Engine Failure Analysis

Other SAE books of interest

Engine Combustion: Pressure Measurement and Analysis
By David R. Rogers
(Product Code: R-388)

Opposed Piston Engines: Evolution, Use, and Future Applications
By Jean-Pierre Pirault and Martin Flint
(Product Code: R-378)

An Introduction to Engine Testing and Development
By Richard D. Atkins
(Product Code: R-344)

For more information or to order a book, contact:
SAE International
400 Commonwealth Drive

Warrendale, PA 15096-0001 USA
Phone: 877-606-7323 (U.S. and Canada only) or
724-776-4970 (outside U.S. and Canada)
Fax: 724-776-0790;

Email: CustomerService@sae.org;
Website: http://books.sae.org

Engine Failure Analysis
Internal Combustion Engine Failures and Their Causes

By Ernst Greuter and Stefan Zima

Translated by Peter L. Albrecht

SAE *International*

Warrendale, Pennsylvania
USA

SAE *International*®

400 Commonwealth Drive
Warrendale, PA 15096-0001 USA

E-mail:	CustomerService@sae.org
Phone:	877-606-7323 (inside USA and Canada)
	724-776-4970 (outside USA)
Fax:	724-776-0790

ISBN 978-0-7680-0885-2
SAE Order No. R-320

Library of Congress Cataloging-in-Publication Data
Greuter, Ernst, 1922-1995.
  [Motorschäden. English]
  Engine failure analysis : internal combustion engine failures and their causes / by Ernst Greuter and Stefan Zima.
    p. cm.
  Includes bibliographical references and index.
  ISBN 978-0-7680-0885-2 (alk. paper)
  1. Internal combustion engines. 2. Structural failures. 3. Machine parts—Failures. I. Zima, Stefan. II. Title.
  TJ785.G74513 2012
  621.43--dc23   2012008127

Information contained in this work has been obtained by SAE International from sources believed to be reliable. However, neither SAE International nor its authors or translators guarantee the accuracy or completeness of any information published herein and neither SAE International nor its authors or translators shall be responsible for any errors, omissions, or damages arising out of use of this information. This work is published with the understanding that SAE International and its authors are supplying information, but are not attempting to render engineering or other professional services. If such services are required, the assistance of an appropriate professional should be sought.

To purchase bulk quantities, please contact:
SAE Customer Service

E-mail:	CustomerService@sae.org
Phone:	877-606-7323 *(inside USA and Canada)*
	724-776-4970 *(outside USA)*
Fax:	724-776-0790

Copyright of the original German edition by Vogel-Verlag und Druck GmbH & Co. KG, Würzburg (Germany). All rights reserved.

Translated by Peter L. Albrecht

Visit the SAE Bookstore at
books.sae.org

Table of Contents

Preface to the Second English Edition ix

Preface to the First English Edition xi

Chapter 1 - Introduction 1

Chapter 2 - The Engine 9

2.1   Properties and peculiarities 9
2.2   Operating conditions 11
        2.2.1   Engine concepts 11
        2.2.2   Engine power output and power reduction 14
2.3   Operational behavior of engines 20

Chapter 3 - Failure—Definitions and Concepts 41

Chapter 4 - Causes of Failure 47

4.1   Wear and tear 47
4.2   Technical defects (product defects) 49
        4.2.1   Design flaws (planning flaws) 49
        4.2.2   Materials defects 52
        4.2.3   Manufacturing defects 53
4.3   Operating errors 53
        4.3.1   Overloading 53
        4.3.2   Changing operating conditions 54
        4.3.3   Operating errors 55
4.4   Humans as the cause of failures 59

Chapter 5 - Explanation of Failures 61

5.1   Type of failure 61
        5.1.1   Failures from mechanical loading 61
        5.1.2   Overload failure 62
        5.1.3   Fatigue fractures 65
        5.1.4   Thermal damage 71
        5.1.5   Failure through corrosion in aqueous media 73
        5.1.6   Failure through tribological loading 76
5.2   Failure analysis 86
        5.2.1   On-site inspection 86
        5.2.2   Securing damaged parts 86

5.2.3　Determining damage-relevant data of a machine
installation 87

5.2.4　Course of events 88

5.2.5　Exact description of damage...............................90

Chapter 6 - Engine Failures 91

6.1　Overview...91

6.2　Crank train failures..99

6.2.1　Pistons 99

6.2.2　Piston rings..178

6.2.3　Connecting rods.....................................190

6.2.4　Crankshafts..203

6.2.5　Crank train bearings 215

6.2.6　Engine oil 275

6.3　Crankcase and ancillary components 284

6.3.1　Crankcase..284

6.3.2　Crankcase damage and failure 287

6.3.3　Cylinders, cylinder liners, and cylinder jackets............291

6.3.4　Cylinder damage 299

6.3.5　Cavitation 305

6.3.6　Cylinder heads..309

6.3.7　Cylinder head damage 312

6.4　Valve train...318

6.4.1　Valve springs...322

6.4.2　Valves 325

6.4.3　Camshaft and cam followers 342

6.4.4　Timing belts, chains, and gears.................347

6.5　Fuel injection and ignition systems........................369

6.5.1　Diesel engine mixture formation and combustion........369

6.5.2　Fuel injection systems 381

6.5.3　Fuel injection system damage...................392

6.5.4　Glow plugs 405

6.5.5　Otto-cycle engine ignition and combustion...................412

6.6　Filters 427

6.6.1　Fundamentals of filtration........................427

6.6.2　Air filters 432

6.6.3　Oil filters 440

6.6.4　Fuel filters 451

6.7　Heat exchangers and heat transfer devices...............454

6.7.1　Shell and tube heat exchangers 459

6.7.2　Heat exchanger damage 463

6.8   Turbochargers...477
    6.8.1   Turbocharger damage...484
    6.8.2   Lubrication inadequacies...491
    6.8.3   Turbocharger housing leaks...494
    6.8.4   Turbocharger operation in zero pressure regime..........495
    6.8.5   Noise complaints...496

Chapter 7 - Preventing Combustion Engine Damage 497

7.1   Preliminary remarks...497
7.2   Introduction..498
7.3   Loss statistics..499
7.4   Advice for the prevention of damage by product faults..............502
    7.4.1   Planning and design...502
    7.4.2   Fabrication and assembly...504
7.5   Advice for loss prevention by operational faults.........................506
7.6   Engine cooling..508
    7.6.1   Information on cooling water treatment.......................508
    7.6.2   Cooling water shortage..510
    7.6.3   Examples of damage incidents.......................................510
7.7   Engine lubrication...511
7.8   Engine fuel...512
7.9   Combustion air...513
7.10   Maintenance and inspection...514
    7.10.1 Maintenance...514
    7.10.2 Inspection...514

Appendix...517

List of Acronyms..525

References...527

Bibliography...535

Illustration Credits..553

Index...557

About the Authors...567

Preface to the
Second English Edition

Combustion engines (Otto and Diesel cycle) are found in a multitude
of forms in motor vehicles, ships, aircraft, construction and agricultural
equipment, stationary machinery, etc. Their wide range of types, sizes,
unique and varied operational conditions (steady-state as well as
nonstationary), as well as the multitude of design details (i.e., pistons,
cylinders, cylinder sleeves, cylinder heads, bearings, valves and valve
guides, crankshafts, camshafts, etc.) results in a wide spectrum of
possible engine failures, whose causes are often difficult to recognize
and evaluate.

This reference work offers consistent and comprehensive assistance
in determining the cause of engine damage, and effective remedies.
With clear text, richly supported by photographs and illustrations,
the authors present a selection of engine failures, along with their
investigation and evaluation, and provide guidance for preventive
measures. Necessary background knowledge of the basic engineering
relationships and processes is presented in a condensed format.
The book interprets the large bandwidth of possible engine failures
for automotive engineers (with emphasis on engine technology),
hands-on specialists—as well as students and teachers, appraisers,
engine consultants, assurance companies, and for laymen (e.g.,
vintage car enthusiasts) who are interested in engine technology—
in a comprehensive, readily understandable manner, free of
manufacturer bias.

In view of the significant and continuing success of the German edition,
now in its third edition, and the fact that there is no comparable text on
the international market covering the topic of engine failure analysis,
SAE International has published this English-language edition. I
have followed the development of this book from its very beginnings
under the primary authorship of Ing. Ernst Greuter †, before Dr.-Ing.
Stefan Zima assumed the task of creating a significantly expanded and
amended second edition. While the English-language edition was in

preparation, Dr. Zima passed away unexpectedly. As an editor at VDI Verlag, I had worked with Dr. Zima for more than a decade, cooperating on the publication of several books. In that time I came to know him as an expert in the field of engine technology.

Compared to the German edition, several sections have been revised and expanded. Wherever possible, older illustrations have been replaced by current material. The contents of the "Filtration" section were brought up to date; for this I would especially like to thank filter manufacturer MANN+HUMMEL for proofreading and correcting the text, and making available the latest illustrations. Compared to the German edition, the extensively subdivided structure of this volume has been arranged more clearly. In addition, a closing chapter, "Preventing combustion engine damage," has been added.

Special thanks go to Lyle Cummins who encouraged Dr. Zima to start work on an English edition. I would like to express my thanks to the translator of this work, Peter L. Albrecht of Costa Mesa, California, whose engineering knowledge, patience, and diligence have resulted in a linguistically clean translation, adapted for English-speaking readers. I would especially like to express thanks to Alena Zima, widow of the author, for her support in the extensive task of relabeling the illustrations.

Siegfried Binder

Publishers Note: SAE International is grateful to Siegfried Binder, who was instrumental in helping to complete the formatting and organization of this edition upon the sudden passing of Author Stefan Zima.

Preface to the
First English Edition

As the first edition of *Motorschäden* (the German original of this volume) is already out of print and largely unavailable, it would appear that a second edition is in order. As the original author, Ing. Ernst Greuter, has passed away, the publisher, Vogel Verlag, has asked me to revise the book for a new edition, expanding the material of the original and bringing it up to date. It has been my pleasure to carry out this task.

Engine failures, like other failures, are occurrences to be avoided at all costs, to be prevented before the fact. Yet experience teaches us that this is not always possible. Although engines have become ever more reliable, with longer service lives, failures still happen. It is the purpose of this book to examine their causes, to give an overview of the wide spectrum of engine component failures, and to examine their underlying causes.

This book is intended for all who deal with engine failures under operational conditions, in repair shops, shipyards, engineering consultancies, insurance companies, and technical oversight organizations, as well as the research and development departments of engine and component manufacturers.

"Pure theoreticians" may also benefit from study of engine failures; they demonstrate that even the theoretically impossible can—and will—happen. This realization should also be imparted to students who place their faith in computational methods.

Engine failures are the result of multitiered, interconnected, and interdependent conditions, effects, and situations. To recognize these, and to determine their causes and remedies, it is necessary to understand the design and manufacture of engine components, their functions, and how these functions depend on other components and in turn affect other parts. To this end, this book examines the design and function of the most critical engine components, as well as their relevant physical and technical properties. Only when one has a good grasp of what happens inside an engine, and the conditions under which engines operate, can one understand engine behavior and the resulting problems and failures. Conceptually, component damage may be organized in an array of causes and effects. In the process, it becomes apparent that not only does the same damage appear on different components, but also quite different types of damage can appear on any given component. Similarly, the same failure may be rooted in any of several causes, and a single cause may manifest itself in various different failure phenomena. To provide a better overview, the most important failure mechanisms have been catalogued. These are followed by discussion of typical failures of functional systems and components, including specific examples whose failure mode or appearance is deemed instructive.

The quality of photographs varies greatly. Failed components often exhibit low contrast, the color range usually consists of shades from light gray to deepest black, and photographs documenting failures are often taken at the scene, under unfavorable lighting conditions, by photographic amateurs. At the other end of the quality spectrum are the outstanding photographs taken in research laboratories, and it is not unusual for such images to possess extraordinary aesthetic appeal. In keeping with their request for anonymity, the sources of many photographs are not credited. Naturally, a field as wide-ranging as that of engine failure cannot be contained within the covers of a single book; a bibliography lists additional literature.

The earlier German edition of this book concentrated on description and analysis of motor vehicle engine failures. This edition expands coverage to encompass general-purpose engines found in heavy commercial vehicles, railway locomotives and vehicles, electrical generators, prime movers, and marine engines. This was done to satisfy a practical need for such information, and because general-purpose engines have been and remain technical precursors for automotive engines. Examples of this may be found in super- and turbocharging, charge air intercooling, and multivalve technology. Problems and failures encountered in these engines are to be expected in automotive powerplants as well.

New additions include Chapters 1 through 5 as well as Sections 6.2.6 and 6.8. The detailed discussion of diesel engine governors found in the first edition has been deleted, as has the section on clutches.

This book is based on practical, real-world experience. For this reason, the author of the first edition, Ing. Ernst Greuter, relied on appropriate reference materials and publications. Similarly, I would like to express my thanks to the following firms and institutions, who have generously provided the material for this book in the form of information, references, and even actual failed components. I am especially grateful to employees of these organizations for informative discussions and advice.

ABB Turbosystems

ADAC e.V.

Allianz-Zentrum für Technik GmbH, Ismaning, Bavaria

BEHR GmbH & Co. KG, Ludwigsburg, Germany

BERU Aktiengesellschaft, Hannover

BorgWarner Turbosystems GmbH, previously Aktiengesellschaft Kühnle, Kopp & Kausch

Caterpillar Marine Power Systems, Hamburg (formerly MaK Maschinenbau GmbH)

ContiTech Antriebssysteme GmbH, Hannover

Cummins Diesel Deutschland GmbH, Gross-Gerau, Germany.

Daimler AG (formerly DaimlerChrysler AG), Stuttgart-Untertürkheim

DEUTZ AG, Cologne

Eckert, Fischer, Sahm & Partner, Engineering Consultants

Federal-Mogul (formerly Glyco-Metallwerke, Wiesbaden, and Goetze-Kolbenringe, Burscheid, Germany)

iwis motorsysteme (formerly Joh. Winklhofer & Söhne GmbH & Co. KG, Munich)

Maschinenfabrik Alfing Kessler GmbH

Mr. Paul Klaver, Klaver Engines and Engineering, Rolde, The Netherlands

Kolbenschmidt Pierburg AG, Neckarsulm, Germany

MAHLE GmbH, Stuttgart, Germany

MAN Nutzfahrzeuge Aktiengesellschaft, Nuremberg works

MANN+HUMMEL GmbH, Ludwigsburg, Germany

Miba AG, Laakirchen, Austria

MTU Motoren- und Turbinen-Union Friedrichshafen GmbH

SCHERDEL GmbH, Marktredwitz, Bavaria

TRW Automotive GmbH, Barsinghausen, Germany

Zeppelin Baumaschinen GmbH

ZF Friedrichshafen AG

Zollern BHW Gleitlager GmbH & Co., Braunschweig, Germany

I would also like to thank the students of the Giessen-Friedberg Technical University for interesting failed components; the unique combination of youthful exuberance and old vehicles and engines found at the university level has provided many truly spectacular examples of engine failure! My thanks also to my wife Alena for her understanding and patience—and for numerous CAD-generated illustrations and schematics. The publisher of the German edition, Vogel-Verlag, is also owed a debt of gratitude for their careful and excellent production of this book.

Stefan Zima

Chapter 1

Introduction

Internal combustion engines have been with us for more than 100 years. In that time, hundreds of millions of engines have been built. One might assume, therefore, that engine technology is so well understood that few or no failures would be encountered. Reality, however, is quite different. To a far greater degree than other machinery, engine development is driven by failures of every type. If one considers the comparatively unproblematic operation of, for example, electrical machines, one cannot avoid the impression that combustion engines are plagued by basic, systemic faults, and this despite more than a century of effort by an ever-growing number of engineers, technicians, and specialists throughout the world to develop and improve combustion engines. What makes combustion engines so susceptible to damage and failure, that entire books, monographs, and indeed standards addressing failure of engines and their components have been, and continue to be, published?

The underlying causes are in the nature of engines themselves, in the complexity of engine functions with their multitude of often obscure interactions, and in the various operating conditions under which engines are expected to operate.

The combustion engine is a self-contained heat engine which converts energy contained in fuel into heat, and thence a part of that heat into mechanical energy. To achieve the same effect using other types of machines would require a firebox, a boiler, a condenser, and a turbine, with the total package no larger than the combustion engine they would replace.

Because conversion to heat—combustion—takes place within the engine ("internal combustion"), the powerplant's precision components are directly subjected to combustion gases and their corrosive and abrasive products. High gas pressures within cylinders, crucial in determining engine power and efficiency, must be maintained by moving seals— the piston rings—whose function in turn requires hydrodynamic lubrication. This breaks down at top and bottom dead center, at exactly the point where piston rings, at ignition top dead center, are required to withstand high gas pressures and temperatures. Moreover, the piston rings must conduct a large portion of the heat transferred into the pistons to the cylinder walls. Most pistons are made of aluminum alloy, whose melting point just below 600°C is far below the peak temperature of more than 2000°C encountered in the combustion chamber. Added to this, the sliding members of the engine—pistons and cylinders—are made of different materials with dissimilar expansion coefficients, for which compensating measures must be applied in both design and operation. Power is transmitted through bearings by lubricating films whose thickness amounts to only a few thousandths of a millimeter. Central components of the engine are subjected to high thermal, mechanical (indeed, dynamic mechanical), and tribological loads.

Engines consist of many parts, mechanically and functionally joined, which must work together faultlessly. These parts are individually subjected to a multitude of effects and influences whose totality is often largely unpredictable and upon which development engineers at various manufacturers can have only limited influence. The engine is a link in a chain which can be no stronger than its weakest link. Faults in engine accessories, power transmission, design, and operational deficiencies in a vehicle or vessel have retroactive effects on the engine. Therefore, engine failures are often indicators of faults, weaknesses, and functional failures within the entire propulsion system and the machine powered by that system.

The external operating conditions experienced by engines are no less unfavorable than internal conditions. Vehicle engines have no rigid foundation, vehicle frames or unit bodies deform, and engines are continually subjected to vibration and severe impacts. Deformation of a soft ship's foundation by ocean swells or the vessel's cargo loading may lead to bearing damage or indeed crankshaft breakage. Engines are subjected to varying ambient temperatures, with daily as well as seasonal variations; they operate in all climate zones from arctic cold to blazing desert heat. There is hardly any other type of machine which is subjected to continuous, often abrupt load changes: cold start, rapid loading to full load, sudden unloading, long periods at idle, stop-and-go operation at low engine temperatures, overloading, and shutdown from full throttle.

Further problems arise from substandard fuels; from dust, sand, moisture, or salt in combustion air; with inadequate filtration of air, fuel, and lubricants; with inadequate coolant maintenance; and with improper maintenance and operation by technically unqualified personnel. Imagine how machine tools, textile mills, or packaging machinery would react to such operating conditions.

The individual key points in engine development:

- Increase in absolute power output
- Increase in specific output (power/weight; power/displacement; power/package volume)
- Decrease in fuel and lubricant consumption; i.e., improved efficiency
- Reduced manufacturing cost
- Increase in service life and extension of service intervals
- Simplified service and repair
- Reduction of pollutant and noise emissions

These points demand measures which are to some degree mutually exclusive, so that compromises must be made which, depending on the importance of certain requirements, are made at the expense of others.

Modern engines consist of considerably fewer components than engines of bygone times, thanks to integration of functions, cost reduction, elimination of fault sources, simplification of service and repair, but also restructuring of entire functional groups from mechanical to electronic operation (mixture formation, ignition, and engine control). Associated with these changes is a development trend toward higher mechanical, thermal, and tribological loads, which must be borne by smaller cross sections and reduced masses. Development cycles are continuously being shortened, so that in view of costs in development, manufacturing, and operation, certain limitations are accepted. If the process were not subjected to these self-imposed limitations, many a fault, defect, or failure would not be tolerated.

Whenever the causes of a fault or failure were recognized and eliminated, the demands on power, operational behavior, and service life were raised to the point where new faults would arise. The result has been a continuous race to keep up with the latest failure modes. Depending on the state of development of an individual engine, the weakest points would trade places; bearings, pistons and piston rings, crankshaft, fuel injection system, valves and valve springs, cylinder heads, and head gaskets would repeatedly take their turn as problem sources. There is hardly a functional group whose problems have not impeded engine development. It soon becomes apparent that engine development, which inevitably runs up against the limits of what is technically possible, is itself the root cause of engine faults and failures.

Engine failures are attributable to a broad spectrum of physical, chemical, and electrochemical processes which appear in various combinations within and among individual functional groups. Consequently, engine failures usually do not have a single cause but rather several, any one of which is difficult to predict, and whose combination is accordingly even less likely—and yet these failures occur, if nothing else as an affirmation of Murphy's Law: "If anything can go wrong, it will." The causes of failures are therefore often of an indirect nature, so that failures develop in roundabout ways, which makes failure determination exceedingly difficult.

In evaluating failures, the question of whether a large component (e.g., crankshaft) or a small component (e.g., bearing or piston ring) was the root cause is immaterial, because the end result—engine stoppage—is the same. Failures can occur in an avalanche or cascading manner: seized piston ring, seized piston, engine damage, and engine failure. The results are sometimes dramatic. Often, the degree of damage bears absolutely no relation to its root cause.

Machine component failures have always been a driving force for development (and remain so to this day). They are unmistakable reminders of weaknesses or faults in materials, design, manufacturing, assembly, and operation. Moreover, they may point to design or operational deficiencies of the entire machinery installation. Corrective measures initiated by failures are primarily directed toward elimination of their causes in the future. Often, such "active" fault prevention has not been possible, because the cause(s) could not be clearly determined, or, for whatever reasons, the underlying causes could not be altered. In such cases, one had to be satisfied in minimizing the effects of such a fault, and making them bearable; in effect, a form of "passive" failure prevention.

For these reasons, failures have played a decisive role in the development of engine technology. They served as a learning experience, entirely in keeping with the old adage that we learn from our mistakes. In the next stage, attempts were made during development to deliberately cause engine failures, to make use of the insight gained and to eliminate such failures from engine operation.

Understandably, one strives to eliminate all possible failure modes during an engine's conception, design, and testing phases. But there are difficulties with this, as described above. After more than a century of engine design, development, and construction, we still have failures that stem from the following:

- Our knowledge of physical and technical processes is limited, even today.

- On the other hand, the advancement of knowledge and experience has led to forced lightweight design, higher engine loading, and lower-cost manufacturing—the cause of many an engine failure.

- It is within the nature of new and advanced development that one crosses the boundaries of what is known and ventures into unexplored engineering territory.

- The chance meeting of multiple conditions, each of which is harmless by itself, but which, together, cause a failure, is individually highly unlikely; moreover, there are so many possible combinations of influential parameters that any attempt to eliminate them beforehand is practically impossible.

- Despite all best efforts, human shortcomings in engine manufacturing, assembly, operation, and maintenance cannot be eliminated completely.

- The engine must be developed and manufactured in keeping with economically justifiable, i.e., affordable, costs; for cost reasons, one cannot push development as far as might be technically possible or desirable. Cost pressures are counterproductive in many different ways, and not only in engine technology:

- Development times have been shortened appreciably. Increasingly, computer simulation takes the place of experiment and field testing, which is possible in many cases—but most definitely not in all cases!

- More work needs to be done with fewer personnel. This forces ever more extreme concentration on "the business at hand." Often, there is a shortage of staff and time to conduct indirectly vital work, such as thorough documentation of failures, their causes and remedies, and, in development work, the important "peek over the garden fence." The result is a loss of experience and a narrowing of the engineering field of view. Rather than the best solution, it is the most readily available, the easiest to realize, and, seemingly, the lowest-cost solution that is ultimately chosen.

- Early retirement of older, experienced workers in all branches of development, manufacturing, and engine operation interrupts the flow of knowledge and experience.

Now, again, failures due to long-known causes reappear, after their solutions are negated for reasons of simplification and rationalization, because the reasons for these solutions are no longer familiar to a younger generation of engineers and technicians.

Still, experiences gained over decades, and increasingly effective instrumentation applied to development methods, have led to the fact that failures which delineate the limits of engine loadability are becoming ever rarer from the engine operator's point of view. These have effectively been transferred to the experimental departments and, in recent times, largely replaced by experimental and theoretical pre-development. These considerations put engine failures in perspective, especially when one considers how long and how reliably most engines operate.

The bandwidth of engine failures is large; it ranges from irretrievable total destruction of the engine, whether from external forces (Fig. 1.1)

Fig. 1.1 Total failure of an aircraft engine after a crash from an altitude of 1000 meters.

Fig. 1.2 Total failure of an engine as a result of a seized piston.

or internal failure (Fig. 1.2), through all intermediate levels, to virtually invisible load and wear patterns.

This book limits itself entirely to engineering failures; that is, it excludes failures arising from external forces, whether unintentionally through accidents, or deliberately through acts of war.

Chapter 2
The Engine

2.1 Properties and peculiarities

Colloquially, the term *engines*—machines used to generate propulsive forces—is understood to mean internal combustion piston heat engines (DIN 1940).

The following are characteristic of reciprocating-piston combustion engines (Fig. 2.1):

- Variable working volume
- Internal combustion
- Discontinuous (intermittent) mass throughput
- Small mass flows
- Working fluid with high pressures and temperatures

These properties constitute the principal advantages, but also the inherent disadvantages of combustion engines. Advantages are:

- Good efficiencies that are based on the high temperatures and pressures of the engine's thermodynamic process. Such efficiencies are also the cause of many engine failures. The effective thermodynamic efficiency of Otto-cycle engines is about 35%; that of diesel engines, between 40 and 55% (the latter for large two-stroke diesel engines). In principle, combustion engines also have good efficiency at part load, because the operational behavior of piston engines (in contrast to flow machines) is largely independent of engine speed.

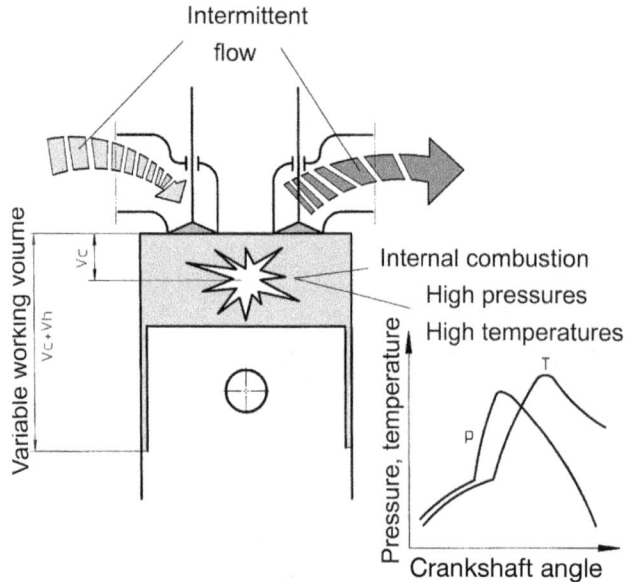

Fig. 2.1 Characteristics of combustion engines (schematic).

- Combustion engines are built to cover a large performance range, in terms of output power range as well as engine speed/torque gradation. The smallest engines—for model airplanes and model cars—develop upwards of 0.07 kW. Large engines for marine and stationary power applications produce up to 66,000 kW. Small engines primarily generate their power through high rotational speeds, while large engines rely on high torque for their power.

A fundamental disadvantage of combustion engines is their discontinuous mode of operation. For kinematic reasons, this has a twofold effect on reciprocating piston engines, first through inertia effects (mass forces and moments), and second through intermittent (interrupted in time) mass throughput. On the other hand, these disadvantages are requirements for the operation of combustion engines, because only discontinuous operation permits engine operation with peak temperatures in excess of 2000°C.

Energy conversion within an engine is the result of different, interconnected processes:

- Gas exchange
- Mixture formation
- Combustion
- Power transmission
- Heat transfer

The engine is a complex system of mutually interacting technical components, whose effects are based on thermodynamic, gas flow, dynamic, tribological, and control technology processes. These processes are effectuated by functional groups:

- Crankcase with cylinders, cylinder heads, and oilpan
- Crank train: piston and related components, connecting rods, crankshaft, bearings, lubricant
- Valve train: camshaft, lifters, pushrods, rocker arms, valves, lubricant
- Mixture formation: fuel injection system (diesel or Otto cycle), carburetor (Otto cycle)
- Cooling system: heat exchangers, pumps
- Filters for air, fuel, and lubricant
- Engine control
- Gears, chains, or drive belts

Individual functional groups consist of components which, as the last links in the chain, are the sites where actual engine damage occurs: pistons, piston rings, bearings, crankshaft, camshaft, valves, etc.

2.2 Operating conditions

2.2.1 Engine concepts

"The" or "a" typical engine as such does not exist; rather, there exists a multitude of different configurations, differing greatly in size, type, application, mode of operation, and operational behavior, and therefore, in keeping with the theme of this book, a similar broad range of failures. Determinant for an engine concept is its intended application, which generally establishes size (cylinder bore, stroke, number of cylinders), working process (four-stroke, two-stroke), working cycle (Otto or diesel), mixture formation and combustion process (e.g., direct or indirect injection), cooling (water or air), crank train layout (trunk pistons, crosshead), and the overall design configuration.

Vehicle engines, especially Otto-cycle engines, are such a dominant engine type that most people think of motor vehicle engines when they hear the word "engine." Even though the great majority of engines are used to propel motor vehicles, the bandwidth of engines for general applications—propulsion of commercial vehicles of all types including railway and marine applications, electrical power generation, and construction equipment—is much larger, in terms of dimensions, power spectrum, design variety—and in its failure modes.

Due to superordinate physical laws of mechanical similitude, engine dimensions (bore, stroke, number of cylinders) are not mutually independent; large engines run at lower speeds, small engines are able to rev higher. Therefore, engine speed range is a reliable index for diesel engine size and type (Otto-cycle engines are strictly high-speed engines.) Distinction is made between low-speed, medium-speed, and high-speed engines (Table 2.1).

Table 2.1
Engine Speed Range of Various Diesel Engine Types

Engine speed range	Working cycle	Cylinder bore	Piston stroke	Number of cylinders	Power output per cylinder
rpm	–	mm	mm	–	kW
Low speed 250 – 75	2 - stroke	260 – 980	980 – 2660	4 – 12	400 – 5800
Medium speed 1000 – 400	4 - stroke	200 – 640	300 – 900	5 – 20	100 – 1940
High speed 2500 – 1000	4 - stroke	130 – 250	140 – 300	4 – 20	35 – 370
High speed passenger car 4000 – 3000	4 - stroke	70 – 90	80 – 100	4 – 8	up to 25

Not only power, size, and speed are interrelated, but also working principle, process, and cooling type. High absolute power can only be achieved with large cylinder dimensions (bore, stroke), while high specific output (power/displacement, power/engine mass) is achieved through working cycle frequency (engine speed, number of strokes) and specific work.

The engine concept ultimately determines operating mode, operating time, utilization, operational behavior, and life expectancy of individual engine components. Engines for general-purpose applications are uniformly diesels; they may also be designed as (natural) gas or dual-fuel engines. Medium and large engines, specifically those used for marine propulsion, are powered by fuel oil. Two-stroke engines use a crosshead configuration to guide the cranktrain, with physical separation of cylinder and crankcase spaces.

Engines are always part of a powerplant installation, whose function is therefore also influenced by the concept, execution, and individual components of the installation. This is understood to include engine mounts, engine alignment, power takeoff at the engine, power transmission, auxiliary drives, cooling type and cooling system, engine preheat and keep-warm system, intake system and air filtration, engine room ventilation, fuel system, lubricating oil system, exhaust system, and noise damping. These peripheral considerations, necessarily different for nearly every application, are the cause of many problems and failures affecting the engine itself.

The driven machinery determines the engine's power and load profiles by mapping torque to engine speed, in accordance with the engine's purpose in propelling a vehicle, a ship, or electrical power generator. In principle, distinction is made (Fig. 2.2) between full-load, generator, or propeller curves.

- **Full-load curve**
  The engine is operated to develop full torque potential (full load) at various engine speeds. Power is a linear function of engine speed. Operation along the full-load curve is characteristic of heavy vehicles, specifically construction equipment.

- **Generator curve**
  The engine is operated at constant speed, because generators are expected to provide constant voltage or frequency; these are dependent on speed. Power is controlled through torque modulation.

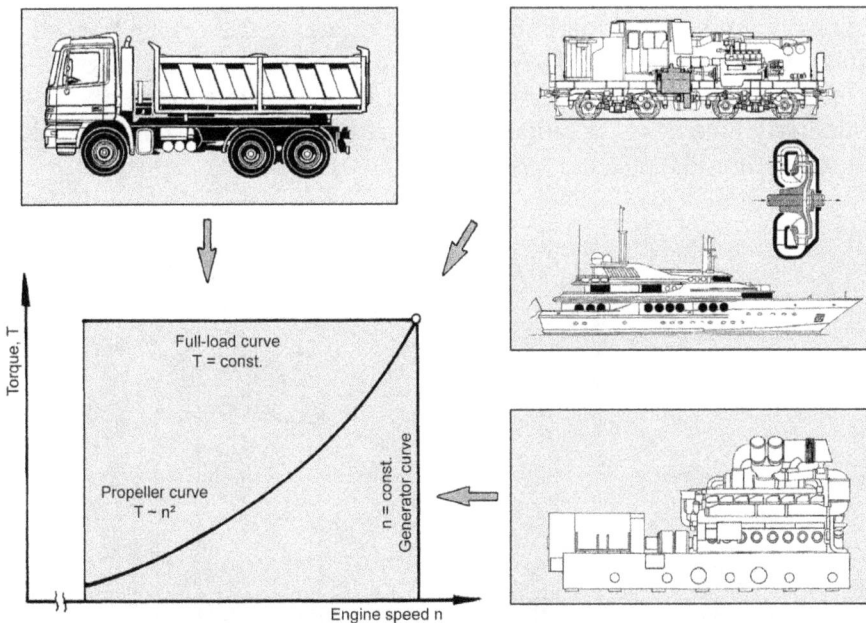

Fig. 2.2 Propeller, full-load, and generator curves.

- **Propeller curve**
  Torque is a function of the square of engine speed; power is a function of engine speed cubed. Such behavior is typical of flow machinery. For marine vessels, power goes as speed to the mth power ($P \approx n^m$). The exponent m depends on the vessel type and size, and ultimately also on its speed. For large, slow vessels, m = 3, while for fast ships m = 3.5 to 3.7. In practice, these curves appear slightly different, because exhaust turbosupercharging increasingly imposes the turbocharger's flow machine characteristics on the engine. The engine map is so restricted in its upper power range that it can no longer satisfy the power demand of the full-load curve (Fig. 2.3).

- **Motor vehicle operation**
  Driving resistance of a motor vehicle (while driving in a straight line on a level surface at constant speed) is composed of rolling resistance and aerodynamic resistance. Rolling resistance increases roughly linearly with speed, while aerodynamic resistance increases with the square of speed. Power rises as speed to the power of 2.6 to 2.8. The engine torque/engine speed relationship is matched to the vehicle's operational requirements by means of torque and speed converters in the drivetrain—mechanical, hydraulic, or electric transmissions. For passenger car engines, this is achieved with three to eight (mechanical) gear stages ("speeds" or "gears"); for heavy commercial vehicles, even more discrete stages are required (Figs. 2.4 and 2.5).

2.2.2 Engine power output and power reduction

Engine power is a function of the amount of fuel which can be burned in its cylinders. Because air is needed for combustion, the air mass within the cylinders is the deciding factor in power development, particularly when one considers that air takes up about 10,000 times the volume of (liquid) fuel.

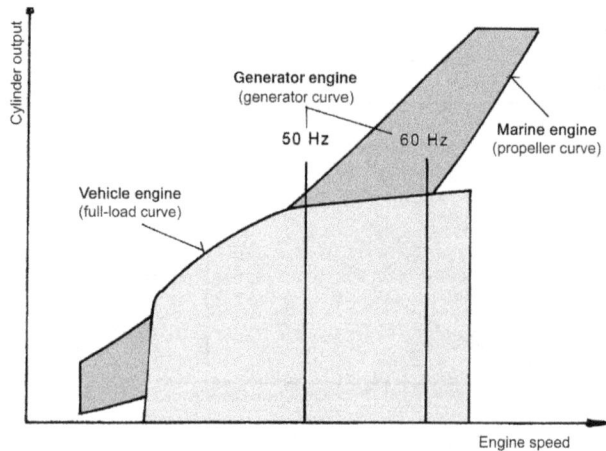

Fig. 2.3 Engine operating regimes under various operating conditions.

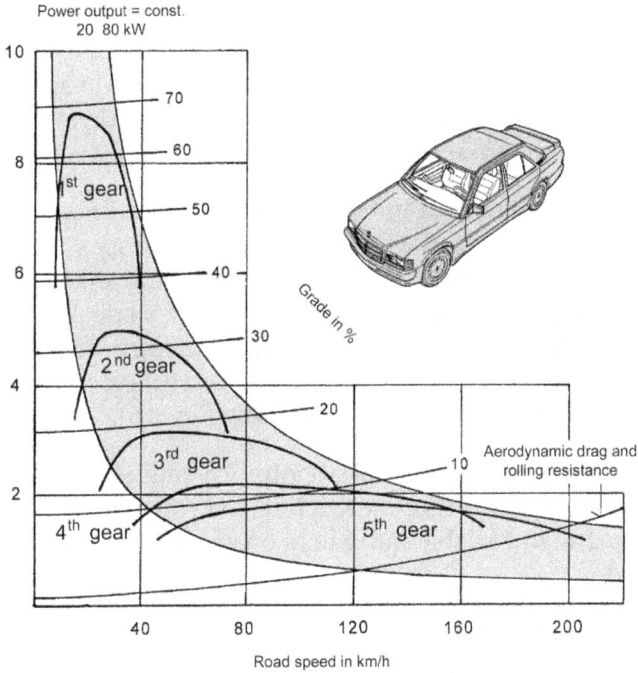

Fig. 2.4 Tractive effort vs. road speed diagram for a passenger car.

Power output = const.
20 80 kW

Grade in %

Aerodynamic drag and rolling resistance

1st gear
2nd gear
3rd gear
4th gear
5th gear

Road speed in km/h

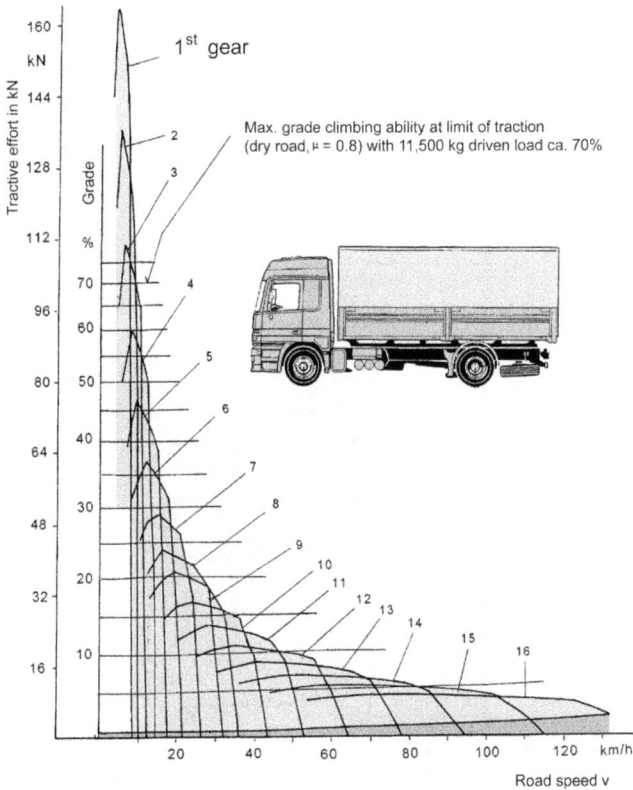

Fig. 2.5 Tractive effort vs. road speed diagram for an 18 ton commercial vehicle, v_{max} = 131 km/h (81 mph).

Tractive effort in kN

1st gear

Max. grade climbing ability at limit of traction
(dry road, μ = 0.8) with 11,500 kg driven load ca. 70%

Grade %

Road speed v

15

The laws of thermodynamics tell us that complete energy conversion is not possible. Consequently, part (indeed the greater part) of the energy released in combustion is rejected to the environment as heat—with the exhaust stream, coolant, and a small portion as radiation and convection (heat carried by the smallest particles in a flow).

If one establishes the energy contained in fuel as the 100% baseline, one obtains the corresponding percentages for individual energy streams, including the usable heat (mechanical energy at the crankshaft output end). Such an energy budget may also be applied to a specific performance point; the relationship between individual energy streams changes as a function of engine speed and torque. A *Sankey diagram* offers a graphical representation of such relationships (Fig. 2.6).

Although *power* is a clearly defined physical quantity, in engine technology, the concept of power is more restrictive and must be defined more precisely, due to variable influences on engine output. These influences include:

- The duration of power delivery. It is obvious that a high-performance engine (e.g., a racing engine) is expected to deliver power for a shorter period of time than engines for commercial vehicles or marine vessels.

- Time intervals between, and modalities of, manufacturer-specified maintenance tasks

- Atmospheric conditions—ambient barometric pressure, air temperature and humidity—under which power must be delivered

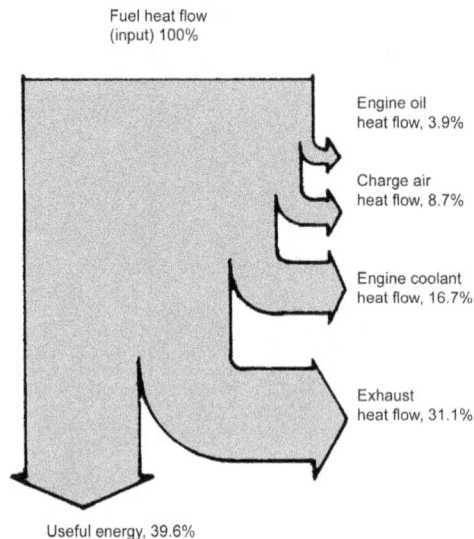

Fuel heat flow (input) 100%

Engine oil heat flow, 3.9%

Charge air heat flow, 8.7%

Engine coolant heat flow, 16.7%

Exhaust heat flow, 31.1%

Fig. 2.6 Sankey diagram (heat budget) of a high-speed diesel engine.

Useful energy, 39.6%

- The power consumed by auxiliary devices which are unavoidable and dependent on the engine application. For motor vehicle engines, these are:
    - Induction and exhaust system
    - Fan and water pump or cooling air blower
    - Fuel pump
    - Injection pump for diesel engines
    - Unloaded alternator or generator

For every citation of output power, the corresponding engine speed must be specified. The superordinate concept of power, according to DIN ISO 3046 T1, is

- **Declared power**
  The value of the power, declared by the manufacturer, which an engine will deliver under a given set of circumstances. In some applications, the declared power is named "rated power." Further distinction is made for location (within the engine) where the power measurement is made:
    - Indicated power: the total power developed in the working cylinders as a result of the pressure of the working medium acting on the pistons.
    - Brake power: the power or the sum of the powers measured at the driving shaft or shafts.

- **Duration of developed power**
    - Continuous power: the power that an engine is capable of delivering continuously, between the normal maintenance intervals stated by the manufacturer, at the stated speed and under stated ambient conditions, the maintenance prescribed by the manufacturer being carried out.
    - Overload power: the power that an engine may be permitted to deliver, with a duration and frequency of use depending on the service application, at stated ambient conditions, immediately after operating at the continuous power.
    - Fuel stop power: the power that an engine is capable of delivering during a stated period corresponding to its application and stated speed, and under stated ambient conditions, with the fuel limited so that this power cannot be exceeded.

- **Type of power specification**
    - ISO power: the power determined under the operating conditions of the manufacturer's test bed and adjusted or corrected as determined by the manufacturer to the standard reference conditions specified in Section 6.

- ISO standard power: the continuous brake power which the engine manufacturer declares that an engine is capable of delivering, using only the essential dependent auxiliaries, between the normal maintenance intervals stated by the manufacturer, and under the following conditions:
    a) At a stated speed at the operating conditions of the engine manufacturer's test bed.
    b) With the declared power adjusted or corrected as determined by the manufacturer to the standard reference conditions specified in Section 6.
    c) With the maintenance prescribed by the engine manufacturer being carried out.
- Service power: the power delivered under the ambient and operating conditions of an engine application.
- Service standard power: the name given to the continuous brake power which the engine manufacturer declares that an engine is capable of delivering, using only the essential dependent auxiliaries, between the normal maintenance intervals stated by the manufacturer and under the following conditions:
    a) At a stated speed at the ambient and operating conditions of the engine application.
    b) With the declared power adjusted or corrected as determined by the manufacturer to the stated ambient and operating conditions of the engine application.
    c) With the maintenance prescribed by the engine manufacturer being carried out.

It is worth noting that power declarations per ISO 3046 T1 specifically mention maintenance—an obvious indication of the significance of maintenance, not only for reliability and longevity, but also for power development itself. Engine manufacturers therefore provide specific instructions for the scope of, and intervals between, maintenance procedures, dependent on the respective service conditions. So, for example, section [2-1] says

"at present, in the European region, a W5 [maintenance] is recommended every 10,000 hours, and a W6 every 30,000 hours. Maintenance intervals outside Europe reflect operating conditions that are more difficult in every way . . ." (see Fig. 2.7).

From the power equation, it is apparent that power, being a function of air density, is dependent on atmospheric conditions; specifically:

- Air pressure, as determined by barometric altitude and current weather conditions (high/low pressure zones)

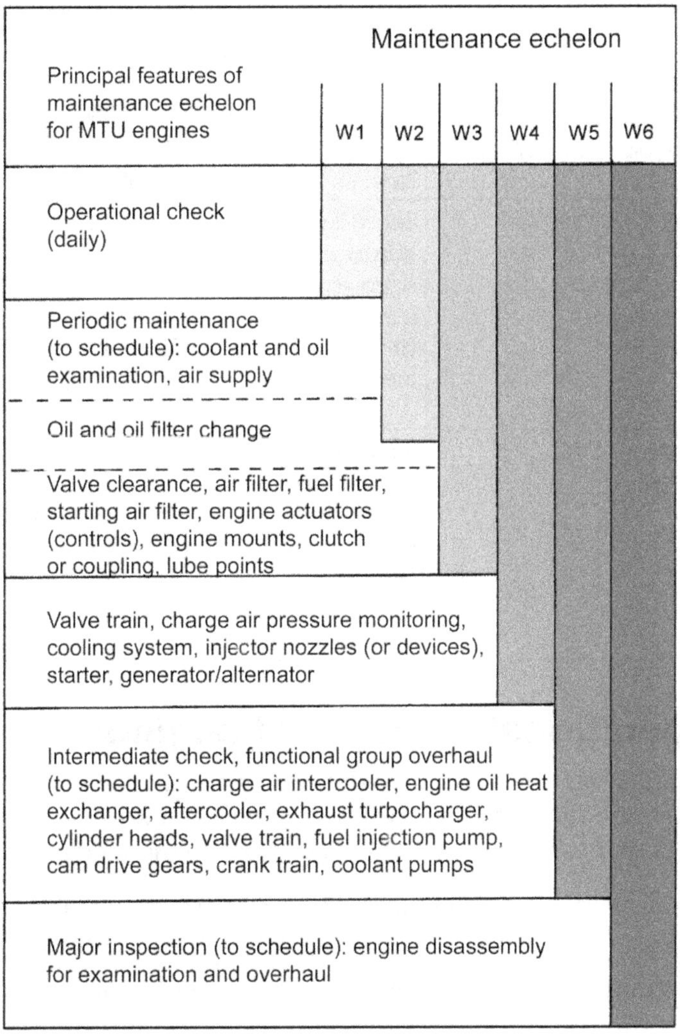

	Maintenance echelon					
Principal features of maintenance echelon for MTU engines	W1	W2	W3	W4	W5	W6
Operational check (daily)						
Periodic maintenance (to schedule): coolant and oil examination, air supply						
Oil and oil filter change						
Valve clearance, air filter, fuel filter, starting air filter, engine actuators (controls), engine mounts, clutch or coupling, lube points						
Valve train, charge air pressure monitoring, cooling system, injector nozzles (or devices), starter, generator/alternator						
Intermediate check, functional group overhaul (to schedule): charge air intercooler, engine oil heat exchanger, aftercooler, exhaust turbocharger, cylinder heads, valve train, fuel injection pump, cam drive gears, crank train, coolant pumps						
Major inspection (to schedule): engine disassembly for examination and overhaul						

Fig. 2.7 The MTU Friedrichshafen maintenance program encompasses "maintenance echelons" W 1 through W 6 (W from the German "Wartung" = maintenance). Source: MTU

- Air temperature—affected by geographic latitude, barometric altitude, and season
- Relative humidity

Therefore, whenever an engine is being operated under nonstandard atmospheric conditions, power must be corrected to the applicable standard atmospheric conditions (power correction); see Table 2.2.

Table 2.2
Reference Conditions for Engine Power

Engine type	Standard	Reference conditions	
Motor vehicle engines	DIN 70020 Part 6	Air temperature	Tr = 298 K (25°C)
		Barometric pressure	pr = 100 kPa
General purpose engines	ISO 3046-1	Barometric pressure (total pressure)	pr = 100 kPa
		Air temperature	Tr = 298 K (tr = 25°C)
		Relative humidity	Φr = 30%
		Charge air coolant temperature	Tcr = 298 K (tcr = 25°C)
Main and auxiliary engines for marine vessels	IACS	Barometric pressure (total pressure)	px = 100 kPa
		Air temperature	Tx = 318 K (tx = 45°C)
		Relative humidity	Φx = 60%
		Sea or bulk water temperature (at charge air intercooler inlet)	Tcx = 305 K (tcx = 32°C)

2.3 Operational behavior of engines

Combustion of mixture in the cylinder causes a rise in pressure which, acting on the piston, moves the engine's crank train. Internal combustion engines operate in a non-steady-state manner, as a result of

- Gas pressure history of the thermodynamic process (induction, compression, ignition, combustion, expansion, and exhaust)
- The nonuniform motion (kinematics) of the crank train

Accordingly, gas and inertia forces vary over the course of a working stroke (Fig. 2.8).

The crank train accomplishes three functions: it controls the course of the thermodynamic process, converts the reciprocating motion of the piston into rotary motion, and converts gas pressure acting on the piston into torque. As a result of its kinematics, crank train components experience nonuniform motion, which induces inertia effects (inertia forces and inertia moments). Gas pressure and inertia forces account for all other forces acting on the crank train (see Fig. 2.9); individual forces act on various engine components (see Fig. 2.10).

The above considerations apply for a single load condition, for example for rated power at rated rpm, and the corresponding torque. However,

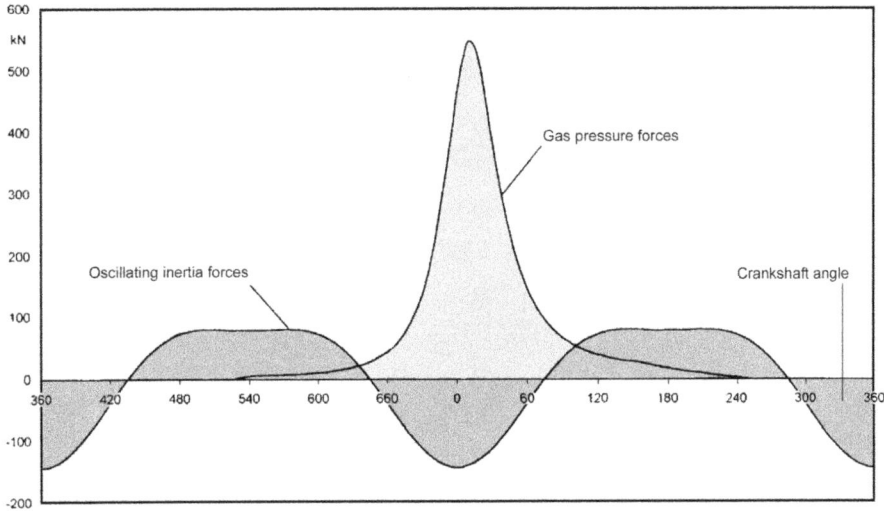

Fig. 2.8 Gas and inertia forces over the course of an entire (four-stroke) cycle.

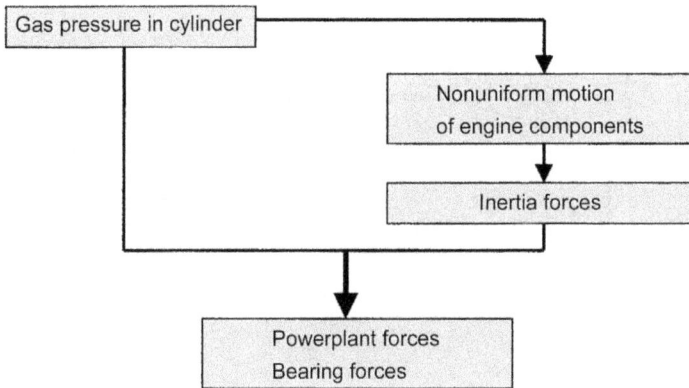

Fig. 2.9 Flow chart: engine internal forces.

engines operate at various speeds and loads (as characterized by torque τ or mean effective pressure p_{me} or specific work w_e). Correlation between the possible power points of an engine, in terms of torque as a function of engine speed, is known as the *engine map*. The limits of an engine map are defined by a lower idle speed, maximum torque, and the breakaway curve (Fig. 2.11).

If the engine is being driven by the vehicle, for example on a downgrade or when a ship is reversing engines, then the engine map expands into the negative range ("retarding curve").

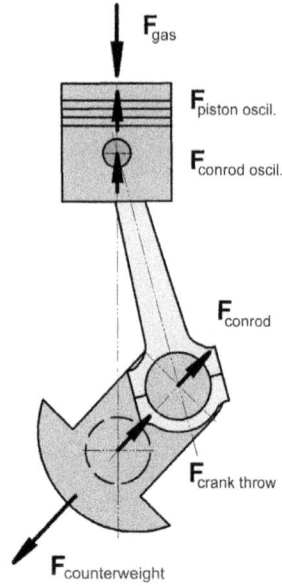

Fig. 2.10 Crank train forces.

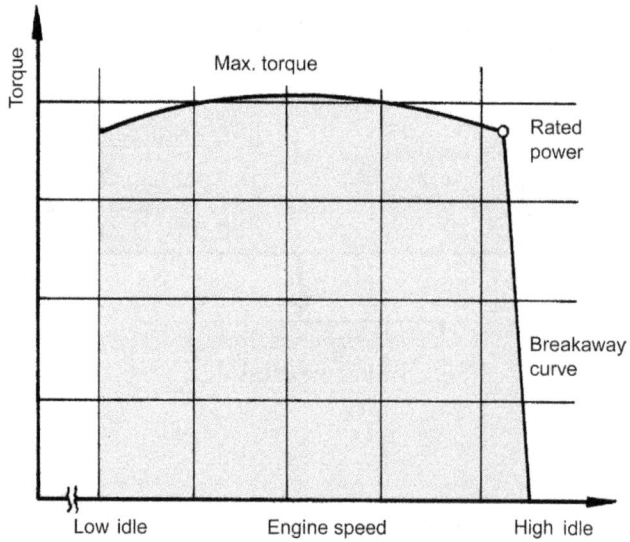

Fig. 2.11 Engine map (schematic).

The engine map illustrates the relationship between physically and mechanically related quantities such as torque, engine speed, and power. This relationship may be

- of a direct nature, such as the dependence of power on torque and engine speed ($P = \tau \cdot 2\pi \cdot n$), or

- of an indirect nature, such as when so many factors affect a quantity that its behavior can no longer be traced back to one or two initiating factors. Examples: dependence of cylinder injection pressure, or efficiency, on torque and engine speed.

To represent engine behavior, engine speed is plotted on the abscissa (horizontal, or x-axis) while torque is plotted on the ordinate (vertical, or y-axis). Every point on the engine map is uniquely defined by the intersection of an engine speed line and a torque line. In this representation, curves of constant engine power are represented as hyperbolas. For individual intersections of torque and speed, all engine properties of interest are measured or calculated, and plotted: speed on the x-axis, torque on the y-axis, and the appropriate property on the z-axis, for example absolute fuel consumption. In this way, we obtain three-dimensional surfaces (Fig. 2.12).

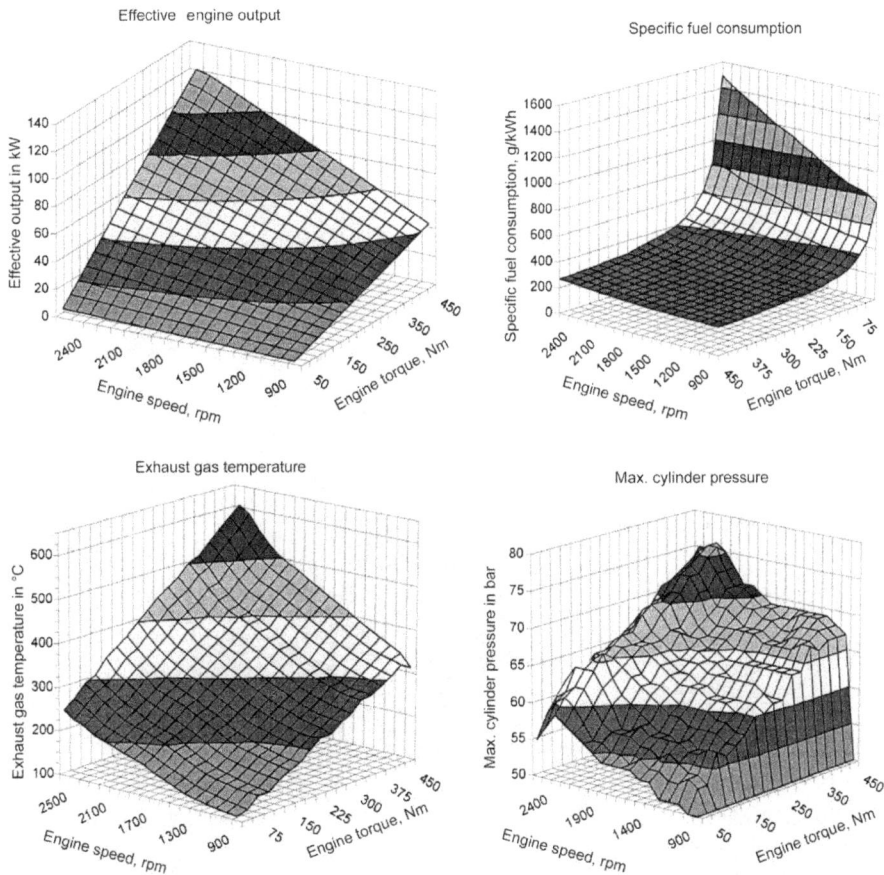

Fig. 2.12 Three-dimensional maps for a diesel commercial vehicle engine. Engine characteristics plotted as functions of engine speed and torque. Cylinder bore ca. 125 mm.

Because it is difficult to work with three-dimensional representations, we project the property of interest onto the speed/torque plane. The measured quantity at the intersection of speed and torque lines is recorded, and all points with this same value are connected (of course, this is carried out by a computer program). In this way, we obtain lines or curves of identical engine behavior, analogous to contour lines on a topographic map. The most familiar examples of this are plots of lines of equal specific fuel consumption, b_e, which due to their similarity to the ridges on certain types of seashells are also known as conchoid curves (Fig. 2.13).

Other measured properties are treated similarly. In the course of an engine development program, but also in association with all other development stages, relevant engine properties are measured and evaluated in map form. As an example of the scope of such investigations, following are the measured parameters and their evaluation for the case of a high-speed diesel engine:

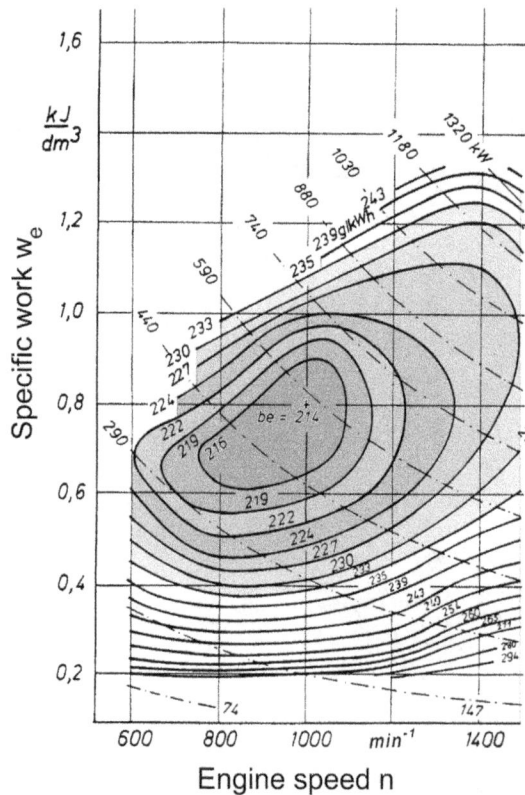

Fig. 2.13 Map of engine specific fuel consumption (conchoid curves).

- Absolute fuel consumption
- Specific fuel consumption
- Smoke number
- Exhaust gas temperatures upstream and downstream of turbine
- Turbine rpm
- Boost pressure
- Peak cylinder pressure
- Air flow rate
- Specific air flow rate
- Total air mass flow
- Volumetric efficiency
- Begin of injection
- Pressure rise rate
- Ignition begin and ignition delay
- Injection duration
- Crankcase blowby volume
- Heat rejected through oil heat exchanger
- Fuel injection quantity
- Charging angle
- Exhaust volume flow rate
- Heat rejected through engine coolant
- Heat rejected through engine coolant and exhaust turbocharger coolant
- Heat rejected through turbocharger coolant

In conjunction with component development, certain specific behaviors are measured, such as bearing, piston, or valve temperature, torsional vibration and component vibration, noise and pollutant emissions, etc. Such maps, because they illustrate targeted subtle engine behavior, may prove to be valuable aids in tracking down underlying causes of engine component failures (Figs. 2.14 and 2.15).

For load changes within an engine map, the relationship between gas and inertia forces changes, along with engine speed and torque. The gas pressure history, especially peak pressure (ignition pressure) is roughly directly proportional to specific work (torque), while inertia forces increase with the square of engine speed. Therefore, the engine and its components are subjected to two different types of load changes (Fig. 2.16):

- High cycle changes due to the working cycle
- Low cycle changes due to shifts from one engine operating point (speed, torque) to another

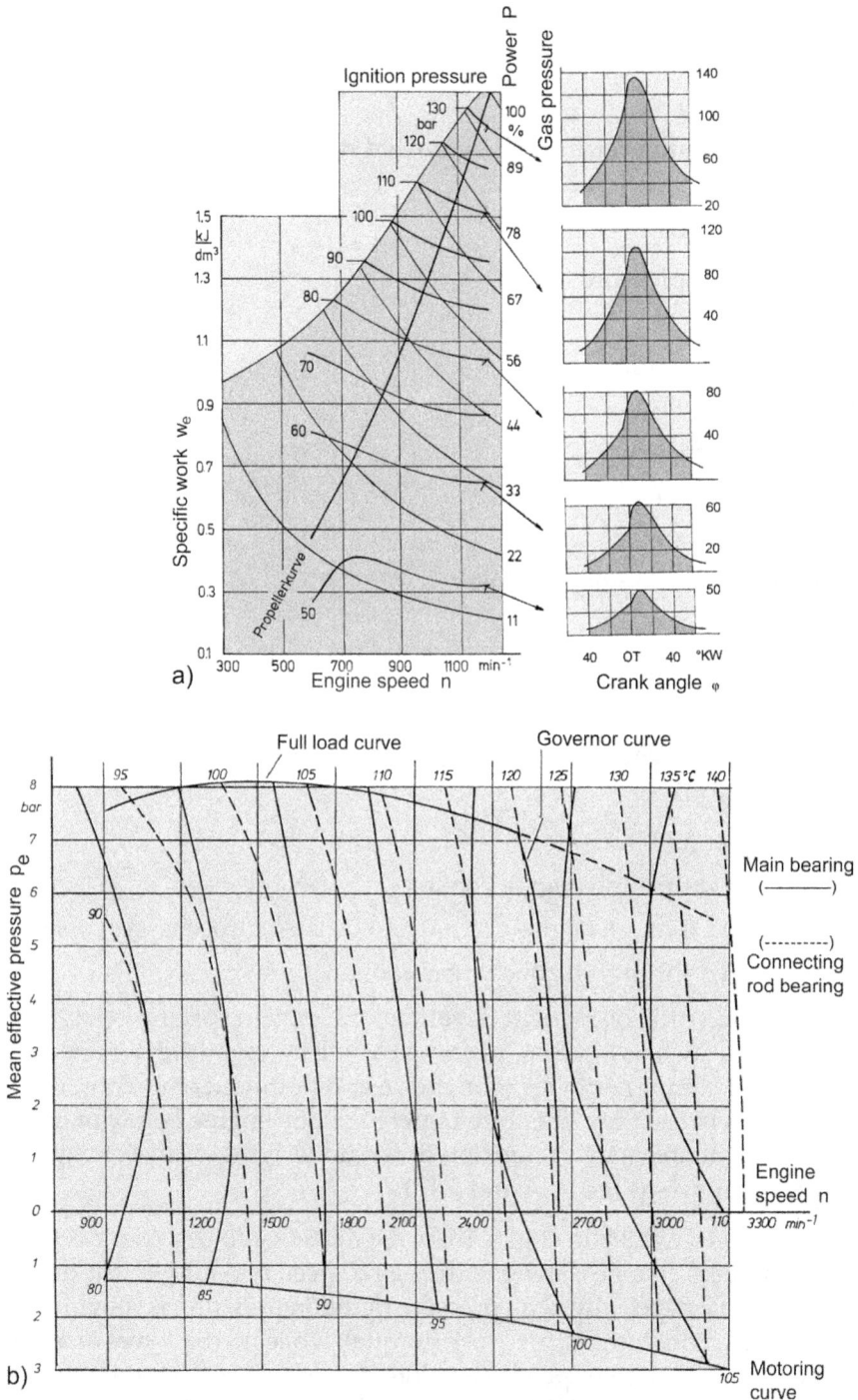

Fig. 2.14 Engine maps represent specific properties of engines, their components, and functional groups. Examples: (a) ignition pressure map of a supercharged four-stroke diesel engine, operating on a propeller curve; (b) bearing temperature map of a four-stroke diesel commercial vehicle engine, measured 1 mm below surface.

High-cycle loading is the cause of many engine component failures, because materials are much more sensitive to dynamic than to static loads. The significance of low-cycle loading was not recognized for a long time, but eventually operational experience with engines in intermittent operation showed how unfavorable this could be, even for applications with relatively low utilization.

Different load collectives, with corresponding effects on component loading, also arise as a function of the type of machinery that the engine is driving, because gas and inertia forces on engine components change along with speed and torque. The number of load changes is a function of engine speed and utilization. "Utilization" can be understood to mean:

- *Time utilization*; in other words, operating hours per year. This is entirely dependent on the type of engine and its particular application. Assuming an average speed of 50 km/h, a passenger car engine that covers 20,000 km per year is in operation for all of 400 hours per year. For commercial vehicles, utilization is strongly dependent on service conditions: long-haul, local service, delivery service, construction sites, etc. Accordingly, distance covered annually may amount to 200,000; 90,000; 50,000; or 10,000 km. Marine engines operate for 6000 to 7000 hours per year.

- *Proportion of full load, part load, and idling* in the engine's operating time. In [2-2], the maximum residence time for passenger cars is at about 20% of full load (torque) and below 50% of maximum speed (Fig. 2.17).

Commercial vehicle engines and general-purpose engines have very different load profiles, calling for different engine families, as indicated by engine manufacturers [2-3, 2-4].

- Engines used to power machinery such as pumps, compressors, or construction equipment are subjected to a relatively high full-load proportion in their total running time (Fig. 2.18).

- Engines for railway locomotives (switchers, road engines, powered railcars) enjoy relatively low utilization in Europe (in contrast to the United States). For example, industrial engines (switching and local service) spend 75% of their operating time in the lowest load range [2-5]. Even for road engines, the full-load proportion is low (Fig. 2.19).

- For ships, utilization is high [2-6]: conventional ferryboats spend about 40% of their operating time at full power, hydrofoils and fast ferries (water jet propulsion) as much as 80% (Fig. 2.20).

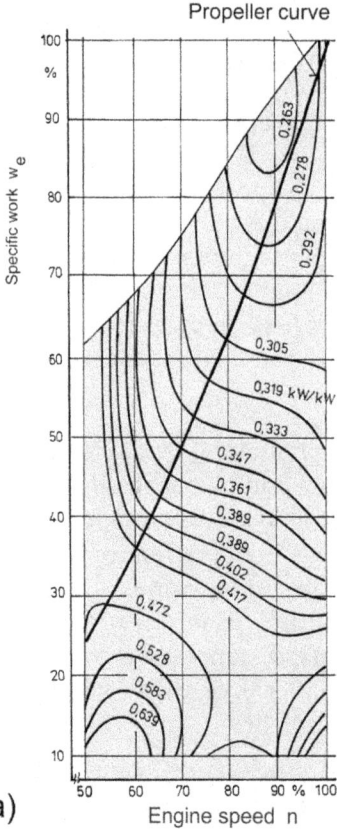

a)

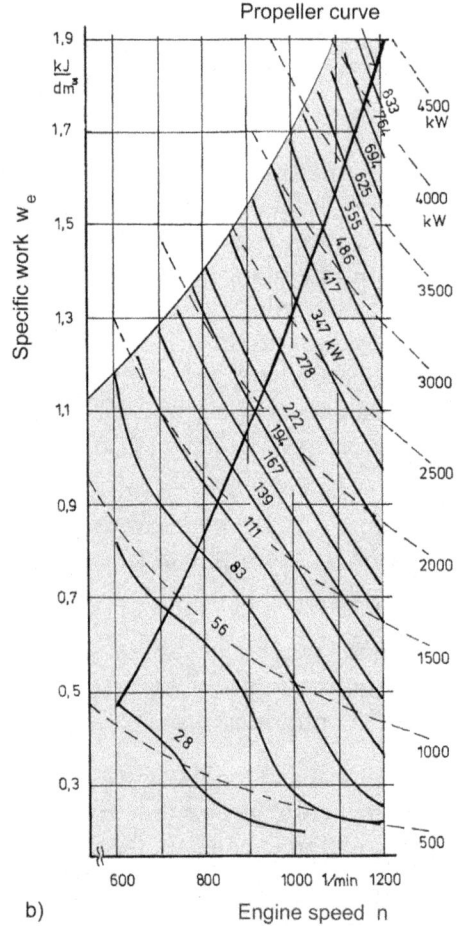

b)

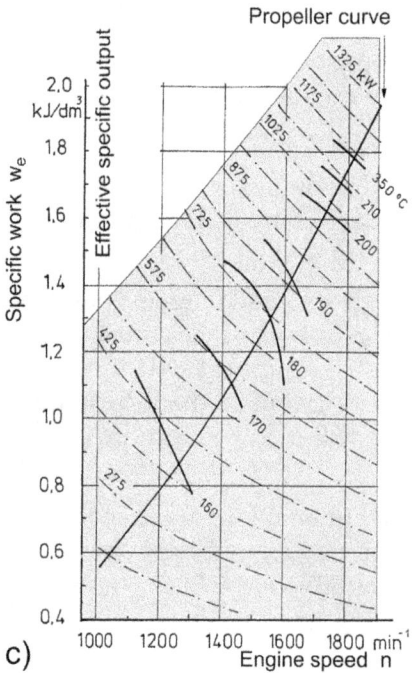

c)

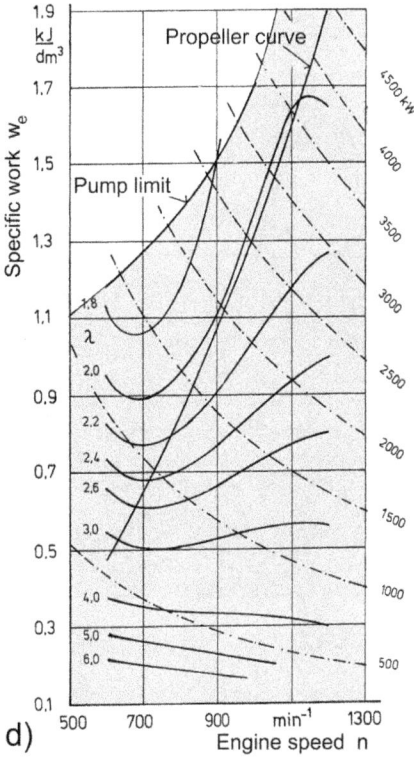

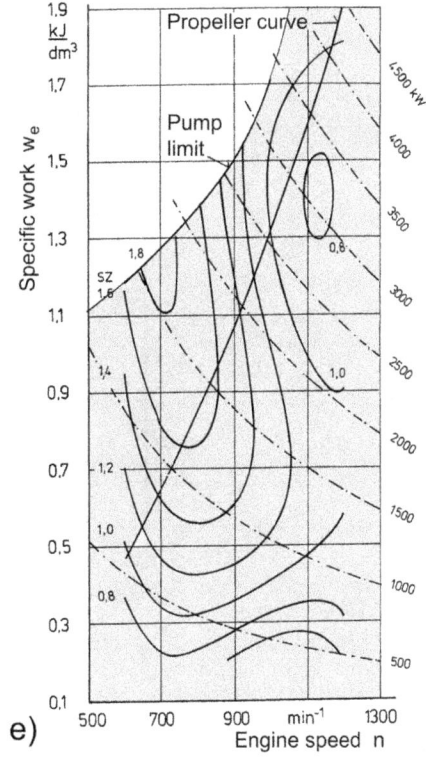

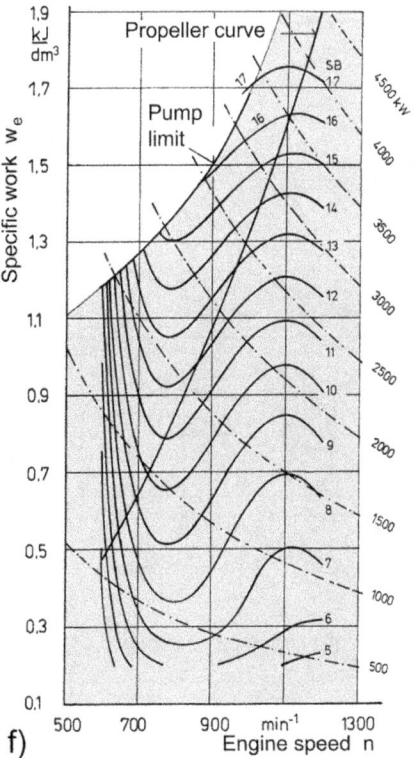

Fig. 2.15 Maps of specific engine behavior, using a four-stroke diesel as an example: (a) Heat lost to engine coolant, relative to engine output; (b) heat lost in charge air intercooler; (c) cylinder head temperature at exhaust valve seat; (d) overall relative air/fuel ratio; (e) smoke number; (f) begin of injection in crankshaft deg. before TDC.

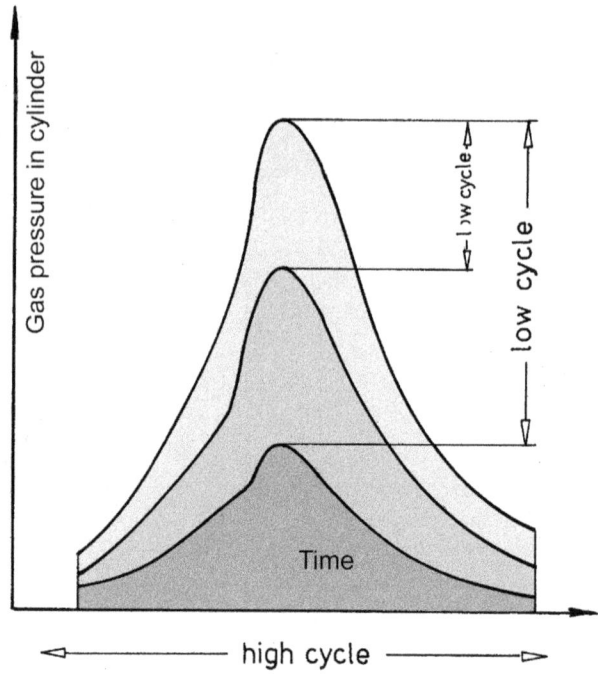

Fig. 2.16 Load types: high cycle vs. low cycle.

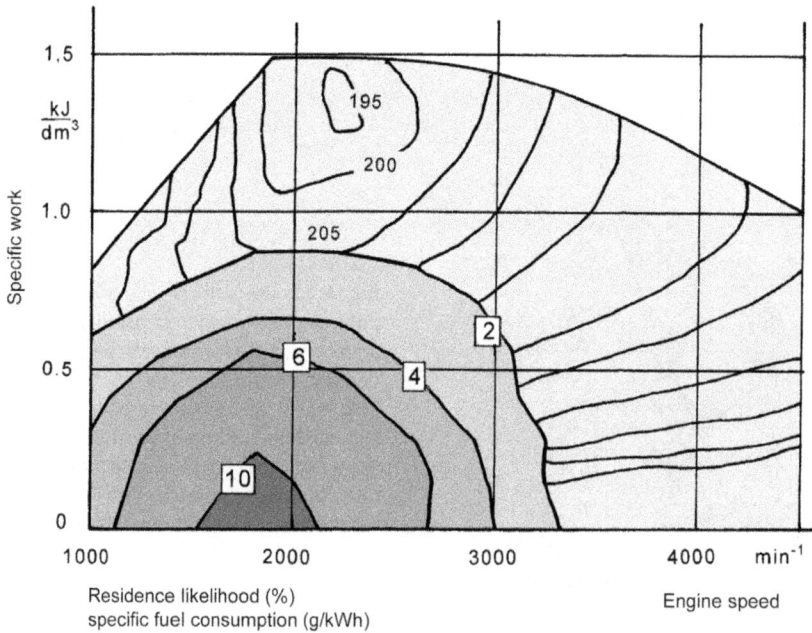

Residence likelihood (%)
specific fuel consumption (g/kWh)

Engine speed

Fig. 2.17 Residence likelihood in fuel consumption map of a passenger car engine on a cross-country trip. Source: IVK

Now, *power* is an overarching concept, composed of engine speed and torque. Depending on the service profile, other allocations of these two parameters may result. The results of research on a medium-weight commercial vehicle in delivery service are illustrated in the following, with the relative frequencies of road speed, engine torque, and engine speed (Fig. 2.21) and, delving into deeper detail, engine coolant and oil temperatures (Fig. 2.22).

The engine speed/torque contribution to power, and duration of the respective power development, determine the load collective for engine components, and so influence the susceptibility to failure and failure frequency (Fig. 2.23). This is reflected by engine manufacturers'

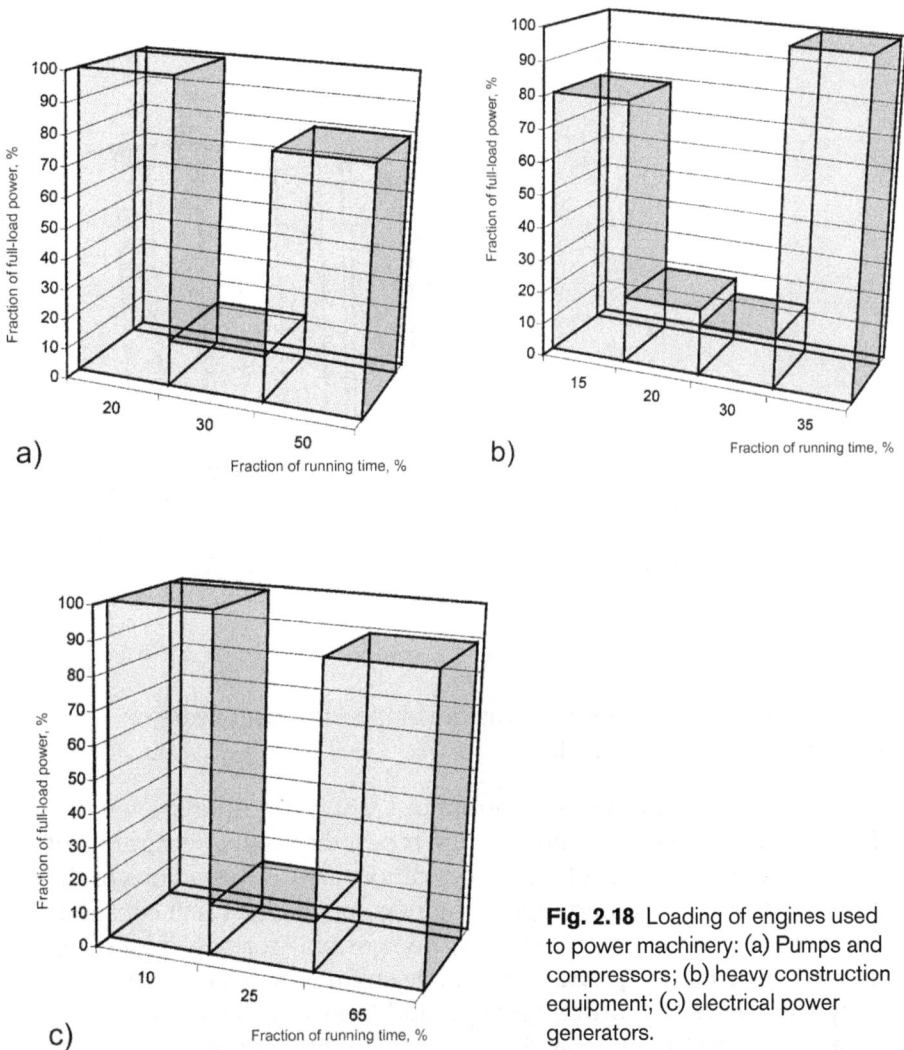

Fig. 2.18 Loading of engines used to power machinery: (a) Pumps and compressors; (b) heavy construction equipment; (c) electrical power generators.

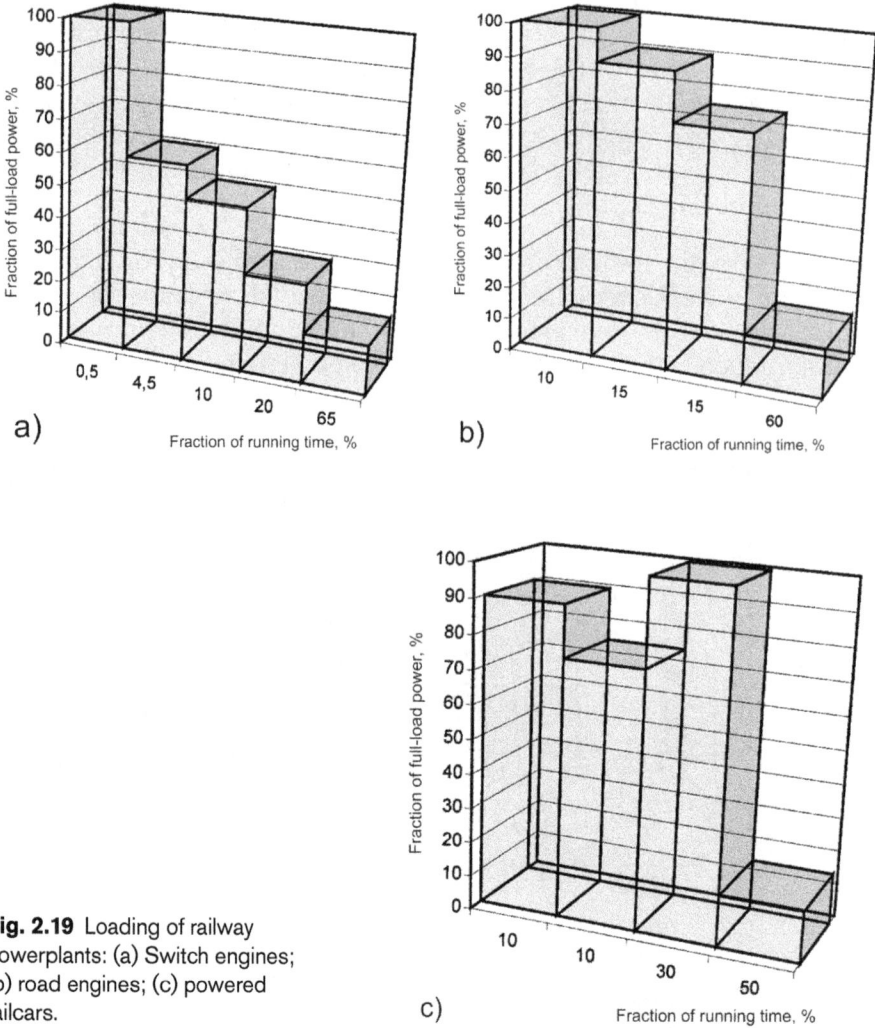

Fig. 2.19 Loading of railway powerplants: (a) Switch engines; (b) road engines; (c) powered railcars.

recommended service intervals, as developed through years of operational experience [2-6].

Different engines experience different operating conditions; even with identical applications and engine types, there may be significant differences. This impacts objective conditions, as do the actions of the vehicle driver or machinery operating personnel. One must consider that operating conditions which may be "normal" for one engine type will be unacceptable for another. Vehicle engines are started and driven at any ambient temperature; medium and large engines would not tolerate such treatment. In such situations, preheating of coolant and lubricant are indispensable.

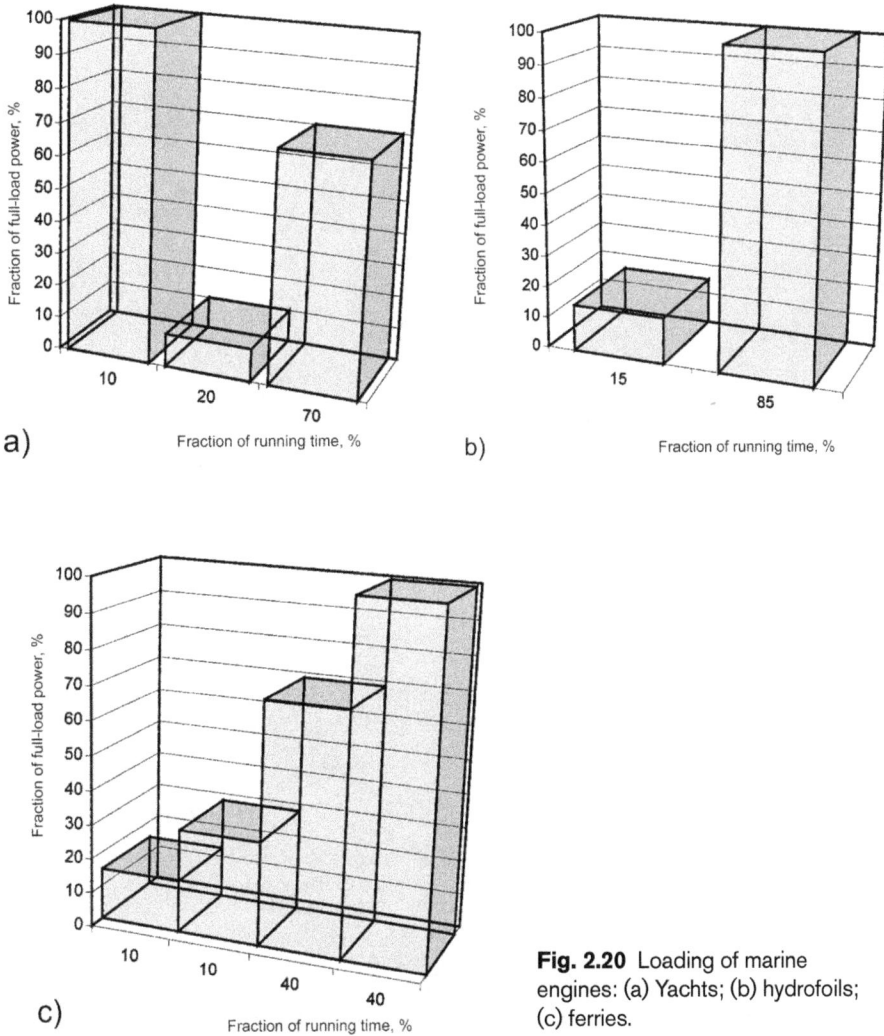

Fig. 2.20 Loading of marine engines: (a) Yachts; (b) hydrofoils; (c) ferries.

Objective conditions, i.e. conditions arising from the situation at hand, are defined not only by the load collective of torque and engine speed, but also by other factors. For motor vehicles, these include starting frequency, operating time in the low-load regime, or sudden load changes, as well as the effects of climate and topography [2-7, 2-8]:

- High-speed freeway or interstate service:
  Long-distance travel over great distances, steady speed at high engine speed up to maximum engine speed, rapid and large engine speed changes, only limited idling phases.

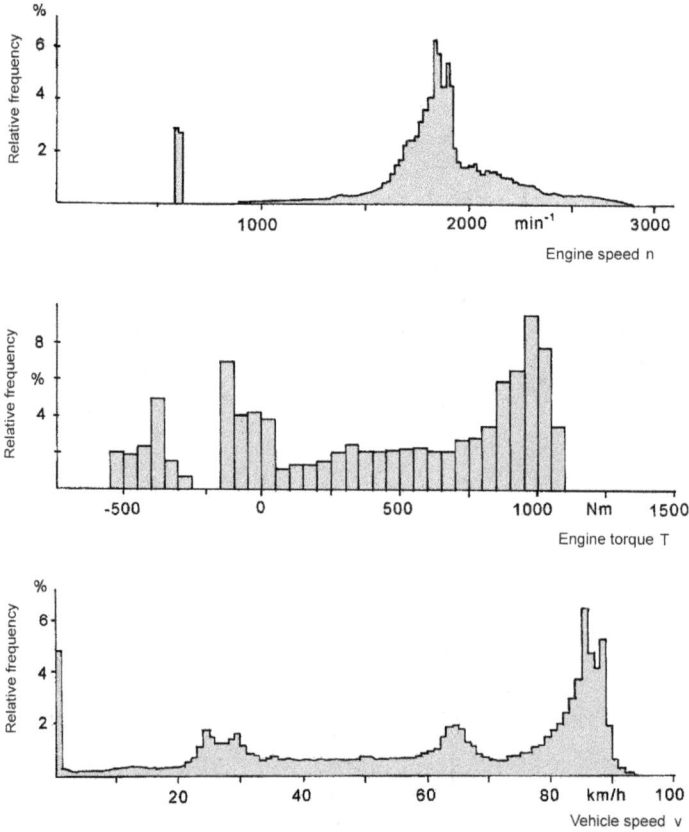

Fig. 2.21 Relative frequency of engine speed, engine torque, and vehicle speed for a diesel-powered commercial vehicle in distribution service. Source: Daimler

- Freeway congestion or freeway exits:
  Idle and low engine speeds with previously heated engine and lubricant.

- Highways:
  Medium engine speeds, slowly changing engine speed over limited range.

- Mixed city and residential-area traffic:
  Flowing traffic, frequent gear changes.

- Extreme short-range travel:
  Cold engine, shutdown phases with frequent starts and large fraction of idling, usually low engine speeds.

European long-distance traffic may pass through zones of ambient temperatures ranging from more than +30°C to less than −10°C; in the United States, these temperature limits are stretched even farther. Uphill

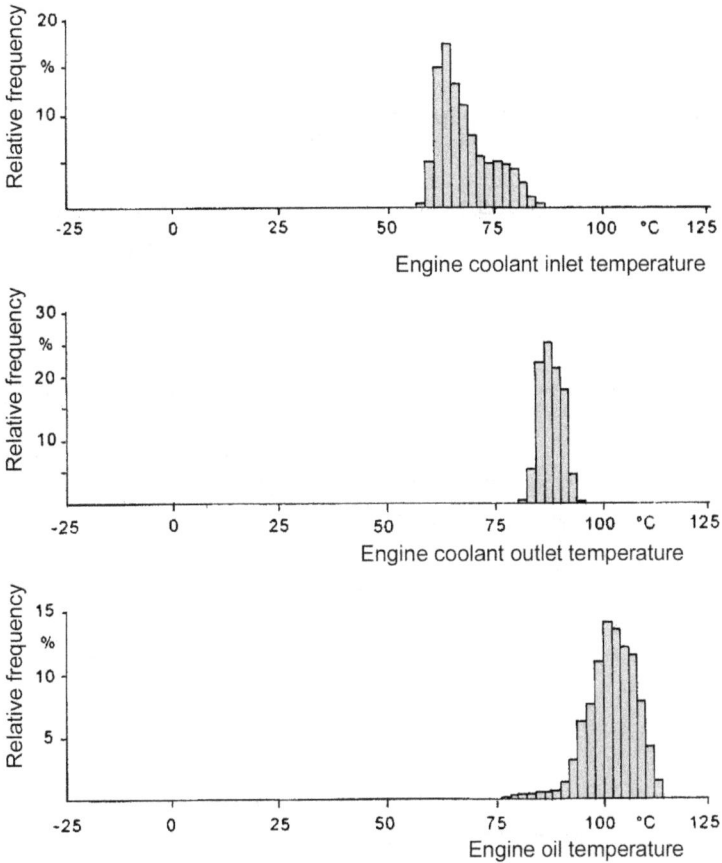

Fig. 2.22 Relative frequency of engine coolant and engine oil temperatures for a diesel-powered commercial vehicle in distribution service. Source: Daimler

grades of Alpine passes represent significantly more difficult service, especially for commercial vehicle engines. Such are the operating conditions which impose their loads on the engine and its components, and escalate the wear rate; for example, in the valve train (Fig. 2.24), increase fuel consumption and accelerate oil aging.

In countries of the Middle and Far East, in Central and South America, in Africa, and in Asian countries, engines are subjected to added difficulties by

- Continuous overloading; for example, in countries with appropriate topography (e.g., Argentina), tractor-trailers of up to 50 metric tons (55 U.S. tons) total weight may be pulled by engines developing only 125 kW (170 hp)

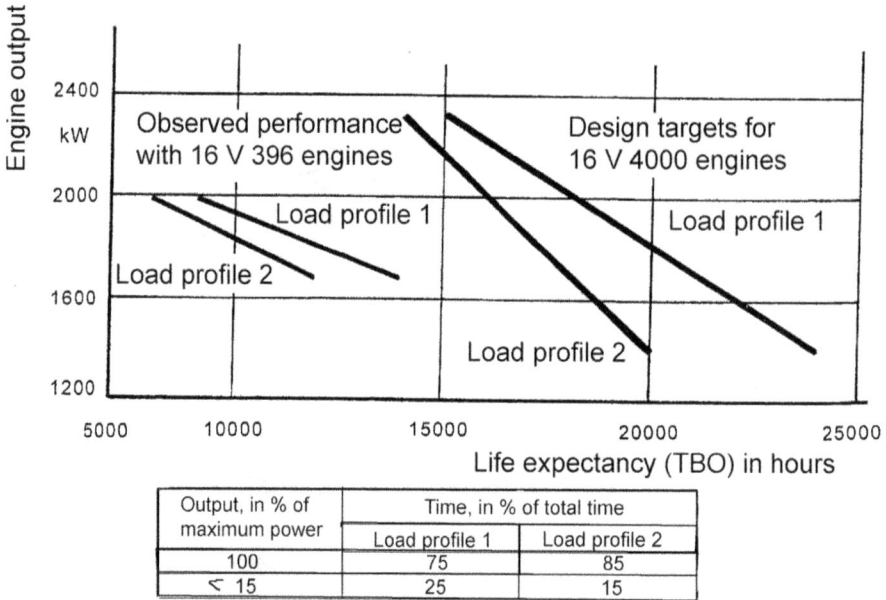

Output, in % of maximum power	Time, in % of total time	
	Load profile 1	Load profile 2
100	75	85
≤ 15	25	15

Fig. 2.23 Life expectancy as a function of load profile and power output for two engine families. Old: 16-cylinder V engine, 396-series (left); new: 16-cylinder V engine, 4000-series (right). Source: MTU

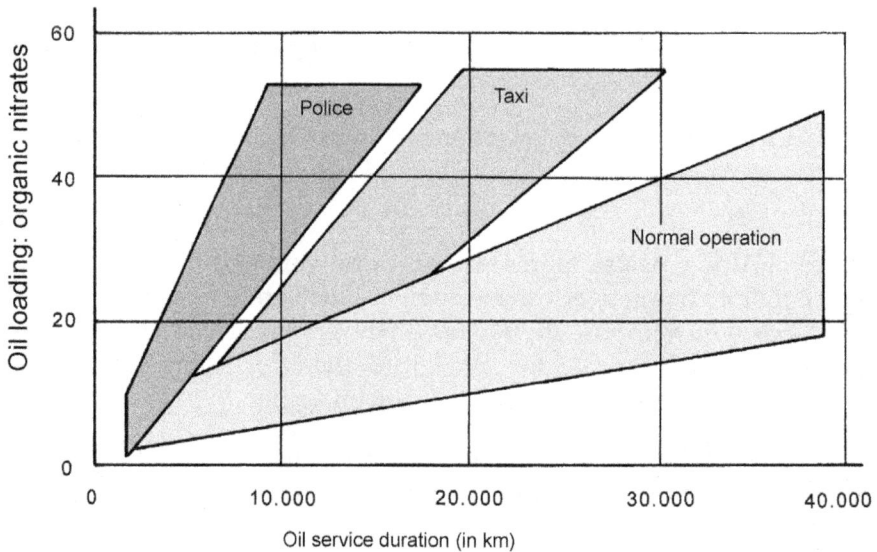

Fig. 2.24 Engine oil loading as a result of unfavorable operating conditions (organic nitrates, as a function of oil service duration).

- Poor road conditions, with corresponding high airborne dust concentration (Fig. 2.25)
- Obsolescence of engines and vehicles
- Dust, water, and salt in induction air
- Inferior fuel quality (particularly in marine applications)
- Improper engine storage before placing in operation (Fig. 2.26)
- Inadequate, often necessarily improvised maintenance

These factors—type, extent, and duration of loading; boundary conditions associated with the overall machine installation; and operating conditions—determine the mechanical, thermal, electrochemical, and tribological loads on individual engine components. These lead to corresponding demands which may result in wear, deformation, cracks, and fractures. The results may range from operational malfunctions and damage to complete engine failure, with possible serious consequences. To prevent predictable damage, engines are fitted with various protective measures. These protect the engine against overspeeding (which would lead to self-destruction), overloading, or other impermissible or dangerous operating situations.

Fig. 2.25 A Bolivian highway. The grid visible in this photograph is a wire mesh mounted in front of the windshield as protection against rocks thrown up by oncoming traffic.

Such protective measures include governors, emergency air shutoff valves, crankcase relief valves, decompression valves, and many others:

- *Decompression valves*, installed in cylinder heads to reduce overpressure, such as may be encountered in large two-stroke engines during crash-stop maneuvers when an engine, still turning in the normal direction, has its rotation suddenly reversed. By opening these valves during starting, a bypass port vents the cylinder head. If water has collected in the cylinder, this will be apparent as it is blown out. Moreover, decompression valves make it easier to turn an engine over during maintenance and repair operations.

- *Emergency air shutoff valves* for emergency engine stops. The air shutoff valve is mounted in the intake tract and, in the event of an emergency, cuts off the flow of induction air. This permits more rapid engine shutdown than interrupting fuel supply.

- *Crankcase relief valves* (blast valves, crankcase relief doors, crankcase safety covers) serve as safety valves in the event of crankcase

Fig. 2.26 Replacement engine for a power-generating station in Calafate, Argentina, was subjected to Patagonian sandstorms for an extended period before it was installed.

overpressure. These permit rapid venting of pressure resulting from an oil vapor explosion, protecting the crankcase against damage (Figs. 2.27 and 2.28).

The primary protective function of a governor is to prevent engine runaway in the event of sudden unloading. Its secondary protective function is to limit engine filling if charge air temperature exceeds acceptable values. Limits for coolant or oil temperature, or oil pressure, can also trigger filling limitation or filling cutback.

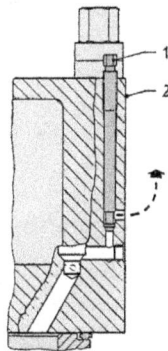

Decompression valve.

A vent passage incorporating a valve seat is drilled into the cylinder head, and sealed by a threaded valve. To open the decompression valve, the valve is unscrewed and the tapered valve head lifted from its seat, allowing air from the cylinder to escape through the decompression valve passage.

1   Decompression valve
2   Cylinder head

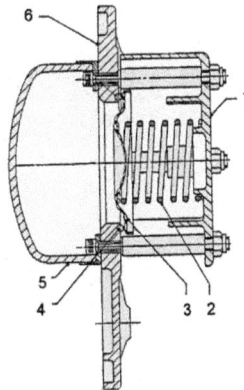

Crankcase relief valve.

A crankcase relief valve consists of a housing, inspection cover, valve head and flame arrester, installed in a crankcase inspection port. A flame arrester to prevent flames from exiting the crankcase is bolted to the inside of the inspection port. If overpressure exceeding the valve opening pressure is created within the crankcase, the valve head is forced against the spring, thereby opening the valve. When pressure drops, the spring reseats the valve head and closes the valve.

Fig. 2.27 Engine protection equipment (1). Source: MTU

1	Housing	4	Sealing ring
2	Spring	5	Flame arrester
3	Valve head	6	Inspection cover

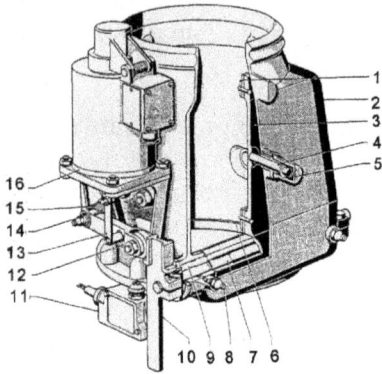

Emergency air shutoff valve.

The emergency air shutoff valve is installed in the intake air ducting, and interrupts supply of combustion air to the cylinders. With no air to support combustion, the engine comes to an immediate stop. This stops an engine more rapidly than can be achieved by cutting off fuel supply. In the event of engine overspeeding (overrevving), the engine monitoring system actuates a solenoid which releases the emergency air shutoff valve, which is then pulled shut by the air flow. This also trips a limit switch which reports the situation to the monitoring system.

1	Sealing ring	7	Shaft	13	Link
2	Housing	8	Grease fitting	14	Pin
3	Flap	9	Lever	15	Torsion spring
4	Pressure pin	10	Bearing	16	Solenoid
5	Spring	11	Microswitch		
6	Flap seat	12	Lever		

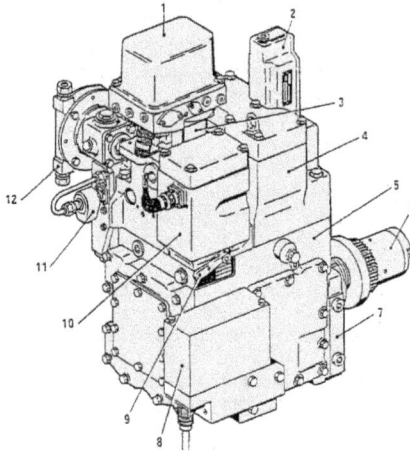

Engine governor.

1 Electro-hydraulic target value control unit
2 Engine filling sensor
3 Target value adjusting unit
4 Governor output
5 Governor body
6 Torsionally elastic governor drive
7 Oil pump
8 Electronically controlled fill limiter
9 Shutoff lever
10 Electric shutoff device
11 Idle speed increase device
12 Backup actuation via Telefax device

Fig. 2.28 Engine protection equipment (2). Source: MTU

Chapter 3
Failure—Definitions and Concepts

According to the *American Heritage Dictionary*, a *failure* is "the condition
. . . of not achieving the desired end or ends; of being insufficient
or falling short; a cessation of proper functioning or performance;
nonperformance of what is requested or expected; a decline in strength
or effectiveness." A shorter definition may be found in [3-1]: "Failure:
a loss of functional ability and necessary or desired properties. Failure
is the result of a process—the *damage process*." This definition draws
attention to the connection between a result—failure—with the
causative agent—the damage process. Now "damage" is an overarching
concept for a multitude of subsidiary concepts, which are listed in a
guideline, VDI 3822 (Section 1, p. 3) ("Failure analysis—Fundamentals,
terms, definitions—Procedure of failure analyses"), and in Table 3.1.

Failure is the result of a damage process which may be initiated by one
or more causes. These causes may lie within the damaged component
itself, as primary failure, or in other components whose functional
failure, fault, or damage affects the damaged component. Often, the
cause is completely disproportionate to the ensuing damage.

Independently of the causes, processes, appearances, and type of
failure, and of the component, functional group, or machine in question,
failures are often explained by a wide variety of physical, chemical, and
electrochemical processes, as summarized in VDI 3822. Ultimately, all
failures of machine components may be traced back to these processes.

- Failure through mechanical loads
    - Overloading failure
    - Yield failure

Table 3.1
Relevant Terms and Definitions

Term	Definition
Failure	changes in a component by which its intended function is impaired considerably, or rendered impossible
Previous failure	failure that previously occurred on a component or in a plant
Primary failure	failure that occurred first, being the cause of further failures
Consequential failure	failure that is triggered by a previous damage on the same or a different component
Recurrent failure	repeated occurrence of a similar failure
Failed item	component or sample thereof, subject to the failure
Reference item	component or sample thereof, used for comparison with the failed item
Failure location	Place of failure on the component, encompassing the primary failure area
Failure symptoms	external condition of the failed item
Failure characteristics	characteristic features of a failure
Failure hypothesis	The probable type of failure and assumed failure time history serve to establish a failure hypothesis and are typically used to guide the sequence of examinations. The failure hypothesis is influenced decisively by experience.
Failure time history	development of a failure over time
Failure mechanism	individual or combined events at the material technology level, leading to a failure
Type of failure	allocation of a failure to a specific failure mechanism
Failure cause	sum total of influences triggering a failure
Failure analysis	systematic examinations and tests to identify failure time history and failure cause
Failure correction	corrective action taken to prevent recurrence of a specific failure
Failure prevention	preventive action taken to prevent the occurrence of failures

- Brittle failure (fracture)
- Fatigue failure
- Fatigue failure with ultimate yield failure
- Fatigue failure with ultimate brittle failure

- Failure through corrosion in aqueous media
  Corrosion types without mechanical loading
    - Uniform surface corrosion
    - Shallow pit corrosion
    - Pitting
    - Crevice corrosion
    - Corrosion due to varying ventilation
    - Galvanic corrosion
    - Corrosion by deposits (contact corrosion)

- Selective corrosion
- Acid corrosion (cold end corrosion)

Corrosion types with added mechanical loading
- Stress crack corrosion
- Fatigue corrosion

Hydrogen-induced corrosion
- Bubble formation
- Crack formation, interior cracks
- Stress crack corrosion
- Corrosion by pressurized hydrogen below 200°C

Corrosion by microbiological processes
- Microbiological corrosion under aerobic conditions
- Microbiological corrosion under anaerobic conditions

- Failure through thermal loading

Hot fracture
- Hot overloading failure
- Hot fatigue failure due to varying mechanical loads at high temperatures
- Failure due to changing temperatures (thermal fatigue failure)

Hot cracking
- Welding stress cracking
- Creep cracking
- Cracking due to changing temperatures
- Grinding cracking
- Hardening cracking
- Hot cracking
- Soldering cracking
- Thermal deformation

Thermal surface damage
- Discoloration, tarnishing
- Scaling
- Hot vapor oxidation
- Hot water oxidation
- Melting corrosion
- High temperature corrosion
- Burning
- Fusing

Functional faults caused by deposits
- Boiler scale, water scale
- Mineral deposition
- Silica deposition

Damage by diffusion processes
- Inward diffusion by materials from the environment
- Outward diffusion by steel alloying elements

- Failures by tribological loading
  Sliding abrasion
    - Hydrodynamic sliding abrasion
    - Elastohydrodynamic sliding abrasion
    - Sliding abrasion with mixed friction
    - Sliding abrasion with solid-body friction, technically dry friction
  Rolling abrasion
    - Hydrodynamic rolling abrasion
    - Elastohydrodynamic rolling abrasion
    - Rolling abrasion with mixed friction
    - Rolling abrasion, technically dry rolling contact
  Fatigue abrasion
    - Fatigue wear with mixed friction
    - Fatigue wear, technically dry
  Abrasive wear
    - Scouring abrasion
    - Sliding grain abrasion
  Flow abrasion (erosion)
    - Hydroabrasive erosion
    - Solid particle erosion
    - Gas erosion
    - Drop erosion
    - Cavitation erosion

Damage processes may operate at different speeds:

- Explosively; e.g., oil vapor explosion in a crankcase
- From impact; e.g., hydraulic locking, or crankcase damage from a broken connecting rod
- Exponentially; e.g., bearing, piston ring, or piston seizure
- Long-period fatigue due to prior damage

The speed at which a failure develops to some extent determines whether the failure will be noticed in time to permit intervention.

In evaluating engine failures, one must differentiate between fate or the history of an individual engine and the engine type, and such considerations must always be made in the context of so-called "normal" engine behavior; i.e., the current state of the engineering art.

In the case of an engine type generally known for its satisfactory behavior, individual examples may draw attention to themselves by their frequency of faults and failures. (In the auto industry, this was once associated with the concept of "Monday morning cars."[1]) The cause of failure lies in the chance meeting of multiple failure causes. This "residual risk" is taken into account, for example in antifriction bearing calculations, by application of a "survival rate" (90%, 99% . . .). For commercial vehicle engines, according to [3-2], a "B10 life expectancy" is defined, which specifies that for an anticipated service of one million kilometers (621,000 miles), only 10% of the engines will need overhaul or replacement. This is in contrast to system-related defects; e.g., as the result of design flaws or faults due to inadequate testing.

"Normal" failure behavior of an engine may be represented by a so-called "bathtub curve," named for its shape (Fig. 3.1). The first portion of the curve—a straight drop-off—characterizes the *break-in phase* with chance failures, caused for example by material flaws, or manufacturing or assembly defects. The adjoining region of the *utilization phase* exhibits a constant failure rate; e.g., through operator error, dirt, vibration, loosening of fasteners, hydraulic locking, or maintenance errors. This transitions to a third region of progressively rising failure rate, the *wear failure phase*. This rise in failure rate is marked by wear, aging, and fatigue failures. Appropriate maintenance measures may shift this region to the right; that is, the onset and increase of failures can be delayed.

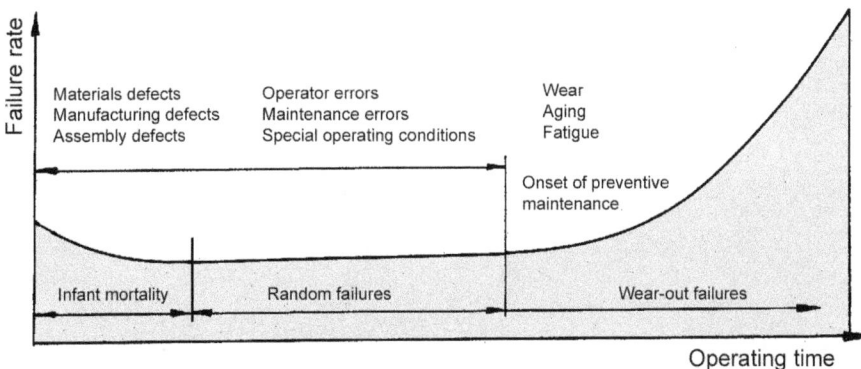

Fig. 3.1 Bathtub curve: idealized illustration of failure rate as a function of operating time.

---

[1] This concept was based on the assumption that workers showed up for work after the weekend, suffering from a lack of sleep, and therefore went about their jobs with less care than usual.

A failure is the result of a process, which, originating from one or more causes, takes a specific course, determined by the conditions and situation at hand. This first makes itself known as a functional fault which, if timely intervention is not made, may result in engine failure and further damage.

Chapter 4
Causes of Failure

4.1 Wear and tear

The task of machines is to convert energy into other forms, as well as transmit and modify forces, moments, and motion. In fulfilling these tasks, loads are applied to machine components, which results in mechanical, thermal, tribological, and electrochemical demands on these parts, as a function of their shape, dimensions, nature, materials, and manufacturing. These demands result in component wear. This is understood to mean an undesirable reduction in serviceability as a result of physical and/or chemical influences. The extent of such influences is affected by the nature of the machine's operation; e.g., by its utilization (operating hours per year) as well as the driven distance relative to rated power (idle, part load, full load) and by external conditions or limitations; e.g., state of maintenance, dust, moisture, and salt content of the ambient air.

"Wear and tear" can manifest itself as simple wear, aging, corrosion, biological damage to material, as well as breakage. Wear and tear are unavoidable accompaniments to machine operation. DIN 31051 [4-1] succinctly states that

". . . wear and tear are the price to be paid for use of the equipment. The equipment cannot be operated without incurring wear and tear . . ."

The rate of wear and tear (wear and tear over time) is the deciding factor for operational safety, life expectancy, and operating economy of machines. For this reason, wear and tear are counteracted by wear and tear reserves—added material, the loss of which, through wear and tear, does not impair the function and operational safety of the component.

For example, if manufacturing tolerances result in bearing clearances of 0.200 to 0.250 mm, then specification of an operational wear limit of 0.300 mm provides a wear reserve of 0.100 mm. Additionally, the useful life of engine components is extended by providing sufficient material for re-machining or re-working, often in several stages (Fig. 4.1).

Of course, wear and tear rates differ for various engine types (passenger car, commercial vehicle, general purpose). One publication [4-2] cited the following wear rates for large four-stroke heavy oil engines:

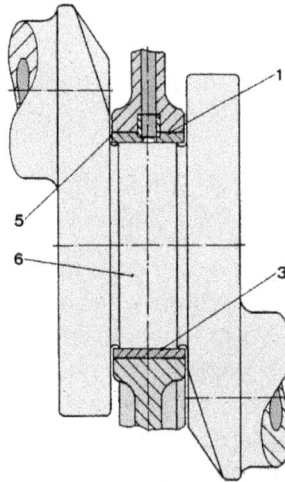

No.	Desig-nation	New condition	Remachining		Dimension		Clearance		Oversize		Wear limit
		Nominal dimension	Stage	Nominal dimension	lower	upper	min.	max.	min.	max.	
1	Crankcase bore dia.	180^{H6}	1	180.50^{H6}	0	+0.029			0.204	0.252	0.200
			2	181.00^{H6}							
			3	181.50^{H6}							
3	Main bearing ID	168.142	1	167.942							
			2	167.742							
			3	167.542							
			4	167.342							
			5	167.142							
			6	166.942							
3	Crankshaft journal dia.	168_{h6}	1	167.8_{h6}	0.025	0	0.142	0.203			Clearance 0.230
			2	167.6_{h6}							
			3	167.4_{h6}							
			4	167.2_{h6}							
			5	167.0_{h6}							
			6	166.8_{h6}							

Fig. 4.1 Operating wear limits of engine components, in this example a crankshaft main bearing. For simplification, only wear points 1 and 3 are shown in the table. Source: MTU

96% of all valves required minor regrinding at 5000 to 6000 operating
hours;
70% of all ring grooves showed less than 0.01 mm wear per 1000
operating hours;
80% of all cylinder sleeves showed wear of 0.03 mm per 1000 operating
hours.

The consequences of unacceptable wear and tear are operational
faults or damage to components and functional groups. Because
of functional relationships and dependencies within a machine or
equipment installation, such faults manifest themselves cumulatively.
Often, various faults and external influences place additional loads
and greater demands on components. These faults and influences may
include technical flaws, operating errors, and external influences. The
boundaries between these influences are not always sharply delineated.

4.2 Technical defects (product defects)

4.2.1 Design flaws (planning flaws)

Design flaws, in the sense of a defective or unsatisfactory engine
concept, may be largely eliminated by the current state of the art. Basic
engine configurations are firmly established and well proven. Today,
only single-acting engines with conventional crank train and valves are
produced, employing a four-stroke cycle or, in the case of very small
and very large engines, a two-stroke cycle. Engines are exclusively
built by vehicle manufacturers or dedicated engine manufacturers,
while most engine components in turn are sourced by specialist
suppliers who act as the interface for know-how, and collect and apply
experience from all branches of engine technology. Engines demand
comprehensive and thorough development. Thanks to electronics
and electronic data processing, the instruments of computation,
simulation, experimentation, and measurement technology have made
great progress. This, and the experience gained through operation
of hundreds of millions of engines, form the foundation of "correct"
engine conception and design.

Nevertheless, faults may creep into the layout of individual
components: in design assumptions for the loads to be borne; in
execution inappropriate for the task; inadequate dimensioning;
unsuitable shape, fits and tolerances, materials and heat treatment, or
manufacturing processes. Such faults usually do not directly impact the
central functional groups of an engine, such as crank train, valve train,
etc., but rather the engine peripherals. Also, the consequences of such
design faults are usually only apparent after long or even very long

service, because the causes of short-term failures can be brought to light during the engine development phase.

An example for faults and damage during introduction of a new line of marine engines is drawn from a manufacturer's publications [4-3]:

". . . during the first 2750 operating hours (consider that in an automotive application, with an average speed of 50 km/h, this represents a distance of 137,500 km, or 85,400 miles!) two main problems arose, which resulted in extended downtime:

- Turbocharger failure as a result of defective layout of the turbine blade damping wire
- Exhaust valve failure, as a result of metal fatigue, caused by corrosion

To date, a series of minor modifications have been carried out, all based on experience gained on board. The following are worthy of mention:

- Change cylinder sleeves from two-part to one-piece design. We had four cases of cracked flame rings, which did not recur with the new design.
- Change studs for combined indicator and safety valve; the strength of these bolts proved to be inadequate.
- Change the fuel pump stop cylinders, as air leaks occurred.
- Several stages of turbocharger modifications.
- Modification of injection pump rocker arms due to inadequate lubrication.
- Change fuel pump delivery valve spring due to material defect.
- Injection pump modification due to design problem with clamping screws . . ."

Engines must meet certain requirements in terms of output power range and transmission layout, as well as the type of service and power definition; the right motor for the right job! Underpowered engines will be overloaded, while overpowered engines will operate uneconomically at part load. Improper engine application, improper matching to operating conditions; e.g., in power transmission characteristics and the machinery to be powered, can cause faults and failures. Maintenance intervals and scope must be matched to the type of service and loading.

Many faults and failures are the result of mistakes in planning the machine installation, for which there are countless opportunities for error. This applies less to the engines of standard production passenger cars and commercial vehicles than to general-purpose engines; i.e., stationary operation, marine, rail, and special vehicles. For this reason, engine manufacturers provide comprehensive planning materials (installation guidelines, Ship Owner's Guide, projections of engine installations, etc.), which cover virtually all aspects of an engine installation: type of cooling, cooling system layout, engine preheat, engine keep-warm operation, combustion air system, exhaust system, fuel system, lubrication system, electrical system, engine operation and monitoring, engine mounting, engine room configuration, engine room ventilation, noise damping, piping, power transmission, engine starting, and rigging the machinery set. Another important point is in planning for maintenance. If service operations are made difficult by design or installation conditions, this will lead to their not being carried out with the required frequency or care, which in turn increases the likelihood of a fault. In one example, a light truck in distribution service, the driver's seat had to be raised to gain access to the dipstick and check engine oil level. For the (usually younger) drivers of this vehicle type, this became too tedious and cumbersome, especially since there was a good chance of soiling the seat; oil level checks were not performed regularly, with the consequence that this vehicle type soon stood out in the failure statistics.

Increasingly, imponderables of engine installation are avoided by providing the customer not with a "naked" engine, but rather an installation-ready module, e.g., for buses or railcars. This concept was realized decades ago with the propulsion package of the "Leopard" main battle tank. This avoids not only installation errors, but also prevents vehicle manufacturers from using unsuitable components, such as heat exchangers and filters in the powerplant installation. Another advantage of this concept is that external test runs may be conducted on power modules. For buses, such a module consists of a framework that carries engine, transmission, air filter, oil tank for the hydrostatic fan drive, power steering fluid tank, engine oil tank with automatic replenishment, coolant recovery tank, radiator with hydrostatically driven fan, air conditioning compressor, and a second alternator, all in a functional, compact package (Fig. 4.2).

This ensures that only proven, matched components are used, and individual installation during the manufacturing process is simplified. This reduces the chance of errors, and permits rapid module replacement for service and repair. These tasks are much easier to perform on an uninstalled module. Overall, this ensures optimum embedding of the engine within the vehicle concept, and good integration of accessories.

Fig. 4.2 Bus propulsion module. Source: MAN

Even with a good planning concept, the devil is often in the details: coolant plumbing, manifolds, shrouds, valves , gates—improperly located, or improperly designed—can be the source of faults and failures.

On medium and large engines, fuel type plays a vital role for operational behavior and susceptibility to failure: gasoil, diesel fuel, MDF, and heavy oil. The worse the fuel ("worse" being synonymous with "cheap"), the more one must contend with the possibility of operational faults.

4.2.2 Materials defects

Materials faults, through use of unsuitable materials, pairing of incompatible materials, or substitution of materials are, fortunately, rare.

The fact that materials faults occur at all is due to the difficulty in making absolutely faultless materials, and the difficulty in testing materials used in engine components three dimensionally; that is, throughout the entire volume of the component. Such materials faults can arise in all phases of the manufacturing process: contamination of the source material, shrink holes, inclusions, bubbles, and separation in castings; lamination in rolling; cracks and folds in forging; structural defects, heat treating defects, or bonding or plating failures in coated components (bearings, piston rings). The price advantages of gray-market parts are due in no small part to the use of inferior, or at least unsuitable materials.

4.2.3 Manufacturing defects

Manufacturing defects may result from

- Nonadherence to dimensions, tolerances, fits, and clearances
- Not achieving the required surface finish
- Mechanical surface damage through turning, milling, and boring: scoring, scratches, chatter marks, nicks, and notches
- Dividing errors on gear teeth
- Heat treating defects
- Tears in cold forming operations

The surface structure of tribologically loaded components is especially critical, such as the formation of the outer layer in honing operations. Manufacturing defects in highly stressed sections such as fillets, chamfering of oil passages, etc. can be the initiators of fatigue failures.

In assembly, misalignment, lack of concentricity, and, above all, improper torqueing of bolted connections may be recognized as the causes of faults and failures, along with errors in rigging and balancing the machine installation. Manufacturing defects are also the cause of the high failure rate of gray market replacement parts.

4.3 Operating errors

4.3.1 Overloading

Overloading—imposing loads beyond permissible limits, and the resulting impairment of functional ability [4-4] represents a "classical" cause of failures. Depending on the physical process involved, distinction is made between mechanical, thermal, chemical, corrosive, and tribological overloading. Engines are overloaded when they are asked to produce more power than they were designed to deliver, or when that power is called for under more difficult conditions such as lower barometric pressure, high induction air temperature, and high coolant and lubricant temperatures. We are all familiar with the sight of overloaded trucks crawling up a grade, announcing their load condition by their low road speed and billowing black exhaust smoke. Excessively high engine loading and overloaded trailers represent an overloading condition often observed in vehicle operations outside Europe. In marine operations, improper matching of engine, propeller, and ship may lead to engine overloading.

For an engine, overloading means that too much torque is being demanded. Because its power is limited, engine speed must drop ("lugging"). But this is only permissible to a limited degree, because exhaust gas temperature will climb as a result of decreasing excess air, and therefore the temperatures of thermally sensitive components such as pistons, cylinder heads, and exhaust valves will rise. Higher temperatures decrease material strength, reduce clearances, and change (either increase or decrease, depending on the situation) interference and press fits. Furthermore, lugging operation and associated high ignition pressures and pressure gradients lead to higher crank train forces. Deformation increases loads on individual components, just as it compromises their operation in conjunction with other components. Mechanical overloading leads to mechanical damage to components, in the form of (unacceptable) elastic ("reversible"), inelastic; i.e., plastic ("irreversible") deformation, to cracks and fractures. Higher crank train forces also give rise to higher bearing forces; higher pressures must be built up in the bearings' lubricating film to offset these higher bearing forces. Consequently, even tribological loading increases.

Engines are not only overloaded externally, "by the gas pedal," but also internally, by engine-internal irregularities such as cold starts with long ignition delays, excessive injection advance, in the case of gasoline engines early ignition or ignition knock, clogged filters, and dirty heat transfer devices. In an engine, mechanical overloading is associated with thermal and tribological overloading, because higher torque means higher gas pressures, which in turn are the result of increased energy conversion in the combustion chamber. More power therefore means more heat, higher component and exhaust temperatures. A familiar example is combustion knock, which, as a result of high flame propagation speeds and gas turbulence, increases heat transfer from combustion gases to the components that make up the combustion chamber to the point where these components are thermally overloaded. In diesel engines, poor spray formation from injection nozzles is often the cause of thermal damage. Thermal overloading leads to hot cracking, burnoff, and burnout, and directly to component failure due to decreased material strength at high temperatures.

4.3.2 Changing operating conditions

Marine engines are operated along the propeller curve, which is a function of the type, size, and form of the vessel. Engine and (if present) transmission must be matched to one another so that the engine can meet the ship's power demands with respect to torque and engine speed. Marine growth on the hull will increase its surface roughness, which moves the propeller curve to the left, toward higher torque.

Fouling of heat transfer equipment (heat exchangers, intercoolers) reduces cooling performance and raises coolant temperatures. Clogged air filters increase induction resistance. For motor vehicle engines, ambient temperatures and decreasing air density with altitude represent an added difficulty. Fuel quality makes its own negative contribution, if this does not meet the engine manufacturer's specifications, as is often the case in Eastern Europe as well as in non-European countries.

Today, engine designs are checked for torsional vibrations as early as the concept stage, so that torsional vibration dampers and matching of the clutch or coupling to the engine may be applied to eliminate critical torsional vibration modes. Still, the resonant frequency of a component may coincide with that of the engine or its installation, with concomitant results. Furthermore, such resonant frequencies may shift as a result of wear or increased clearances. If the damping oil breaks down, the torsional vibration damper loses its effectiveness.

Engine and attached accessories are subject to a wide range of shock and vibration. This may result in relaxation ("setting") or burnishing effects in bolted connections, which can lead to loosened fasteners.

An especially dangerous operational anomaly for positive displacement machines, which includes combustion engines, is hydraulic locking. If, as a result of leakage, large quantities of water, fuel, or engine oil find their way into the cylinder, the piston will come against an essentially incompressible fluid. As the crankshaft continues to turn, the connecting rod will be deformed to such a degree that usually the big end will fracture, and often enough in such cases, break a hole through the crankcase wall.

4.3.3 Operating errors

4.3.3.1 Break-in

As a result of their various manufacturing processes, the parts of a new engine exhibit roughness; even within specified tolerances, the shape of mating wear surfaces deviates from the geometrically ideal shape. During engine run-in or break-in, mating sliding surfaces must adapt to one another in terms of their shapes and deformation behavior; rough "peaks" need to be worn down. This is especially true for piston rings and cylinder walls, but also for bearings of any sort. Engine break-in is considered completed when the top piston ring completely seals the combustion chamber. If full load is applied to the engine before this condition is reached, hot combustion gases will blow between the ring and cylinder wall and burn the film of lubricating oil, which will make top ring sealing even worse. The ring will "burn," and wear and oil consumption will increase. In the worst case, ring and piston seizure occurs. Over the years, the break-in regimen for motor vehicle engines has been steadily reduced. For some new designs of gasoline motor vehicle engines, the entire break-in procedure consists of just a few minutes' operation under external power, during which coolant and lubrication systems are checked for function and freedom from leaks. Medium and large engines, however, are subjected to test and acceptance runs at the factory [4.21]; furthermore, the engine manufacturer issues targeted specifications for break-in modalities by the end user (Fig. 4.3).

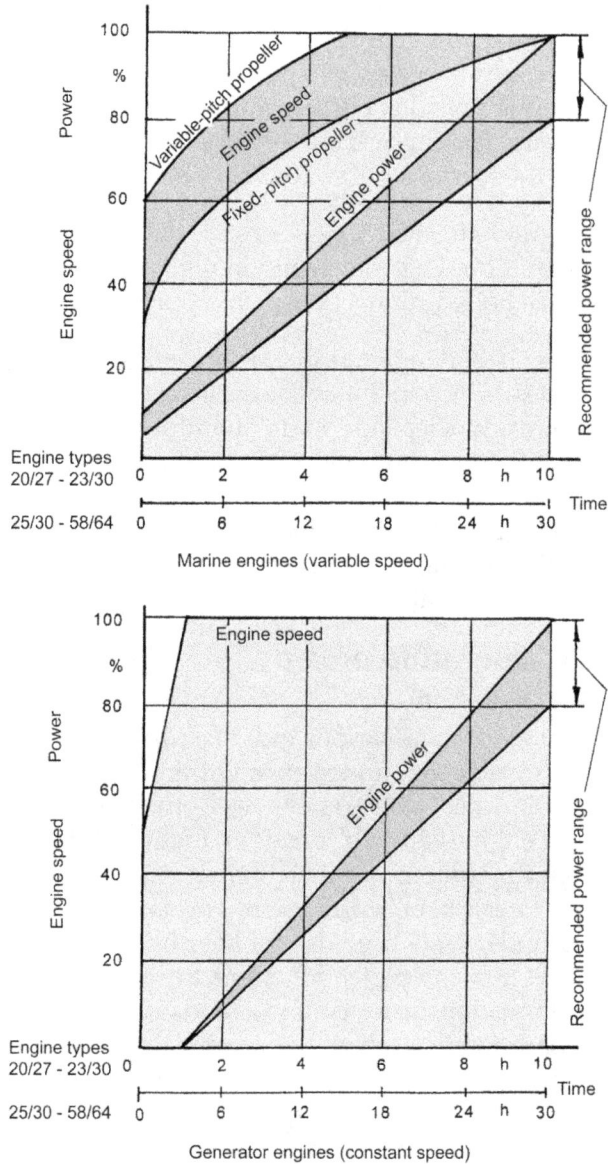

Fig. 4.3 Break-in program for medium-speed MAN B&W four-stroke diesel engines. Source: MAN

4.3.3.2 Cold start

For motor vehicle engines, cold starting ability is a prerequisite, particularly because application of antifreeze prevents coolant freezing. Theoretically, the engine is also able to tolerate operation at very low temperatures. In practice, engines do not take kindly to this, particularly if full load is applied immediately after starting. Lubricating oil is still very viscous, especially since oil temperature exhibits a significant lag behind coolant temperature. The formation of a lubricating oil film in the bearings is problematical, and, with excessive backpressure against the oil pump, the bypass valve in the oil filter will open. Hydrocarbon emissions of cold engines are significantly increased. Low temperatures promote cold sludge formation. Therefore, for temperatures below −15°C (+5°F), special cold-start equipment, such as a flame start system, is required. As a rule, larger engines must be preheated; i.e., coolant and lubricating oil must be warmed up to 40°C (104°F). In locomotives and other heavy vehicles, a temperature monitor in the starter electrical circuit prevents starting if temperatures are below minimums (but these can be circumvented by industrious mechanics).

4.3.3.3 Long idling periods

In general, engines are sensitive to long idling periods because ignition misfires may occur with the small fuel flow rates encountered in idling. The unburned fuel condenses on the cylinder walls, washes off the lubricating film, and dilutes the lubricating oil. Additionally, unburned fuel finds its way into the exhaust system and exhaust turbine, and has a negative effect on exhaust emissions quality. Carbon deposits (coking) may occur at the injector nozzle holes and impair injection spray formation. Poor combustion may result in deposits on the piston crown and in the piston ring grooves. Boost air passages and turbocharger compressor blades may "oil up." Overall, long periods of operation at low loads lead to increased internal engine fouling and contamination. Then, when the engine is brought to full load too rapidly, wear increases considerably; in extreme cases, combustion gases will blow past the rings, and pistons will seize.

4.3.3.4 Working materials, and use of unsuitable working materials

The flow of lubricants and coolant, as well as their temperatures, is monitored by measuring devices (pressure and temperature). On medium and large engines, temperatures and pressures outside acceptable limits will cause governor intervention. Yet engines continue to fail due to lack of oil, lack of coolant, or excessive coolant temperatures. Many such failures could be avoided if an acoustic warning signal were combined with a visual warning, to make drivers and machinery operators more aware of a problem.

The desire to reduce operating costs at any price is the mother of many engine failures:

". . . it would be a mistake to blame lubricant for failures resulting from exhaust valve deposits when an especially parsimonious operator buys used motor oil and mixes it with his diesel fuel, or when the operator of a bucket loader neglects to perform scheduled oil changes and manages to turn oil with a viscosity of 104 mm^2/s at 40°C into something with a viscosity of 500 mm^2/s, with resulting engine damage . . ." [4-5].

4.3.3.5 External forces

The engine is subject to a multitude of effects from the machinery which it powers, as well as from the installation's surroundings. In the case of automotive engines, these may come in the form of mechanical damage as a result of traffic accidents, which, if not recognized, can lead to engine failure. For marine engines, hull deformation may be transmitted through the engine foundation to the crankcase and crankshaft bearings. Lack of antifreeze may cause coolant to solidify, cracking the block. Absent or inadequate air filtration causes increased wear of sliding elements. Clogged or restricted coolant passages result in insufficient coolant flow and engine overheating. Leaking fuel lines can result in engine fires (Fig. 4.4).

4.3.3.6 Engineering progress

It is only an apparent paradox that with continuing development of technology in general, under the blanket designation of "progress," old failure modes, whose causes have long been recognized and eliminated by suitable measures, arise again, and that it is precisely this "technological progress" which serves as a trigger for new, previously unknown failure modes. One must envision that engineering solutions do not offer any finality, but rather in most cases are coupled with preconditions, constraints, etc. As soon as these change, formerly successful solutions no longer satisfy the new requirements, and new failures appear. Greater specific work, and lower specific fuel consumption—combined with higher ignition pressures and "fatter" p-V diagrams—not only subject components to higher loads, but also in new, different ways: altered load collectives and longer operating times at higher utilization on the one hand allow "old" faults to re-emerge, on the other hand also trigger "new" types of faults. High-quality materials no longer satisfy demands. With the introduction and proliferation of, first, electrical technology, then electronics in engine technology, new

Fig. 4.4 Passenger car engine damage as a result of an engine compartment fire. Source: Woidich

failure forms appeared, whose causes might no longer be rooted in components ("hardware") but also in the immaterial ("software").

4.4 Humans as the cause of failures

If one traces an engine failure back to its cause, in more than a few cases it will be found that the root cause is actually a human failure. In principle, human failures are not limited to engine technology itself; they can be found in development, manufacturing, operation, and repair. The working conditions in these fields are very different, which compounds the possibilities for error, as well as the likelihood that these will remain undetected and grow to cause new failures.

In development, the nature of the process requires a redundant approach. Designs are carefully validated by calculation as well as experiment. Manufacturing has numerous built-in control systems to eliminate faults. Nevertheless, as already mentioned, strong cost pressures and one-sided "economical" thinking lead to adoption of a quickly realized, apparently most cost-effective solution, rather than the better engineering solution. The "experience discontinuity" resulting from early retirement of older, experienced workers, extreme deadline pressures, and outsourcing of development and manufacturing leads to faults which, in the past, might have been more easily avoided. In the majority of cases, faith in computers is justified—but not always!

In engine operations, the situation is as follows: as a result of complex and nontransparent situations, inadequate training and knowledge, as well as physical and psychological pressures due to unfavorable working conditions, personnel may be overburdened if rapid yet considered response is required. Although accident-related failures are not the subject of this book, the course of events of an accident, its accompanying conditions, and the time pressure under which its principals (machinery operators, drivers, pilots) needed to act, make clear that the possibilities for appropriate, considered action are limited. In association with the engine failures treated here, various forms of human error are also addressed [4-6]:

- Errors of omission
    - Lack of attention and/or
    - Memory lapses

- Errors of execution as a result of
    - Errors in recognizing conditions, events, processes, irregularities, and faults
    - Errors in interpretation of conditions, events, processes, irregularities, and faults
    - Errors in action as a result of
        inadequate knowledge
        inadequate skill
        inadequate experience

Such errors are compounded, if not actually caused, by

- Fatigue/exhaustion through
    - Excessive working hours
    - Too many tasks

- Unfavorable working conditions due to
    - Heat/cold
    - Noise
    - Vibration

- Overburdening through
    - Situational complexity
    - Situational dynamics (time pressure)

- Lack of motivation, as well as

- Convenience/laziness

Chapter 5
Explanation of Failures

5.1 Type of failure

5.1.1 Failures from mechanical loading

Fractures are "macroscopic material separations as a result of exceeding the bonds within rigid bodies through mechanical loading" [5-1]. Fracture initiation and progression are governed by the failure cause(s). Depending on circumstances, the course of a fracture may be predetermined, or it may develop in very different ways. External loading exerts normal and tangential (shear) forces on components, which, if of sufficient magnitude under the given conditions, result in fracture of the component. Fractures follow a certain pattern; we speak of *fracture mechanisms:*

- Crack formation
- Crack growth
- Crack propagation
- Fracture

Cracks arise during component loading or as a result of intrinsic stresses caused by

- Internal influences
    - Slag, separation, microinclusions, microstructural differences, etc. and/or
- External influences
    - Designed-in stress risers such as stepped shafts or oil passages

- Areas of high force concentration where loads are applied, discontinuous stiffness within component such as transitions to reinforcing ribs
- Surface damage from mechanical processes, assembly, corrosion, fretting, etc.

Cracks represent notches which give rise to highly localized, high stress spikes; these stress spikes deepen the crack. Depending on conditions, a crack may grow very slowly, or with reduction in load may even cease growing (stable propagation stage), or continue to grow, with propagation accelerating along with crack depth. Instances with very high propagation speed (up to the speed of sound) exhibit "explosion-like" brittle fracture. The transition from stable to unstable crack propagation is determined by a critical crack length, which is a function of load condition and load application rate (speed of loading), as well as material temperature. Important factors in crack propagation include:

- Type of load: tension/compression, shear, bending, torsion, surface pressure, bearing stress (Fig. 5.1).

- Stress condition: uniaxial/multiaxial.

- Time/load history: static/dynamic loading (constant, increasing, alternating, impulsive).

- Load application rate.

- Temperature: strength behavior of materials is temperature dependent. In general, strength decreases and deformation increases with increasing temperature. At low temperatures, deformability decreases and materials become brittle.

After VDI Guideline 3822, Page 2, distinction is made between

- Overload failure
  - Ductile overload failure with macroscopically visible deformation
  - Brittle overload failure without macroscopic deformation
- Fatigue failure
  - With ductile residual overload failure
  - With brittle residual overload failure

5.1.2 Overload failure

Overload failures result from rapid or impulse loading or overloading.

Ductile fractures with macroscopically visible deformation result after what is often considerable plastic deformation. For ductile behavior, failure criteria include exceeding a maximum allowable deformation.

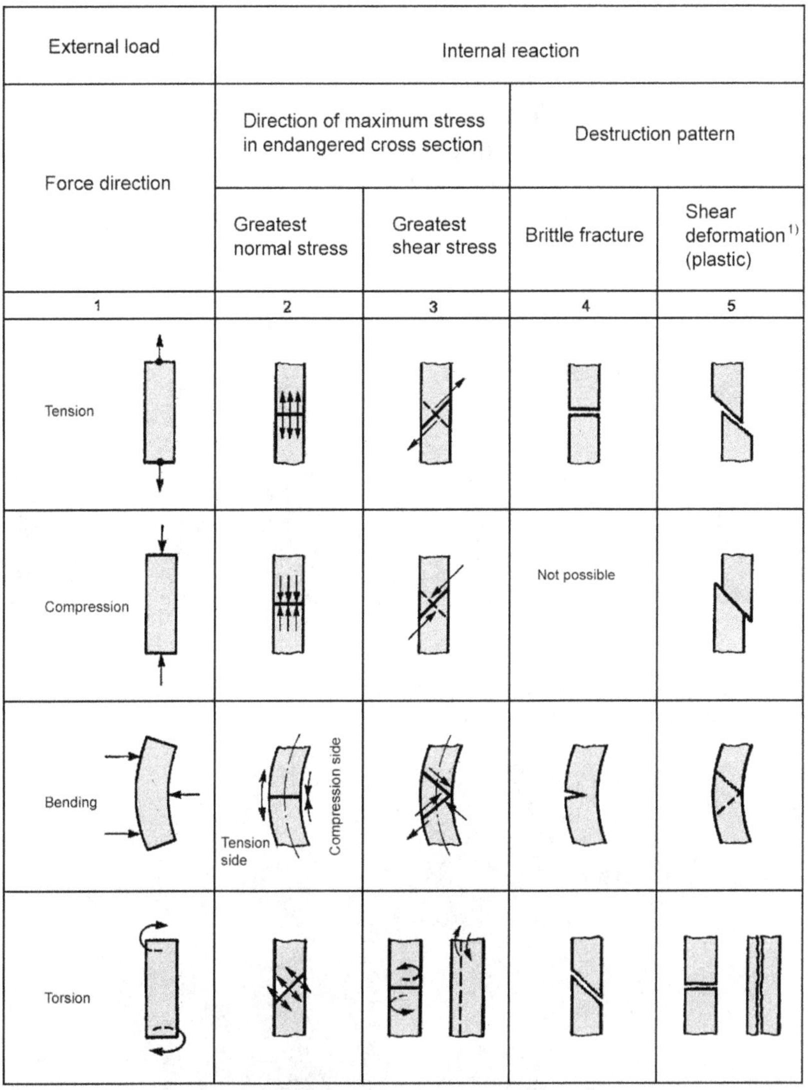

External load	Internal reaction			
	Direction of maximum stress in endangered cross section		Destruction pattern	
Force direction	Greatest normal stress	Greatest shear stress	Brittle fracture	Shear deformation[1] (plastic)
1	2	3	4	5
Tension				
Compression			Not possible	
Bending	Tension side / Compression side			
Torsion				

[1] Fracture takes place in the deformed zone after deformation.

Fig. 5.1 Fracture mechanisms for various external loads (schematic). (After Pohl, Das Gesicht des Bruchs metallischer Werkstoffe)

Initially, component function may only be compromised if the failure occurs suddenly. Along with the appearance of external "necking," there may be internal constrictions at material inhomogeneities such as voids, inclusions, or precipitates. After the fracture occurs, these may be recognized by the "honeycomb" appearance of the fracture surface [5-2].

Brittle overload failures without macroscopically visible deformation are characterized by fracture perpendicular to the direction of normal stress. The fracture surface has a coarse texture (Figs. 5.2 and 5.3). Influential parameters include stress condition and the rate of load application. If the fracture arises in part due to shear fracture as a result of shear loads, it may be observed that the shear fracture jumps from shear layer to shear layer, forming steplike fracture surfaces (Fig. 5.4).

Fig. 5.2 Brittle fracture of a cast iron (GGG-50) guide shoe, after impulse overloading. The fracture initiated at the upper edge. Source: Allianz

Fig. 5.3 Reflection electron micrograph (1000 X) of a brittle overload fracture. Source: Allianz

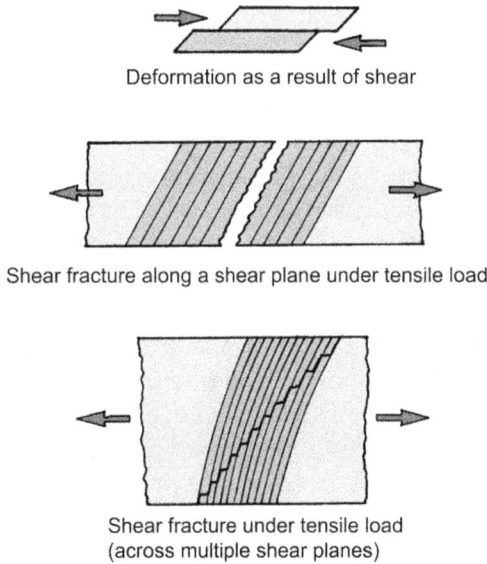

Deformation as a result of shear

Shear fracture along a shear plane under tensile load

Shear fracture under tensile load
(across multiple shear planes)

Fig. 5.4 Schematic of sliding fractures.

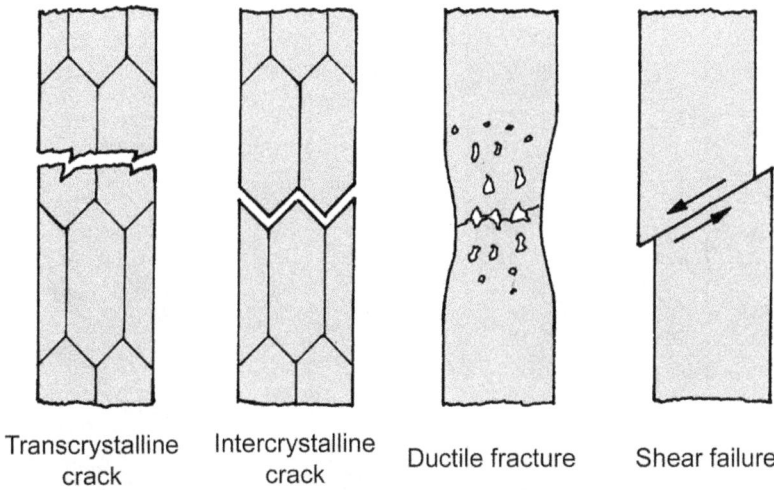

Transcrystalline crack   Intercrystalline crack   Ductile fracture   Shear failure

Fig. 5.5 Overload failure fracture forms. (After Czichos: "Materials" in: Hütte, 30th edition)

Brittle overload fractures are promoted by multiaxial load conditions, notches, and abrupt section changes, limited deformability of the material, and microstructural irregularities, as shown in Fig. 5.5.

5.1.3 Fatigue fractures

Fatigue fractures are the characteristic failure mode for dynamically loaded engine components. They are initiated by (often microscopically small) surface cracks, or material inhomogeneities (e.g., nonmetallic inclusions immediately below the surface, or deeper within the component). In the latter case, the fracture propagates in the form of a "fracture lens" (Fig. 5.6).

The initiators of such cracks are local stress concentrations caused by material defects, manufacturing defects, design flaws, as well as operating influences. Starting from such cracks, the fracture proceeds in stages as loads are applied and released, in general with load changes; this is recognizable by the so-called "beach marks" (so called because they resemble wave marks on a beach), or "rest lines" because the fracture does not proceed at a constant rate, but rather *rests* between periods of propagation; one may regard the "rest lines" as similar to tree growth rings. The development of fatigue fractures depends on damage precursors (incipient cracks), and the level and duration of loads. Prerequisites for creation of rest lines/beach marks are changes in component loading, as may result from engine operation though periods of starting, idling, part load, full load, etc. Under low loads, crack propagation halts, only to start again as load is increased. Because

| Fracture initiation on surface | Fracture initiation immediately below surface | Fracture initiation within material ("fish eyes") | Multiple fracture initiation sites |

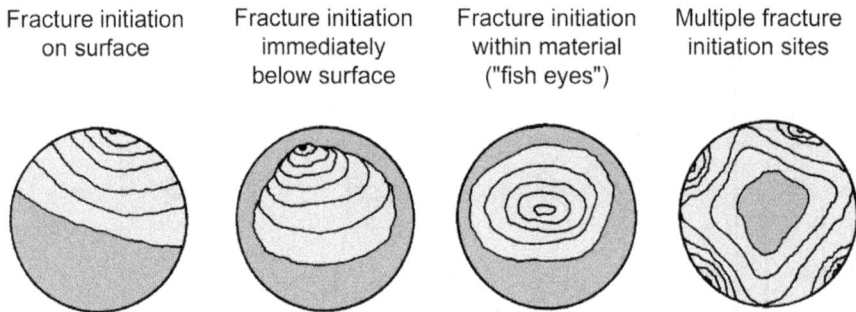

Fig. 5.6 Points of origin for fatigue fractures.

Fig. 5.7 Fatigue fracture of two bolts, M20 8.8, due to dynamic overloading as a result of insufficient tightening torque. Source: Allianz

the fracture propagates slowly, the lines near the point of origin are closer together. As the fracture progresses, the spacing between lines increases because crack propagation accelerates as the load-carrying cross section decreases. Propagation of rest lines may be forced by local stress concentrations, and they may have multiple points of origin, with the rest lines intersecting and merging (Figs. 5.7 and 5.8).

In terms of rest lines, ductile and less ductile materials differ insofar as the former materials show more distinct rest lines early on, initially concave, later, beyond a certain distance from the point of origin, appearing convex. Less ductile materials show convex rest lines from the outset, with more rapid crack propagation because of their internal notch sensitivity. If the crack has propagated far enough, the remaining load-carrying cross section is no longer able to carry the load; it breaks suddenly, as a so-called overload failure. The failure history may be recognized at the fracture surface by

- The smooth fracture surface of the actual fatigue failure surface, with limited crack propagation rate. The stepwise progression of the crack may be recognized by the rest lines.

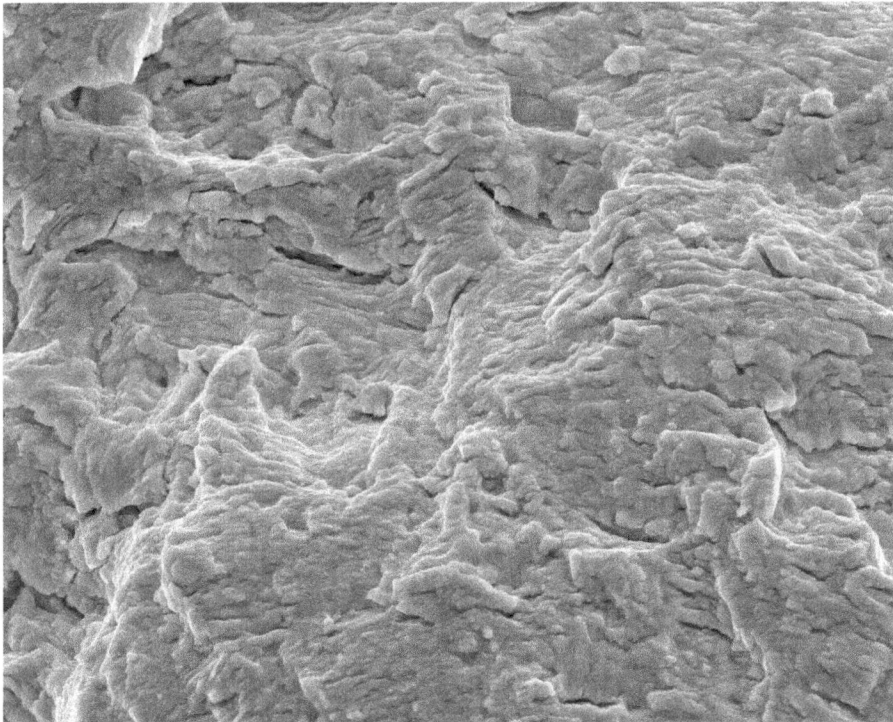

Fig. 5.8 Reflection electron micrograph (2000:1) of a fatigue failure. Source: Allianz

- Rougher zone of more rapid crack propagation.
- Very rough fracture surface of the remaining cross section, as a result of brittle or ductile overload failure (Fig. 5.9).

The size, shape, and location of the final fracture surface within the component cross section depend on the load magnitude and nature. A small final fracture zone is an indication of high static safety margin, a large final fracture zone indicative of smaller static safety margin or low amplitude of the cyclic loading (Fig. 5.10).

The fracture surfaces are smoothed by their relative motion, and take on a velvety to rough appearance. Smooth fracture surfaces therefore indicate slow crack propagation, because the fracture surfaces are burnished by rubbing against one another. Often, however, the fracture surfaces are altered by secondary influences, so that their interpretation may be difficult. Fatigue fractures of engine components are caused primarily by loading in tension, in tension and compression, rotating bending, and torsion (Fig. 5.11).

Recognition and interpretation of torsional fatigue fractures is more difficult than that of other fracture forms. This is caused by the fact that

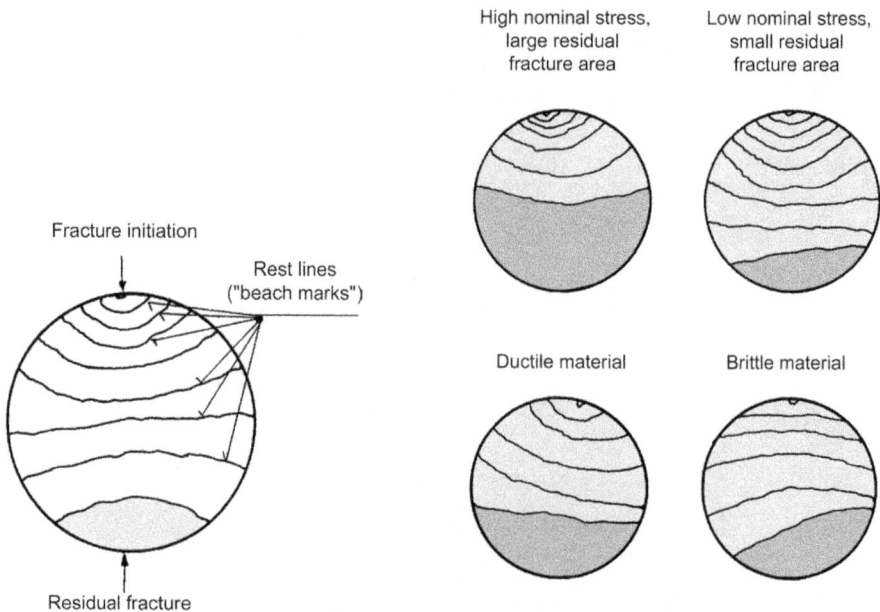

Fig. 5.9 Schematic of a fatigue failure.

High nominal stress, large residual fracture area

Low nominal stress, small residual fracture area

Ductile material

Brittle material

Fig. 5.10 Rest line appearance and residual fracture zone for fatigue fractures with various boundary conditions.

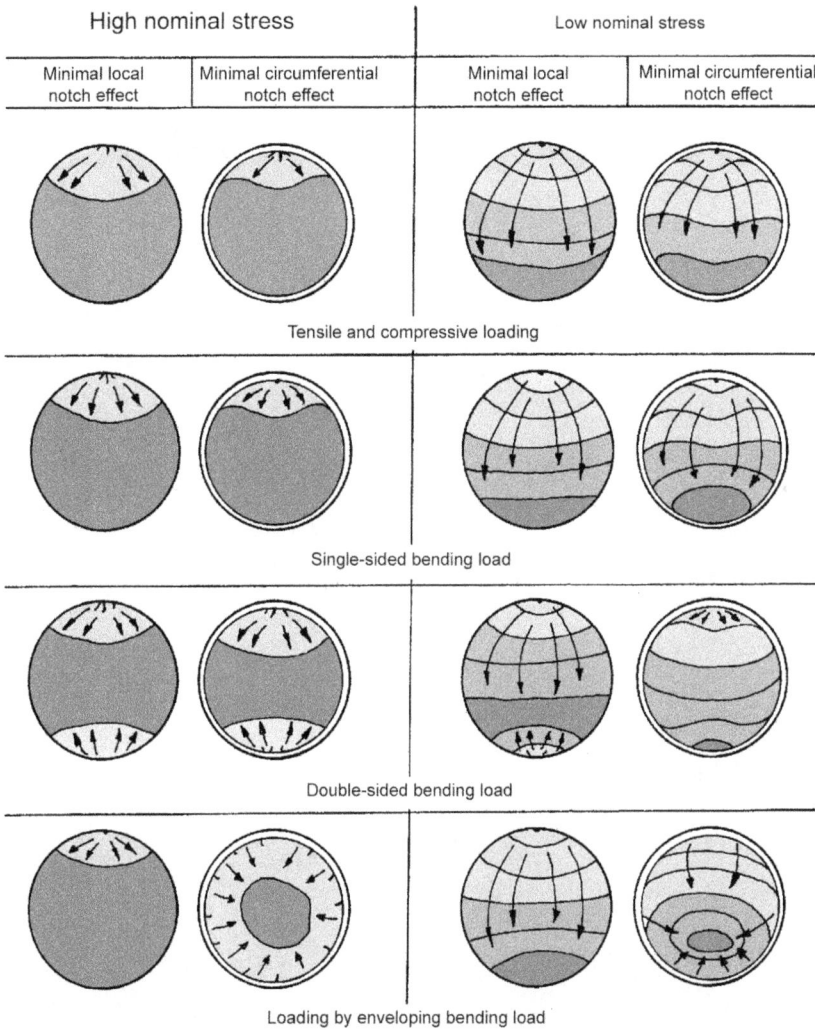

High nominal stress		Low nominal stress	
Minimal local notch effect	Minimal circumferential notch effect	Minimal local notch effect	Minimal circumferential notch effect

Tensile and compressive loading

Single-sided bending load

Double-sided bending load

Loading by enveloping bending load

Fig. 5.11 Morphology of fatigue fractures (after Tauscher, Berechnung der Dauerfestigkeit).

torsional fractures may run longitudinally, transversely, or take on a spiral path (Figs. 5.12 and 5.13).

The last of these results in the "classical" torsional fracture, with the fracture surface at a 45° angle (Fig. 5.14). This fracture appearance is primarily encountered in more brittle materials (as may be readily demonstrated with a piece of blackboard chalk).

A valuable aid in making the stress condition of a component visible, by means of the resulting strain, is the so-called brittle coating method. The part to be examined is coated with a material that adheres firmly

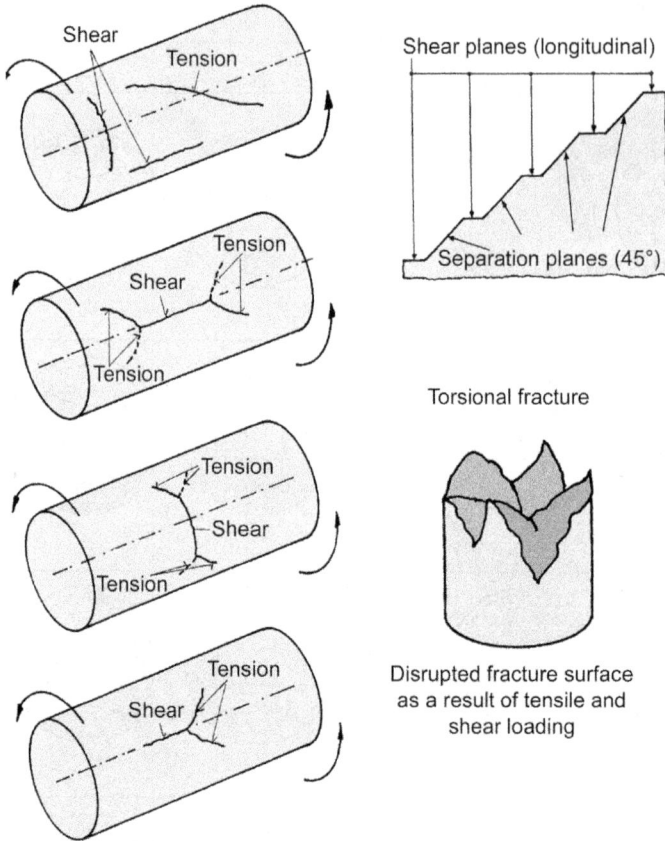

Shear
Tension

Tension
Shear
Tension

Tension
Shear
Tension

Tension
Shear

Shear planes (longitudinal)

Separation planes (45°)

Torsional fracture

Disrupted fracture surface
as a result of tensile and
shear loading

Fig. 5.12 Schematic of torsional fractures.

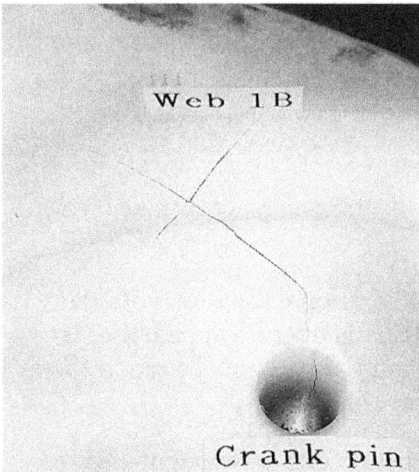

Web 1B

Crank pin

Fig. 5.13 Torsional fracture of a crankpin.

Fig. 5.14 "Classic" torsional fracture of a crankpin on a modular cast iron (GGG 70) passenger car crankshaft.

Fig. 5.15 Strain lines on a crankpin subjected to torsional loading. The strain lines run at a 45° angle.

enough to participate in the component's elastic deformation, but which is also sufficiently brittle to crack at stress levels well below the component's elastic limit, perpendicular to the direction of greatest stress (Fig. 5.15).

The point or points of origin of torsional fractures are less obvious. Although the fracture usually runs along one or two preferred planes, additional cracks may arise in various forms; e.g., in the form of stairsteps. If torsional loading changes direction, a large number of fracture surfaces will occur. The appearances of fracture surfaces are dependent on the type and level of load, and on the material.

5.1.4 Thermal damage

Engine components are subjected to various and varying temperatures, ranging from room temperature to peak combustion temperatures of more than 2000°C. Individual components operate at various temperatures and are accordingly subjected to thermal loads, the source of a wide range of failures: hot fractures, hot cracking, hot deformation, thermal surface damage, functional faults caused by deposits, etc. Hardest hit are components adjoining the combustion chamber: cylinder head and valves, cylinders, crankcase or cylinder sleeves and of course pistons, additionally exhaust passages and exhaust-driven turbines (Fig. 5.16). Thermal overloading can also result from friction

Exhaust valve

combustion		
chamber side	500	- 580 °C
seat side	340	- 430 °C
valve stem	180	- 260 °C

Injector nozzle

nozzle tip	180 °C

Cylinder head

combustion		
chamber side	210	- 260 °C
water jacket side	120	- 140 °C
valve webs	260	- 300 °C

Cylinder

piston at TDC	200	- 320 °C
top ring at TDC	140	- 180 °C
top ring at BDC	100	- 110 °C

Piston

top ring groove	160	- 220 °C
wrist pin boss	180	- 190 °C
wrist pin boss	100	- 110 °C
(built-up piston)	80	- 100 °C

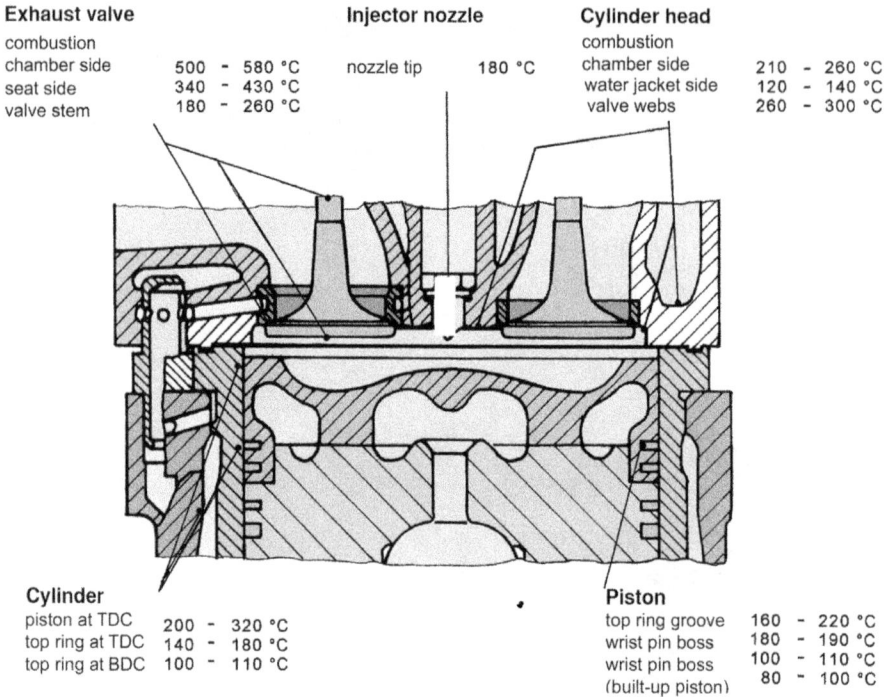

Fig. 5.16 Temperatures of components bordering the combustion chamber (example: medium-speed diesel engine). Source: MAN

processes (e.g., in bearings) or excessive operating fluid temperatures (e.g., intercooler temperature).

There are no clearly defined temperature limits for the concept of thermal loading. This is understood to mean temperatures clearly above room temperature; [5-3] mentions temperatures above 600 K (327°C). Specifically, high temperatures have the following effects on components:

- Local temperature differences within components lead to temperature gradients, which obstruct thermal expansion and therefore cause thermal stresses. Elevated and high temperatures may cause changes to material structure, with far-reaching effects on material properties.

- At elevated temperatures, materials begin to undergo plastic deformation under static loads; they "creep."

- Rapid application of thermal loads—thermal shock—combined with restricted thermal expansion can result in hot cracking. Cylinder heads of medium and large engines are especially susceptible, as they are subjected to sudden thermal shocks in going from starting with compressed air, to full load, to full stop.

- Temperature changes caused by engine speed and load changes represent LCF (low cycle fatigue) loading, which can result in cracks and fractures in the region below 10^5 load cycles.

- Temperature changes in the course of engine operating strokes represent HCF (high cycle fatigue) loading. This affects the surfaces of components bordering the combustion chamber, such as cylinder head and pistons, subjected to rapid shifts between high combustion temperatures and inrushing cold air.

- Parts subjected to thermocorrosive loading include above all exhaust valves, especially in engines operating with MDF (marine diesel fuel) or heavy oil.

- Thermoabrasive damage may be observed on piston rings even under normal operating conditions, even though wear may not be very far advanced. For this reason, manufacturers of large engines provide operators with documentation to assist in visual inspection of piston rings that have been in service.

5.1.5 Failure through corrosion in aqueous media

Corrosion is the electrochemical reaction of a material with its surroundings, in which an anodic reaction between metal and metal ions results in electrolytic removal of metal (DIN EN ISO 8044, formerly DIN 50900, Part 1). This metal removal may damage components to the extent that their function is impaired and/or their service life is compromised. In an engine, parts subjected to corrosion primarily include those surrounded or penetrated by coolant: cylinder sleeves or crankcase, water pumps, coolant lines, and heat exchangers and radiators, but also valves, bearings, and other internal components.

When two electrically conductive, connected metallic materials of differing electrical potential, or a material with an electrical potential gradient, is brought into contact with an electrolyte, galvanic elements are formed (*corrosion elements*), in which the less noble metal forms the anode, the more noble the cathode. Corrosion proceeds as an oxidation of the attacked metal, and reduction of the attacking medium. In aqueous solutions, both processes may occur simultaneously but at different locations on the metal surface. Sensitivity of a metal to corrosion depends on its position in the electrochemical series. Metals unsusceptible to corrosion are termed noble metals, while those sensitive to corrosion are termed less noble or base metals. The corrosion process is determined by the material, the attacking medium, and the electrolyte. Electrolytes usually consist of solutions of salts, acids, bases, or other liquefied (including molten) materials. Corrosion is greatly aided by the presence of atmospheric oxygen; dirt, soot, salt, etc. contained in ambient air also promotes corrosion. The behavior of corrosion products is of great significance. If these are dense enough to hinder electrolyte access to the base material, they function as corrosion protection (*passivation*), e.g., aluminum oxide, an extremely resistant coating that forms on bare aluminum. The results of corrosion are metal removal, material penetration, and cracking. Distinction is made between

- **Surface corrosion**
  Depending on circumstances, the entire surface is more or less evenly attacked and material removed. Because it is easily recognized, surface corrosion is regarded as relatively less dangerous, especially since it can be alleviated through protective measures.

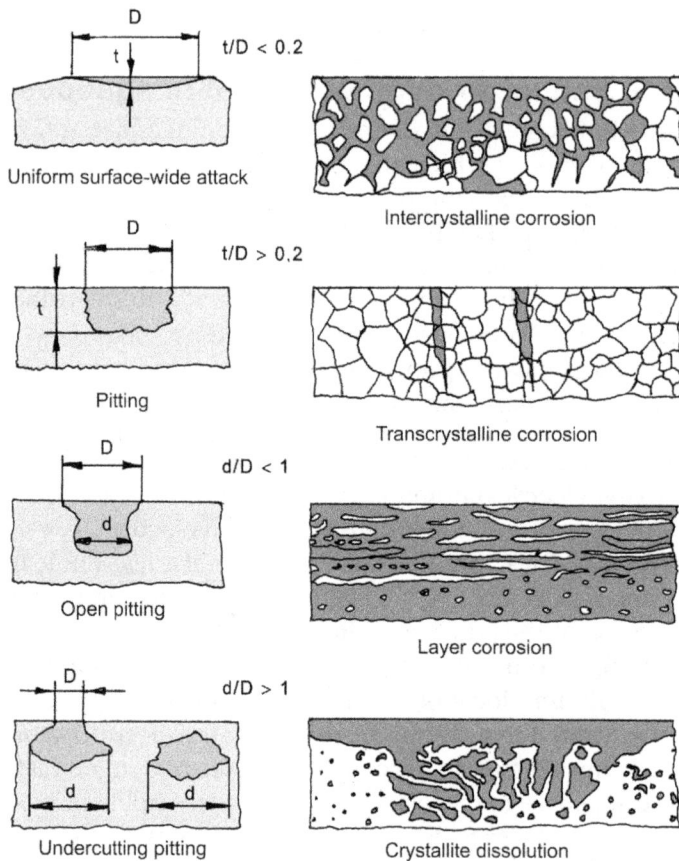

Fig. 5.17 Corrosion manifestations (after Wiederholt).

Uniform surface-wide attack — $t/D < 0.2$

Pitting — $t/D > 0.2$

Open pitting — $d/D < 1$

Undercutting pitting — $d/D > 1$

Intercrystalline corrosion

Transcrystalline corrosion

Layer corrosion

Crystallite dissolution

- **Selective corrosion**
  In selective corrosion, certain areas are preferentially attacked:
  - *Local corrosion* (pitting) occurs when only one component of an alloy is attacked, or in the case of local contamination of material surfaces, below which corrosive attack may progress into the material. Because little may be detected in the way of corrosion products, this type of corrosion is regarded as more dangerous.
  - *Intercrystalline corrosion* takes place at grain boundaries in alloys and is associated with severe reduction in material properties.
  - *Transcrystalline or intracrystalline corrosion,* with corrosive attack, takes place within crystal grains (Fig. 5.17).

A corrosion process familiar to everyone is rusting of ferrous materials, recognizable by the color and structure of the corrosion products, shown in Table 5.1.

**Table 5.1
Degrees of Corrosion (source: [5-4])**

Color	Degree of corrosion
Light brown rust	Appears within 2 weeks in open air, in the presence of humidity and rainy weather; does not penetrate metal, may be removed at room temperature by relatively nonreactive means (diluted hydrochloric or muriatic acid or sulfuric acid)
Red rust	Numerous deep surface pits; may be removed within an hour by chemical means
Dark brown rust	Relatively deep surface pits; may be removed by chemical means only after several hours of treatment

Corrosion of engine components poses a hazard because the juxtaposition of corrosion and mechanical loading may give rise to dangerous manifestations:

- **Stress crack corrosion**
  arises in components subject to tensile loads, as corrosive attack promotes growth of an already present crack. Even more dangerous is

- **Fatigue crack corrosion**
  as cyclic loading accelerates the process.

- **Fretting corrosion**
  occurs when two components are joined under pressure and undergo miniscule relative motion. The materials are especially prone to react at points where the peaks of surface irregularities deform and break away. The reaction products cause stresses, which can result in fatigue failures.

- **Erosive corrosion**
Solid particles in flowing media preferentially damage protective surface layers and so initiate corrosion.

- **Cavitation corrosion**
The same effect may be initiated by material disruption by imploding gas bubbles in a fluid stream.

As undesirable as corrosion may be in general, it does serve useful purposes in manufacturing, e.g., in the case of electrochemical milling, in which metallic material is locally removed by electric current in an electrolyte [5-5].

5.1.6 Failure through tribological loading

DIN 50323 defines *tribology* as "the science and technology of surfaces interacting in a relative motion. It comprises the entire field of friction and wear including lubrication. It includes corresponding interfacial interactions between solids as well as between solids and liquids, consistent materials or gases."

Distinction is made between

- Solid body wear

- Wear under conditions of mixed or boundary lubrication

Wear is the "progressive loss of material from the surface of a solid body, because of mechanical causes . . . (i.e.) contact and the relative motion of a solid, liquid, and gaseous counter body" [DIN 50320]. In machine operation, wear is as undesirable as it is unavoidable. Indeed it may even be used to achieve desirable effects, for example in writing (pencil/paper, chalk/blackboard). Written characters on paper or slate are nothing but visible traces of adhesive wear. Another example of "desirable wear" is the break-in process, in which contact surfaces of machine parts are matched to one another by the wearing down of surface roughness. Various wear mechanisms are employed in manufacturing processes.

Preconditions for the appearance of wear are:

- **Two wear bodies** (main body and counterbody)
In the case of solid bodies, material microstructure, strength, hardness, shape, and surface geometry play important roles.

- **Relative motion between main body and counterbody**
In an engine, these are predominantly: sliding, rolling, pushing, and flowing—continuously, or oscillating, or intermittently. Parameters

affecting tribological loading include size, direction, and time history of the relative motion.

- **The presence of a normal force between the wear bodies**
- **An intermediate material between the wear bodies**
  Solid particles, fluids, or gases (Fig. 5.18).

Tribological loads lead to various types of wear, such as sliding, rolling, vibration, or scouring abrasion; stream impact wear; erosion; cavitation; or droplet impact erosion. These are based on mechanisms which individually or in various combinations are responsible for the type of wear at hand: shearing, elastic, and plastic deformations, and boundary layer processes. Specifically, these involve:

- **Surface disruption**
  The forces at the contact points of sliding bodies cause material fatigue, as a function of load amplitudes and number of load cycles. Cracks and pits are formed, which lead to fretting (Fig. 5.19).

- **Abrasion** (from the Latin *abradere*, to scratch or scrape away)
  If one wear partner is appreciably harder and rougher than the other, the harder material will plow its way through the softer. The same effect may be seen if a hard foreign particle is trapped between the two wear partners. As it is carried along, it draws furrows in the surface of the softer wear partner. Depending on the microscopic processes involved, distinction is made between microplowing, microcutting, microfatigue, and microcracking [5-6]. The results are scratches, grooves, pitting, and waves.

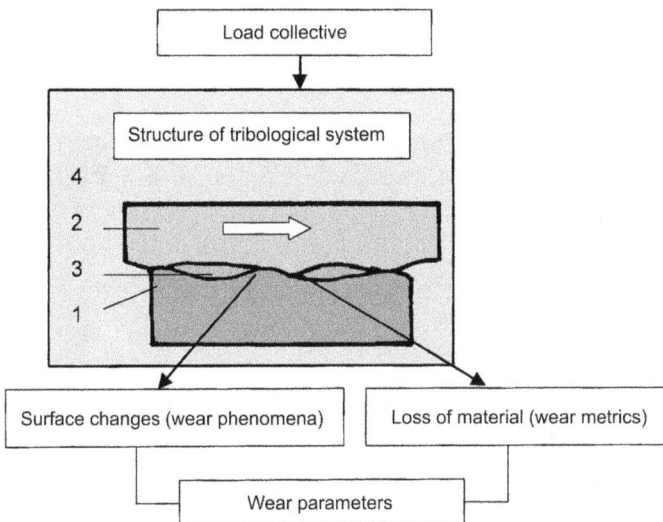

Fig. 5.18 Representation of a tribological system: 1. Body; 2. Counterbody; 3. Intermediate material; 4. Surrounding medium.

- **Adhesion** (from the Latin *adhaerere*, to stick to, to hang on to something)
  When asperities touch one another, high local pressures are generated, which disrupt protective surface coatings. "Cold welding" or seizing creates local contact bridges; with relative motion of the contact partners, the high tensile strength of these bridges causes separation not at the actual points of contact but rather in the adjacent volume of one of the partners.

- **Tribochemical reactions**
  Reactions between wear bodies and the surrounding medium and/or an intermediate substance result in reaction products in the form of layers and particles, which tend to break out and form hard wear particles (Fig. 5.20).

One form of wear which appears at many points within an engine is fretting corrosion, the result of tiny relative motion between two components connected through a prestressed joint, such as bearings and crankcase bores, crankshaft and drive flange, all manner of shaft and hub connections, roller chains, etc. All of the above mentioned wear mechanisms can take part in fretting corrosion, which explains the different appearances that this form of corrosion can take: shiny bare metal, adhesive material transfer, abrasion by tiny wear particles, and fatigue cracking. Fretting corrosion damages both wear partners, but with differing failure progression.

Apart from the break-in process, wear is an undesirable, but unavoidable process. Particles are released from bodies and counterbodies as a result of wear processes, and these in turn intensify wear. The decisive factor is the wear rate; i.e., the speed at which wear develops:

- **Degressive**
  Break-in processes, in which roughness from the manufacturing process is smoothed, and the area of contact between sliding partners is increased

- **Linear**
  Normal operation, in which wear increases steadily but at a very low rate

- **Progressive**
  Self-amplifying, accelerated wear, which quickly leads to functional faults and resulting failures (Fig. 5.21).

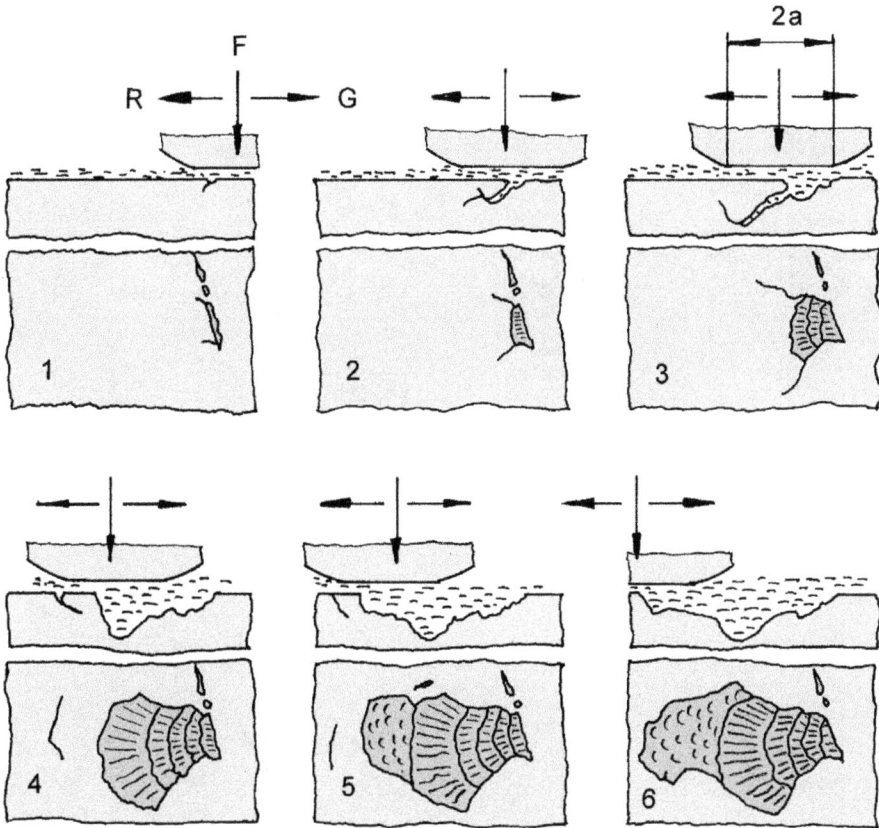

Fig. 5.19 Pit formation (schematic; after Niemann and Bötsch).

(1) Compressive and tangential loading of tooth flanks. Additional notch stresses at scoring lines of surface roughness; incipient cracks form here.

(2) Oil, which has penetrated under pressure and with capillary action, is trapped in the contact zone and, as it is rolled over, exerts bursting pressure which promotes crack growth.

(3) Continuous opening and closing of the incipient crack result in fatigue and material breakouts: pitting takes place.

(4) The oil in the crack only exerts bursting pressure until the crack is so long that oil is able to escape beyond the contact zone. Until this point, pits have a typical triangular appearance. The sickle-shaped appearance of pitting formation represents the rest lines ("beach marks") typical of fatigue failure.

(5) As pit formation continues, pit shape and breakouts, as well as the pit bases, become irregular.

(6) Enlargement of the breakout by compressive and tangential loading, because in the rolling direction, one side of the pit edge is unsupported due to missing material.
F, normal force; G, direction of sliding motion; R, direction of rolling motion; 2a, width of compressive area.

Abrasive wear

Adhesive wear

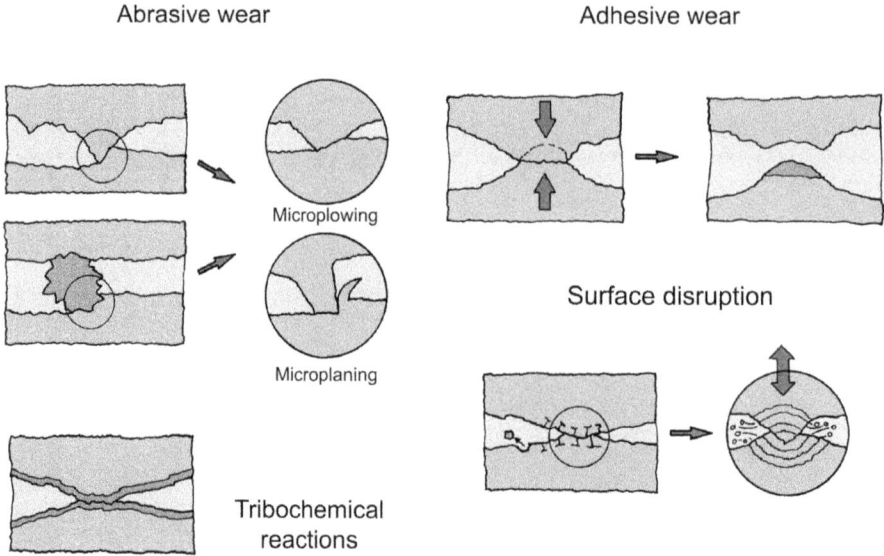

Microplowing

Microplaning

Surface disruption

Tribochemical
reactions

Fig. 5.20 Schematics of basic wear mechanisms.

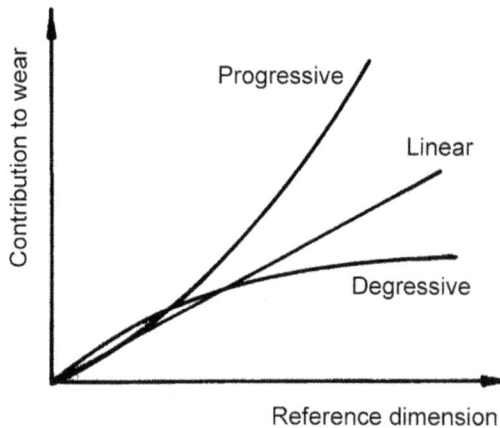

Fig. 5.21 Wear rates.

In engines, wear is primarily caused by

• **Boundary and mixed lubrication** (incomplete separation of main
  body and counterbody)
  The typical wear mode in the boundary lubrication regime is fretting
  corrosion. It occurs on dry as well as oil lubricated contact surfaces.
  Short oscillatory motion (in the range of 0 to about 1/3 mm), as may
  be caused by vibration, shock, and elastic deformation, result in local
  abrasion and, in the presence of air, in oxidation processes. Given
  the small relative motions, wear particles are not transported away,

and continue to react with the basic materials. The resulting extreme pressures cause the lubricant to lose viscosity.

- **Fluid friction** (complete separation of main body and counterbody)
- **Cavitation**
  Cavitation occurs anywhere the pressure on a fluid drops below its vapor pressure. The results of such bubble formation include:
    - Damage to adjacent surfaces
    - Loss of hydrodynamic properties (primarily affecting flow machinery)
    - Noise generation

These vapor bubbles are unstable. If they enter a region of higher pressure, hydrostatic pressure, amplified by surface tension, causes them to implode, which, near walls, may not proceed symmetrically. The vapor bubble constricts on the side opposite the component surface, and produces a fluid stream (microstream) directed at the interior of the bubble. This narrows down to a needle-like jet and impacts the wall surface at high speed (500 to 1000 m/s). The material is initially plastically deformed and (cold) worked (incubation phase), then fatigues and is disrupted. Fine cracks are formed, which lead to pitting of the material surface. At first, the damage rate increases rapidly (acceleration phase). After distinctly visible damage has been created (rough surface, pitting), the damage rate decreases, presumably because the pits are filled with fluid which acts as a protective buffer (delay phase). Next, material damage increases steadily (steady-state phase). Over time, the process continues its attack, deep into the material. In the case of acoustic cavitation, material damage takes place at the cavitation site; in the case of flow cavitation, damage occurs more or less downstream. Cavitation can destroy any metal, and cannot be prevented by use of more resistant materials. Cavitation is usually associated with erosion [5-7, 5-8]. In engines, (flow) cavitation may occur in bearings, in the fuel injection system (pumps, lines, nozzles), and as acoustic cavitation at cylinder sleeves (Figs. 5.22 and 5.23).

- **Erosion**
  If solid bodies are exposed to fluids containing particles, for example lubricants or fuel containing foreign particles, or a gas flow carrying particles (exhaust gas with combustion products), the result is erosion, primarily an abrasive form of wear, in which material is carried away by repeated impact of solid particles against the component surface. The degree of erosion is a function of particulate volume in the fluid, particle size, density, shape, hardness, and structure. The effect of impact goes as the square of velocity, and up to the fourth power of velocity. The effect of impact angle varies,

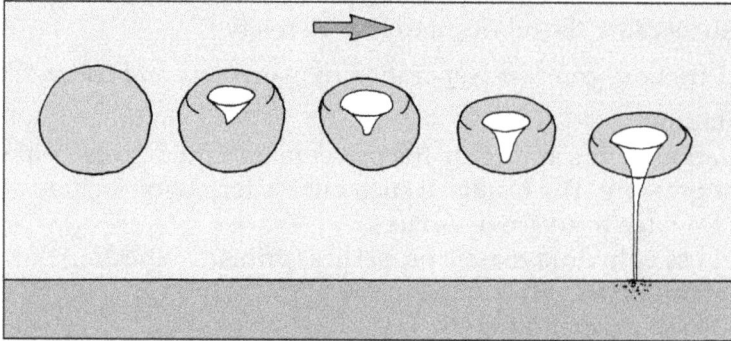

Fig. 5.22 Implosion of vapor bubbles during cavitation (schematic).

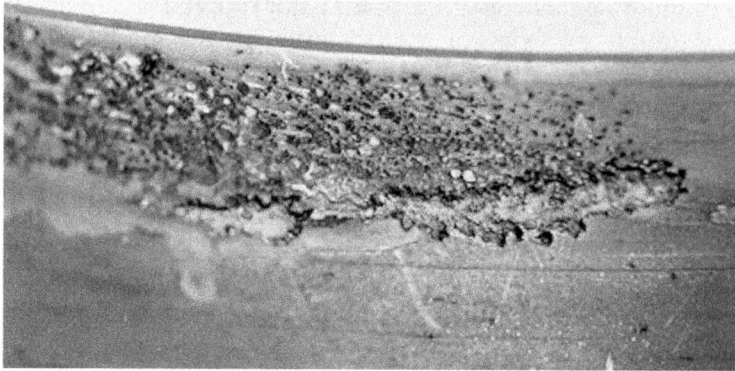

Fig. 5.23 Cavitation damage to a journal bearing.

depending on the material; ductile materials exhibit higher wear with a flat angle of impact, while brittle materials are more susceptible to normal impact [5-9]. Erosive wear is usually observed in the area of cross section changes and changes in flow direction (Fig. 5.24).

- **Droplet impact wear** (a form of "reverse" cavitation) is a primarily a problem in steam turbines and aircraft powerplants. It is seldom encountered in internal combustion engines. One example may occur during combustion knock, as fuel droplets are hurled against the piston crown by highly turbulent gas flow. Pressure (as a function of impact speed) is generated at the point of droplet impact. If the fluid flows away radially at high speed, the piston material is subjected to shear forces. Repeated stress by individual fluid impacts causes severe destructive pitting [5-7], see Figs. 5.25 and 5.26.

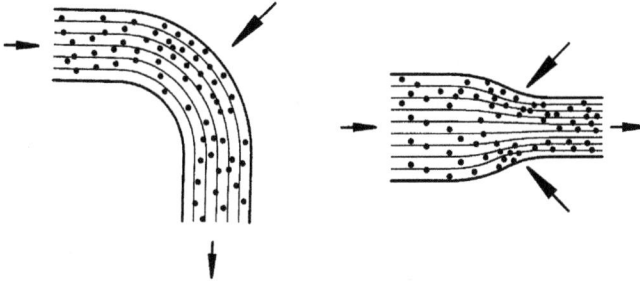

Fig. 5.24 Preferred areas for erosive attack.

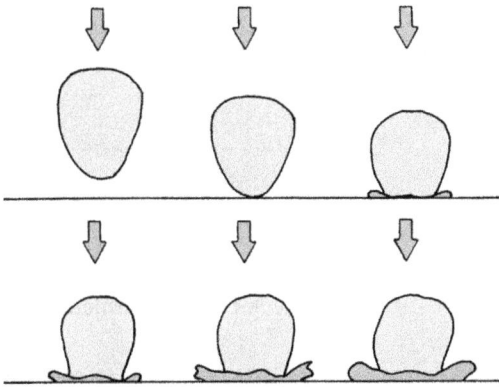

Fig. 5.25 Droplet impact (schematic).

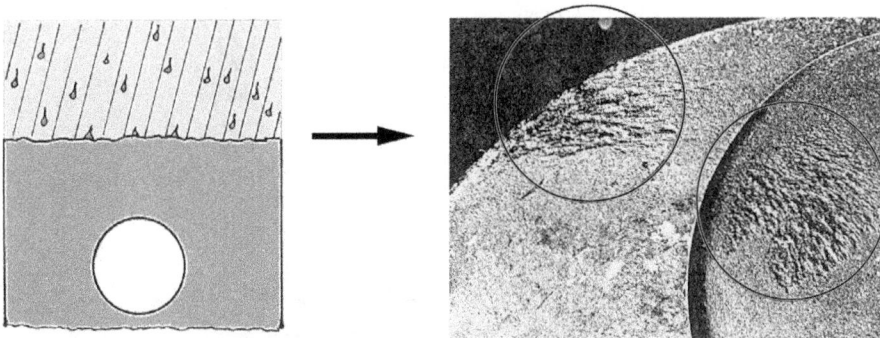

Fig. 5.26 Droplet impact and its effects.

The effects of wear (Fig. 5.27) are:

- Reduction in cross sectional area

- Surface alteration

- Compromised functionality as a result of:
    - Increased clearances
    - Reduction of interference
    - Compromised geometry and kinematics

In an engine, wear, as progressive loss of
material from a surface of a solid body,
manifests itself in various forms.

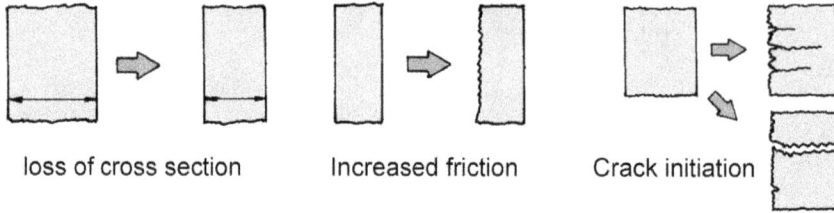

loss of cross section   Increased friction   Crack initiation

Changes to surfaces

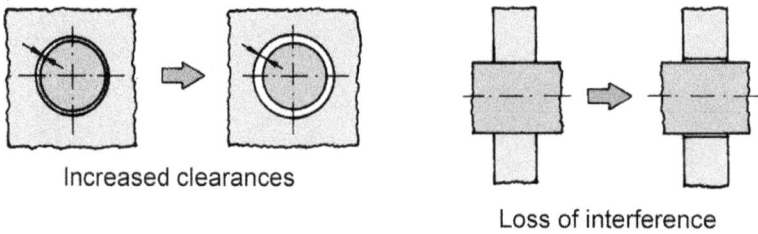

Increased clearances

Loss of interference

Direct compromise of functions

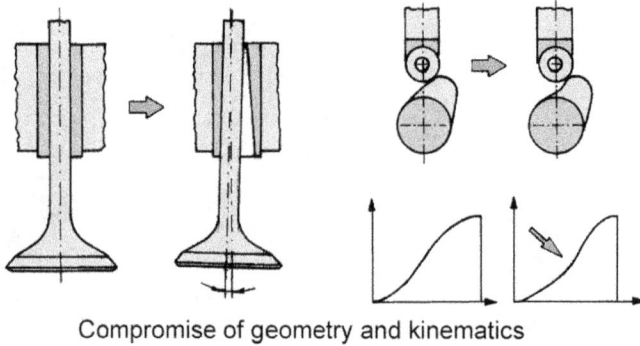

Compromise of geometry and kinematics

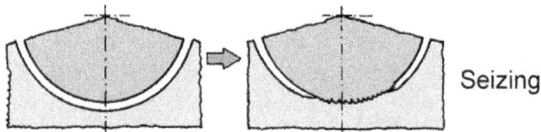

Seizing

Fig. 5.27 The effects of wear.

This leads to:

- Increased friction
- Seizing/galling
- Force and fatigue failures

As far as the engine is concerned, wear affects include

- Transmission of motion (crank train, valve train)
- Restricted motion (torsional vibration damper)
- Power transmission (crank train, valve train, gear train, press fits such as crankshaft to crankshaft flange, gears to shafts, etc.)
- Material transport (fuel injection lines, coolant passages)
- Information transfer (electrical contacts)

The most important engine wear areas are the crank train and valve train (Fig. 5.28).

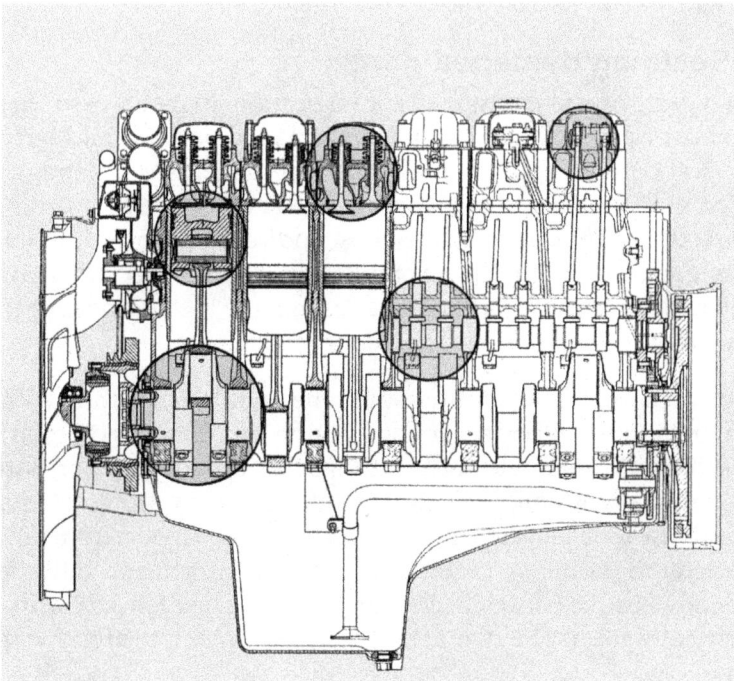

Fig. 5.28 Internal combustion engine wear zones.

5.2 Failure analysis

5.2.1 On-site inspection

The first step in any failure analysis is visual inspection of the damage to view the damaged engine installation, if possible on-site. Although this is often not possible, it is nonetheless desirable. Inspection of the machine installation or vehicle with a damaged engine provides vital background information which would otherwise be unavailable; inspection offers insight into the conditions under which the engine had been operating: condition of the machinery, installation situation, routing of ducts, lines and plumbing, air filter condition, dirt, leaks, traces of leaks, oil puddles, carbon deposits, rust, general impression of operating personnel, storage of operating fluids and spare parts, and many other details.

In the case of vehicle engines, the condition of the vehicle provides important clues: external visual impression, tire condition; e.g., in the case of a passenger car, is it a "sporty" car with corresponding equipment, does it have a trailer hitch, etc. Such apparently incidental details provide information regarding the manner in which the engine is operated, and the types of loads to which it is subjected.

5.2.2 Securing damaged parts

Damaged parts must be removed for inspection, if at all possible without incurring additional damage. Damaged parts should be secured. It should be noted that all other components which might have interacted with the damaged component(s) should also be examined, those directly and, if possible, also those indirectly involved. In the case of a damaged bearing, the entire crank train may be affected, along with the crankcase. It is always advantageous to have the engine sent to the manufacturer or a repair shop for examination.

Damaged areas are first examined with the unaided eye, then with a powerful magnifying loupe. Damaged parts should be photographed, along with their immediate surroundings. Because engine parts are generally coated with oil, they usually offer poor, low-contrast images; they should instead be carefully cleaned and illuminated from all sides. Attention should be given to shadows, contrast, and reflections. Additionally, simple hand sketches are made, as needed, to clarify the damage photograph. The damage is described, using the following guidelines:

- Type of damage
- Site of damage

- Appearance of failure: color, structure, direction of propagation, etc.

- Evidence of previous repairs

For transportation, damaged parts should be packed in such a manner that the damaged areas (e.g., the fracture surfaces) are not "falsified" by transport.

5.2.3 Determining damage-relevant data of a machine installation

Detailed, precise documentation is just as important as careful securing of a damaged engine or its components and damage examination and analysis. Initially, documentation ascertains basic data, followed by a description of the course of events. This should include interviews with the driver/operator or operating personnel present when the failure manifested itself, to record their observations. Such verbal evidence should be recorded as soon as possible after the event. They may provide valuable preliminary clues to possible causes of failure (and possibly indicate hidden associated peripheral damage to the engine and installed equipment). All of these records are ultimately documented in an examination report (Section 5.2.4). The report concludes with an extensive description of the failure, which should as a rule examine in great detail the examinations which have been conducted; e.g., tolerances, material pairings, fits and clearances, material quality, etc., which in specific cases may be supplemented by laboratory reports (Section 5.2.5).

The basic information which should be recorded at the outset of any examination are:

- Short description of the machinery installation

- Operator, previous owner

- Engine:
  manufacturer, engine type, model, year built, total operating hours, peripherals, operating mode, operating conditions, maintenance (oil changes, coolant maintenance), and previous damage. The latter is of particular importance because it is often observed that engine failures occur after previously conducted overhauls and repairs. An engine's history often provides important clues to the cause(s) of failures.

- Damaged parts:
  manufacturer's marks, manufacturing date, forging number, casting number, group number, part number, drawing number

5.2.4 Course of events

How did the engine failure manifest itself?

- Leaks
- Loosening (of parts, fasteners)
- Imbalance
- Vibration
- Noise
- Deformation
    - Cracks
    - Fractures
- Locking

How did operating personnel react to the engine fault, and what were the consequences of the failure?

The written failure report contains the following data:

- Engine type
- Engine serial number
- Purchase order number
- Purchaser
- Operating time
- Reason for delivery
- Removed from what vehicle or machinery
- Previous operating location
- Number of performed maintenance procedures
- Previous maintenance
- Date of failure
- Date of removal
- Total operating hours or miles (km)
- Operating time or distance since last inspection
- delivery condition (including packing)
- Parts missing as received
- scope of damage
- Determination of damage
- Cause and progression of failure

Following is a sample report form for a crankcase examination:

Component	Pieces	Evaluation	Conclusion
Crankcase Number Version			
Crankcase (upper portion)			
Coolant chambers			
Piston fit (top)			
Piston fit (bottom)			
Cylinder sleeve flange mating surface			
Main bearing bore			
Cylinder head bolt			
Main bearing bolt			
Camshaft bearing			
Tappet bore			
Crankcase (lower portion)			
Cover			
Cover with crankcase ventilation			
Cam box cover			
Oil spray nozzle for piston cooling			
Oil tank			
Cover			
Oil filler neck			
Oil dipstick			

Individual component evaluations may, for example, be classified as follows:

1 Normal
2 Rework
3 Reject
4 Measurement form
5 Missing as delivered
6 Replacement part
7 Replaced in favor of improved parts

5.2.5 Exact description of damage

- Type of damage
  wear, crack, fracture (orientation, shape, condition of fracture
  surface), corrosion, cavitation, fatigue, aging

- Damage pattern
    - Visual assessment (macroscopic damage assessment):
      appearance of damaged part(s) as well as surrounding area,
      coatings, lubricant remnants, manufacturing details
    - Examination of
         dimensions and tolerances
         manufacturing quality
    - Laboratory tests
         metallographic examination
         microstructure examination
         chemical analysis
         heat treatment
         strength properties
         ultrasonic examination
         reflection electron microscope examination
         energy dispersive X-ray spectroscopy examination
    - Computational and theoretical examination
    - Statistical evaluation of failure-relevant parameters

- Damage report
    - Scope
    - Overview of results
    - Procedure:
         cause of damage
         damage process
         damage pattern
         type of damage
    - Results of examination
    - Opinion

Chapter 6
Engine Failures

6.1 Overview

Engine failures, like failures of other types of machinery, as undesirable as they may be, nevertheless serve to increase knowledge and are part of the development process. For this reason, failures are deliberately introduced in the course of engine development, to establish failure modes, causes, and relevant parameters of characteristic failures, and to anticipate and avert such failures in the future. The history of failures may be traced like a thin red line through the entire history of engine development.

In the early days of combustion engines, problems were experienced with nearly all functional groups and components: seized piston rings, seized pistons, seized bearings, fatigue failures, and thermal overloading as a result of combustion knock were part of the everyday experiences of drivers, pilots, and machinery operators. The switch to light-alloy pistons in World War I and thereafter gave rise to such an abundance of problems that new piston designs were created in response: controlled-expansion pistons, built-up pistons, articulated pistons, etc. In the 1920s, bearings became the major obstacle to aircraft engine development, ultimately overcome by the development of steel-lead-bronze bearings by Allison in the United States. Combustion knock in gasoline engines hindered continued development. This was remedied by the addition of tetraethyl lead as a gasoline antiknock additive. High rates of cylinder wear, especially in tractor engines, were greatly diminished with the development of pearlite castings by tractor manufacturer Lanz. In the late 1920s/early 1930s, as compressorless fuel injection enabled fitting of diesel engines to motor vehicles, initially in heavy, then lighter commercial vehicles and then finally in passenger cars, excessive piston ring groove wear was countered by incorporation of a ring carrier, invented by Ernst Mahle. The next problem to rear its head was crack-initiated fractures in highly stressed pistons for motor vehicle and rail diesel engines, as well as aircraft engines.

Power increases for aircraft engines, spurred by the arms race of the 1930s, led to new materials, shapes, and manufacturing techniques for piston rings. Fatigue failures of intricately shaped parts such as pistons, connecting rods, crankcases, but also shafts and fasteners, put the concept of "shape stability" into the engineering consciousness; the importance of small-scale shaping, i.e., detail design, was recognized. Not only components, but also engine operating materials showed their weaknesses. Although the importance of lubricant as a machine element, specifically as a medium for power transmission, was increasingly recognized, the available engine oils could no longer meet ever more rigorous demands, above all in diesel engines. Addition of additives raised the load capacity, in stages, of engine oils, as did oil testing, classification, and specification. The enormous wear rates of piston rings and cylinders in motor vehicles, tank, and aircraft engines—of the German as well as British and American combatants— in the North African campaign of World War II led to the use of chrome-plated piston rings.

As "dieselification" of rail transport began in Germany in the 1950s, engineers became aware of the importance of coolant preparation and handling: cavitation and corrosion attacked the outer walls of cylinder wet liners and crankcase water galleries. In the 1950s, hot cracking of piston crowns marked the end of indirect injection for commercial vehicle diesels. The 1950s also saw a spate of valve spring failures. Improvements to materials, manufacturing, and above all testing methods provided relief. Cavitation in fuel injection lines forced modification by means of pressure relief; e.g., by constant pressure or throttle valves in the injection pumps. The need to power marine diesels with heavy oil once again led to built-up pistons, a concept that was abandoned for four-stroke engines in the early 1950s with the introduction of single-piece aluminum pistons. Piston ring flutter and ring burning in the 1960s and 1970s, and cylinder liner glazing in the 1980s underscored that tribology was and remains a central factor in engine design. Bearing problems as a result of supercharging and high-pressure supercharging led to development of new bearing configurations ("rillenlager" or grooved bearings, sputtered bearings).

In the 1990s, low-cycle fatigue failures developed into an important field of research and development.

Given the great bandwidth of engines—from lawn mowers to the largest marine diesels—vast differences may be seen in concept, layout, and engine design philosophy in terms of operating time and life expectancy, operating mode, operating conditions, and maintenance modalities. Naturally, these differences also affect failure types and failure modes.

- Medium and large marine engines must be repaired and overhauled onboard; the ship's onboard personnel is supported by the engine manufacturer's specialists or specialized shops and shipyards. On the high seas, marine engines must be brought back online, even after suffering extensive damage, with onboard tools and methods, by any means possible. If necessary, an individual cylinder may be taken offline by disconnecting its fuel pump. Depending on the nature of the failure, the failed cylinder may be motored along for vibration reasons, even though it is working against compression. In extreme cases, the crank train of the offending cylinder must be removed.

- Minor failures of land-based engines; i.e., those used to power electrical generators, machinery, construction equipment, etc. may be repaired on site—which then forms the initiator for the next failure. In the case of extensive failures, the engines are taken to a shop or repair facility, or are returned to the manufacturer. Operators of large engine fleets, such as the military, railroads, or Post Office, operate their own repair facilities. Vehicle engines are either brought to a repair shop, where minor damage is repaired, or, in the event of major failures, engines are sent to the manufacturer, and the vehicle is fitted with an exchange engine.

- Motor vehicle engines are often repaired in a do-it-yourself fashion. In such cases, specialist knowledge, skill, and experience—as well as attitudes of responsibility and honesty—of those conducting the repair may cover a broad spectrum, ranging from exemplary to criminal. Improper and—increasingly—deceptive, superficial engine repairs, mainly on engines for older vehicles, are the subject of countless legal disputes. Motor vehicle adjusters can all tell stories about these. One example: "Upon examination of the failure symptoms . . . it may be assumed that the connecting rod bearing failure was caused either by prior damage at the time of the sale, or an assembly error at the time of installation. In view of the inexpert treatment of the cylinder walls, piston ring wear, and excessive clearance between piston and cylinder, engine failure in some form, even without the connecting rod bearing defect, was almost certainly predestined . . . it cannot be ruled out that the engine block may have much higher mileage (in excess of 200,000 km), and the pistons were removed from a different engine in the course of an improper repair . . ."

Even the conditions involving engine repair—cleanliness, required facilities, fixtures, and tooling—affect engine operation. Another cause of failures lies in modifications, as they are often performed on engines in an effort to increase performance (real or imagined). Inexpert "tuning" ranges from scooter engines, through passenger car engines, to fiddling with the governors of engines in commercial service, and, most recently, intervention in engine management systems ("chip tuning").

Manufacturing defects are becoming increasingly rare. More frequent are faults introduced in repair, and installation after repair. Many failures have their root causes in errors and deficiencies in the process of reinstalling engines, or failure to heed the engine manufacturer's detailed requirements, instructions, and suggestions.

The wish, indeed the drive, to keep operating costs down, leads many engine operators to feed their engines with fuel of inferior quality. "Savings" are also often taken at the expense of operating fluids—oil types and oil change intervals, as well as coolant care. Many failures can be explained by details of engine operation itself, for example when engines are overloaded, when they are asked to provide (too much) power for too long a time, or under unfavorable environmental conditions (altitude, temperature). Cooling system and filtration are often found to be simultaneously victim and perpetrator. Carelessness in filter replacement may have expensive consequences, and small radiator and coolant system leaks may result in serious damage. All of this is reflected in engine failure statistics, as reported by major insurance organizations [6-1]; see Fig. 6.1.

It is obvious from these statistics that the engines of earth-moving machinery are most strongly affected, making up 51% of the failures. This is attributable not only to engineering reasons, but also to statistics: because insured damages are involved, only damage to insured engines is reflected in the statistics, and insured engines represent a larger portion of the earth-moving fleet than in other application fields. Also, engines in earth-moving equipment operate in extremely dusty

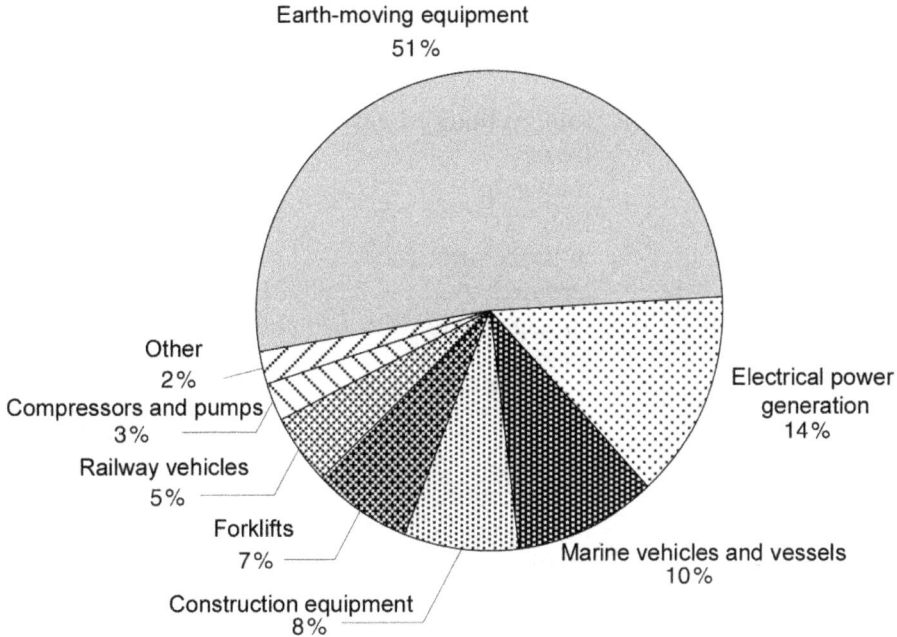

Fig. 6.1 Combustion engine failures by application. Source: Allianz

conditions, are subject to hard usage, and their operating hours have a high full-load component. Maintenance and repair work is usually carried out on site, under imaginably less than ideal conditions.

Most engines are installed in vehicles, so it is naturally of interest to see what types of failure are encountered in motor vehicle engines. Major automobile clubs operate emergency road services for the benefit of their members. For example, the German auto club ADAC keeps breakdown statistics [6-2] and publishes these annually, with classification by the nature of the failure, based on reports filed by its emergency service technicians. All vehicles less than ten years old are evaluated. It is apparent that the predominant contributors to breakdowns are engine peripherals, albeit these are usually minor faults or failures. Internal engine failures are not diagnosed in detail, as such vehicles are taken to repair shops. As on-site, that is, roadside, failure diagnoses are made for the purpose of quickly determining whether the failure can be repaired at the scene to allow the vehicle to drive away under its own power, or, if this is not possible, be towed, failure descriptions are given in very general terms, yet provide a good overview of failures, faults, and damage to motor vehicle (passenger car) operations:

animal damage	foreign body in part	nonfunctioning
bent	frozen	overcharged
blocked	hangs up	overheated
broken	iced	rusted
broken off	in limp-home mode	seized
burned	jammed	short
burned through	leakage current	slipped off
burst	leaking	soiled
charred	loose	too little
chatter	loosened	too little pressure
choked off	lost	too much
clogged	malfunction	torn
contact	misadjusted	unhooked
corroded	misfire	won't hold
damaged	missing	worn off
difficult movement	moist/wet	
discharged	noises	

A remarkable type of motor vehicle engine damage that has become increasingly common over the past decade is gnawing damage by small animals. Martens, beavers, porcupines, and other animals are adopting human cities and settlements as their living spaces, and have shown a preference (as yet unexplained by animal biologists) for plastic electrical insulation and radiator hoses, which leads to electrical shorts, operational faults, and loss of coolant. Ignition leads account for 38%, coolant hoses for 25% of such damage. In Germany, one remarkable fact is that damage by martens is mostly encountered south of a line between Koblenz, Frankfurt, and Coburg [6-3]. It is not known if there are geographically localized concentrations for this type of damage in other countries.

Table 6.1 is an excerpted list of faulty or damaged parts from the German auto club ADAC breakdown statistics. Automotive assistance services in other countries keep similar statistics, with largely identical results. Description of the fault/damage and/or localization to the affected component are likely generalizations. Nevertheless, just from the large number of afflicted engines, such statistics are rather revealing. The note *no details* indicates that the damage involves the actual part; visually apparent damage is more precisely described (e.g., *attachment, studs, gasket*, etc.)

Table 6.1
Excerpt from ADAC Breakdown Statistics, 2007 (Source: (6-2))

Affected component	Detail	Fault/cause
Engine	No details, studs/bolts, ground connection to bolt, attachment, gasket, shaft seal, pressure hose, V-belt, ground connection, control unit fuse, control unit	Worn down, jammed, frozen, overheated, charred, soiled, defective, animal damage, noises, misadjusted
Engine mounts	Studs/bolts, attachment	Torn out, broken, cracked, seized, damaged
Bearings/crank train	No details, studs, attachment, seal	Seized, jammed, leaking, damaged, foreign bodies in part, loose
Fan belt pulley, flywheel, starter ring gear	Attachment	Loose, imbalanced, damaged, lost, noises, worn down, bent
Crankcase ventilation	Pressure hose	Clogged, iced, soiled, incorrectly installed, animal damage
Tensioning system	No detail, bolt/stud	Broken, detached, loose, stiff
Distributor drive	No detail, bolt, stud, attachment	Seized, damaged, loosened, broken off
Timing chain tensioner	No detail, attachment	Broken, misadjusted, foreign body in part
Timing chain	No detail, attachment	Broken, misadjusted, foreign body in part
Timing belt	No detail	Worn out, malfunction, slipped off, loose, stiff, jammed, seized, charred, burned, improperly installed, damaged
Timing belt tensioner	No detail	Loose, damaged, noises, broken off
Timing belt pulleys	No detail, attachment	Loose, broken off, broken out, worn down
Camshaft drive	No detail, shaft seal, stud/bolt	Twisted, chatters, misadjusted, cracked, loose, damaged, worn down
Valve train	No detail, attachment, seal, shaft seal	Detached, worn down, blocked, cracked
Camshaft	Stud/bolt, attachment, seal, timing belt	Damaged, worn down, burst, soiled, corroded, misadjusted
Rocker arm, rocker arm shaft	No detail, attachment	Loose, defective, worn down
Lifter (solid/hydraulic)	No detail, attachment	Broken off, seized, noises, insufficient pressure
Valve	No detail, attachment, seal	Broken off, seized, jammed, warped, burned
Valve lash	No detail	Misadjusted, noises
Cylinder head	No detail, stud/bolt, attachment, seal	Loose, cracked, burned through, cracked, leaks
Spark plug thread	Attachment, seal	Cracked, damaged, burned through, broken out

Table 6.1 (continued)

Affected component	Detail	Fault/cause
Valve cover	Screw/stud, seal	Twisted off, missing, damaged, leaks, slipped off
Intake manifold	Screw/bolt, attachment, seal, pressure hose	Loose, lost, worn down, charred
Exhaust manifold	Attachment, seal	Cracked, loose, leaks, burned, damaged
Exhaust muffler	Bolt, stud, attachment, seal	Broken, cracked, burst, detached, loose
Catalytic converter	Attachment	Broken, cracked, burst, detached, loose
Turbocharger	Attachment, bolt/stud, hose clamp, hose clamp low pressure, shaft seal, actuator motor	Blocked, loose, noises, damaged, leaks, broken, insufficient pressure
Turbocharger oil line	No detail	Broken off, loose, burst, leaks
Oil drain plug	No detail	Damaged, missing, defective
Oil cooler	No detail	Cracked, burst, clogged
Oil line	No detail, bolt/stud	Worn down, loose, lost, leaks
Oil pump	No detail	Cracked, damaged, leaks, seized
Oil pressure switch	No detail, attachment, seal, light bulb, ground wire, wiring, plug, switch	Leaks, malfunction, defective, broken, cracked, loose, moist/wet
Oil filter	No detail, attachment	Clogged, loose, won't hold, cracked, leaks
Cooling system	No detail, attachment, seal, hose clamp, pressure hose	Iced, cracked, burst, malfunction, clogged, leaks, overheated, animal damage
Coolant pump	No detail, attachment, seal, drive belt	Leaks, imbalance, slips, clogged, damaged
Thermostat	No detail	Locked, jammed, rusted, defective
Fan	No detail, attachment, fan belt, ground connection, plug	Nonfunctional, loose, dirty, lost, worn out, slips, shorted, incorrectly installed, corroded

ADAC statistics for 2007 tell us 47.5% of vehicle breakdowns can be traced to the engine and its surroundings. Contributions of the engine and various peripherals to breakdown statistics are shown in Table 6.2.

If one traces failures by type of machine, in other words the engine's field of application, back into the engine itself, one recognizes that the following functional areas are susceptible to faults:

- Geometry and kinematics of crank train and valve train
- Mixture formation (gasoline and diesel engines)
- Ignition (gasoline engines)

Table 6.2
Contribution of Engine and Engine Peripherals
to Vehicle Breakdowns (Source: [6-4])

Ignition	29.7%
Fuel system	13.9%
Fuel injection	13.1%
Cooling	17.7%
Engine	25.7%

These, too, can be statistically documented [see also 6-1, 6-2]. Pistons and cylinder sleeves, connecting rods, crankshaft, and bearings are the components most frequently affected by engine failures (Fig. 6.2).

6.2 Crank train failures

6.2.1 Pistons

The piston is the central component of the so-called *piston machine*, which encompasses not only engines, but also pumps and compressors. In an engine, pistons perform the following functions; see also Fig. 6.3:

- Provide a variable working space (cylinder) boundary

- Seal the working space

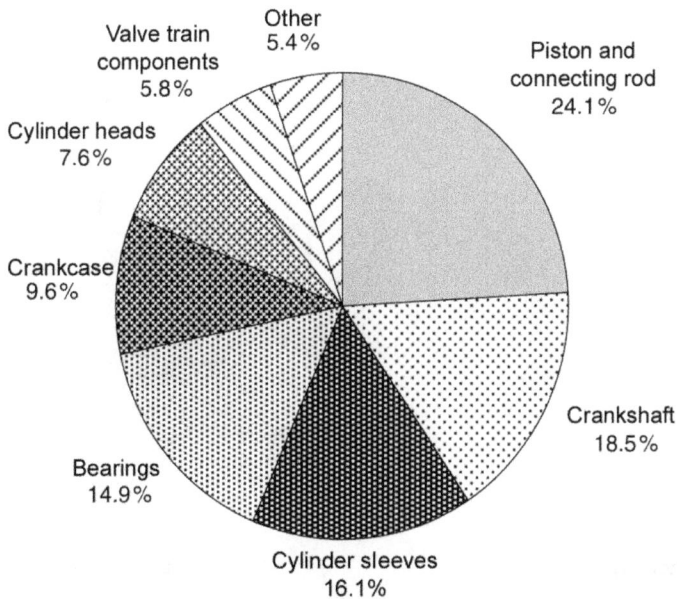

Fig. 6.2 Combustion engine failures by location. Source: Allianz

- Transmit power to and from the working fluid (gas)
- Transfer (remove) heat
- Guide the crank train (in trunk piston engines)
- Support gas transfer during induction and exhaust strokes (four-stroke engines)
- Control gas transfer (two-stroke engines)
- Support mixture formation (by means of suitable shape for the combustion chamber side of the piston crown)

The basic structure of a piston is that of a hollow cylinder, closed at one end. Individual areas of the piston perform various special functions:

- The piston head, including crown and wrist pin bosses, transmits power.
- The ring zone, including the fire land, performs the sealing as well as heat transfer functions.
- The piston skirt guides piston motion.

The "performance" of a piston, the loads and stresses to which it is subjected, and its sensitivity to its operating conditions, are readily apparent upon closer inspection of its functional requirements.

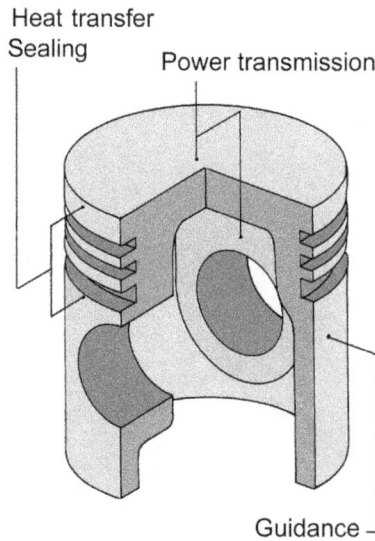

Fig. 6.3 Functional zones of a piston.

Because the function of piston engines is predicated on the fact that the working space of combustion gases within the cylinder varies in volume during the course of a stroke, the piston, as a power-transmitting element, represents the movable boundary of the combustion chamber. It can only perform this function if a hydrodynamic lubricating film is present between piston skirt and cylinder wall, as well as between wrist pin and wrist pin bosses. This lubricating film is also a prerequisite for the sealing function of piston and piston rings; without it, the high gas pressures of combustion could not be maintained. To enable the piston to withstand the extreme combustion temperatures (even though they are present only briefly), it must be able to transfer heat to the cylinder and/or coolant, primarily through its piston rings. In addition, the piston crown is called upon to support mixture formation. To this end, it is given a thermally disadvantageous, discontinuous shape, with large heat-absorbing and small heat-rejecting (edges of valve pocket or bowl) surfaces. In the case of two-stroke engines with exhaust ports rather than valves, the piston serves as a slide valve and is subjected to high thermal loads from the high-speed flow of exhausted combustion gases. Along with power transmission to and from the connecting rod, the piston's crank train guidance function imposes additional burdens. To meet these demands, the smallest details of the piston must be developed with great care.

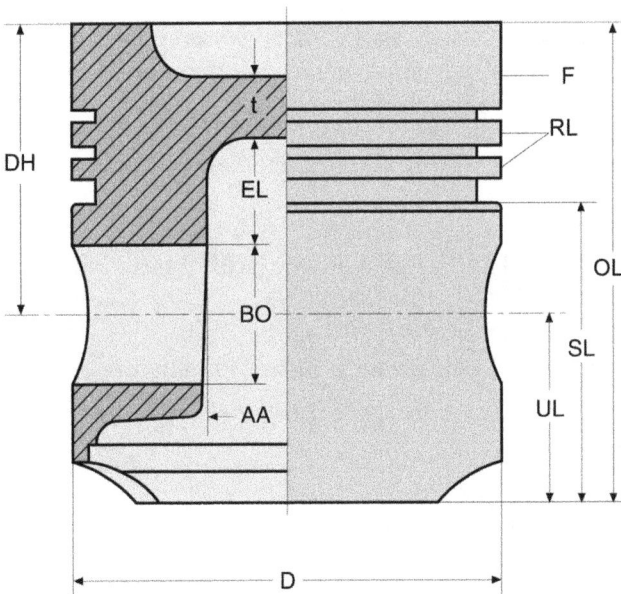

Fig. 6.4 Basic construction of a piston, with designations: F, Fire land; t, Crown thickness; RL, Ring lands; DH, Deck height; EL, Expansion length; OL, Overall length; BO; Wrist pin bore (wrist pin diameter); SL, Skirt length; LL, Lower length; AA, Wrist pin boss clearance; D, Piston diameter. Source: MAHLE

This is evidenced by the fact that every single element of the piston is designated by its own unique technical terminology (Fig. 6.4).

The large span covered by combustion engine design requires a wide palette of piston configurations. Depending on engine size and type, pistons are differentiated by type, shape, principal dimensions, and material. The most important types (Fig. 6.5) are:

- Smooth or full-skirt pistons (nomenclature used by individual piston manufacturers varies) with various weight-saving features

- Controlled-expansion pistons, with cast-in steel bands, rings, or sleeves to control thermal expansion

- Built-up pistons, two- or three-part pistons with a steel upper section and aluminum or spheroidal (ductile) cast iron lower part(s), upper and lower sections rigidly joined but separable

- Articulated pistons, two-part pistons with steel upper and aluminum or gray cast iron lower parts, joined by the wrist pin

- Monoblock pistons, single-part spheroidal (ductile) cast iron pistons

Pistons of any one type may also differ in their principal dimensions. Depending on piston type, size, and application, piston manufacturers provide guidelines for piston dimensioning. Characteristic piston dimensions are (nominal) diameter (D), to which all other dimensions are referenced; e.g., wrist pin diameter (BO/D). Piston materials include aluminum-silicon alloys, primarily eutectoid but also hypereutectoid alloys. For higher thermal, mechanical, and corrosive loads (heavy fuel oil), steel (piston heads) or spheroidal cast iron (piston heads, monoblock pistons) is used. Piston materials are:

- Aluminum-silicon alloys with minor alloying elements copper, nickel, and magnesium
    - Eutectoid alloys with 11 to 14% silicon
    - Eutectoid alloys with 11 to 14% silicon and added heavy-metal content
    - Hypereutectoid alloys with about 18% silicon
    - Extremely hypereutectoid alloys with 24 to 25% silicon

- Aluminum-copper alloys, with about 4% copper and 2% nickel

- Ferrous materials
    - Lamellar graphite gray cast iron
    - Spheroidal graphite gray cast iron
    - Malleable iron
    - Forged steel

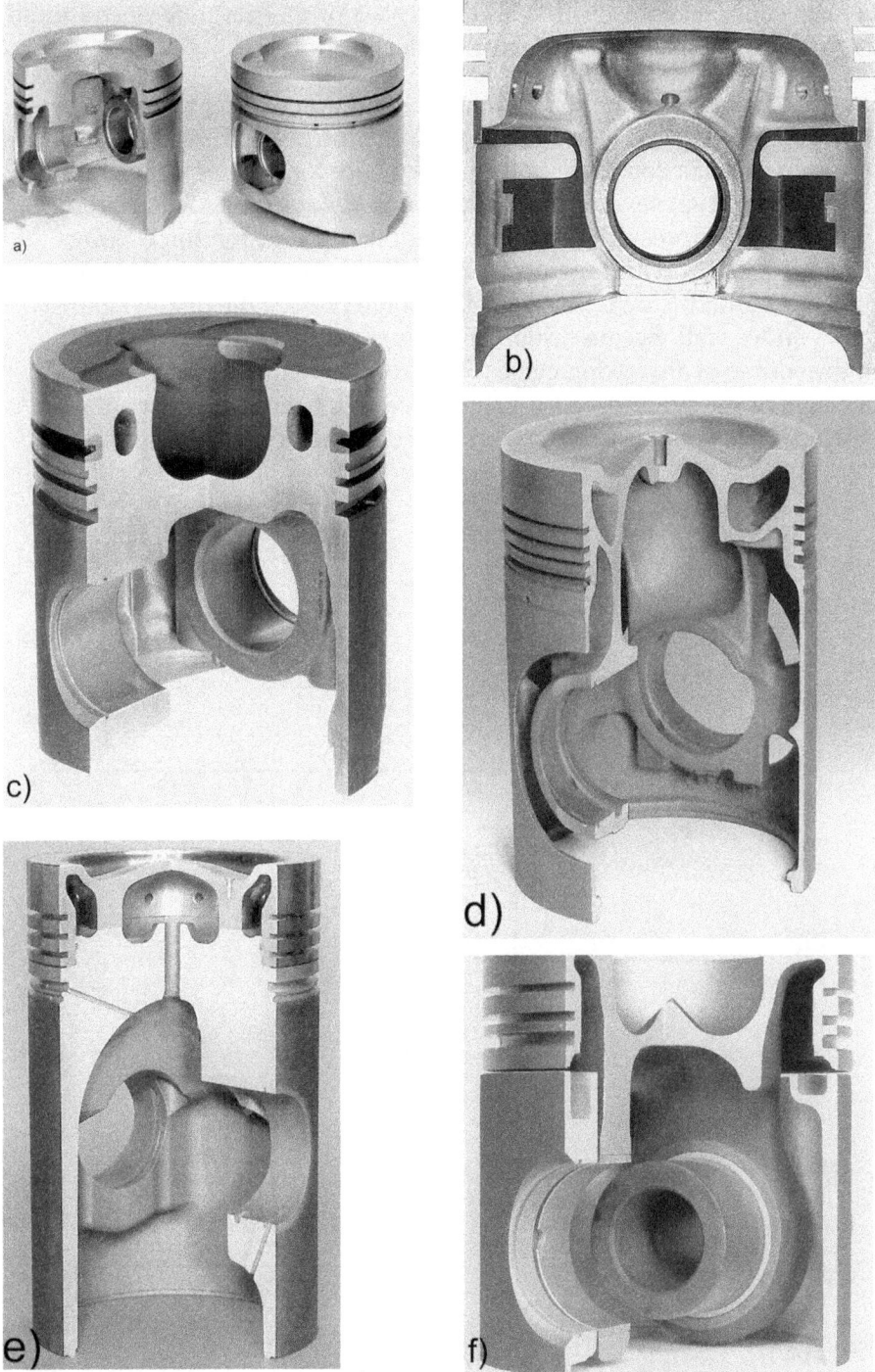

Fig. 6.5 Piston configurations: (a) Full skirt piston; (b) controlled-expansion piston; (c) piston with oil cooling gallery and cast-in ring carrier; (d) monoblock piston; (e) built-up piston; (f) articulated piston. Source: MAHLE

In terms of exact composition and materials properties, reference should be made to the piston manufacturers' specifications.

Gas pressure acting on the piston crown, and oscillating inertia forces of the connecting rod, combine to generate the piston force F_p. Because, in the course of a complete cycle (shown here for a four-stroke engine), these change sign several times, the piston is subjected to alternating tensile and compressive forces. Resolution of the piston force along the connecting rod axis results in an additional component, the normal force acting on the cylinder wall. This force presses the piston against the cylinder wall. Because this, too, changes direction several times in the course of a working cycle (Fig. 6.6), the piston is forced from one side of the cylinder to the other. Because of the required piston clearance, this results in an impact of the piston against the cylinder

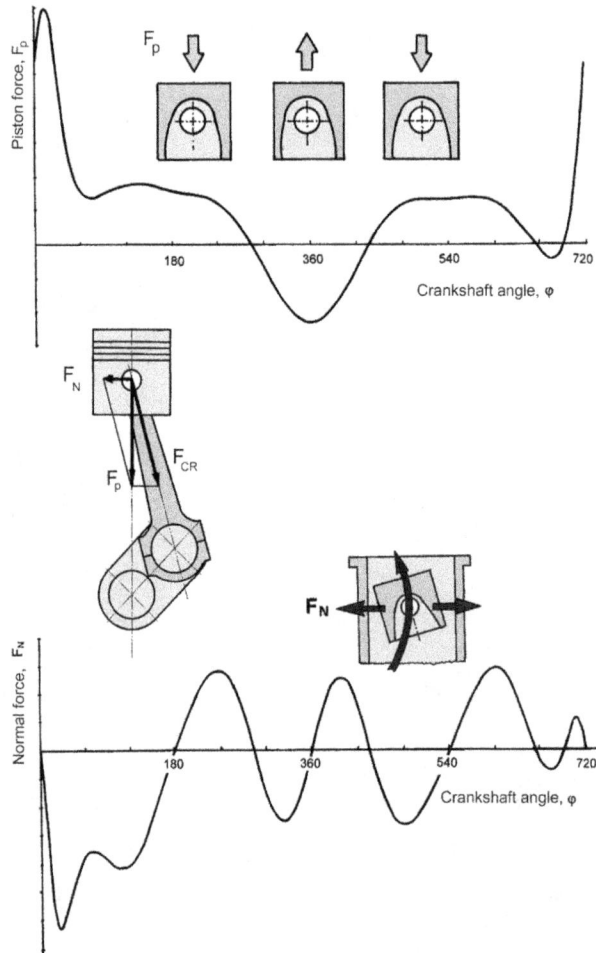

Fig. 6.6 Piston and normal force history over one engine cycle.

wall, especially around the point of ignition top dead center, with a force that is acoustically perceptible as *piston slap*.

Special piston designs have been developed to suppress piston slap: controlled expansion pistons, with cast-in steel bands, rings, or sleeves. These control elements modify the piston's thermal expansion in a desirable way, and so hold piston clearance largely constant. Another possibility for "softening" the side-to-side reversal is to offset the wrist pin from the cylinder axis (Fig. 6.7). This moves the connecting rod axis from the cylinder centerline by a few millimeters, either in the major thrust direction, or in the minor thrust direction. Displacement in the major thrust direction causes early reversal of the piston, at a time when the normal force has less effect on the piston. In the process, the piston, as a result of its rocking motion, first contacts the cylinder with its "soft" lower section (piston skirt), which additionally reduces the impact. One therefore speaks of "offset for noise reduction." For motor vehicle diesel engines, "thermal offset" is employed, an offset in the direction opposite the normal force. This keeps the piston (within its clearance) closer to the cylinder centerline, which has a beneficial effect on piston ring sealing function and therefore helps to prevent the accumulation of carbon deposits on the fire land.

Because piston slap excites vibrations in the cylinder sleeve, the outside wall of the sleeve may experience cavitation from the surrounding coolant.

Gas forces on the piston crown are transmitted through a comparatively small area of the wrist pin bosses, through a lubricating film only a few

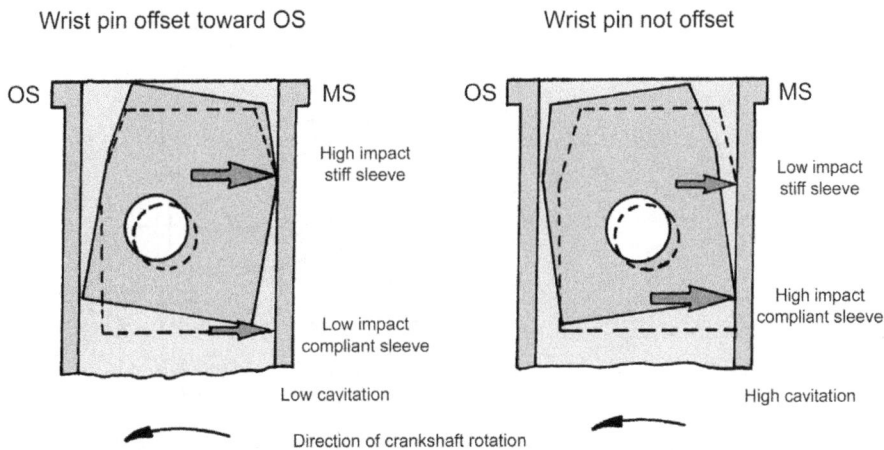

Wrist pin offset toward OS   Wrist pin not offset

OS ⌐   ⌐ MS   OS ⌐   ⌐ MS

High impact
stiff sleeve

Low impact
stiff sleeve

Low impact
compliant sleeve

High impact
compliant sleeve

Low cavitation   High cavitation

Direction of crankshaft rotation

Fig. 6.7 Piston offset and its effect on contact reversal after ignition top dead center: MS, major thrust side; OS, minor (opposite) thrust side. Source: MAHLE

thousandths of a millimeter thick, to the wrist pin and, again through a lubricating film, to the wrist pin bushing and the connecting rod. In the process, force transfer changes direction twice, which causes high local stresses (Fig. 6.8).

Gas pressure changes the piston shape. The center of the piston crown is pushed inward, and is actually bent over the stiff wrist pin. The wrist pin bends as well, and in addition is deformed into an oval cross section. Because the wrist pin is made of high-strength material, these deformations are passed on to the piston. The result is radial compressive stresses and circumferential tensile stresses (peripheral tensile stresses), which act on the piston's wrist pin bosses. In particular, high stresses are imparted to the walls just above the wrist pin, to the point where these areas are plastically deformed, which somewhat diminishes the peak stresses. Transfer of normal force to the cylinder wall presses the thrust side of the piston skirt inward.

Because piston forces change over the course of a power cycle, the piston is dynamically loaded, which represents a considerable intensification of working conditions. Added to this are thermal loads imposed by combustion gases (Fig. 6.9). Maximum gas temperatures are in excess of 2000°C (~ 3600°F), and are dependent on the working cycle, combustion principle, and engine load condition.

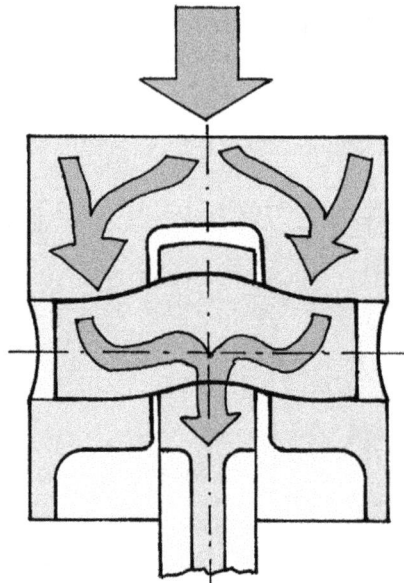

Fig. 6.8 Schematic of force transfer in piston.

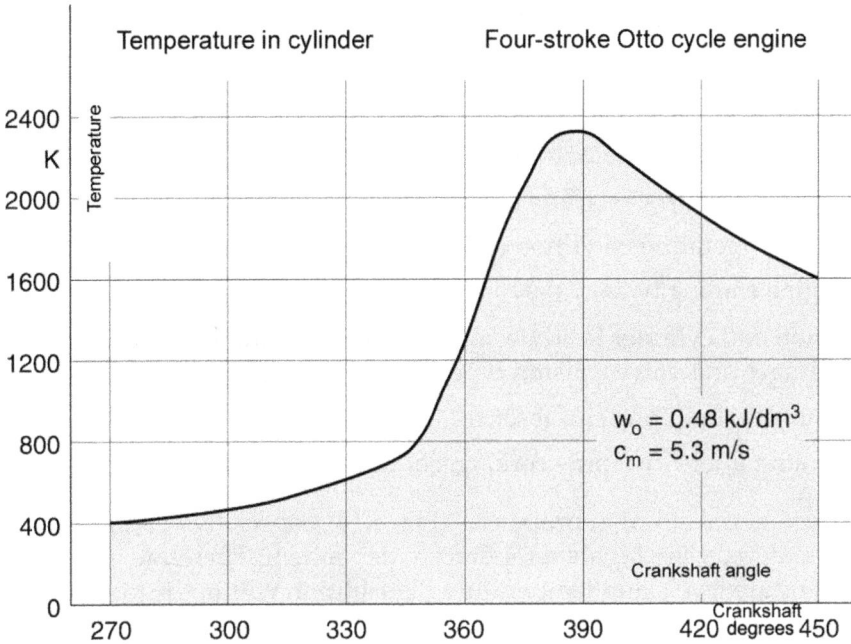

Fig. 6.9 Working gas temperature history.

Exhaust temperatures; i.e., gas temperature after the expansion stroke and during the exhaust stroke, are in the range of 500 to 900°C (~ 900 to 1600°F) while the temperature of inducted charge (air or air/fuel mixture) is 50 to 100°C above ambient temperature. Because of its thermal inertia, the piston, like other combustion chamber parts, does not take part in these temperature fluctuations. The amplitude of temperature variation on the piston crown amounts to only a few degrees Kelvin and rapidly decreases as one moves inward.

Heat entering the pistons is removed, on the one hand through the piston rings and piston skirts, and on the other hand by coolant (cooling oil) and from the piston inside surfaces to the crankcase air (see Table 6.3).

Table 6.3
Piston Heat Flow (Source: [6-5])

	Uncooled	Oil spray cooling	Cooling oil gallery
Cooling oil	-	45°K	68°K
Piston rings	62°K	41°K	18°K
Crankcase air	24°K	8°K	8°K
Piston skirt	14°K	6°K	6°K

Temperature levels and temperature distribution inside a piston (Fig. 6.10) are dependent on:

- Working process (Otto or diesel engine)
- Working cycle (two- or four-stroke cycle)
- Combustion process (direct/indirect)
- Engine operating condition (idle, part load, full load)
- Engine cooling (water/air)
- Piston and cylinder head design (location and number of gas passages and valves, piston type, piston material)
- Piston cooling (present/absent)
- Cooling intensity (spray cooling, cooling gallery, cooling coils, etc.)

Piston material strength properties (Fig. 6.11), especially of aluminum alloy materials, are highly temperature dependent. Therefore, interior temperature levels and temperature distribution within the piston determine permissible loads.

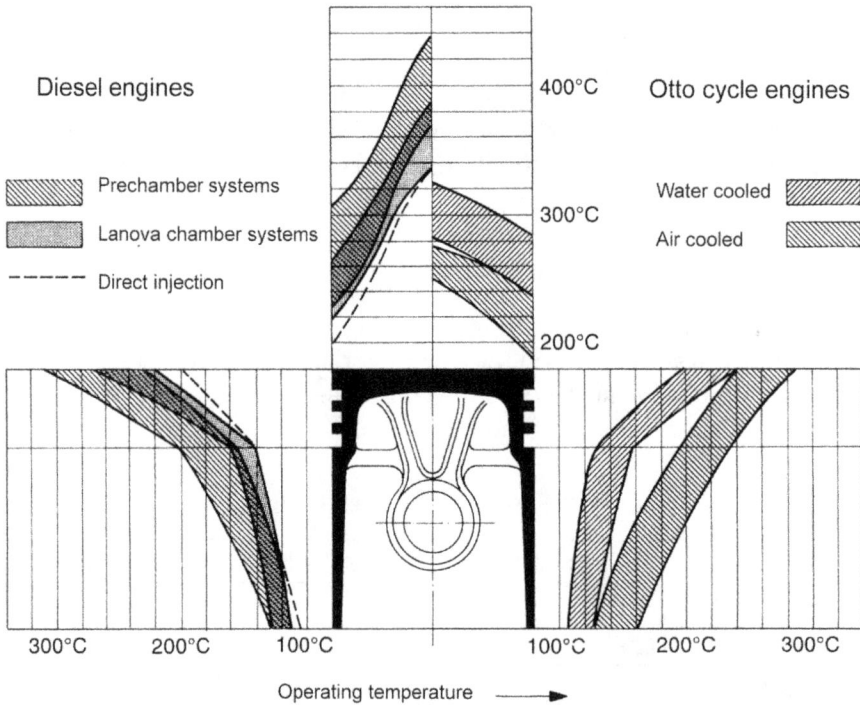

Fig. 6.10 Piston temperatures. Source: MAHLE

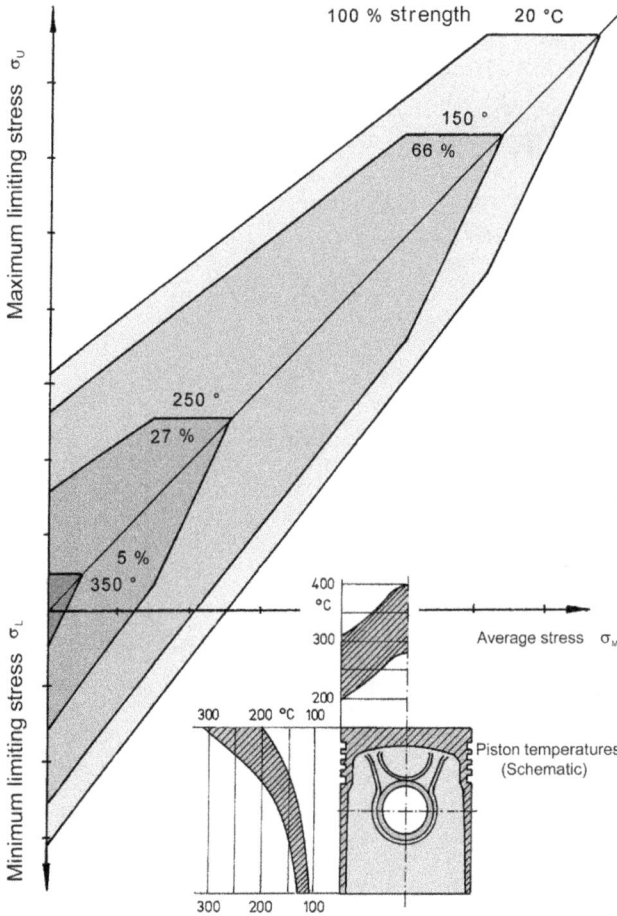

Fig. 6.11 Qualitative fatigue strength diagram (Smith diagram) of a eutectoid AlSi piston alloy (forged) at various temperatures.

Because pistons deform and expand in response to forces, above all gas forces, and gas temperatures, this change in shape must be taken into account in design to ensure freedom from seized pistons at operating temperature. This is accomplished by imparting a shape to the piston that differs from the ideal circular cylinder. The piston must be installed with a cold clearance (Fig. 6.12) which reflects the expected deformation.

Cold piston clearance is composed of the difference between the engine cylinder and an imagined round cylindrical piston (the nominal clearance) as well as deviation of the piston from this ideal circular cylinder shape. The actual piston shape (piston contour) deviates from the ideal circular cylinder in the axial direction (cone, barrel shape) and in the circumferential direction (oval shape). This compensates for piston thermal deformation as a result of temperature history, and mechanical deformation from gas and inertia forces.

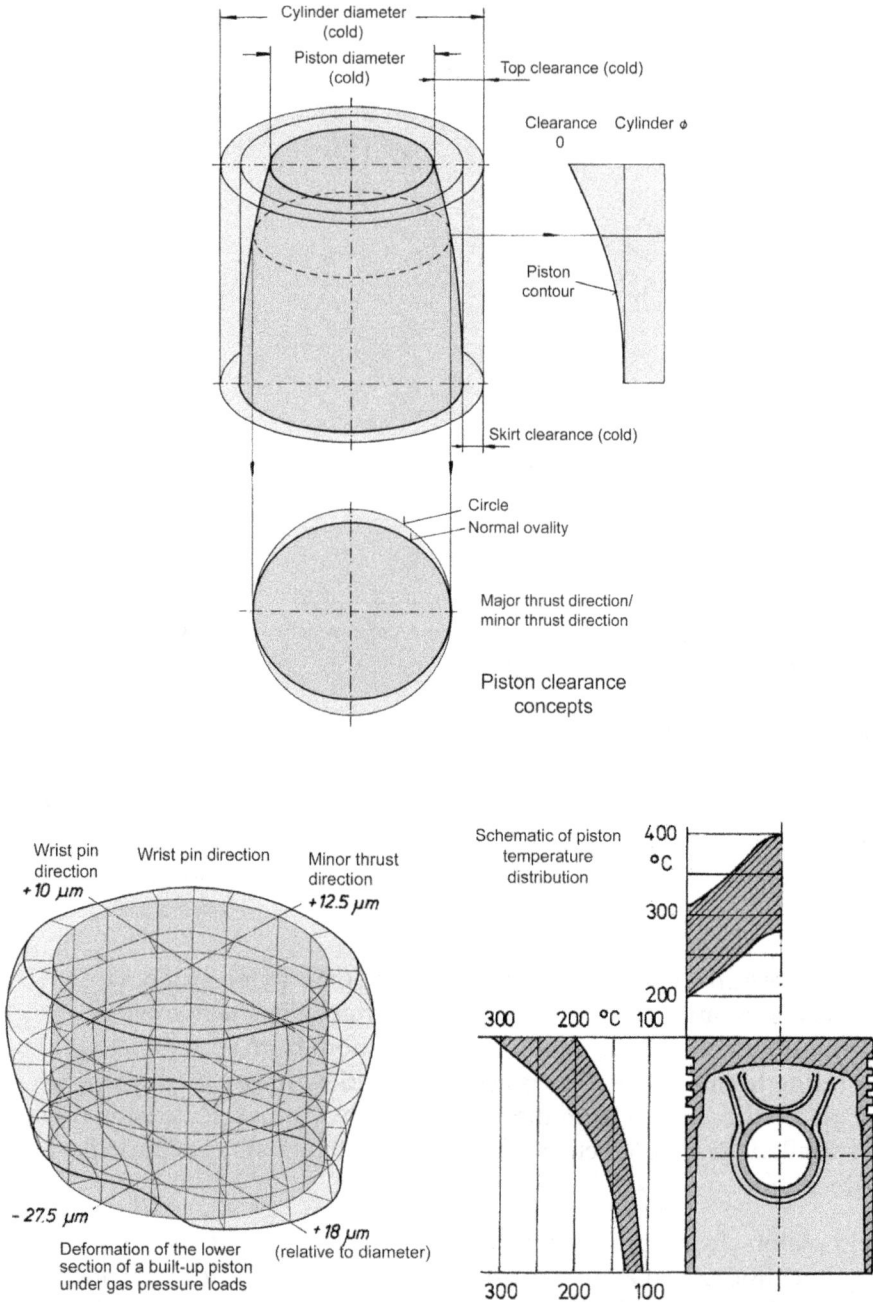

Fig. 6.12 Piston clearance schematic. Cold piston clearance is composed of the difference between the engine cylinder and an imagined round cylindrical piston (the nominal clearance) as well as deviation of the piston from this ideal circular cylinder shape. The actual piston shape (piston contour) deviates from the ideal circular cylinder in the axial direction (taper, barrel shape) and in the circumferential direction (oval shape). This compensates for piston thermal deformation as a result of temperature history, and mechanical deformation from gas and inertia forces.

Piston damage or failure (this also applies to damage/failure to other engine components) may be regarded from several different perspectives [6-6]:

- Damage/failure symptoms on the component, in this case, piston
- Engine symptoms
- Failure consequences
- Causative agent(s)
- Factors affecting the damage/failure

The starting point of a failure analysis is the failure symptom on the piston; e.g., charred piston crown. This in turn has its counterpart on the engine; e.g., the cylinder sleeve. The failure was initiated by engine symptoms such as combustion irregularities or high-speed knock. Results or consequences are localized melting of the fire land, and piston and ring seizing. Overarching causes in this example might include overloading, faulty mixture formation, ignition, or injection, influenced by improper engine operation and maintenance.

Ultimately, piston failures can be traced to problems in kinematic and tribological conditions, to thermal, mechanical, as well as tribological overloading, which in turn are the result of engine overloading through too much power, for too long a duration, to engine faults, unfavorable operating conditions, external influences, and inadequate care and maintenance. There is a continuous spectrum ranging from normal operating behavior, to minor faults, to complete failure.

A single physical agent can cause various failures, just as a single failure type may have various causes. The differences in failure symptoms are in such cases minor and are usually masked by the degree of damage. In this way, the most common failure, that of a seized piston, may be the result of excessive temperature levels as a result of overloading, lubricant or fuel starvation, problems with mixture formation and combustion, or manufacturing and assembly defects.

Table 6.4 provides an overview of the tightly woven web of piston failures, their causes, and influencing factors, arranged by piston zone and type of failure.

- *Faulty kinematic relationships*
  The kinematic relationships between engine components, in this case the crank train and valve train, are carefully matched to one another. High compression ratios and large ports, with long valve duration, mean that clearance between pistons and valves is very

small. For this reason, especially in the case of diesel engines, it is vital that the correct quench height (Fig. 6.13) be observed. Improper cylinder head gaskets or excessive gasket "set" reduce piston to valve clearances.

Wrist pin bearing wear or damage can increase piston stroke, eventually damage the valve train, and, in extreme cases, result in seized valves with damage to valve train and crank train kinematics, and impact of the valve heads against the piston crown. Cocked pistons may result in stuck piston rings. Carbon deposits in the ring grooves may impede piston ring movement.

- *Thermal, mechanical, and tribological piston overloading*
  If power output and/or duration exceed an engine's permissible limits, pistons will be thermally, mechanically, and tribologically overloaded. Also, improperly adjusted engine parameters such as ignition, mixture, fuel injection, coolant and lubricant temperatures, manufacturing and assembly defects, external influences, inadequate care and maintenance have a negative effect on piston operating conditions, leading to overloading.

A systematic overview of possible causes of piston failure, arranged by affected piston area, is given in Table 6.5. Additionally, Sections 6.2.1.1 to 6.2.1.4 cover characteristic piston failures after [6-7], according to type of failure and the piston areas in which they appear.

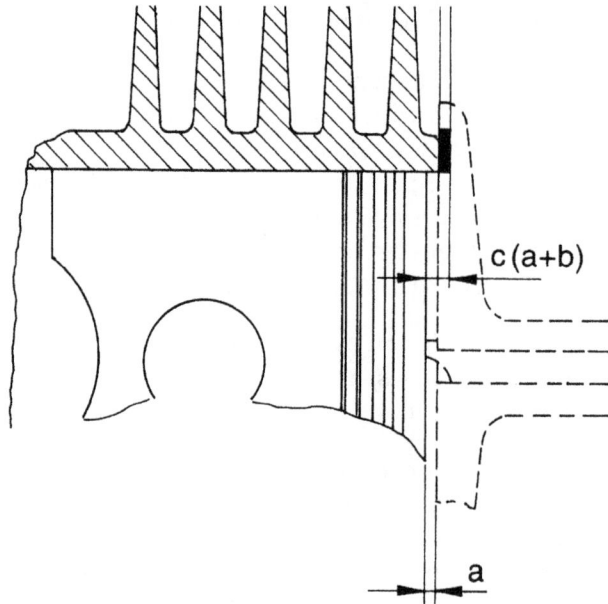

Fig. 6.13 Quench height: a, piston to deck head space at top dead center; b, cylinder head gasket thickness; c (= a+b), total quench height. Source: MAHLE

Table 6.4
Piston Damage Symptoms

Symptom	Cause
Contact pattern (Figs. 6.14, 6.15, 6.16)	When starting or stopping the engine, and with load changes and associated changes in piston and cylinder deformation, it is inevitable that the piston will at times operate in the mixed lubrication regime. This leads to a minimal removal of material, as is also evident in other sliding bearings. The areas where the piston bears are more or less clearly visible, especially if the piston has been coated with a layer of surface protecting material (e.g., bonded graphite coating). These areas are referred to as the contact pattern.
Wear (Fig. 6.17)	After long service, measurable wear (material removal) is apparent at load-carrying areas, but these do not impair piston function. Affected zones include the skirt, particularly on the major thrust side; wrist pin bores; and ring grooves.
Scuffing	If operating conditions deteriorate, e.g., as a result of oil starvation or added thermal or mechanical piston deformation, scuffing (gouging, scoring) may be apparent on the piston skirt. This is understood to mean obvious wear marks as a result of "hard" load carrying.
Seizing	Further deterioration of conditions then lead to piston seizure: mixed lubrication raises the temperature of the mating sliding surfaces, lubricant in the contact zone evaporates, local dry friction leads to mating surfaces welding together, followed by the weld bridges being torn apart. Seizing is self-amplifying and leads to severe damage to the component. Affected zones may include piston skirt, ring lands, and fire land, along with the wrist pin bearing surfaces. Because seizing is an interactive process involving two sliding surfaces, the cylinder wall is also affected. Seizing may also be the result of excessive piston temperatures, insufficient clearance, faulty assembly, lubrication and coolant problems, etc.
Crack initiation	Thermal as well as mechanical overloading may initiate microscopic cracking.
Cracking	From crack initiation, cracks grow at varying speeds. Under thermal loading, such cracks occur primarily in the piston head; in direct injection diesel engines, the edge of the bowl is susceptible. By contrast, under mechanical loading, the wrist pin bosses are susceptible to hub cracking (starting inside the wrist pin bore) and cracking on the outer face above the bore.
Fracturing	Cracks continue to develop, ultimately leading to fatigue fractures. Characteristic fracture failures are split pistons and hub-crevice fractures.
Melting	Melting is caused by combustion irregularities.
Burn through	Melting expands until eventually a hole is burned through the piston.
Erosion	Erosion occurs when fuel droplets impinge on the piston surface at high speed, causing mechanical damage. This is especially noticeable in association with combustion knock.
Corrosion	Corrosion may be caused by combustion products. Salt content of ambient air may also have a corrosive effect.
Carbon deposits	Improper mixture formation, as well as inadequate piston ring sealing, result in deposits on the piston and in the ring grooves. Over time, these "bake on" and solidify into very hard coke (oil carbon) deposits. These deposits polish the cylinder walls to a mirror finish (bore polishing), lead to abrasive wear, and may, if pieces break off, compromise piston ring play. Moreover, heavy carbon deposits may interfere with valves and hamper heat transfer.

Fig. 6.14 Various piston contact patterns for a series of V-configuration diesel engines.

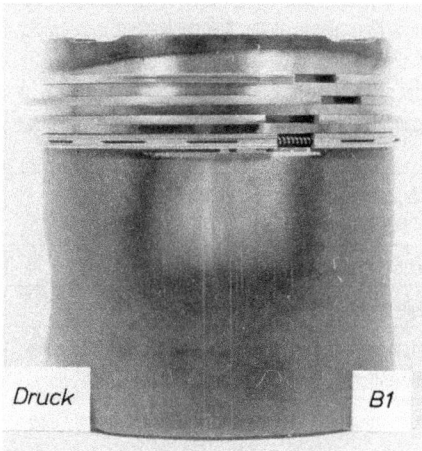

Fig. 6.15 Pronounced contact pattern on the pressure side of a diesel piston.

Fig. 6.16 Emphasis of contact pattern by means of equidensity photography.

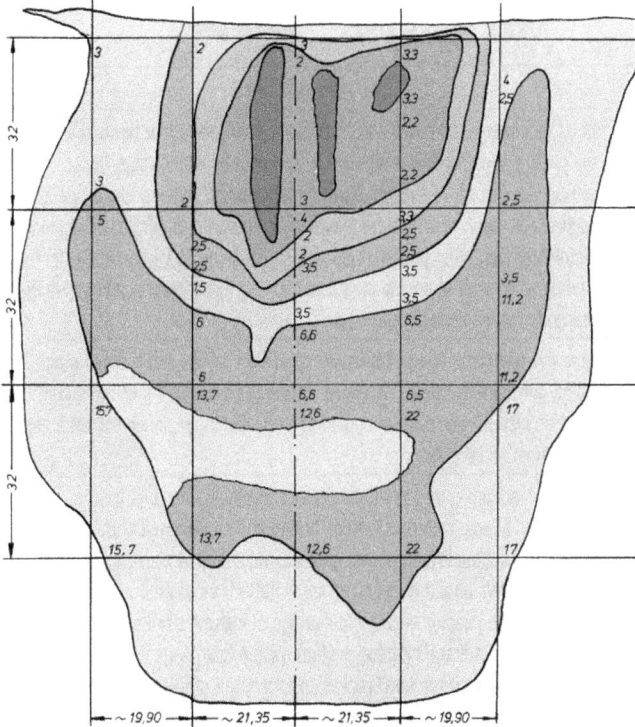

Fig. 6.17 Wear measurement of the piston in Fig. 6.16.

6.2.1.1 Piston failures in the skirt area

Piston skirt seizing

Appearance

The piston skirt shows several similar areas of scuffing/scoring (Figs. 6.18, 6.19). These appear on both the major thrust side and the minor thrust side. Matching evidence of seizure will also be found on the cylinder wall, exactly opposite the damaged piston areas. The surface of the affected areas transitions from highly polished contact areas to areas of dark, relatively smooth scuffing. The piston ring area is undamaged.

Possible causes

- Insufficient clearance between piston skirt and cylinder wall, either through improper dimensioning or because later distortions arising from engine operation caused excessive constriction.

- The cylinder bore diameter was too small.

- The cylinder head was torqued excessively or unevenly (warped cylinders).

Table 6.5
Piston Failure Areas

Affected area of piston	Cause of failure
Skirt	Faults and interruption of the hydrodynamic lubricating film through overloading, fuel flooding, excessive oil scraping by piston rings, "edge bearing" as a result of excessive piston deformation, cylinder distortion as a result of too much interference (sleeves) or uneven cylinder head torqueing (piston/cylinder, wrist pin/wrist pin bores), polishing by carbon deposits, fuel flooding as a result of engine installation errors, insufficient piston installation clearance and oil dilution, corrosion
Piston head and wrist pin bosses	Excessive gas pressures and pressure gradients, forces exerted by foreign bodies, valve interference, excessive carbon deposits, hydraulic lock, axial thrust on piston as a result of angled engine installation, incorrect wrist pin and retainer installation
Piston head, fire land, ring zone	Thermal overloading as a result of combustion faults resulting from combustion knock, incorrect ignition timing and fuel mixture (Otto cycle), inadequate fuel quality; for diesel engines, uncontrolled combustion, with improper timing, added to distorted spray pattern as a result of carbon deposits on the injector nozzle and damage to injector needle, local thermal overloading on indirect injection engines, piston bowl edge cracking on direct injection engines, piston bowl recession as a result of excessive temperatures

Fig. 6.18 Seized piston. Source: KS

Fig. 6.19 Seized piston. Source: KS

- The nominally plane surfaces of cylinder head or crankcase/ cylinder block were uneven. Threads in the tapped stud holes or cylinder head bolts were contaminated and distorted. The bolt head mating surface had fretting corrosion or was unevenly oiled during installation.

- An improper or unsuitable cylinder head gasket was installed. Scale, calcification, dirt in the coolant galleries, cooling system, etc. might have caused uneven heating and resulting cylinder warping (defective thermostats). The engine was subjected to heavy loads before it was fully warmed up.

Piston seizing as a result of overheating

Appearance

Severe scoring is visible from the piston head downward, increasing toward the bottom of the skirt. The surface of the seized areas is dark colored, deeply scuffed, and partly torn open. Seized areas extend all the way around the piston. The ring area also shows seizing all around its circumference, with the marks becoming less severe toward the skirt (Fig. 6.20).

Possible causes

- Thermal overloading from the combustion chamber side has heated the piston head to the point where piston clearance to the cylinder wall has been eliminated. The piston operated with mixed and boundary lubrication, which ultimately caused a combination of clearance and dry-running seizure around the entire circumference of the head.

Fig. 6.20 Seized piston as a result of overheating. Source: KS

- General lack of clearance between piston and cylinder can be eliminated as the cause, because despite severe seizing around the head, the lower skirt area, with its smaller clearances, shows less damage.

Engine-related causes
- Combustion faults leading to overheating of components bordering the combustion chamber, and/or
- Cooling system defects (coolant circulation, piston cooling).

One-sided piston skirt seizure (possibly with overheating)
Appearance
As may be seen in Fig. 6.21, one side of the piston head (it could be the major thrust side, or just as easily the minor thrust side) shows severe, dark-colored seizure marks, with a severely torn surface. There is even a transverse crack visible across the seized area. The area of the skirt opposite the seized area is completely undamaged, which, in the early stages of such a failure, also applies to the ring area.

Possible causes
- Localized overheating of one-half of the cylinder has caused the oil film on this side to collapse completely, resulting in dry seizure.
- Lack of clearance can be eliminated as a cause, because despite severe seizing, there are no corresponding marks on the other side of the piston.
- At the time of the failure, coolant circulation had failed completely, either as a result of air bubbles in the cylinder area or in the water pump.
- Influential factors might have included a broken fanbelt, defective thermostat, or water pump failure.
- Oil starvation as a result of a failed oil pump.
- Severe scale deposits in the cooling system.
- Incorrect ignition timing in Otto-cycle engines (early or late ignition).
- Faulty injector nozzles.

One-sided seizure marks due to oil starvation
Appearance
Figure 6.22 clearly shows a one-sided skirt seizure. This is on the major thrust side of the piston. Obvious by comparison is the good contact pattern on the opposite, undamaged side.

Possible causes

- During its downward movement, the side of the piston that is more heavily loaded during the working stroke (the major thrust side) has insufficient lubrication. Lack of lubricating oil between piston and cylinder leads to local seizing, with overheating and welding, until, in short order, the entire contact surface on the major thrust side has seized. Causes could include: too much scraping effect from the oil control ring, low oil level (especially when climbing or descending grades), or oil dilution.

- This failure, caused by oil starvation, is recognizable by the fact that the piston was apparently working well before the failure. The still-visible contact pattern on the side opposite the major thrust side is well formed, and the piston shows no sign of overheating.

- Insufficient piston ring lubrication leads to ring seizure, recognizable by their burnt working faces. In cases where piston seizure is a result of ring seizure, this is usually recognizable by pronounced evidence of seizing in the ring area. Low oil pressure in the lubrication system, due to a defective oil pump, may have caused this failure.

Fig. 6.21 One-sided piston skirt seizure. Source: KS

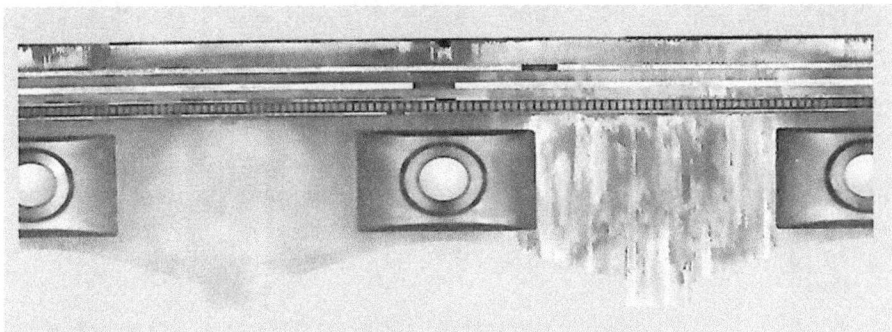

Fig. 6.22 Piston seizure due to oil starvation. Source: MAHLE

Engine-related causes

- Oil starvation or oil dilution

- Unsuitable oil control rings

- Defective cooling system including thermostat, fan belt, scale, water pump

- Injection system in the case of diesel engines

- Fuel flooding in the case of Otto-cycle engines, defective cold start equipment, filter system

Piston seizure (major thrust side and minor thrust side)

Appearance

The piston skirt area shows pronounced rubbing and scoring on both the major and minor thrust sides (Fig. 6.23). The areas are shiny and to some extent appear polished, and are concentrated toward the bottom of the skirt. The ring area and all piston rings are in good condition.

Possible causes

- The piston shape is formed in such a way that at operating temperature, the entire skirt surface bears against the cylinder wall. If, as shown here, both sides (major and minor thrust sides) show seizure marks beginning at the bottom of the skirt, the conclusion is that piston clearance is insufficient as designed. If the piston crown does not show any carbon or varnish deposits, any effects from overheating as a result of combustion problems can be ruled out. This failure occurs very early (in nearly new condition), as thermal expansion is restricted due to inadequate clearance. Engine cooling faults (lack of coolant, defective thermostat, etc.) and resulting overheating could also cause insufficient clearance.

- The coefficient of expansion of aluminum is twice that of cast-iron cylinders. Also, rapid topping up of cold water may result in

Fig. 6.23 Piston seizure on thrust and opposite sides. Source: MAHLE

insufficient clearance, as the cylinder is rapidly cooled and shrinks, while the piston is still hot.

- If seizing marks are unevenly distributed around the piston circumference, the seizure may have been caused by a warped cylinder. Local polishing of the cylinder wall points to this scenario.

- In the case at hand, it must be assumed that the cylinder bore was too small, after the cylinder had been overhauled.

Engine-related causes
- Incorrect cylinder or piston dimensions

- Improper cooling system function, defective thermostat, defective water pump, etc.

- Defective belt tensioner (water pump drive belt)

Diagonal seizing marks on skirt near wrist pin boss
Appearance
The piston in Fig. 6.24, with a fairly flat area around the piston pin bores, has only seized in the transition between the skirt and this flat area. These seizing marks appear to be entirely diagonal. The skirt area on both the major and minor thrust sides is relatively free of seizing marks. Highly polished surfaces are visible immediately adjacent to the seizure marks.

The connecting rod is difficult to rotate on its wrist pin. Lateral seizure traces are visible in the piston pin bore (Fig. 6.25).

Possible causes
This situation applies strictly to pistons associated with connecting rods that have clamped or shrink-fit (interference-fit) wrist pins; i.e., the wrist pin is firmly held in the connecting rod and is only free to move in the piston. Due to the limited range of oscillation of the connecting rod, and therefore the wrist pin, its lubricating conditions are critical. In normal operation, oil is supplied by drilled passages, or radial or axial lubricating grooves. However, placing a new or overhauled engine in operation can be problematical if connecting rod and wrist pin, as sliding, mating components, are not yet lubricated. By the time oil manages to find its way through the bearing gap, the wrist pin has already seized in its bore. The resulting added heat causes additional piston expansion in the wrist pin area, in other words, around the wrist pin bosses. With further overloading, the lubricating film on the cylinder wall is squeezed away, and the result is seizing.

Another hazard is excessive interference fit between connecting rod and wrist pin; this could cause oval deformation of the wrist pin

Fig. 6.24 Diagonal seizure marks on piston skirt. Source: MAHLE

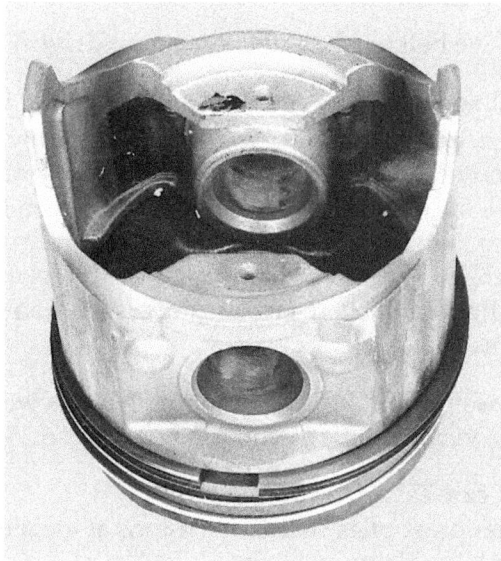

Fig. 6.25 Seizure marks in wrist pin bore. Source: MAHLE

(reflecting the cross section of the connecting rod upper end). If this oval deformation is transferred to the piston, the result could be hard contact or even seizing. It is important that wrist pins and wrist pin holes in connecting rods be well lubricated before assembly.

- The new or overhauled engine had too much "shelf time" before being placed in service.
- Faulty engine assembly practices.
- Inadequate tolerances in wrist pin/connecting rod end.
- Misaligned connecting rod.

Crooked piston—asymmetrical contact pattern

Appearance

Figure 6.26 shows one-sided piston contact against the cylinder wall. The fire ring is blackened by carbon deposits above one wrist pin bore (at left and right side of the photograph), yet is relatively clean above the other wrist pin bore (center of photograph). The skirt contact pattern is asymmetrically displaced and, despite piston ovality, the contact patterns at the edge of the skirt join on only one side, below the wrist pin bore (at edges of photograph).

Possible causes

- Piston angularity due to wrist pin axis non-orthogonal to connecting rod, nonparallel connecting rods, or incorrect crankshaft bearings

- Misaligned connecting rods

Consequences

Under these conditions, the piston rings will have difficulty breaking in, with resultant compression and power losses. Moreover, hot combustion cases can blow past the rings and destroy the oil film on the cylinder wall. Dry running and seized pistons will result. The skewed piston orientation will cause the rings to flutter during their oscillating motion. This will cause pumping action, with high oil consumption. This piston orientation also imposes axial thrust on the wrist pin. The wrist pin retainers may wear, or even be forced out.

Seized pistons as a result of distorted (warped) cylinders

Appearance

The piston shown in Fig. 6.27 has a sharply defined pinched zone at the bottom of the skirt, with several smeared areas, in part bright, shiny, and without deep scoring. The cylinder has a circumferential shiny edge near the head gasket. This is a case of localized clearance reduction, usually encountered on engines with individual cylinder sleeves.

Fig. 6.26 Skewed piston. Source: MAHLE

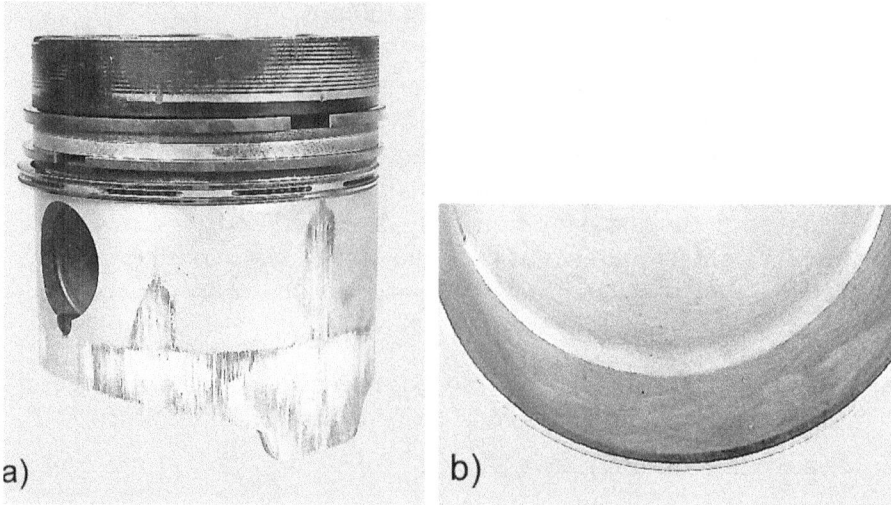

Fig. 6.27 Pinched piston. Source: MAHLE

Possible causes

In the case of wet liners, excessively tight fits in the block spigots or too-thick sealing rings may constrict the cylinder. Such faults can also occur in resleeved cylinder blocks. Obviously, in boring out the block or in machining the outside diameter of the sleeve, a step was created, which led to sleeve deformation.

Localized seizing marks are usually caused by warped cylinders and by improper cylinder head torqueing, an especially dangerous situation in the case of air-cooled finned cylinders.

For these reasons, it is advisable to first install wet liners without sealing rings to check fit and warping. Cylinder blocks may have cylinders whose lower ends are tighter due to uneven wear of the honing stones, or if the honing tool is inserted too far or not far enough. Cylinder bores should always be measured at several different levels, and the measured workpiece must be at room temperature.

Engine-related causes
- Cylinders or sleeves (liners) are installed improperly
- Cylinder or sleeve bores are too small
- Defective mounting surfaces, seal ring grooves, incorrect seal ring thickness and/or diameter
- Cylinder block and/or cylinder head are not flat, possibly warped, wrong cylinder head gasket or unsuitable head gasket configuration

Piston wear caused by dirt

Appearance

Both sides of the piston show dull contact surfaces (Fig. 6.28). Piston rings have excessive gap, and in addition show excessive radial wear and razor-sharp corners. The edges of the ring grooves show considerable axial wear. The piston skirt has a "sanded off" appearance. Symptoms included high oil consumption, combined with increased blowby and oil dilution by fuel. In addition, the engine did not develop full power and exhibited poor starting, especially in cold weather.

Possible causes

The appearance of the piston surface suggests heavily contaminated oil. If piston ring wear (predominantly in the axial direction) decreases from the top ring downward toward the oil control rings, contamination probably entered through the intake passages. However, if wear is greatest at the lower piston rings (oil control rings), and the piston skirt also shows heavy wear, the cause may be found in contaminated oil. If the piston only shows scoring due to dirt, and the rings—especially the oil control rings—are worn (and axial wear of the top compression ring is minimal), the problem was caused by cylinder honing. Either the cylinder bores were not properly rinsed after honing, or the honing marks were smeared over, with smeared and folded metal rubbed away. This can result in burned rings.

Engine-related causes

- Contaminated engine oil, oil filter, and oil strainer
- Contaminated air filtration system, possibly a bypass in the air supply
- Intake manifold or passages defective or improperly sealed

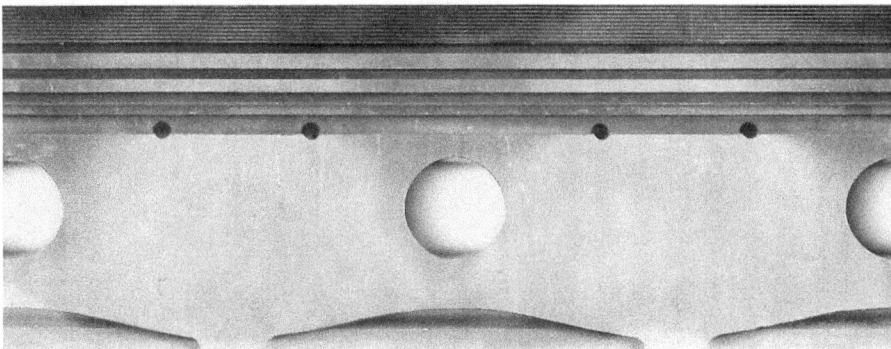

Fig. 6.28 Piston wear caused by dirt.

- Contaminated engine components (inadequate cleaning after machining/honing)
- Poor workmanship or improper machining of cylinder bores or sleeves
- Boring or honing debris remaining in engine

The piston in Fig. 6.29 also shows a milky-gray skirt contact pattern, with more or less prominent vertical scoring marks. A large number of foreign bodies have embedded themselves over the entire running surface of the piston and the fire land. The sharp, oil-scraping edges of the piston rings have a "fringed" appearance, and a "beard" is forming. The radial thickness of the piston rings has been appreciably reduced by this wear. The sides of the compression rings, especially the top ring, as well as the sides of the ring grooves are heavily worn and no longer parallel in the axial or radial directions.

Long-term dry seizing

Seizing of this type (Figs. 6.30, 6.31, 6.32) may also occur even with adequate piston clearance. The lubricating film between piston and cylinder wall, already locally compromised by excessive heat or fuel flooding, has broken down completely. The result is mixed or boundary lubrication at these contact points, which quickly forms seizing areas with torn-open surfaces. Similar damage appears as a result of oil starvation.

Appearance

Along with a uniformly well-formed contact pattern, the upper skirt area shows local scuffing, one-sided or on both sides. The fire land may also show similar areas of scuffing.

Fig. 6.29 Piston wear caused by dirt. Source: KS

Fig. 6.30 Long-term dry seizing. Source: KS

Fig. 6.31 Long-term dry seizing. Source: KS

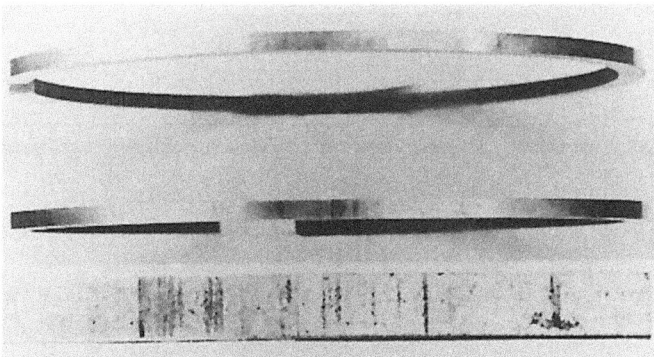

Fig. 6.32 Long-term dry seizing. Source: KS

Possible causes

After extended, previously unproblematic engine operation, lubrication in the upper cooling regime has broken down, locally and in limited areas—possibly as a result of fuel flooding or overheating (e.g., from excessive blowby).

Engine-related causes

- Combustion problems as a result of fuel injection faults

- Overheating due to exhaust system problems

Piston scuffing as a result of fuel flooding (1)

Appearance

Figure 6.33 clearly shows one-sided, streaky, narrow piston wear areas, and considerable scuffing along the entire skirt length. The piston rings may also exhibit burned spots.

Possible causes

The oil film on the cylinder walls was washed away by excess fuel.

- Piston and piston rings run dry. Scuffing first appears on the major thrust side; finally, seizing appears.

- Excess fuel is often the result of improper carburetor operation; either the automatic choke failed to shut off soon enough, or a hand-operated choke remained engaged too long.

- A defective fuel injection system (cold start enrichment), or ignition misfiring as a result of damaged spark plugs in individual cylinders, can lead to fuel deposits on the cylinder walls. At higher power levels, the resulting oil dilution usually leads to piston damage.

Engine-related causes

- Defective fuel injection system, incorrect ignition settings, defective spark plugs

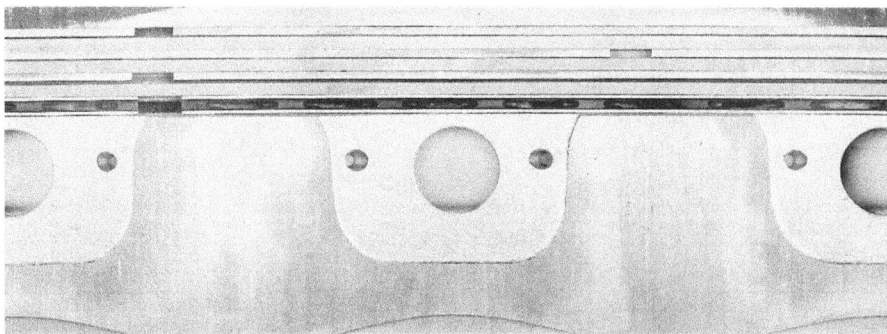

Fig. 6.33 Piston scuffing as a result of fuel flooding. Source: MAHLE

- Oil dilution as a result of fuel flooding

- In diesel engines, defective injection system and/or nozzles.

Piston scuffing as a result of fuel flooding (2)

Appearance

All piston rings show clearly visible wear, but this is still within tolerable limits. With increasing operating time, scuffing appears on the piston skirt, characteristic of dry running as a result of fuel flooding (Fig. 6.34).

Possible causes

- As in Fig. 6.33

- Fuel flooding as a result of combustion faults always leads to compromised oil films, forcing the pistons to run under mixed lubrication. This causes wear in the piston ring area.

- Familiar "fuel scuffing" happens only when the oil film is so badly affected by fuel flooding that the piston runs dry.

- Continued lubrication failure leads to considerable piston ring, ring groove, and cylinder wear.

- Initially, the piston skirt is little affected, because the crankshaft provides oil with adequate lubricating properties.

- Once wear particles from the upper part of the piston are mixed with lubricating oil, and the oil loses its lubricating abilities as a result of fuel dilution, wear spreads to all sliding parts of the engine. Ultimately, the crankshaft bearings are also affected.

Fig. 6.34 Fuel flooding.
Source: KS

Piston skirt corrosion

Appearance

Piston skirt pitting is characteristic of this form of damage (Fig. 6.35).

Possible causes

- The pitting occurs because of intercrystalline corrosive attack. The cause is usually use of an unsuitable solvent for removal of carbon deposits, combined with too long an exposure time.

- Storing engines or vehicles for years, with unsuitable preservatives, can lead to damage of this type.

6.2.1.2 Ring and fire land damage

Damage as a result of skewed piston due to jammed piston rings

Appearance

Piston rings have streaky seizure marks over their entire circumference, possibly with heat discoloration. Moreover, there is one-sided heat discoloration in the wrist pin direction, on the fire land and first ring land (Fig. 6.36).

Possible causes

Connecting rods not aligned or poorly aligned, and possibly twisted

Consequences: As a result of a skewed piston due to twisted connecting rods, the jammed piston rings are no longer able to seal properly. Hot combustion gases can flow past the rings and destroy the oil film. The rings run dry and seize. This generates heat which causes discoloration. Further damage includes seizing in the ring and upper skirt area as a result of piston heating. Stuck piston rings, with identical consequences,

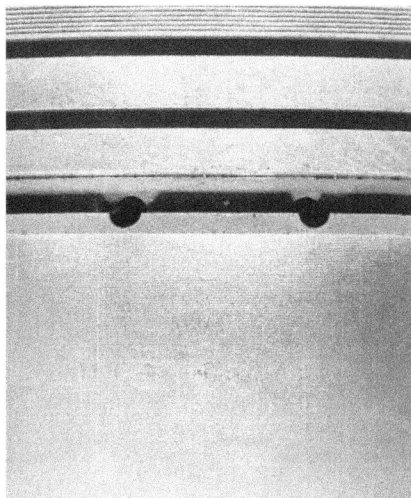

Fig. 6.35 Corrosion.
Source: MAHLE

may also be caused by carbon deposits, dirt, or, in the case of improper assembly, by metal shavings and damaged ring lands.

Burned piston rings

Appearance

Compression rings show severe seizing on their entire circumference, along with burn marks. The cylinder bore is also damaged as a result, with long gouges and the beginnings of seizing marks (Fig. 6.37).

Possible causes

Burned rings (see also piston rings)

Damage of this type begins with local oil starvation between piston ring and cylinder wall, with associated ring seizing. The minimal oil film that normally remains on a cylinder wall is completely scraped away by the piston rings. This permits local metal-to-metal contact, followed by welding as a result of frictional heating and irregularities on the sliding surfaces, as well as material cracking. Such damage occurs primarily in the break-in period, under heavy loading, when the piston rings have

Fig. 6.36 Skewed piston. Source: MAHLE

Fig. 6.37 Burned piston ring. Source: MAHLE

not yet achieved their full sealing function. Increased blowby of hot combustion gases heats the piston skirt, so that piston seizing may be observed along with burned rings.

Burned rings as a result of aggressive oil control rings, or excessively smooth cylinder walls
In this type of damage, the pistons are usually still in flawless condition, that is, there is little or no sign of overheating (discolored lower ring section and upper skirt section). Excessive oil scraping as a result of badly fitting rings, not in keeping with specifications, is known to lead to inadequate oil supply to the cylinder walls, with burned rings as a result. When piston rings are replaced without honing the cylinder walls, only the asperities of the fresh, not yet broken-in rings bear against the smooth cylinder wall, and these find insufficient lubricating oil. Installation of new piston rings must always be accompanied by re-honing of the cylinder bores to ensure the required roughness for ring break-in.

Engine-related causes
- Improper honing (unsuitable honing pattern or honing structure; see also under "cylinder")
- Unsuitable piston ring package
- Lubricating or cooling problems
- Oil contamination

Piston burning (Otto-cycle engines)
Appearance
Piston head material behind the rings has been burned away. The piston skirt has not experienced primary seizing; one sees that at worst, some piston material from the burned area has been smeared onto the skirt (Fig. 6.38).

Possible cause
Pre-ignition
Burned piston heads in Otto-cycle engines are the result of pre-ignition on pistons with flat crowns and large quench areas. Lack of clearance can be eliminated as a cause, as in the early stages, despite high temperatures caused by burning, no seizing is visible on the skirt. The quench zones of the piston head are heated so intensely by pre-ignition that the softened material is carried away, from the fire land, past the rings, all the way to the oil ring groove, by inertia forces and by combustion gases forcing their way into the ever-growing burned areas.

Fig. 6.38 Burned piston. Source: KS

Pre-ignition is caused by:

- Incorrect spark plugs
- Insufficient or nonexistent valve clearance, damaged or leaking valves
- Improper fuel
- Glowing deposits on piston crown, cylinder head, valves, and spark plugs
- Soft carbon deposits, which form in the combustion chambers of high-performance engines during extended city driving
- Diesel fuel in gasoline
- Oil in combustion chamber due to excessive oil consumption
- High engine and high intake air temperatures, etc.

Seized fire lands on diesel engine pistons

Appearance

The piston head shows signs of seizing, primarily in the fire land area, and opposite a heavily carboned impact zone for a fuel stream from the nozzle (Figs. 6.39 and 6.40). The surface of the seized area is rough and torn; to some extent larger pieces of piston material have been torn out.

Fig. 6.39 Seized fire land. Source: KS

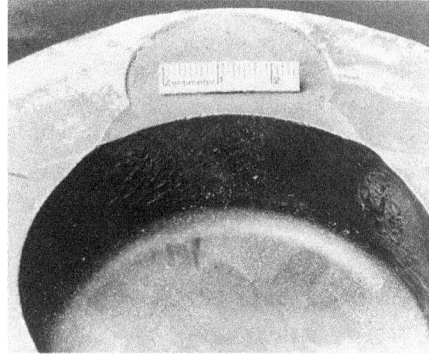

Fig. 6.40 Seized fire land. Source: KS

Possible cause

Unvaporized fuel from the injector has penetrated all the way to the cylinder wall, where it has diluted the oil film to the point where the piston is running completely dry.

Due to dry running, the piston material in this fire land area has seized so severely that the piston material has literally welded itself to the cylinder wall, resulting in several large pieces being torn out of the piston head.

Engine-related causes

- Dripping, in other words, defective injector nozzles
- Hanging injector needle caused by distorted nozzle holder (possibly as a result of overheating damage)
- Incorrect injection timing and resulting added heating
- Burned prechamber or swirl chamber
- Defective fuel filtration system

Ring zone damaged by broken piston rings (1)

Appearance

Large craters have been hollowed out of the ring lands or fire land. The surfaces of these craters are shiny and scoured smooth. The ring in the hollowed-out groove is broken (Fig. 6.41). The edges have been forced outward and worked away by rubbing against the cylinder wall, forming a sharp edge.

Possible causes
Broken piston rings: the craters are the result of broken piston rings. Ring breakage can be traced back to possible assembly defects, insufficient ring gap, or ring flutter (see "piston rings").

For two-stroke pistons with pinned rings (to prevent rotation), such craters often appear next to the securing pins. The cause is broken ring ends. These are caused by rings with insufficient or no gap, pressing against these pins, or ring ends which were scratched or cracked during assembly.

Ring groove wear is increased by overheating in the ring zone, because hot combustion gases can blow past the broken rings or through the craters.

Ring area damaged by broken piston rings (2)
Appearance
If pistons are overheated, wear increases due to loss of material strength. The risk of piston seizing increases because the lubricating film has been burned away.

Great care should be taken when installing pistons. A ring compressor should always be used, and the piston should under no circumstances be tapped or hammered into the cylinder. Piston ring gap must be checked and measured. For pinned or doweled rings, care should be taken to ensure that the rings are free to move. For single-cylinder,

Fig. 6.41 Broken piston rings. Source: MAHLE

especially single-cylinder two-stroke engines, the cylinder must not be rotated during assembly, as the cutaway ring ends could break off if they are pressed against the dowels or port edges. The port edges should always be well deburred and must not have sharp corners.

Possible causes
- Broken rings—defective locating dowels
- Insufficient ring gap

Severe ring and ring groove wear

Appearance
Due to excessive wear, the piston rings have too much axial clearance (Fig. 6.42), especially in the top groove. Usually the rings show severe radial wear; ring gap has grown to several millimeters. In the shaded areas of Fig. 6.42, piston material has been worn away. Such pistons are usually removed due to excessive oil consumption or loss of engine power.

Possible causes
In addition to normal wear in the course of long service life, various other causes come into consideration.

In principle, such extreme wear is caused by lubrication problems. Incorrect carburetor adjustment, defective automatic choke, fuel injection problems, excessive choke operation, long idling times, cold starts and short trip service, ignition misfire in only one cylinder, etc. all promote cylinder wall wetting by fuel, which dilutes or washes away

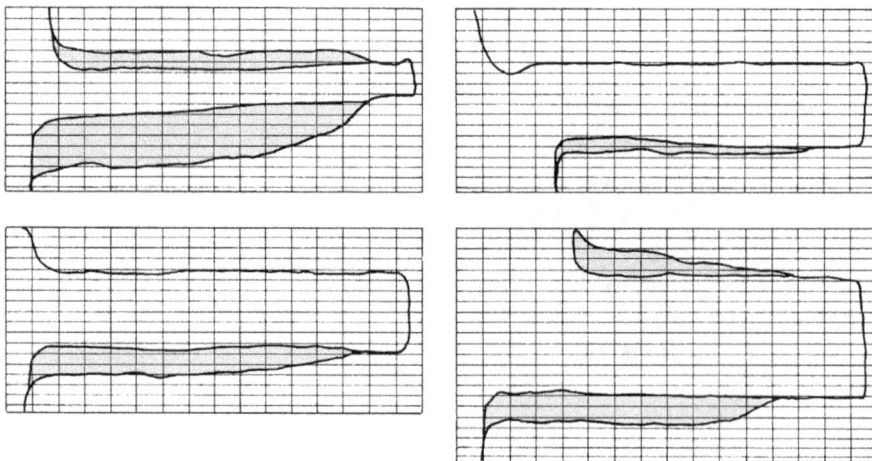

Fig. 6.42 Ring gap wear. Source: MAHLE

the oil film, and so leads to extraordinary wear. This also applies to diesel engines if unburned fuel impinges on the cylinder walls due to starting problems or other reasons.

Dirt or foreign bodies carried by combustion air, due to damaged filters or neglected filter service, also cause severe wear of this type. In the case of extreme axial wear combined with radial wear in the first ring groove (especially when compared to the oil control ring), if the piston shows a dull, worn-out contact pattern, one may conclude that the cause is improper filter maintenance. Above all, the wear pattern at the wrist pin provides clear evidence of wear caused by dirt.

If, however, only axial wear is present, the cause is excessive combustion (and therefore piston) temperatures in combination with excessive engine speed. Inertia forces work the rings into the sides of their grooves. Also, ring flutter at high engine speeds causes excessive groove side wear. Beat-out grooves allow ring flutter even at lower speeds, and wear rates accelerate.

Engine-related causes
- Lubrication problems
- Incorrect carburetor adjustment or fuel injection problems, defective cold start system
- Ignition misfire on a single cylinder
- Dirt and foreign bodies in oil, inadequate combustion air filtration

Ring land failure in Otto-cycle engines
Appearance
In this case, the ring lands have broken on one side, in the quench area opposite the spark plug, slightly offset from the exhaust valve. The break (Fig. 6.43) begins at the top edge of the land at the bottom of the groove, and runs into the piston material at an angle. Near the lower edge of the land, the crack changes direction, heads outward, and exits at the lower edge of the land or below it, at the base of the ring groove. The crack may also continue inward, at an angle, through the entire ring area. Longitudinal cracks which limit the lateral extent of the ring land breakage expand downward. In the case of slotted pistons, the entire piston head has been bent downward by about 0.25 mm in the damaged area.

Similar deformation is exhibited by the engine's remaining pistons, even though their ring lands have not yet broken.

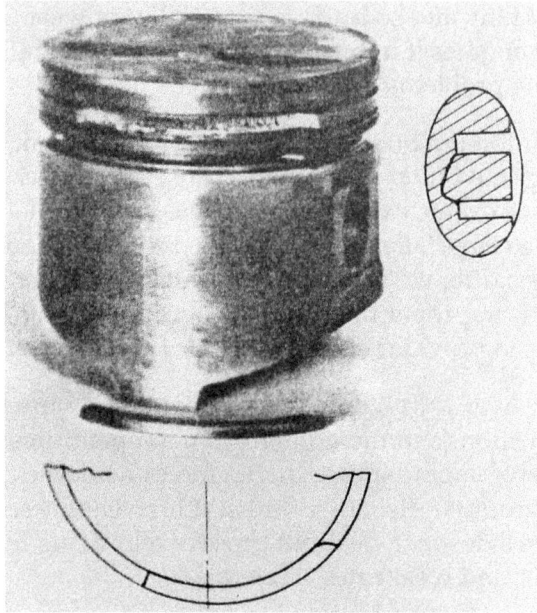

Fig. 6.43 Broken ring land. Source: KS

Possible causes

- Design-mandated narrow ring lands, incorrect ring package, ring groove base radius too small

- Assembly error

- Combustion knock

In Otto-cycle engines, broken ring lands are often encountered in high-compression engines with large quench areas opposite the spark plug. These are always the result of combustion knock. This means that the fuel did not satisfy the engine's octane demand under all operating and load conditions.

Although ring land breakage usually only affects individual cylinders, the remaining cylinders are always severely deformed. Excessively narrow ring lands and insufficient ring groove base radius also contribute to ring land failures.

Ring land failure due to mechanical overloading (1)

Appearance

Figure 6.44 shows that the top ring land has broken over about 1/3 of its circumference. Examination of the fracture surface shows that the crack progressed from top to bottom. There are no visible traces of combustion problems. By contrast, Fig. 6.45 shows a ring land fracture which progressed upward.

Possible causes

- Overloading the ring area as a result of engine overloading ("sporty" driving style)

- Assembly defect in piston installation, wrong cylinder head gasket

- Top piston ring striking ridge after ring replacement (cylinder not machined; ridge not reamed)

This failure occurred after only a short operating time. The engine had been prepared for motorsports ("tuned"). Increased engine performance caused cylinder gas pressure to rise to the point where the ring lands, especially in the major and minor thrust directions, were overloaded to the point of breaking.

Single breakage of the second ring land can also be attributed to the same cause; with the sealing function of the first ring compromised, the ring land below the second ring is overloaded as it experiences nearly the full combustion pressure. It is therefore possible for the second land to break while the first ring land remains undamaged.

While the fracture in Fig. 6.44 runs from top to bottom, the ring land of Fig. 6.45 has broken in the opposite direction. The fracture lines run in such a way that the top of the broken-out section is larger than the bottom. Such damage is usually the result of improper installation, in which the rings were not squeezed by a ring compressor, and the piston, instead of being pushed into the cylinder, was hammered in. This cracks ring lands, leading to eventual breakage in operation. Often, improper installation also damages the piston rings.

Ring land failures, as described above, may also be caused by the top ring hitting a ridge near the top of the cylinder at top dead center,

Fig. 6.44 Broken ring land: mechanical. Source: MAHLE

Fig. 6.45 Broken
ring land: mechanical.
Source: MAHLE

- if the cylinder was not bored out and rehoned in the course of an individual piston replacement, or
- the piston hit the cylinder head or an incorrect head gasket, or
- the crankshaft main bearings are worn out and were not replaced.
- Ring land failures may also occur as a result of worn ring grooves and the resulting ring flutter.
- Hydraulic locking of the cylinder also places considerable stress on the ring area.

Ring land failure due to combustion problems (2)

Appearance
In Figs. 6.46, 6.47 and 6.48, it may be seen that the ring land fracture runs from top to bottom. Figures 6.46 and 6.47 also show considerable erosion damage, caused by combustion problems.

Possible causes
- *Otto-cycle engines*
  - Combustion knock ("ping") leads to rapid pressure rise and simultaneous overheating as a result of turbulent gas flow. High temperatures and high gas pressures overload the ring lands; cracks begin, and propagate downward to develop into fractures. High-speed knock is especially dangerous as it results in excessive heating of the piston head. Causes for such combustion anomalies include:
  - Extreme early ignition (excessive spark advance)
  - Mixture too lean
  - Fuel octane too low
  - Increased compression due to combustion chamber deposits (city driving)

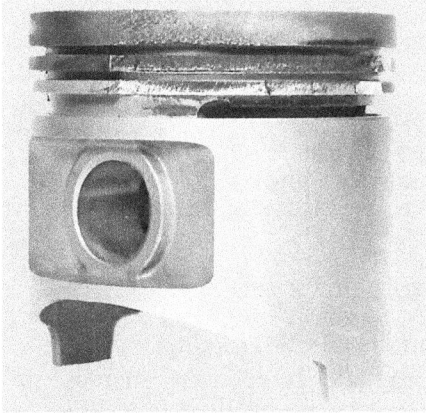

Fig. 6.46 Broken ring land as a result of harsh combustion. Source: MAHLE

Fig. 6.47 Broken ring land as a result of harsh combustion. Source: MAHLE

Fig. 6.48 Broken ring land as a result of harsh combustion. (Source: MAHLE

Elevated induction air temperature (defective air preheating system) may result in combustion knock, as this will demand higher fuel octane.

- *Diesel engines*
    - Here, too, harsh combustion as a result of excessive ignition delay leads to high peak pressures. These result in mechanical overloading of the ring lands. The causes of excessive ignition delay are:
    - Incorrect injection pressure (too early)
    - Excessive injection quantity (poor distribution)
    - Starting aids inject too much fuel
    - Insufficient fuel ignitability (cetane number too low).

Engine-related causes

- Air filtration problems: clogged filters, leaks in the air induction system

- Engine cooling problems

Seized pistons as a result of extreme overheating

Appearance

Fire land seizing and seizing in the ring area (in general, in the upper skirt area) is characteristic of overheating damage (Fig. 6.49).

The piston shape or clearance specifications are designed for normal operating temperatures. Excessive heating and the resulting increase in diameter can result in pinching of the upper parts of the piston. In the case at hand, the melted-off fire land and ring area indicate combustion problems and associated elevated piston temperatures.

Possible causes

- Most local overheating as a result of combustion problems such as knock, or detonation as a result of excessively lean mixtures, incorrect spark plug heat range, incorrect ignition timing, as well as cooling system problems, leads to significant diameter increases and to melting of the piston itself. Such damage is independent of engine operating time.

- In contrast to damage caused by insufficient clearance, the pistons show a well broken in contact pattern, overlaid by the seizure marks.

- Similarly, problems such as piston obliquity or lack of clearance between piston and wrist pin, or connecting rod bearing and wrist pin, may cause local contact between piston and cylinder. This may restrict ring movement.

Fig. 6.49 Seized piston as a result of overheating. Source: MAHLE

- The resulting degradation of cylinder sealing (as a result of piston obliquity) lets hot combustion gases past the rings, which not only heats the piston, but also burns away the oil film on the cylinder walls.

- The same phenomenon may occur while the engine is being broken in, if combustion gases blow past the piston rings, which are not yet sealing completely. The results are seizing in the ring area and in the upper skirt area.

Engine-related causes
- Incorrect ignition timing

- Defective injection system

- Defective cooling system

- Pistons misaligned or rotated

- Faulty fuel system, including filter system

Fire land erosion damage

Appearance

Piston damage in the ring belt and fire land area may appear in the form of erosion damage. Figures 6.50 and 6.51 show erosion damage to a fire land. This piston shows no visible signs of seizing or other damage to the head or skirt. The example shown here serves only as a reference to this type of damage, which may be classified as piston, piston ring, or piston head damage. A more complete treatment with illustrations of the causes of this type of damage, and the necessary checks, may be found in section *6.2.1.3 Piston head failures*, subsection *Erosion damage to piston head and fire land*; see also Figs. 6.77 through 6.80.

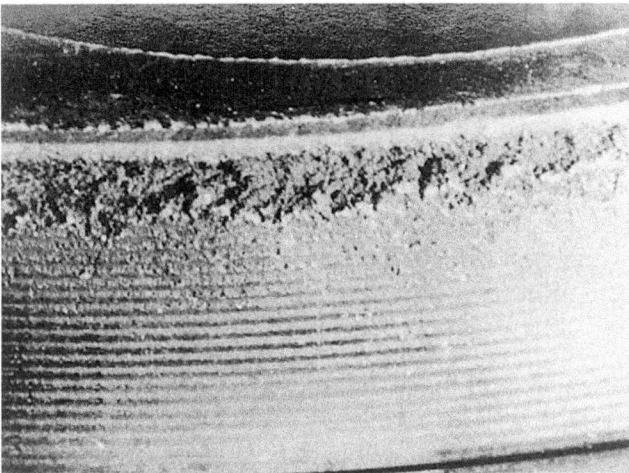

Fig. 6.50 Fire land erosion damage.
Source: KS

Fig. 6.51 Fire land erosion damage. Source: KS

Broken fire land (two-stroke Otto cycle)

Appearance

The fire land in Fig. 6.52 has been broken by force. The piston was not overheated and shows no other damage.

Possible causes

This is without doubt attributable to an assembly error. The piston is from a two-stroke Otto cycle engine; the damage appeared after faulty assembly. During installation into the cylinder, the top piston ring sprung into the exhaust port. Forcible assembly cracked the fire land, resulting in breakage after only a short period of operation.

6.2.1.3 Piston head failures

Piston burn through (holed piston)

Appearance

It is obvious that Fig. 6.53 shows a hole in the piston, as if penetrated by a projectile. The surrounding piston crown surface is mostly covered by melted piston material. Figure 6.54 shows a burned-away fire land and locally penetrated or burned through ring area. The diesel engine piston of Fig. 6.55 has been melted away as far as the piston skirt.

Possible causes

Complete overheating of the piston crown as a result of a defective spark plug, or in diesel engines, defective nozzles, prechamber, or swirl chamber; incorrect ignition settings; defective injection system; unsuitable fuel.

Starting from the damaged area on the piston crown, either the spark plug or, in diesel engines, the injector nozzle or prechamber/swirl chamber throat will be seen to be directly opposite. This establishes

Fig. 6.52 Broken fire land. Source: MAHLE

Fig. 6.53 Burned piston crown. Source: MAHLE

Fig. 6.54 Burned piston crown. Source: MAHLE

Fig. 6.55 Burned piston crown. Source: MAHLE

that these components must have had some influence on the incurred damage. The damage was initiated by combustion problems. In the early stages of this burn-through, overheating softens the piston material, and the piston crown bends downward under cylinder gas pressure.

- *Otto-cycle engines*
  The damage of Fig. 6.53 was caused by a spark plug with too low a heat range.
  Detonation ignited by an overheated spark plug insulator resulted in

a flame that caused local overheating of the piston and melt-through of the piston head.

Other causes such as early ignition timing, excessively lean air/fuel mixture, detonation caused by combustion chamber deposits, defective injection system, and fuel with too low an octane rating lead to charring as shown in Fig. 6.54.

- *Diesel engines*
  The damage shown in Fig. 6.55 was caused by a defective fuel injection system. Excessive fuel quantity as well as dripping nozzles resulted in poor mixture formation. The resulting combustion problems led to locally high temperature peaks and piston burn-through.

Engine-related causes
- Incorrect mixture and ignition settings

- Spark plug with wrong heat range

- Unsuitable fuel

- Fuel injection system problems, incorrect injection timing

Piston damage in large diesel engines with 12, 16, 18, or 20 cylinders is often severe because the friction of a single seized piston is insufficient to brake the engine, especially if full engine power is not being called for. Combustion problems as a result of a poor injector nozzle spray pattern caused the head of an electron beam welded piston to burn through in the cooling channel area. Cooling oil entered the combustion chamber and took part in combustion. The piston fire land "hung itself" against the cylinder liner near top dead center; i.e., it seized so solidly that the piston material, softened by high temperatures, could no longer sustain the loads imposed on it, and tore off. The piston failure was only noticed when, after shutdown, starter torque was no longer sufficient to turn the engine over. Figures 6.56, 6.57, and 6.58 show the full extent of this remarkable piston failure.

Piston burn through (2): Otto-cycle engines

Appearance

Figure 6.59 shows a failure case in which a hole has been burned through a piston head after only comparatively short service (see also Fig. 6.60). The outer third of the hole has been melted out of the piston, the remaining two-thirds on the other hand are a funnel-shaped breakout, widening toward the bottom. The piston skirt as well as the undamaged parts of the piston head often show no sign of thermal overloading, and the decrease in material hardness is within normal limits. Seizing is hardly visible.

Fig. 6.56 Piston without piston crown.

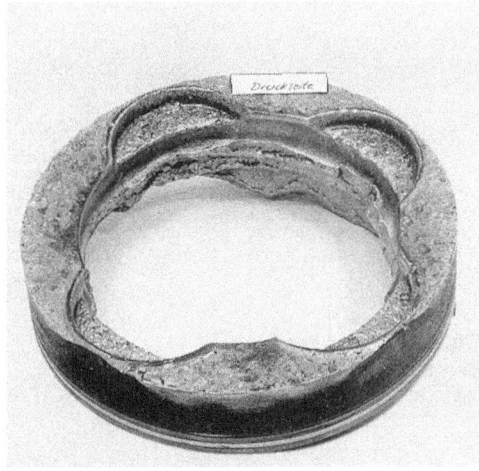

Fig. 6.57 Torn-off piston crown.

Fig. 6.58 Piston in cylinder, with torn-off crown firmly seized in cylinder.

Fig. 6.59 Piston head burn-through. Source: MAHLE

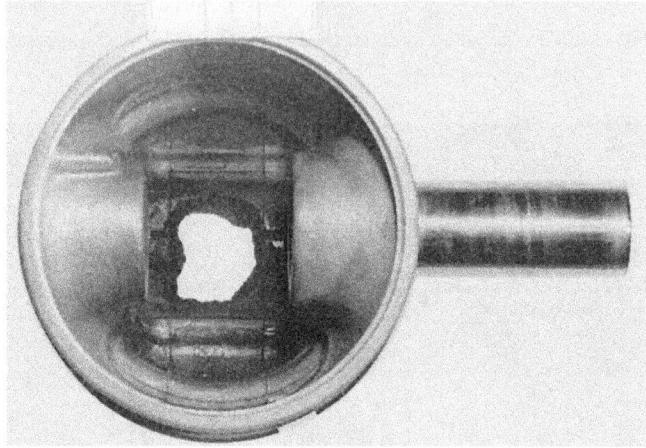

Fig. 6.60 Hole in piston head. Source: MAHLE

Possible causes

Burn through on a high-compression engine with hemispherical combustion chambers. Given certain specific combustion problems, the piston head is rapidly heated at a critical point, causing the material to soften locally. Inertia forces caused by the piston motion as well as rapidly flowing combustion gases carry away the softened aluminum. As a result of progressively decreasing material strength at this location, combustion pressure eventually forces the remaining two-thirds of the piston head thickness, in the shape of a downward-enlarging cone, inward toward the crankshaft.

Burned piston head and fire land

Appearance

As seen in Figs. 6.61 and 6.62, piston head and/or fire land burning has taken place in the direction of one or more injector sprays. Piston head and skirt do not exhibit any seizing, although some burned-off piston material may have been smeared downward.

Possible causes

- Failures of this type occur in direct-injection diesel engines (Fig. 6.61). Prechamber engines are only affected if the prechamber is damaged, causing the prechamber engine to effectively become a direct injection engine (Fig. 6.62).

- Leaking injector nozzles or hanging or jammed injector needles.

- Broken injector springs, or springs with insufficient preload.

- Defective pressure relief valve in the injection pump, etc.

Piston head and fire land melting

Appearance

All three pistons of Fig. 6.63 show various degrees of melting. The rightmost and middle pistons have slight depressions at the edge of the piston crown; in the left piston, the crown has completely melted away. Figure 6.64, like Fig. 6.63, shows advanced damage. Even the exhaust valve stem (left) shows melted material, evidence that pieces of the piston left the engine in a molten state.

Figure 6.65 shows the beginnings of melting on the edge of a diesel engine piston bowl, as well as the piston crown, while the head and fire land areas of Fig. 6.66 are completely destroyed.

Fig. 6.61 Burned piston head. Source: KS

Fig. 6.62 Burned piston head. Source: KS

Fig. 6.63 Melted pistons.

Fig. 6.64 Melted piston.

Possible causes
Extreme overheating

- *Otto-cycle engines*
  Uncontrolled preignition as a result of glowing combustion deposits, and overheated valves as a result of insufficient valve lash and incorrectly installed or damaged cylinder head gasket always results in high peak temperatures, which can reach the melting point of piston material (ca. 570–600°C, or 1060–1110°F). Excessively lean mixture, or fuel with too low an octane rating and excessive ignition advance lead to combustion problems such as knock, which results in the same damage pattern.

- *Diesel engines*
  Here, too, excessively high power output as a result of increased fuel injection quantity, and higher ignition pressure, lead to temperature increases up to the melting point of the piston material. The edge of the piston bowl is the first to melt away, followed by piston seizing, then, finally, piston disintegration. In direct-injection engines with

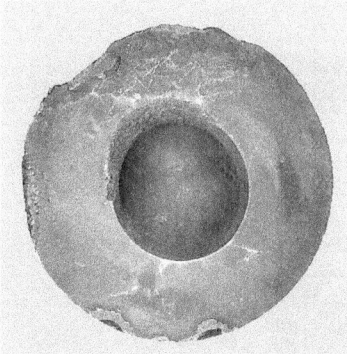

Fig. 6.65 Melted piston.

Fig. 6.66 Melted piston.

damaged injector nozzles, as well as prechamber or swirl chamber engines, unburned or poorly distributed fuel may also lead to the same temperature spikes. The reason is carbon deposits on nozzles, hanging needles, or excessively high fuel injection quantity.

Piston head deformation (valve or cylinder interference)

Appearance

Figure 6.67 shows the impact mark of a valve head on the top of the piston. Displaced material has clearly been pushed up at the edge of the impact zone.

Figure 6.68 shows that the piston head has been deformed by about 5 mm, and has molded itself to the combustion chamber shape in the cylinder head.

Figure 6.69 shows a diesel engine piston, in which the edge of the piston head has been deformed and the fire land has contacted the cylinder wall.

Possible causes

The damage of Fig. 6.67 may have been caused by incorrect valve timing, broken valves, broken valve springs, carbon deposits on the valve stems, insufficient valve lash, damaged connecting rod or wrist pin bearings ("big end" and "little end" bearings). The damage of Fig. 6.68 indicates bearing damage or valve interference.

In Fig. 6.69, carbon deposits, seizing, and clearance changes are recognizable, as well as:

- In the case of the valve head imprint, the cause is obviously not to be found in the piston itself, but rather the failure is traceable to

Fig. 6.67 Piston head deformation. Source: MAHLE

Fig. 6.68 Piston head deformation. Source: MAHLE

incorrect valve timing (incorrect camshaft installation). Other possible causes include broken valve springs or carbon deposits on the valve stem, as well as insufficient valve lash. Another cause may be bearing damage and the resulting increase in bearing clearance, especially connecting rod bottom end bearings, or worn-out wrist pin bushings.

- In the second case, piston stroke probably changed as a result of a loosened connecting rod bolt or bearing damage, allowing the piston to hit the cylinder head.
  This caused piston material to be forced into the combustion chamber; in the quench zones, the ring belt has been forced into the skirt area. In this case, disintegration of the piston was avoided, only thanks to the material's ductility.

- Fire land rubbing (Fig. 6.69) is a phenomenon usually associated with direct-injection diesel engines. Carbon deposits on the piston head (in excess of the clearance space above the piston) result in contact with the cylinder head. This deforms the fire land, forcing it against the cylinder wall. The most severe consequences of such contact are severe seizing or torn-off pistons. Also, as a result of other problems in V, slant, or horizontal engines, fluids (oil or fuel) can collect in the lowest corners of the combustion chamber after shutdown, and cause material displacement when the engine is next started.

Engine-related causes
- Insufficient clearance to cylinder head (diesel engines)

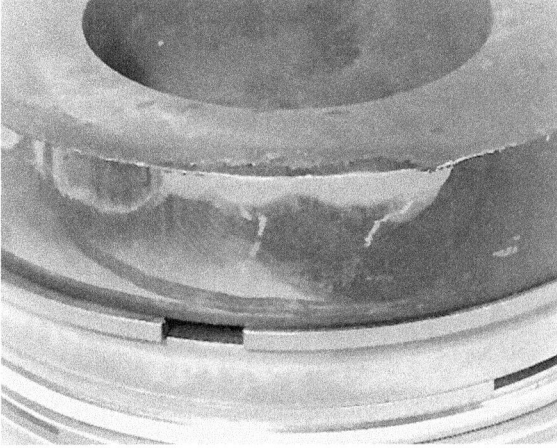

Fig. 6.69 Piston head deformation. Source: MAHLE

- Tolerance problems and impermissible clearances in the crank train and valve train
- Anomalies in the overall valve train
- Formation of oil or fuel deposits

Piston damage as a result of valve interference

Appearance

It is apparent that in every cylinder, a valve has struck the piston crown (Figs. 6.70 and 6.71). The impact depth increases from the front to the rearmost piston. Similarly, piston and ring wear increases from front to back. In all pistons, the oil ring faces are completely worn out.

Possible causes

Valve contact with the piston head.

Impact of the valves against the piston heads interferes with piston motion (and piston/piston ring contact). Ring flutter, increased blowby, and lack of lubrication cause ring wear.

Another possibility is increased piston head to cylinder wall contact. Oil consumption may rise steeply as a result of wear. Wear increases as valve impact depth increases.

Engine-related causes
- Basic engine adjustments (crank train and valve train; timing) not in accordance with specifications, or shifted as a result of wear and/or damage
- Defective timing belt or timing chains, broken valve springs, insufficient piston to cylinder head clearance (diesel engines)

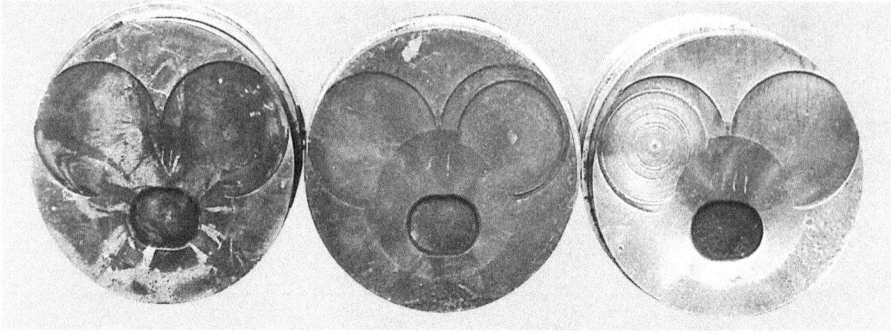

Fig. 6.70 Valve interference. Source: KS

Fig. 6.71 Valve interference. Source: KS

- Damaged crank train and valve train bearings

An extreme case of valve interference occurs when a valve head breaks off ("dropped valve") and falls into the combustion chamber. For the piston, the more dangerous situation is if the valve breaks just behind the valve head, because the head not only drops into the combustion chamber, but also—like a ping-pong ball—bounces back and forth between piston and cylinder head, and, when wedged unfavorably, damages the piston head to the point of breaking through (Figs. 6.72 and 6.73).

Piston head recession

Appearance
The piston head is pressed inward; there are no impact marks visible (Fig. 6.74). Ring belt and skirt are covered by oil carbon deposits.

Possible causes
Here, combustion problems caused high temperatures but without local overheating. Material strength was reduced by this overheating, to the

Fig. 6.72 Piston head completely destroyed by a dropped valve head.

Fig. 6.73 Failed piston with fragments. Source: Scania

Fig. 6.74 Piston head recession. Source: MAHLE

point where the piston head could no longer withstand combustion pressures and so bulged downward. Given further heating, the piston head would have been penetrated. This damage may be the result of improper fuel/air mixture, resulting in elevated combustion temperatures.

Foreign body in combustion chamber and on piston head

Appearance

Figure 6.75 clearly shows particles impacted on the piston head. Presumably, these parts also struck the cylinder head. The magnified view of a section through a particle (Fig. 6.76) shows that the particles consist of tempered gray cast iron, which probably entered the combustion chamber from outside, but might also have come from valve seats, because all piston rings remain undamaged.

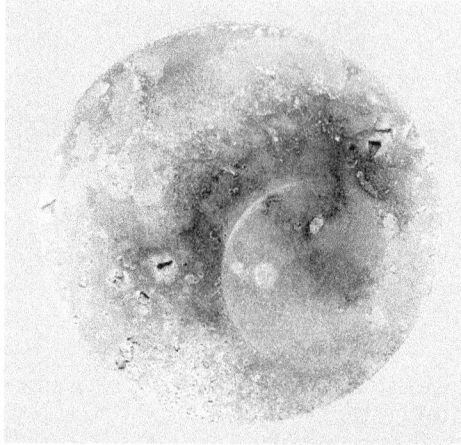

Fig. 6.75 Foreign bodies in combustion chamber. Source: MAHLE

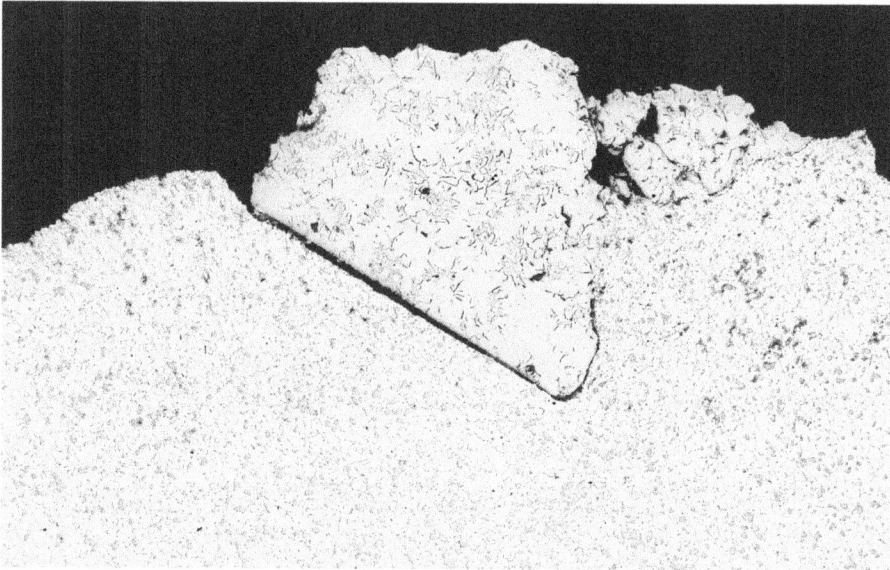

Fig. 6.76 Foreign body in combustion chamber. Source: MAHLE

If only ring freedom is compromised, or only the cylinder walls are damaged and the piston head is not immediately penetrated, such damage is usually not directly detectable.

Possible causes
- Presumably, foreign bodies entering the engine have caused this damage, as neither piston rings nor valve seats show any abnormalities.

- It cannot be ruled out that this damage was caused by foreign bodies that entered the combustion chamber during engine assembly.

Erosion damage to piston head and fire land

Appearance

The pistons in Figs. 6.77 and 6.78 show erosive surface damage in the fire land area. Figure 6.79 shows erosive damage to the piston crown; Fig. 6.80 also shows damage to the edge of the piston bowl.

Possible causes

- *Otto-cycle engines* (see also Fig. 6.77)
  High gas velocities with easily ignitable fuel/air mixtures (low octane numbers), combined with elevated temperatures, lead to erosive surface damage at the edge of the piston crown, fire land, and upper flanks of the top compression ring groove.

- *Diesel engines* (see also Fig. 6.78)
    - Erosion at edge of piston crown, fire land, and upper compression ring flank
      Excessive fuel injection quantity or fuel with insufficient ignitability (cetane number) lead to damage as a result of a "secondary combustion chamber" effect.
    - Piston crown erosion (see also Fig. 6.79)
      Here, too, excess fuel has ignited and formed a so-called secondary combustion chamber. This forms preferentially below the hot exhaust valve area.
    - Erosion at the rim of the piston bowl (see also Fig. 6.80)
      Excess fuel ignites at the hot surface of the piston bowl rim and locally tears out particles of piston material.

Fig. 6.77 Erosive damage to fire land, Otto-cycle engine. Source: MAHLE

Fig. 6.78 Erosive damage to fire land, diesel engine. Source: MAHLE

Fig. 6.79 Erosive damage to piston head. Source: MAHLE

Fig. 6.80 Erosive damage to piston head and rim of piston bowl. Source: MAHLE

Engine-related causes

- Unsuitable fuel

- Mixture formation and ignition problems

- Incorrect fuel injection quantity and injection timing

Notes

Sustained operation with engine knock can result in damage of this type. If knocking is caused by preignition, material erosion may expand over time.

Piston head damage—ablated edge of head and damaged fire land

Appearance

Figure 6.81 (Otto-cycle engine) and Fig. 6.82 (diesel engine) show fire lands that appear to have been chewed away.

Fig. 6.81 Piston head damage. Source: MAHLE

Figure 6.83 shows material displacement at the edge of a diesel engine piston bowl. This image, taken with a scanning electron microscope (see also Fig. 6.84), shows that the cause is not melting, but rather abrasive wear.

Possible causes

Carbon deposits with ensuing damage: in both engine types, carbon deposits in the cylinder bore have resulted in contact with the upper edge of the piston. Carbon may also build up at the edge of the head gasket, in the cylinder head quench area. Piston material is displaced by contact with the carbon deposits, and abrasive wear also occurs. Carbon layer growth is directly proportional to piston wear. In the case of preignition, the piston edge may melt, which accelerates the rate of material removal.

Fig. 6.82 Piston head damage. Source: MAHLE

Fig. 6.83 Piston head damage. Source: MAHLE

Fig. 6.84 Reflection electron micrograph of piston head damage. Source: MAHLE

In swirl chamber engines, carbon and combustion deposits preferentially form at the hot underside of the swirl chamber insert, against which the piston then impacts. In these cases, material removal is very often limited in extent to the outside diameter and hole diameter of the insert. Fire land contact areas are usually indicated by material displacement. In laboratory examination, the piston material shows no local melting.

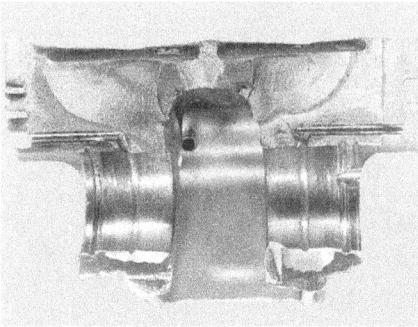

Fig. 6.85 Cracked piston head. Source: MAHLE

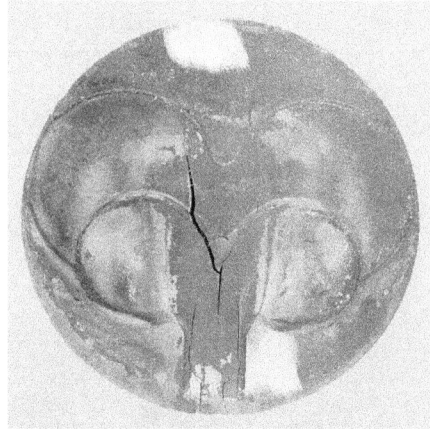

Fig. 6.86 Cracked piston head. Source: MAHLE

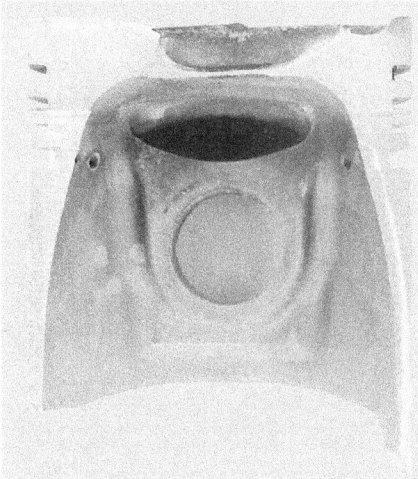

Fig. 6.87 Cracked piston head. Source: MAHLE

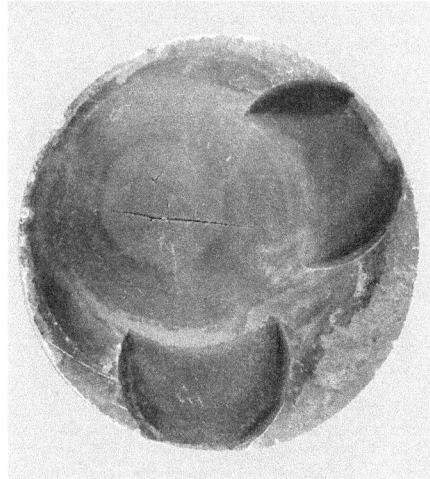

Fig. 6.88 Cracked piston head. Source: MAHLE

Cracked piston heads

Appearance

Figure 6.85 shows the cracked piston of a sports car engine, with the crack running in the direction of the wrist pin axis. Figures 6.86 and 6.87 shows a crack running in the thrust direction, i.e., perpendicular to the wrist pin axis, and Fig. 6.88 shows a piston head with a network of cracks.

Possible causes

Mechanical loading and excessive heating

- *Otto-cycle engine*
  High mechanical loads in sports and racing applications, combined with elevated piston head temperatures, lead to fracture with burn through. The crack direction is parallel to the wrist pin axis.

- *Diesel engine*
  Piston heads of swirl chamber engines (see also Figs. 6.86 and 6.87) are subjected to especially high thermal loads in the vicinity of the chamber throat. Restricted thermal expansion due to temperature differences in the piston head cause plastic deformation at the piston head surface, which may lead to cracking in the engine cool down phase. Gas pressure loads cause the piston head to deform, primarily perpendicular to the wrist pin axis. This promotes crack propagation from the outer edge of the piston head, and can lead to cracking through the entire head cross section.

In prechamber engines (see also Fig. 6.88), the flame hits the piston in the center of a flat bowl. Excessive temperatures lead to piston head cracks. In diesel engines, increasing engine output by raising the fuel injection quantity, or changing the pump governor characteristics, leads to increased piston thermal loading.

In direct injection engines, a distorted injection spray pattern and local piston head overheating can lead to cracking, followed by breakout of entire sections of the piston; see Figs. 6.89 and 6.90.

Piston head cracking (piston head and piston bowl cracks in diesel engine pistons)

Appearance

Starlike cracks have appeared in the flat piston bowl. One crack has developed into the main crack, extending through nearly the entire piston head (Fig. 6.91).

Possible causes

Causes of failure are the same as in Figs. 6.83 through 6.86. Other than these head cracks, there are no other recognizable anomalies, in pistons or piston rings.

Due to thermal overloading, the piston material of prechamber engines is heated by the exit stream from the prechamber, and in direct injection engines, the edge of the piston bowl is heated to the point where this very hot core, surrounded by cooler piston material, is deformed beyond its elastic limit, and crushed. Upon cooling, this area experiences tensile stresses, which lead to stress cracking. If, in addition, stresses from wrist pin bending are superimposed on these thermal stresses, the stress cracks may develop into a greatly enlarged main fracture, resulting in complete breakage and piston failure. Pistons

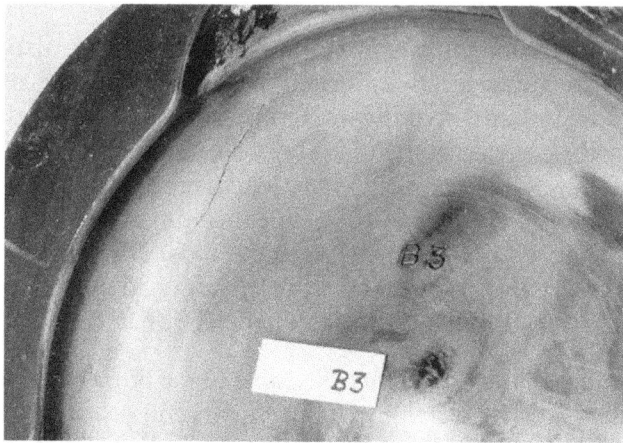

Fig. 6.89 Cracked piston head of a large diesel engine. Source: MAHLE

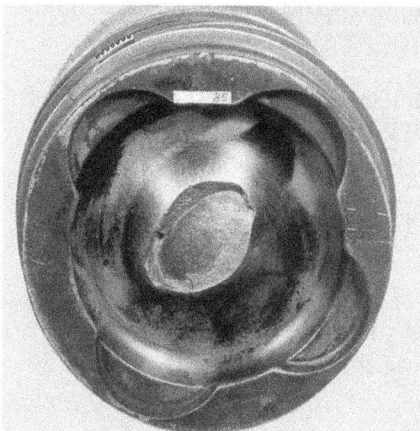

Fig. 6.90 The next stage: breakouts from the piston head of a large diesel engine.

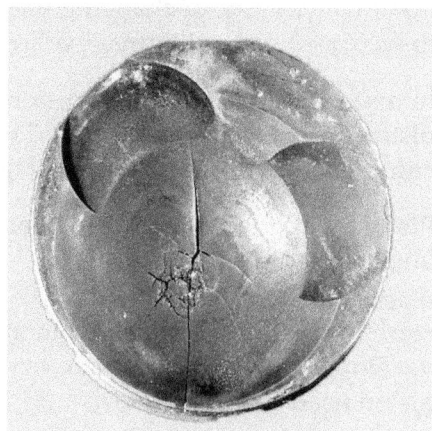

Fig. 6.91 Cracked piston head. Source: KS

with stress cracked heads and a main crack, either in the direction
of the wrist pin axis or perpendicular to it, must be replaced during
intermediate service.

Piston bowl cracking

Appearance

The piston bowl edges of both direct-injection pistons shown are
cracked. The piston in Fig. 6.92 shows deep cracks radiating from the
undercut spherical combustion chamber bowl. The piston in Fig. 6.93
shows short, but wide, gaping cracks at the edge of the piston bowl.

Possible causes

High thermal loads imposed on diesel engine pistons, and large
temperature differences between bowl rims and surrounding material,
restrict thermal expansion. This results in stresses at the bowl rim, and
plastic deformation of the material. Upon cooling, tensile stresses are
created, which lead to cracking. Sharp-edged and undercut bowl rims,
as well as the edges of valve clearance pockets at the edge of the bowl,
are especially susceptible. These cracks can propagate all the way to
the wrist pin bores, and then penetrate the entire piston head. Smaller
cracks also occur in the course of normal loading after long service life,
without resulting in functional problems.

Engine-related causes

Incorrect fuel injection quantity and injection timing. Excessive power,
e.g., as a result of increased engine speed (governor curve altered),
results in temperatures which lead to cracking; engines should only be
adjusted in accordance with the manufacturer's specifications.

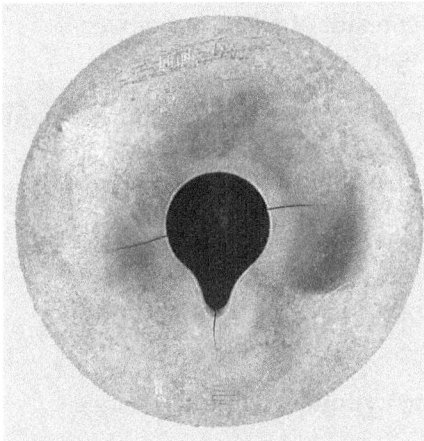

Fig. 6.92 Piston head/piston bowl cracking.
Source: MAHLE

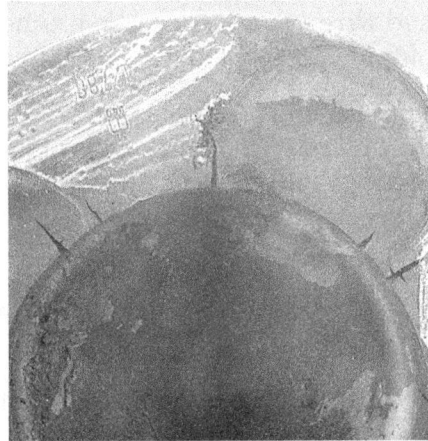

Fig. 6.93 Piston head/piston bowl cracking.
Source: MAHLE

6.2.1.4 Failures in the power transmission area, and wrist pin bearings

Cracks in the wrist pin bores (split-piston fracture and wrist pin boss fractures)

Appearance

In transferring power from the piston head to the wrist pin, there is a concentration of force at the top of the wrist pin bores. In the 1930s, this led to split-piston fractures, with fracture initiation at the inner edges of the wrist pin bosses and the piston split in half vertically. The causes were recognized as inadequately proportioned wrist pins, sharp-edged wrist pin bores, insufficient cold clearance between wrist pin and bore, as well as (especially for diesel pistons) too rigid and, in terms of stiffness, unsuitable piston head reinforcement by means of stiffening ribs. After the causes were recognized, effective remedies were implemented. Similar problems arose again in the 1960s and 1970s, but with an important difference: the fractures began several millimeters inboard of the edge of the wrist pin bores—boss cracking and boss fractures. The causes of these failures were found in greatly increased ignition pressures and pressure gradients, combined with more-compact piston designs. Whenever deformation has to be accommodated in (too) small a space, the result is large stress gradients, which the material is no longer able to withstand. Stronger, but more elastic support for the piston head ("stretch length"), oval-section wrist pins, and reduced wrist pin bending as a result of larger diameter and closer boss spacing, bracing, and wrist pin bushings in the wrist pin bores have proven effective in combating boss fractures.

Although the causes of split pistons and boss fractures, as well as their solutions, are known, such failures may still be encountered if piston and wrist pin are overloaded, either as a result of excessive power output or combustion abnormalities (Fig. 6.94).

The piston in Fig. 6.95 has cracked from the wrist pin bore all the way to the fire land, while the piston of Fig. 6.96 shows a crack which extends to the piston crown and has been burned through. The wrist pin boss supports in Fig. 6.97 are cracked.

Possible causes
- Overloading as a result of incorrect ignition and injection timing
- High temperatures
- Carbon deposits resulting in piston to cylinder clearance changes
- Water entry

With piston overloading, e.g., as a result of horsepower increases without corresponding piston modifications, the wrist pin boss edge cross sections (at the inner or outer faces of the bosses) are overstressed. These cracks initiate at the inside upper edge of the wrist pin bores, or at lubricating oil passages in the boss, under the influence of higher temperatures as a result of excessive power. In contrast to piston bowl or head cracks, these propagate in the direction of the wrist pin axis. In the process, the piston can be split in half. This in turn can lead to burn-through, with accompanying failures such as seizing (as a result of excessive thermal expansion and lack of lubricating oil). Transverse

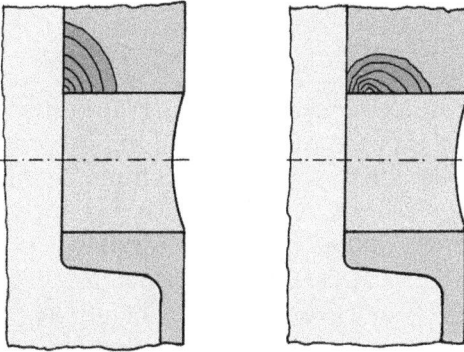

Split piston: Cracks in the headward portion of the wrist pin boss, with crack initiation at the inner edge of the boss. Causes are underdimensioned wrist pins, sharp-edged boss, insufficient clearance between wrist pin and wrist pin bore, as well as unfavorable reinforcing rib configuration with respect to component strength.

Wrist pin boss fracture: Crack initiation several millimeters from the inside edge of the wrist pin boss. Causes are increased piston loading as a result of higher power, with high combustion pressures as well as extremely compact piston designs.

Fig. 6.94 Schematic of characteristic wrist pin boss damage: split piston (historical) and wrist pin boss fracture (modern). Source: MAHLE

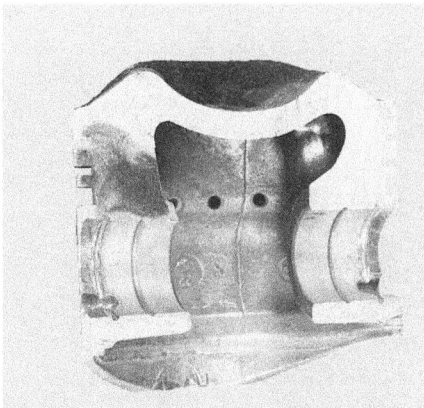

Fig. 6.95 Cracks originating at the wrist pin boss bores. Source: MAHLE

Fig. 6.96 Cracks originating at the wrist pin boss bores. Source: MAHLE

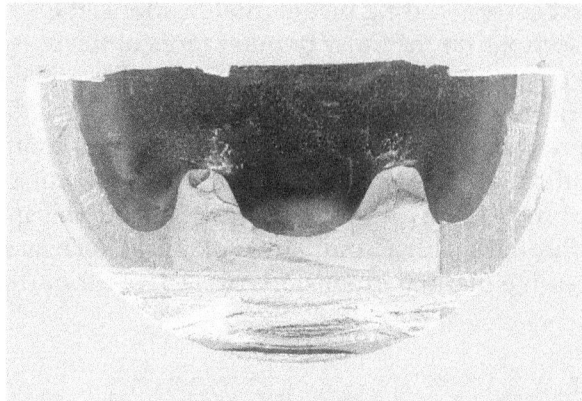

Fig. 6.97 Cracks originating at the wrist pin boss bores. Source: MAHLE

cracks in the wrist pin boss reinforcements (Fig. 6.97), on the other hand, are caused by very high temperatures and high ignition pressures in racing applications. Ultimately, the piston may be completely torn apart. Wrist pins can also cause such cracking. If overloaded, the wrist pin deforms into an oval cross section and causes damage such as that shown in the accompanying photographs. Impact against the cylinder head as a result of carbon deposits or incorrect quench height lead to overloading of the wrist pin bearings and therefore to wrist pin boss cracking or even split pistons. Hydraulic lock (as a result of water, fuel, or lubricating oil collecting in the cylinder) can cause similar damage.

Wrist pin boss cracking inside the wrist pin bores

Appearance

In this case (Fig. 6.98), cracks are visible that do not extend to or have not initiated at the ends of the wrist pin bores. This is more clearly visible in the cut-open piston, Fig. 6.99.

Possible causes

- High temperatures

- Excessive ignition pressures

These wrist pin boss cracks, encountered on diesel engines, begin several millimeters from the inside edge of the wrist pin bores. The cracks appear in the boss reinforcing at high temperature levels and with high ignition pressures. Usually, these are only observed after long operating times and may propagate to the ring belt and to the piston bowl. Here, too, injection pressure, injection timing, and injected fuel quantity should be checked.

Fig. 6.98 Wrist pin boss cracks. Source: MAHLE

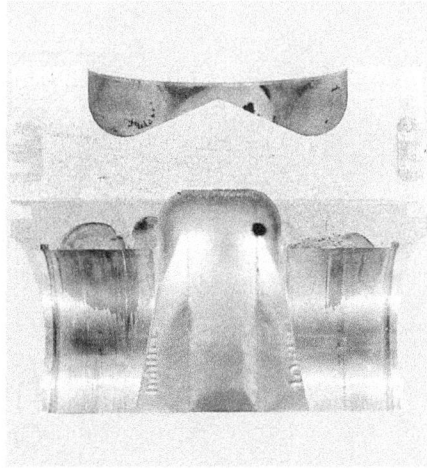

Fig. 6.99 Wrist pin boss cracks. Source: MAHLE

Transverse fracture through the lower part of built-up pistons

Appearance

In built-up pistons, the piston head is made of steel, and joined to a lower part made of an aluminum alloy, gray cast iron, or spheroidal cast iron by means of bolts. Built-up pistons are capable of withstanding high loads and seldom present any operational problems, thanks to their separation of duties—power transmission, resistance to corrosive attack and heat transfer through a cooled piston head, and kinematic location by the lower portion of the piston assembly. Yet even these pistons may experience failures.

Possible causes

Example: In the case of a large natural gas fueled diesel engine (6 MW at 500 rpm), an endoscopic examination of the piston interior after more than 20,000 hours of operation discovered incipient cracks in the reinforcing ribs between piston head and wrist pin bores (Figs. 6.100 and 6.101). The crack was broken open for examination, which showed that this was a case of fatigue cracking (Fig. 6.102). Thereupon the remaining pistons of this engine were examined, and similar cracks detected in more than half of them. Material and manufacturing could not be faulted, leading to the conclusion that the cause of these cracks must be regarded as engine overloading. As a remedy, engine power was reduced.

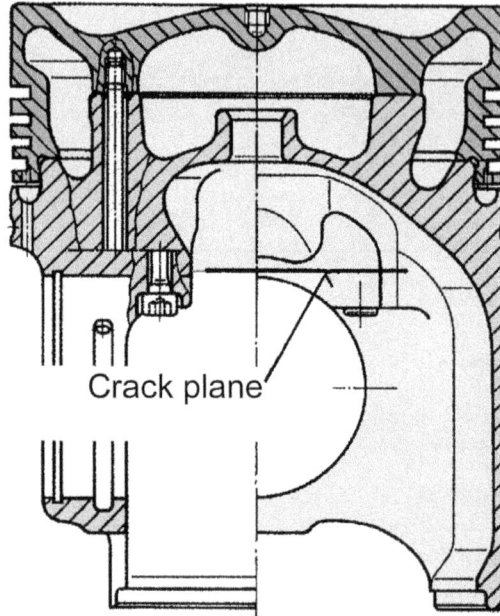

Fig. 6.100 Crack in piston head (schematic). Source: Allianz

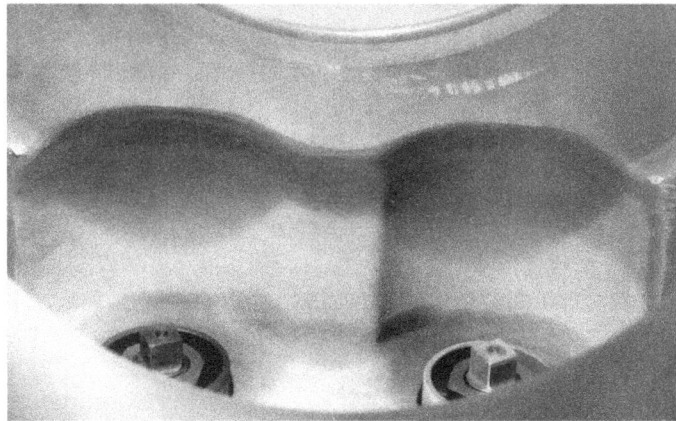

Fig. 6.101 Piston head crack: view inward through the wrist pin bore, crack location marked by dye penetrant (dark color).

Wrist pin bore seizing (1)

Appearance

The picture of the cut-open piston (Fig. 6.103) shows seizing marks in the wrist pin boss, especially in the area of the supporting wall. Because this piston has no groove for a wrist pin clip, the wrist pin must be held in place by the connecting rod (clamped or shrink fit).

Possible causes

Insufficient wrist pin lubrication or tight fit of wrist pin in piston. Both configurations practiced today (floating wrist pin or clamped connecting rod) are treated separately.

- *Clamped or shrink fit connecting rod*
  In this design, the wrist pin is firmly held by the connecting rod; movement is only possible between piston and wrist pin. In contrast to "floating" wrist pins, this oscillating motion through only a few degrees of rotation imposes far higher lubrication demands.

Fig. 6.102 Cracked piston head, after opening crack: fatigue cracking. Source: Allianz

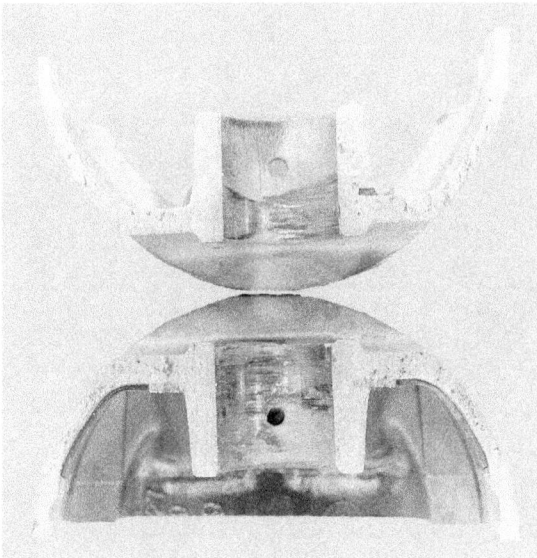

Fig. 6.103 Wrist pin bore seizing. Source: MAHLE

Therefore, provisions for adequate oil supply to the wrist pin bosses, such as grooves and oil passages, must be provided. In the case at hand, inadequate oil supply was not the only cause. Because of the seizing location in the area of the piston wall, one may assume that there was inadequate clearance between piston and wrist pin. Wrist pin bore seizing may result from insufficient play and inadequate lubrication (see Fig. 6.104).

- *Floating wrist pins*
  This configuration normally permits free rotation of the wrist pin in the piston as well as in the connecting rod. If, as a result of insufficient clearance or jamming in the connecting rod, the wrist pin is only free to move in the piston, its clearance and lubricating oil supply may no longer be adequate. This will result in seizing in the wrist pin bores. The heat generated as a result may result in seizing of the wrist pin itself.

Wrist pin bore seizing (2)

Appearance
These seizing marks in wrist pin bores (see Fig. 6.104) have the exact same appearance as those of Fig. 6.103.

Possible causes
Seizing of this type may appear as primary or secondary effects of piston skirt seizing. Because the wrist pin bearing surfaces in the wrist pin bosses are not positively supplied with oil, rather with splash or drip lubrication, seizing in the wrist pin bosses is almost always dry seizing with severely torn surfaces and evidence of equally severe material welding.

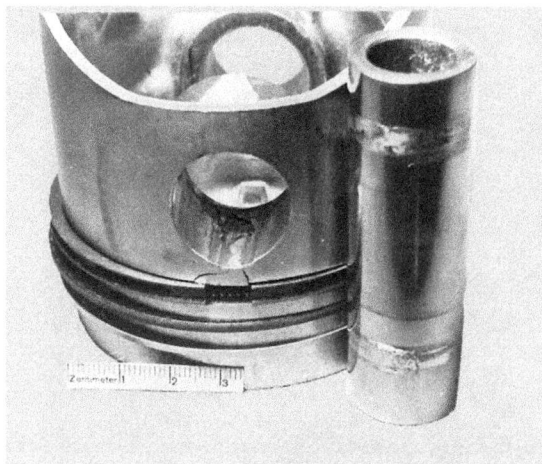

Fig. 6.104 Wrist pin bore seizing. Source: KS

In the case of primary seizing in the wrist pin bosses of floating wrist pins, movement of the wrist pin in the connecting rod bushing is so severely limited (by insufficient clearance or as a result of jamming due to misaligned/warped connecting rods) that the wrist pin is only free to rotate in the piston bosses between limits set by connecting rod oscillation. But floating wrist pins are not designed with sufficient clearance for this motion.

Effects

The results are extreme heating and the resulting collapse of lubrication, with dry running and seizing. Due to heating, the piston skirt expands in the wrist pin boss areas, which causes local lack of clearance, dry running, and seizing. For wrist pins that are shrink fit to connecting rods, clearances in the wrist pin bosses are dimensioned so that there is always sufficient oil film thickness in those areas. When shrink-fit connecting rods are re-used, care must be taken to ensure that the connecting rod bore is not distorted or otherwise damaged. Otherwise, the wrist pin could deform enough in its shrink-fit condition that it has diminished local clearance in the wrist pin bosses, which could easily lead to seizing. When installing a piston with a shrink-fit wrist pin/connecting rod into the engine, the wrist pin bearing area should be adequately oiled so that there is sufficient lubricant for the first few engine revolutions. The piston of Fig. 6.104 shows no significant deposits or wear marks, so that one may conclude that this piston has only run for a brief time. The wrist pin has seized in both bosses, on the upper, loaded sides. The surfaces of the seized areas show a clean, metallic appearance. There is no trace of burned-in oil. In this case, the wrist pin bearing areas were apparently not oiled during installation in the engine.

Broken wrist pin boss

Appearance

The wrist pin bores, near the central plane of the piston, show the beginnings of a typical wrist pin boss edge fracture. The fracture has propagated in a semicircular arc around the point of initiation. (See Fig. 6.105).

Possible causes

Wrist pin boss fracture as a result of overloading and inadequate oil supply. From experience, an initiating crack quickly develops into a split piston fracture, which can propagate through the entire piston head and break a piston into two parts. Wrist pin boss fractures arise from overloading and may be aggravated by insufficient oil supply. Cracks that have been initiated by overloading then propagate even under normal loading, and finally lead to piston splitting.

Fig. 6.105 Broken wrist pin boss. Source: KS

- Combustion problems, in particular harsh combustion as a result of ignition delay
- Excessive or improper application of starting aids in cold starting
- Water accumulation in cylinders while the engine was not operating (alternatively, hydraulic lock due to fuel, water, or oil)
- Inadequately dimensioned wrist pins

Broken or loose wrist pin retainers

Appearance

The piston has been "beaten out" above and below the wrist pin bosses. The surfaces are scrubbed smooth and shiny. The matching cylinder wall areas of pistons whose running areas are damaged this way are usually also damaged and attacked; see Fig. 6.106.

Fig. 6.106 Damaged wrist pin retainer. Source: MAHLE

Possible causes

The damage was caused by a defective piston retaining ring or damaged retaining ring groove. The reciprocating motion of the piston can cause broken-out sections of the wrist pin retaining ring groove to hollow out, with such hollows extending up to the ring belt. Even the piston rings may be worn down by this process.

Broken wrist pin retainers may have several causes.

- Too much deformation in installation could crack the retainer groove or reduce retainer pretension. Subsequent inertia forces then lead to fracturing or gradual loosening, knocking out or even snapping out of the untensioned retainer.
- Axial thrust from the wrist pin against the retainer may cause these to wear or expand.
- The piston is running obliquely (cocked in cylinder) due to crank train problems.

Once the retainer has been forced out of its groove, or has cracked, the ensuing damage cannot be halted. The fragments or even a single part do their destructive work, often damaging the cylinder to the point where boring oversize is no longer possible within the allowable overhaul limits. In such cases, the cylinder must be resleeved, if possible. Furthermore, small fragments can pass through the inside of the wrist pin to beat out the opposite side as well (even though the retaining ring on that side is still firmly seated).

Damaged wrist pin retainers (Fig. 6.107)

Appearance

Here, too, as in Fig. 6.106, the wrist pin boss has been beaten out. Damage as a result of wrist pin axial thrust can be extensive.

Fig. 6.107 Damaged wrist pin retainer. Source: KS

Possible causes

The causes of this damage are similar to those of Fig. 6.106. Here, the very significant axial thrust of the wrist pin has forced out the retaining ring. Assuming they were installed correctly, in operation, the wrist pin retainers—regardless of whether they are wire rings or snap rings—can only be forced or hammered out by wrist pin axial thrust. Axial thrust always occurs if, under operating conditions, the wrist pin axis is not parallel to the crankshaft axis.

This is usually the case if a bent connecting rod causes the piston to run at an angle. Reciprocating motion causes alternating axial thrust, which actually hammers out the retaining ring that happens to lie in the major thrust direction. This popped-out ring wedges between the wrist pin (which is being forced outward), the piston, and the cylinder. There, it is worn down and finally breaks into several fragments. In short order, inertia forces during reciprocating motion cause these fragments to beat out the piston material. Individual fragments wander over to the other side of the hollow wrist pin and initiate similar damage on the opposite side of the piston.

Broken-out wrist pin boss

Appearance

The piston wall around the wrist pin boss has broken out, starting from the wrist pin retainer groove.

Possible causes

Faulty wrist pin installation. The retaining ring was not forced out by the wrist pin during operation, so that further damage did not occur.

The fit between piston and wrist pin is sometimes still designed as a medium force fit. To install the wrist pin, the piston must be heated to 80–100°C (176–212°F). If this is not done carefully, or if there are difficulties in assembly, for example because the wrist pin does not immediately slip into the piston's wrist pin bores, the piston cools down before the process is completed. Now, if the wrist pin is hammered into the wrist pin bosses, the opposite piston wall will be pressed outward and/or severely damaged. Once the engine is in operation, the wall breaks, see Fig. 6.108.

In the case at hand, a floating wrist pin was hammered in by force, because the connecting rod angularity was not correct or the connecting rod was not lined up properly. The retaining ring, already installed on one side, was forced against the outer wall of the retaining ring groove. The piston wall was cracked in a ring around the wrist pin bore, and later, in operation; this piece of material broke away from the piston.

The damage clearly shows that this failure was certainly caused during installation. If the wrist pin had been forced against the retaining ring during operation, thereby causing breakage, the retaining ring would have been forced out of its seat, and even greater damage would have ensued.

Broken wrist pin (1)

Appearance

The wrist pin of Fig. 6.109 has broken transversely; the left, shorter part also shows a longitudinal crack. The wrist pin boss has been destroyed.

Possible causes

Wrist pin overloading from any of several causes.

The source of this fracture is apparently located at the intersection of the two fracture planes. At this point (recognizable from wear marks as the shear point between wrist pin and boss), the case hardening layer has broken out. The wrist pin boss was destroyed after the wrist pin broke.

The fracture, at the intersection of the two fracture planes, is located at the most highly loaded point on the wrist pin. The two fracture planes were created by various forces, which caused a transverse fracture from shear forces, and a longitudinal fracture from oval deformation as a result of wrist pin bending. From these two fractures, it may be

Fig. 6.108 Broken-out wrist pin bore. Source: MAHLE

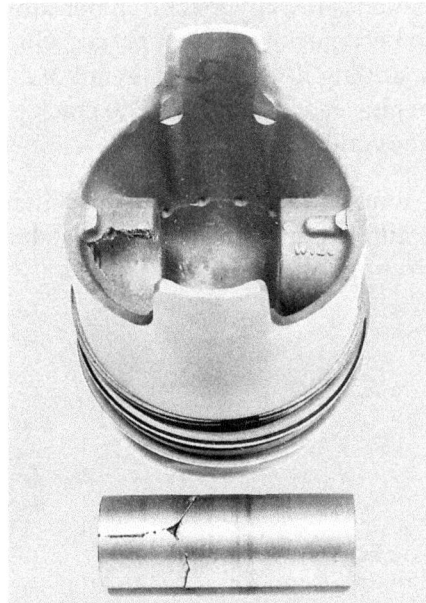

Fig. 6.109 Broken wrist pin. Source: MAHLE

concluded that the wrist pin was overloaded, possibly as a result of hydraulic lock. A material fault cannot be ruled out, but this would require extensive laboratory tests after contacting the supplier of these parts.

Broken wrist pin (2)

Appearance

Figure 6.110, like 6.109, shows a broken wrist pin. This fracture shows a transverse fracture in the transition area from connecting rod to one of the wrist pin bosses. The shorter fragment has also split lengthwise. All fracture surfaces clearly show signs of fatigue failure.

Possible causes

Overloaded wrist pin. Here, too, wrist pin failures could result from overloading, caused by any of a number of possible problems: the combustion process, the effects of foreign bodies, or material defects. This case, however, probably involves improper utilization of starting aids, which resulted in extreme peak firing pressures.

As a result, the wrist pin was deflected and deformed into an oval cross section. With such oval deformation, a longitudinal crack could form at the wrist pin ends as a result of overloading; its origin could be on the outer surface, or at the inner wall of the wrist pin. The crack then propagates as a fatigue failure, toward the center of the wrist pin, until it reaches the area between the connecting rod and wrist pin boss, which is heavily loaded in bending. There, the crack changes direction to become a transverse crack, which finally leads to the entire wrist pin snapping in two. In the event of a material defect, this crack usually begins at an inclusion. The crack propagation is the same as in the case of overloading.

Along with the failures shown here and in Fig. 6.109, wrist pin breakage could, among other things, also be the result of defects caused by improper hardening.

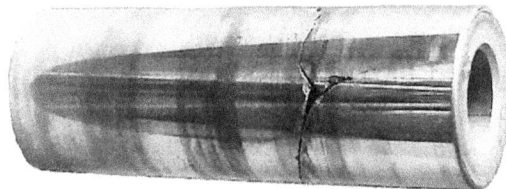

Fig. 6.110 Broken wrist pin. Source: KS

Fig. 6.111 Piston that had been generating excessive noise. Source: KS

Piston noise (running noise)

Appearance

The fire lands of pistons that call attention to themselves as a result of noise show impact marks in their rocking direction (Fig. 6.111). The piston skirt shows more prominent wear marks at the top and bottom than in the center. The fire land shows impact marks in the plane of the wrist pin bosses, above the wrist pin bores.

Possible causes
- Poor piston guidance
- Excessive cylinder bore diameter
- Bent or twisted connecting rods (rods out of plumb)

Running noises may be caused by various influences during engine operation; e.g., a piston that is able to rock due to oversized cylinder bore or piston skirt wear. The rocking motion is excited by connecting rod motion and the resulting shift in piston contact against the cylinder wall; the piston head is forced especially hard against the cylinder wall.

Sideways impact of the piston against the cylinder wall, in the direction of the wrist pin, is usually caused by connecting rods. If a connecting rod is out of plumb (bent or twisted), the piston, in its reciprocating motion, also undergoes an oscillating motion in the direction of the engine's crankshaft axis, which alternately forces the piston against opposite cylinder walls.

Asymmetrical connecting rods or off-center piston support by the connecting rods have the same effect. Here, the amount of connecting rod bearing clearance plays an important role. The smaller the clearance, the better the radial location of the connecting rod, and therefore the smaller the oscillating motion.

Axial wrist pin thrust and alternating side-to-side impact against the wrist pin retainers are always the result of alignment problems between the wrist pin and crankshaft axes. Bent or twisted connecting rods, as well as asymmetric connecting rods, are the most common causes. But excessive connecting rod wrist pin bearing clearance may cause connecting rod oscillation, especially at low engine speeds. This results in the wrist pin impacting the retaining rings.

6.2.2 Piston rings

Piston rings are vital components with a decisive role in the function of an internal combustion engine. They are required to fulfill a variety of tasks, which has resulted in development of a wide variety of configurations. Engines use single-piece, split, self-tensioning rings. The necessary pretension to ensure contact against the cylinder walls is obtained by giving the rings, in their uninstalled state, a larger circumference than that corresponding to the cylinder inside diameter. During installation, the ends of the rings are compressed to the diameter represented by their ring gap; see Fig. 6.112.

Piston rings must fulfill the following functions:

- Seal the combustion chamber to maintain pressure of the combustion gases. Combustion gases must not be allowed to enter the crankcase, nor oil to enter the combustion chamber.

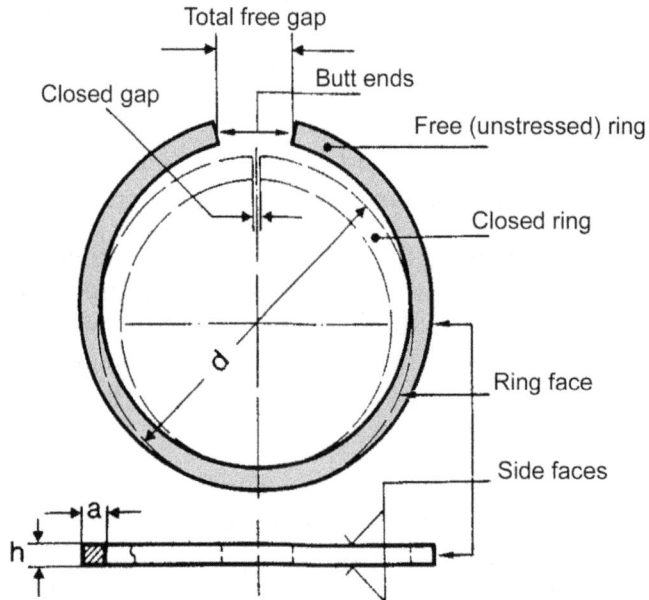

Fig. 6.112 Piston ring terminology: a, radial wall thickness; h, ring width; d, nominal ring diameter.

- Heat transfer. Most of the heat entering the piston is removed through the piston rings.
- Lubricating oil management.

Because these tasks cannot all be met by a single piston ring design, special rings have been developed to address specific functions:

- Compression rings (Fig. 6.113)
- Oil control rings (Figs. 6.114 and 6.115)

Of the latter, steel band oil control rings represent one unique design configuration (Fig. 6.116).

These tasks cannot be separated from one another; compression rings also scrape away oil, or, better said, are intended to meter oil in appropriate quantities. Piston rings, above all the top ring, operate under extremely unfavorable conditions:

- Varying speeds: from high sliding velocity to full stop at top and bottom dead center
- High pressures

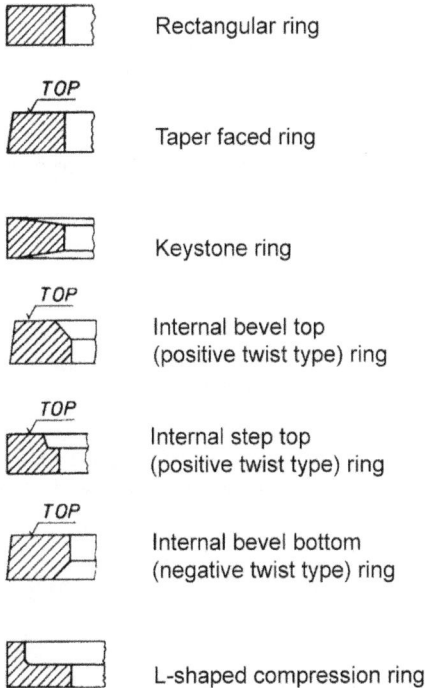

Rectangular ring

Taper faced ring

Keystone ring

Internal bevel top (positive twist type) ring

Internal step top (positive twist type) ring

Internal bevel bottom (negative twist type) ring

L-shaped compression ring

Fig. 6.113
Compression rings.

Scraper ring (stepped)

Scraper ring (Napier)

Slotted oil control ring

Bevelled edge oil control ring

Double bevelled oil control ring

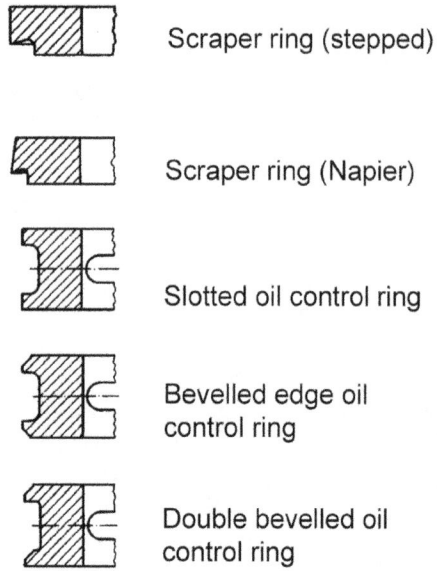

Fig. 6.114 Springless oil control rings.

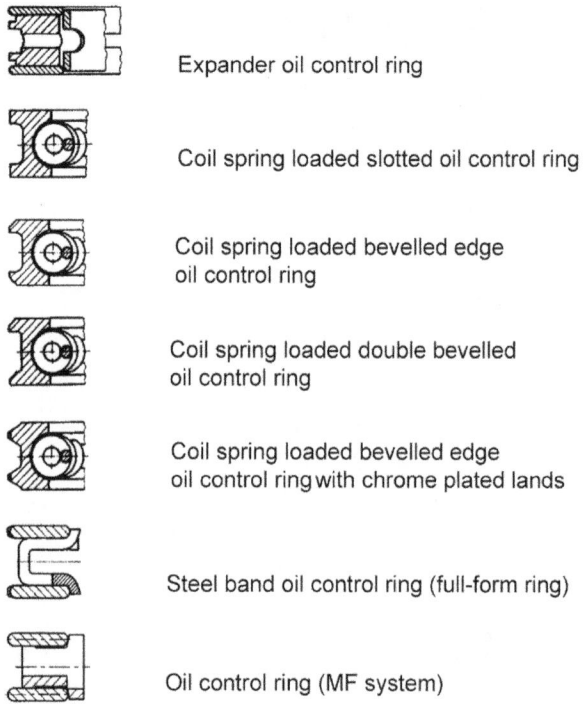

Expander oil control ring

Coil spring loaded slotted oil control ring

Coil spring loaded bevelled edge oil control ring

Coil spring loaded double bevelled oil control ring

Coil spring loaded bevelled edge oil control ring with chrome plated lands

Steel band oil control ring (full-form ring)

Oil control ring (MF system)

Fig. 6.115 Spring supported and spring loaded oil control rings. Source: Federal-Mogul

- High temperatures
- Poor oil supply
- Exposure to aggressive combustion products

The operating principle of piston rings (see Fig. 6.117) is based on the rings forming a system of labyrinth seals, combined with throttling chambers.

To enable them to seal at all, they must first of all fit snugly against the cylinder wall; on the other hand, a lubricating film must exist between ring and cylinder wall. Operating conditions at top and bottom dead center are critical, because at those points the sliding velocity is zero and the hydrodynamic lubricating film breaks down. Added to this, at ignition top dead center, gas pressures and temperatures reach their maximum values. Piston rings must therefore exhibit good sliding and boundary lubrication characteristics. The desired contact pressure distribution (Fig. 6.118) may be achieved by cold forming (hammering, rolling), heat treating (thermal pre-loading), or, as is usually done today, by double contour machining. The desired radial and axial contact pressure distribution of piston rings are determined by the engine operating process and operating conditions.

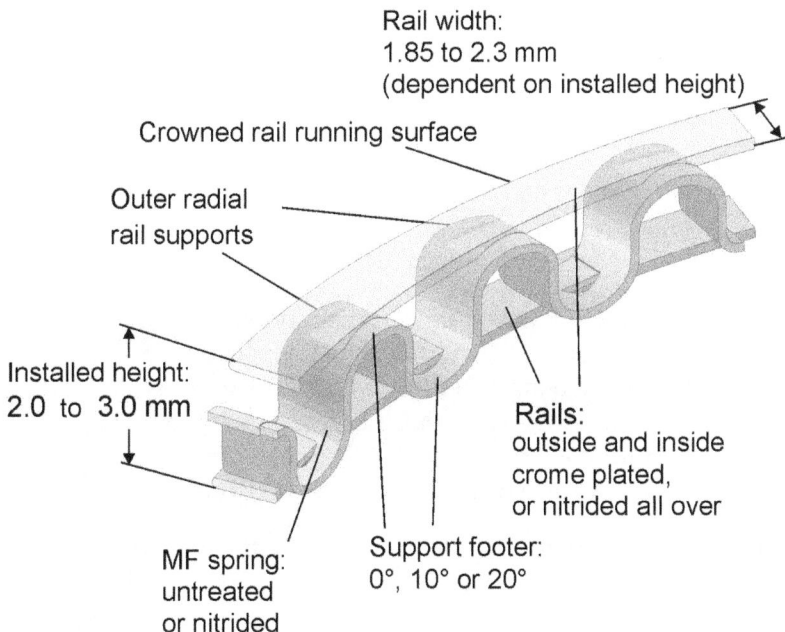

Rail width:
1.85 to 2.3 mm
(dependent on installed height)

Crowned rail running surface

Outer radial
rail supports

Installed height:
2.0 to 3.0 mm

Rails:
outside and inside
crome plated,
or nitrided all over

MF spring:
untreated
or nitrided

Support footer:
0°, 10° or 20°

Fig. 6.116 Steel band full-form oil control ring by Federal-Mogul (ex Goetze). Source: Federal-Mogul

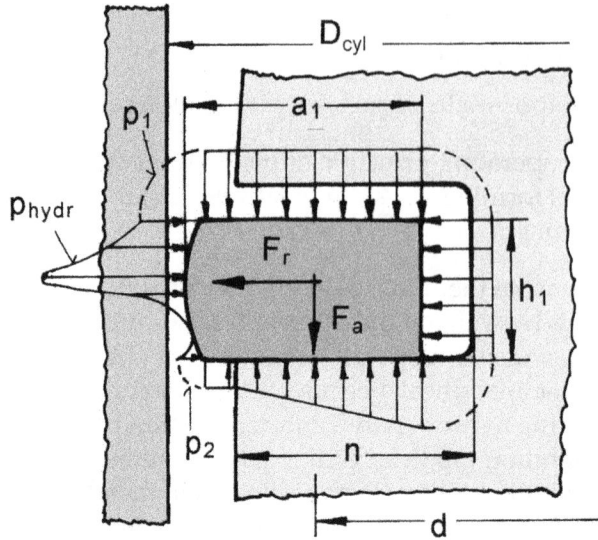

Fig. 6.117 Schematic of piston ring/ring groove/cylinder wall sealing system: a_1, piston ring radial wall thickness; d, mean piston ring diameter; n, ring groove depth; p_1, p_2, gas pressure above and below piston ring; p_{hydr}, hydrodynamic lubricating film pressure distribution on the ring face.

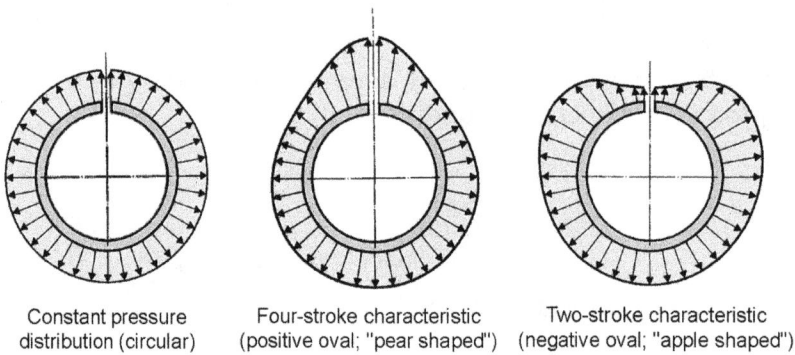

| Constant pressure distribution (circular) | Four-stroke characteristic (positive oval; "pear shaped") | Two-stroke characteristic (negative oval; "apple shaped") |

Fig. 6.118 Piston ring contact pressure distribution.

Piston rings in four-stroke engines have higher contact pressure at their ring ends to counteract ring end flutter, a so-called pear-shaped pressure characteristic. Two-stroke engines use rings with reduced contact pressure at their ends to prevent these (assuming the rings are not positively secured against rotation) from catching in the gas ports—a two-lobed or "apple shaped" pressure distribution.

The radial contact pressure forcing the ring against the cylinder wall is small compared to the gas pressure that is applied to the ring by gas in the piston ring groove. Therefore, the ring face is pressure relieved by appropriate shaping so that pressure from the front reduces pressure acting from the rear (partial and full pressure relief; see Fig. 6.119).

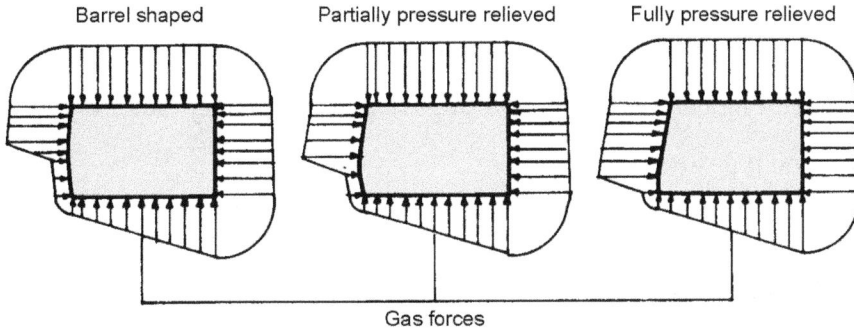

Fig. 6.119 Piston ring pressure relief.

The following demands are placed on piston ring materials:

- Good sliding and boundary lubrication capabilities
- Elastic behavior
- Mechanical strength
- High temperature strength
- High thermal conductivity
- Good machinability/workability

Because of its outstanding running qualities, lamellar cast iron had become a standard piston ring material. Today, however, its strength properties no longer satisfy the demands of high specific output engines, resulting in a switch to spheroidal cast iron (nodular iron) and, ultimately, to steel. These, however, have poor running qualities, requiring coatings on the ring faces. Coated ring faces address the following goals:

- Improved break-in behavior as a result of chemical or thermochemical surface treatment: phosphating and ferro-oxidizing provide layers of phosphate crystals or iron oxide, respectively, which improve ring break-in.
- Ring face armor:
    - Chrome plating: a galvanically applied layer of hard chrome has, in addition to high wear resistance, good running properties, and reduces cylinder wall wear and wear of the other piston rings. Special surface shapes (special lapping, channel plating) provide improved break-in and running properties.
    - Molybdenum coating: melting resistance is high, and, thanks to its porous structure, it functions as an oil reservoir, enabling it to withstand critical operating conditions.

- Metal/ceramic protective layers: these are extremely wear and temperature resistant, but have less layer porosity.

The operating behavior of piston rings depends on many factors:

- Engine type and design
- Combustion principle, combustion process, pressures, and pressure gradients
- Cylinder configuration, cylinder material, and cylinder machining
- Fuel and lubricants
- Ring type, ring material, and ring face
- Operating conditions

Piston rings primarily exhibit the following failures:

6.2.2.1 Faulty assembly

In practice, it is not uncommon to encounter piston ring failures that can be traced back to the assembly process. For installation, the ring must be expanded to the point where it can be slipped over the piston. In the process, the ring is easily overloaded, leading to fracture. The sharp-edged ring fracture surfaces then cut grooves in the cylinder wall. For this reason, rings are installed using special piston ring expanders with a hard stop to limit opening and so prevent overexpansion of the piston ring. As unstressed piston rings have a larger circumference than the cylinder bore, piston installation requires that they be compressed with a piston ring compressor to permit the rings to slide easily into the cylinder. If this is not done, a protruding ring may strike the cylinder liner, damaging both ring and liner.

- *Loss of pretension*
  Reduced or declining ring tension leads to ring butt ends that no longer contact the cylinder wall correctly; the ring no longer seals properly. This may be recognized by black discoloration on the ring butt ends.

- *Wear*
  - Corrosive wear as a result of sulfuric and sulfurous acid in combustion gases condensing as temperature drops below their dew point.
  - Abrasive wear as a result of particulates—dust (oil), carbon, etc.
  - Wear as a result of scored cylinder walls.

  Side face wear allows increased side clearance, which allows the ring to tilt; the ring no longer bears against the cylinder wall correctly, and therefore it experiences increased running face

Fig. 6.120 Ring face wear near the butt end.

wear (Fig. 6.120). The results are local overheating, compromised lubricating film, poor sealing, increased blowby, and cylinder wall scoring.

Example: the chrome layer on the steel bands of an oil control ring (Fig. 6.121) has been locally removed, down to the base material, leaving individual lands still standing in the direction of motion. If both steel bands are laid atop one another in their as-installed orientation, it can be seen that the raised lands match, with identical spacing.

This damage was caused by scored cylinder bores. Such locally limited protrusions are raised by cylinder wall scoring. Compared to normal oil control rings, high contact pressures, radial as well as axial, along the sides of the ring, as well as its configuration, hinder rotation of the individual steel bands. In areas with cylinder scoring, practically no wear of the chrome layer takes place; i.e., ultimately tiny chrome lands are formed, which then prevent ring rotation. Cylinder scoring could arise from a variety of causes. If the cylinder was not bored out correctly in the course of an overhaul (precision boring on appropriate boring or milling machines), existing scoring may not be bored out completely, or cylinders installed without undergoing thorough cleaning may experience their first scoring immediately after the engine is placed back in service. Another cause of scoring is burned piston rings; local lack of lubricating oil results in seized rings. Scoring could result during installation of pistons in cylinders, if the piston rings are damaged (broken) in the process. Given adequate lubrication, seizing marks on the rings may disappear in the course of normal wear, but cylinder scoring remains.

Fig. 6.121 Piston ring wear: ring surface, magnified 50x. Source: MAHLE

6.2.2.2 Burned rings

Ring burning is understood to mean a partial seizing process that leads to increased wear, poor sealing, increased blowby, and oil consumption. Burn marks (Fig. 6.122) are recognizable as axial, streaky changes to the ring face, with clearly discernible circumferential crack structures covering a range of colors from tobacco brown to black. The magnified view shows bright, smeared areas of material that has become plastic or liquefied, around torn out or scraped out furrows, partially filled with brown or dark deposits. Often, these streaks appear in pairs.

Burn marks increase ring wear by as much as a factor of 10,000 over normal [6-8]. They can compromise the piston/piston ring/cylinder wall sealing system after only a few hours of operation. Usually, the top ring is affected most strongly, the lower rings usually less. Burn marks transfer to the piston by way of the cylinder wall. Blowby volume and oil consumption increase significantly; the seizing process intensifies to the point of ring and piston seizing. Distinction is made between early burning, in the course of engine break-in, and old-age burning, after medium to long service life. In the 1960s, burned rings developed into one of the most urgent problems in engine development; at the time, plateau honing provided an effective solution.

Rising engine specific output increased heat flow through the rings, especially the top ring. When added ring loading as a result of a locally

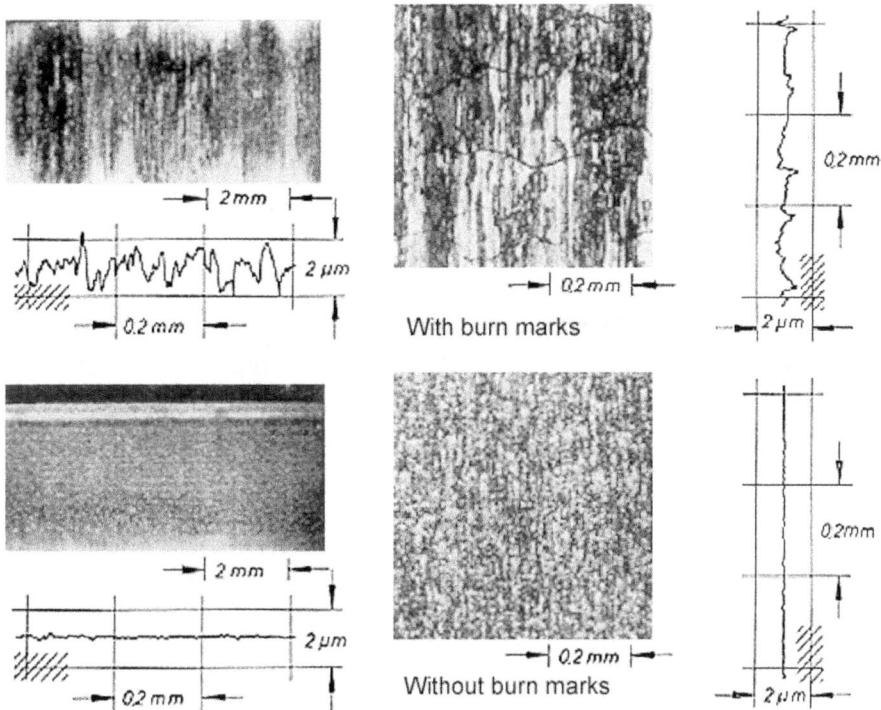

Fig. 6.122 Chrome plated rings, without and with burn marks. Source: Federal-Mogul

compromised lubricating film—regardless of cause—transcends a critical limit, heat flow through the ring is reversed; i.e., heat flows inward from the outer face of the ring, instead of the normal situation, from the inner side of the ring to the outside. This causes the ring to bend more, and local pressure and friction at the ring-to-cylinder contact patch are increased.

As a result of this deformation (see Fig. 6.123), the ring now presses harder against other areas of the cylinder wall, resulting in the typical wear streaks associated with burn marks (see Figs. 6.124 and 6.125).

The actual process of burn mark formation consists of the ring(s) operating under conditions of boundary lubrication. The resulting high friction leads to adhesive and abrasive wear (see Fig. 6.126).

6.2.2.3 Ring flutter

Radial and axial piston ring vibrations are excited by a sudden drop in pressure behind the ring when the ring lifts off of the sides of its ring groove. This causes the ring to momentarily collapse inward. The consequences of ring flutter are heating of piston and piston ring,

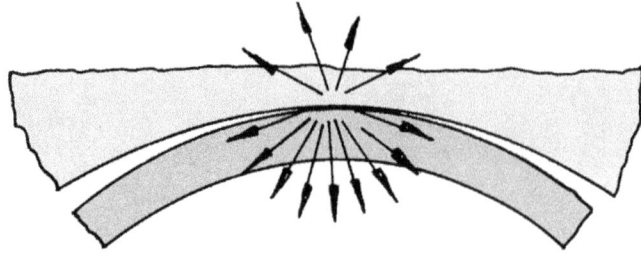

Fig. 6.123 Ring deformation as a result of concentrated heat flow. Source: Federal-Mogul

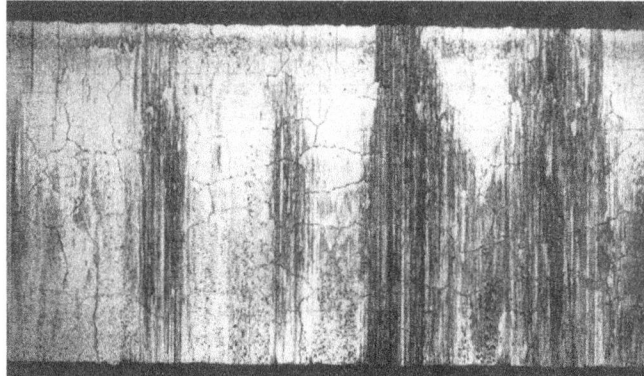

Fig. 6.124 Burn marks on a piston ring face. Source: MAHLE

increased blowby (in large engines, this could result in crankcase explosions), high oil consumption, and loss of power.

6.2.2.4 Ring breakage

Harsh combustion—combustion knock in Otto-cycle engines, cold starts and long ignition delay in diesels—results in high gas pressures and steep pressure rise (pressure gradients), which lead to ring breakage, where the rings, starting from the ring butt ends, break into many small fragments (Fig. 6.127).

Additional causes of piston ring breakage could include

- Insufficient end gap and/or side clearance
- In two-stroke engines, ring ends catching in the ports

6.2.2.5 Stuck rings

In the event of excessive temperatures in the ring grooves, oil can form carbon deposits, which lock the rings into their grooves, prevent any movement, and thereby prevent the rings from performing their sealing function. Before the introduction of piston cooling, stuck rings were a major problem for the piston group, above all affecting high-performance engines (aircraft gasoline and motor vehicle diesel engines).

Fig 6.125 Burned rings on a large diesel engine.

Fig. 6.126 Burned rings and the resulting piston seizing.

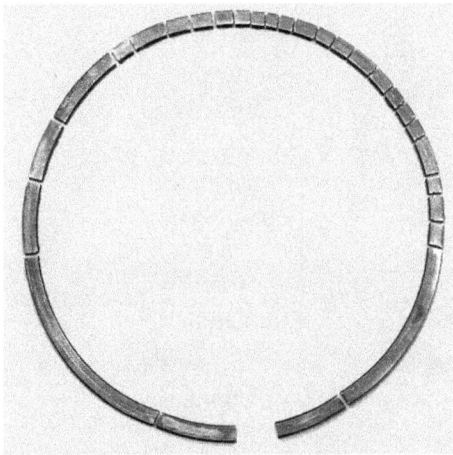

Fig. 6.127 A piston ring, broken into many small fragments.

6.2.2.6 High oil consumption

A certain amount of oil consumption is unavoidable; after all, even the top piston rings and the lower portion of the valve stems are expected to receive adequate lubrication. With every stroke, 1 to 5% of the oil film is replaced [6-9]. One therefore speaks of an "oil budget" for good reason.

When it comes to oil consumption, we face conflicting goals, because some measures that can be taken in terms of ring layout and ring package selection to reduce oil consumption also increase friction losses, and therefore fuel consumption. Fewer rings per piston, shorter fire lands, reduced ring height, higher ring pretension, in general higher forces and temperatures, all serve to worsen tribological conditions and the oil budget. Oil consumption depends on many factors, not all of which are always quantitatively definable. These may include number,

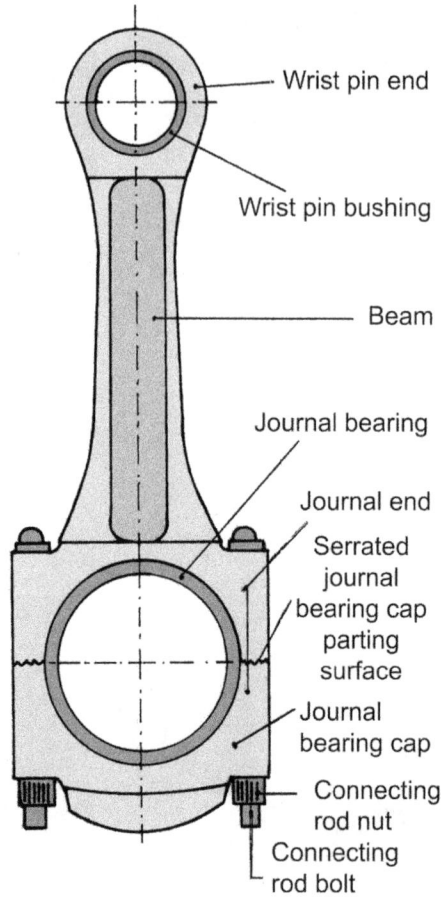

Fig. 6.128 Connecting rod terminology.

Labels in figure: Wrist pin end, Wrist pin bushing, Beam, Journal bearing, Journal end, Serrated journal bearing cap parting surface, Journal bearing cap, Connecting rod nut, Connecting rod bolt

combination, type, geometry, and pretension of piston rings, cylinder wall surface treatment (honing), deformation behavior of piston, and cylinder sleeve. The latter is in turn affected by crankcase stiffness. Oil consumption is determined not only by the piston group, but also by valve stems, with oil able to pass through the valve guides, as well as crankcase ventilation.

6.2.3 Connecting rods

As part of the crank train, the connecting rod is tasked with transferring displacement and forces from the piston to the crankshaft—and vice versa. It also represents the kinematic link between piston and crankshaft. The connecting rod consists of a wrist pin end (the "small end"), a beam, and a journal end (the "big end"). The journal end is made up of a connecting rod head and bearing cap. The wrist pin end is connected to the piston by means of the wrist pin; the journal end is connected to the crankshaft by means of a crankpin journal (Fig. 6.128).

The wrist pin end executes an oscillating (pure reciprocating) motion, while the journal end does a rotating motion, and the beam a swinging motion (Fig. 6.129). Because of increased engine power and the resulting larger crankpin diameters, the journal end will no longer pass through the cylinder bore; accordingly, for high-speed diesel engines, the bearing cap joins the beam at an angle. The added transverse forces on the parting line are often taken up by a keyed interface between bearing cap and beam, in the form of interlocking serrations (Fig. 6.130).

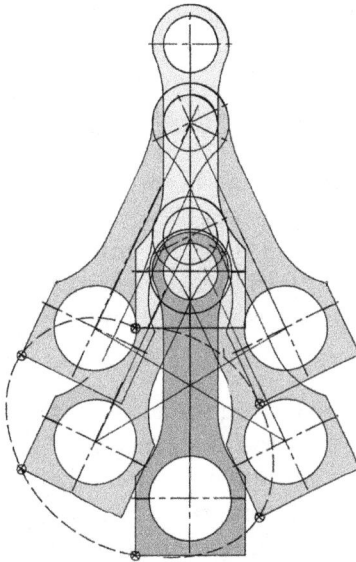

Fig. 6.129 Connecting rod positions during one crankshaft revolution.

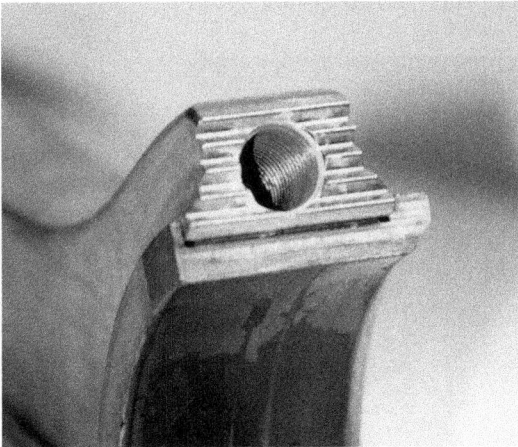

Fig. 6.130 Serrated parting surface of an angled journal bearing cap.

In older designs, the interlocking interface consisted of a key and keyway. Even in straight-parted connecting rods found in motor vehicle engines, modern practice is to take advantage of mechanical interlocking at the bearing cap interface, without the manufacturing complexity of machining serrations; the journal end is "cracked" to yield a "cracked connecting rod" (Fig. 6.131). The fracture surface provides perfect seating of the bearing cap against the beam. Originally only possible with cast connecting rods, cracking is now also applied to forged connecting rods.

Connecting rods of large diesel engines have straight parting surfaces, and are of a "marine" design, in which the connecting rod is bolted to the journal end. This enables the pistons to be pulled without disassembling the journal end bearing. In newer designs, this parting surface is located directly below the wrist pin end, so that it is no longer necessary to remove the heavy connecting rod when pulling a piston. This also reduces the headroom needed to perform this operation.

Because of considerable increase in gas pressures (in motor vehicle diesels, up to 150 bar; medium and large engines have ignition pressures up to 200 bar), modern vehicle engines employ trapezoidal connecting rods, in which the lower part of the wrist pin end, which must take up these high gas pressures, is wider than the upper part, which takes up inertia forces.

As the element responsible for transferring force and motion between piston and crankshaft, the connecting rod is subjected to alternating, high forces. As a moving engine component, it should be as light as possible; with regard to its interaction with the wrist pin and crank pin, it should be rigid. These conflicting requirements force very restricted connecting rod dimensioning, making the connecting rod susceptible to many different types of secondary failures (Fig. 6.132).

Fig. 6.131 Parting surface of a "cracked" connecting rod.

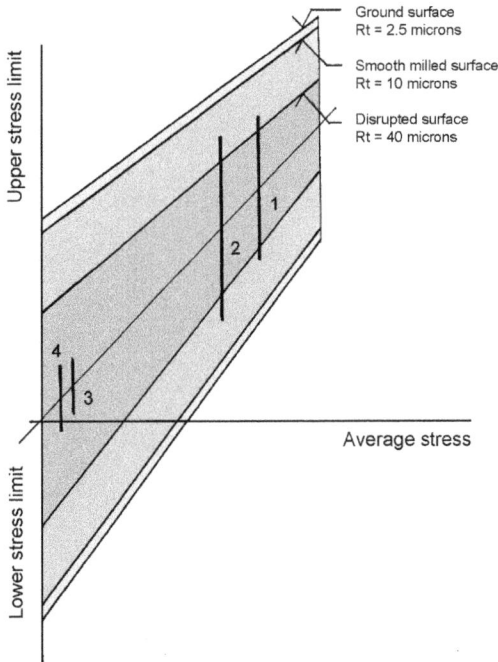

Fig. 6.132 Effect of surface finish on connecting rod load limits: 1, Load in bearing cap fillet during assembly; 2, Load in bearing cap fillet from inertia forces; 3, 4, Load on bearing cap back in transition to fillet from inertia forces.

Connecting rods are usually forged. Passenger car engines also use cast connecting rods. Rods are also made using sintering processes. The connecting rod is loaded in compression (predominantly by gas forces) and tension (predominantly by inertia forces). Moreover, the additional swinging motion of the connecting rod imposes bending loads (Fig. 6.133).

Forces transmitted from the wrist pin, through the connecting rod, to the crankshaft, and the reverse, are borne by lubricating films in the bearings. Forces imposed on the journal and wrist pin ends are therefore dependent on the pressure distribution in the lubricant. This, in turn, is affected by connecting rod stiffness. Under load, the journal and wrist pin ends deform (Fig. 6.134).

- The upward-acting inertia force is counteracted by the lubricating film between the crank pin and the lower (cap side) bearing shell. Forces between the connecting rod and bearing cap are taken up by the connecting rod bolts. The journal bore deforms into a vertical oval, and the rod bolts are bent outward. With insufficient bolt tension, the inside edge of the cap/beam interface would open.

- By contrast, under maximum gas pressure, the beam exerts force through the lubricating film, against the crank pin. The journal bore

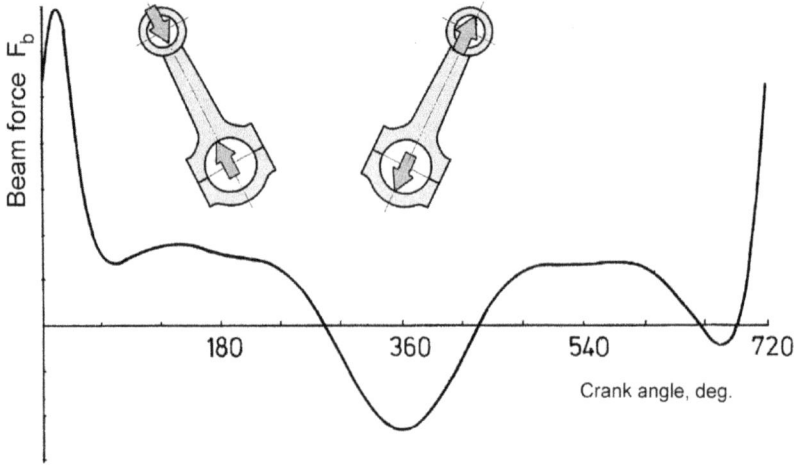

Fig. 6.133 Connecting rod beam force over one complete engine cycle.

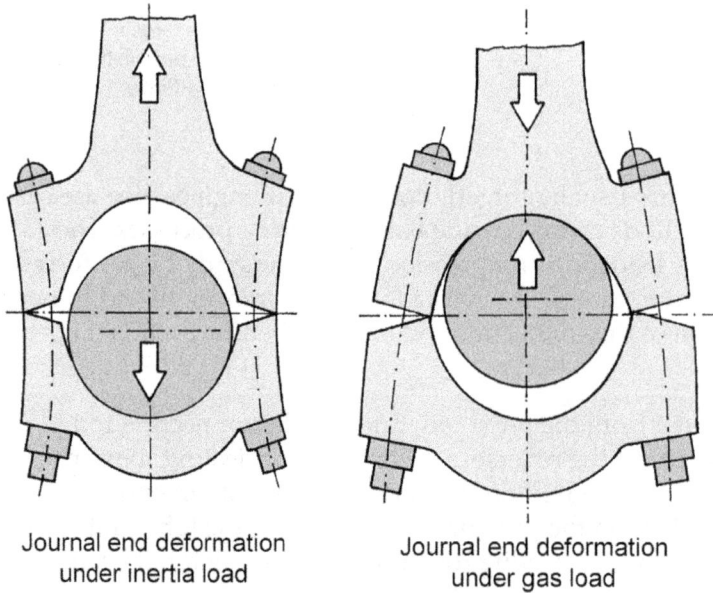

Journal end deformation
under inertia load

Journal end deformation
under gas load

Fig. 6.134 Connecting rod deformation (schematic).

deforms into a horizontal oval, the bolts bend inward, and the outer edges of the cap/beam interface tend to open.

These deformations impose considerable bending loads on the journal and wrist pin ends. Assembly forces are even larger than operating forces, because after all, these must be able to withstand operating forces. Therefore, bolting the bearing cap to the beam takes on special significance: the interface must not be allowed to spread open, and sufficient bearing interference (tight fit of bearing in the journal bore) must be assured.

Because bolt tension determines the stress condition of the journal end, and therefore the shape of the bore, care must be taken that the connecting rod bolts are tightened in the same manner at every subsequent reassembly as in their initial factory assembly. If the bolts are tightened excessively, the journal bore will deform into a horizontal oval; if not tightened sufficiently, into a vertical oval (Fig. 6.135). The achievable bolt tightening accuracy is dependent on the tightening method. If tightening to a specific torque, the portion of torque converted to bolt tension depends on friction conditions in the threads and at the beam/bearing cap interface. As roughness is smoothed with every tightening, thread friction decreases, so that a higher portion of torque is converted to bolt pretension. As a result, the journal bore deforms into a horizontal oval. This explains why journal end bearing

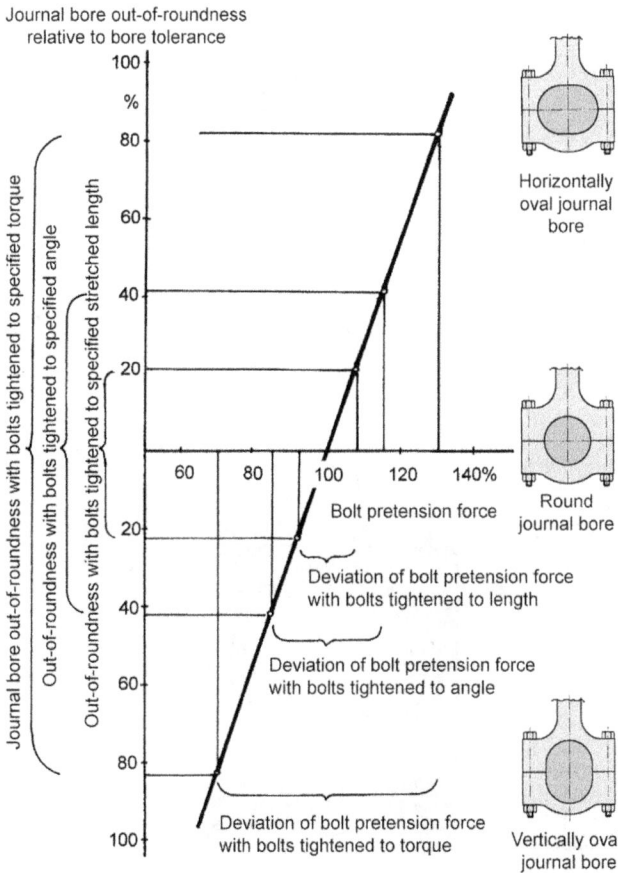

Journal bore out-of-roundness relative to bore tolerance

Horizontally oval journal bore

Round journal bore

Bolt pretension force

Deviation of bolt pretension force with bolts tightened to length

Deviation of bolt pretension force with bolts tightened to angle

Deviation of bolt pretension force with bolts tightened to torque

Vertically oval journal bore

Journal bore out-of-roundness with bolts tightened to specified torque

Out-of-roundness with bolts tightened to specified angle

Out-of-roundness with bolts tightened to specified stretched length

Fig. 6.135 Journal bore out-of-roundness as a function of bolt tension.

seizing seems to occur preferentially after engine overhaul. Better results are obtained by tightening by angle, but the smallest scatter in bolt pretension is obtained by tightening to length, because bolt length is directly proportional to bolt pretension.

The most heavily loaded areas of straight-split connecting rods are the fillets where the beam transitions to the journal and wrist pin ends (Fig. 6.136). In angle-parted connecting rods, the upper part of the blind-threaded holes are often the source of fatigue failures.

Because of its function as a linking element between piston and crankshaft, connecting rod failures usually engender serious consequences. The connecting rod is, in effect, trapped between hammer and anvil, and is deformed and shattered accordingly.

Failures in which the connecting rod is severely deformed, even bent, underscore the astonishing ductility of high-quality heat-treated steels (Fig. 6.137).

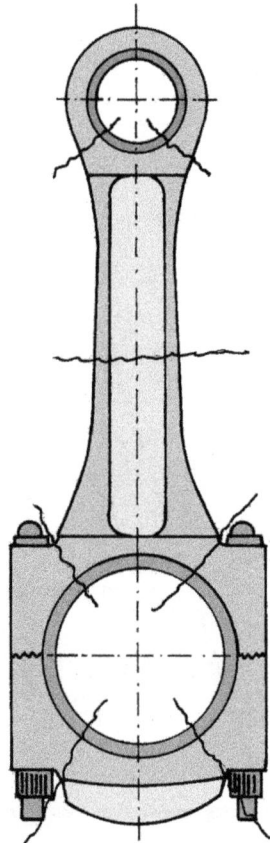

Fig. 6.136 Connecting rod fracture zones.

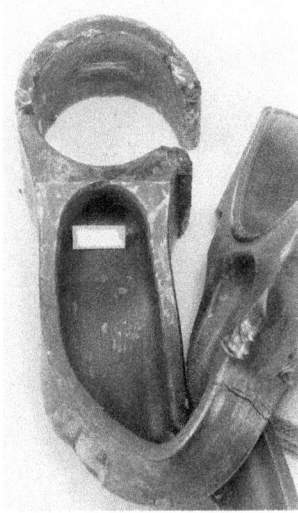

Fig. 6.137 Extremely deformed connecting rod of a large diesel engine.

6.2.3.1 Manufacturing defects

The greater the load on a piston, the more sensitive it becomes to machining marks, notches, or nicks (see Fig. 6.132).

Example a: After an operating time of 725 hours following engine overhaul, the bearing cap of a diesel engine connecting rod broke. The total operating time of the engine (and rod) was 8700 hours. The fatigue failure (Fig. 6.138) was initiated by a ground-in groove in the already highly stressed fillet. The fillet was probably nicked and damaged while grinding the flat for the bearing cap nut (Figs. 6.139 and 6.140). In view of the otherwise long service life of the rod, this damage probably occurred during the most recent engine overhaul.

Example b: After a component operating time of 36,500 hours in a diesel engine, a connecting rod broke below the wrist pin end, resulting in a thrown rod that penetrated the crankcase (Fig. 6.141). The cause of the fatigue failure in the highly stressed transition from beam to wrist pin end was impact damage by some foreign object, incurred during engine operation (Fig. 6.142). The fracture surface showed a fatigue fracture followed by force fracture of the remaining cross section, starting in the area of the flange edge, and extending across roughly half of the shaft cross section (Figs. 6.143 and 6.144). Impact damage was visible on the upper surface of the flange, as well as the extended axis of the wrist pin bearing. The assumption that the failure involves depressions around impact damage that occurred prior to the fracture was borne out by the uneven, disturbed fracture progression at the point of origin, as well as smoothing of the slightly raised fracture edges.

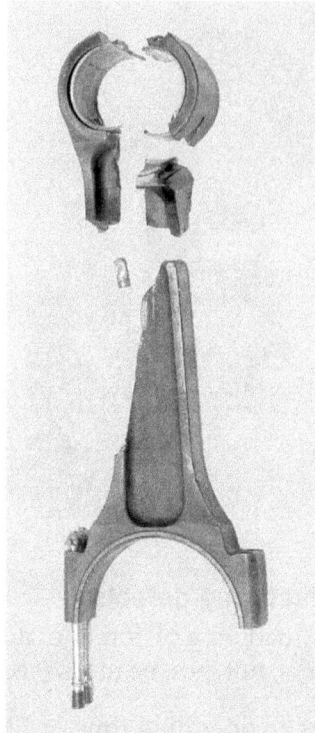

Fig. 6.138 Connecting rod fatigue failure.

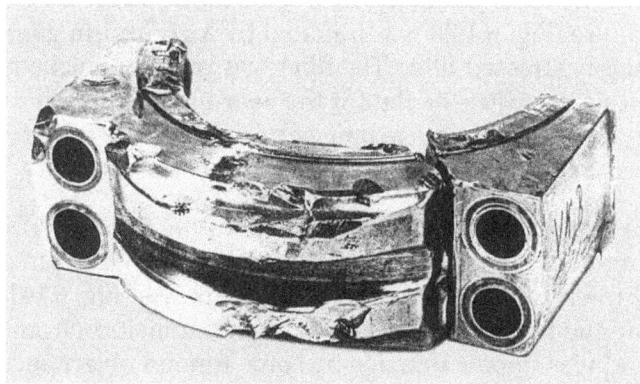

Fig. 6.139 Connecting rod bearing cap with crack through fillet at nut surface.

6.2.3.2 Fretting corrosion

Example a: After 20,000 hours of operation, a locomotive engine was removed and disassembled for overhaul. During overhaul, a crack was detected in a connecting rod (Fig. 6.145). The crack initiated at the journal end, stretched across the entire width of the connecting rod, to within about 6 cm of the fillet to the bolt head mating surface. Furthermore, traces of severe fretting corrosion were visible near

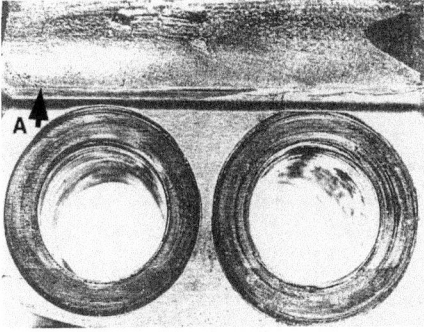

Fig. 6.140 Site of fracture initiation (A) at a nick in the fillet, caused by grinding.

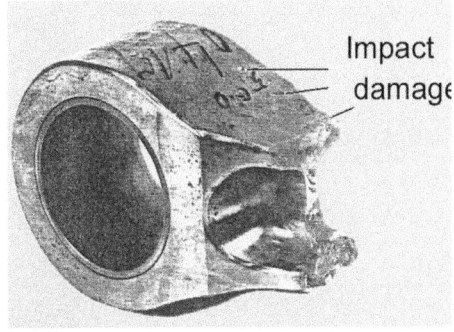

Fig. 6.141 Fragment of broken connecting rod.

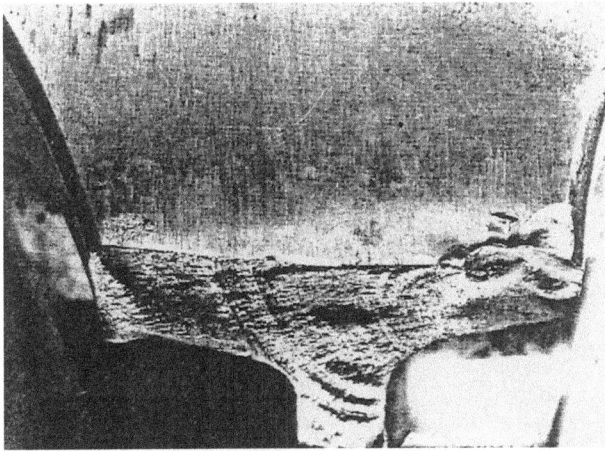

Fig. 6.142 Magnified view of connecting rod fragment, showing impact damage on upper side of flange near fracture initiation site.

the connecting rod parting line. The fracture surface showed a well-advanced, slowly propagating fatigue crack. The crack initiated at a small projection, N (see Fig. 6.146), at the connecting rod journal end, typical of fretting fatigue cracks. The ultimate cause of this fatigue fretting failure was insufficient bearing insert pretension; for subsequent engines, this pretension had been increased.

Example b: A connecting rod bolt broke after 750,000 km of operation in a locomotive engine (Fig. 6.147). The fracture began with a fatigue fracture surface F1, about 2 mm wide, oriented about 45° to the bolt axis. This continued into a fatigue fracture surface F2, perpendicular to the bolt axis, ending with a small residual fracture surface F3 (Fig. 6.148). The fact that the initial crack began at 45°, typical of fretting fatigue fracture, suggests that the crack began as a result of severe fretting corrosion at one of the pilot diameters. After engine disassembly at the indicated mileage, the bolt was re-installed in a

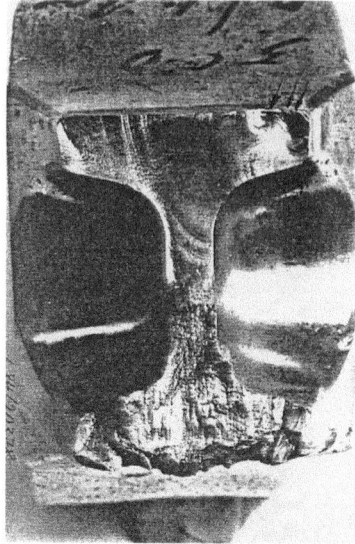

Fig. 6.143 Fracture initiation area with fatigue failure fracture surface and adjacent force-failure fracture surface.

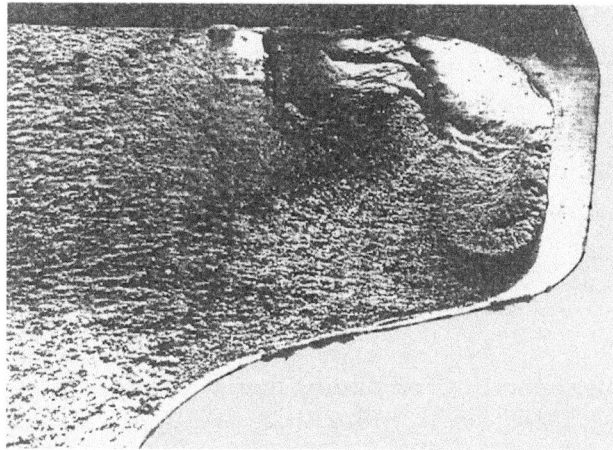

Fig. 6.144 Magnified view of fracture initiation area of the previous illustration.

different orientation, which changed its loading. This led to a fatigue fracture surface running in a different direction (F2) and ultimately to complete failure.

6.2.3.3 Assembly defects

The most common mistake in engine assembly is incorrect tightening of fasteners. The consequences are:

- Spreading of the mating surfaces, followed by bolt fracture (Fig. 6.149); the connecting rod, no longer guided by the crankshaft, is severely damaged and in turn damages the crankshaft, crankcase, and piston

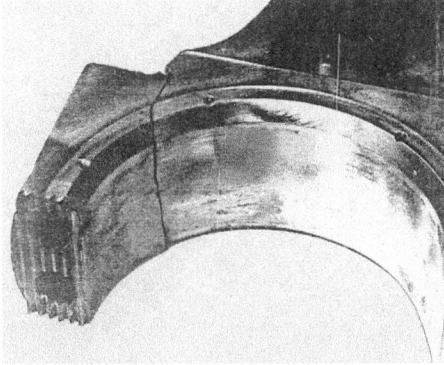

Fig. 6.145 Crack propagation at the journal end of a connecting rod; traces of fretting corrosion at the bore surface.

Fig. 6.146 Exposed fatigue and laboratory-induced residual force fracture, with crack initiation site marked at N (A: laboratory force fracture; B: fatigue crack).

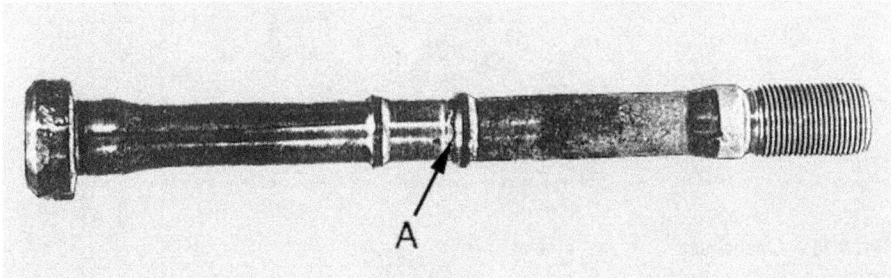

Fig. 6.147 Fractured connecting rod bolt; crack initiation site at A.

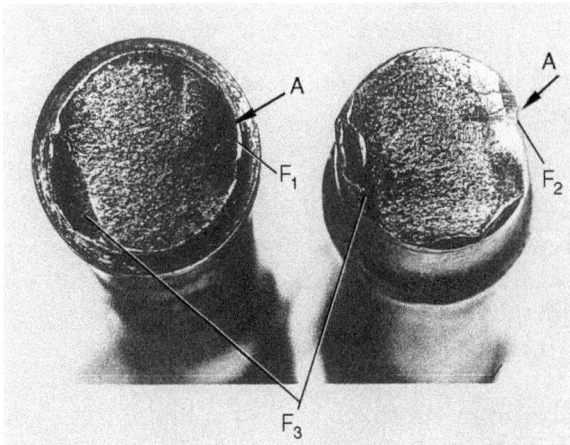

Fig. 6.148 Fracture surfaces of a bolt shaft, with crack initiation site at A, primary fretting fatigue fracture surface F_1, secondary fatigue fracture surface F_2, and residual fracture surface F_3.

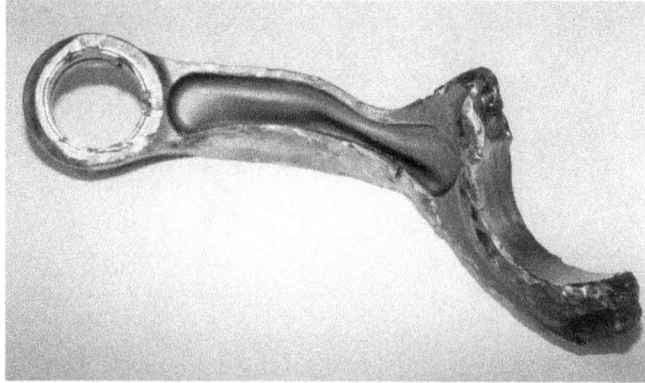

Fig. 6.149 Connecting rod failure as a result of connecting rod bolt fracture.

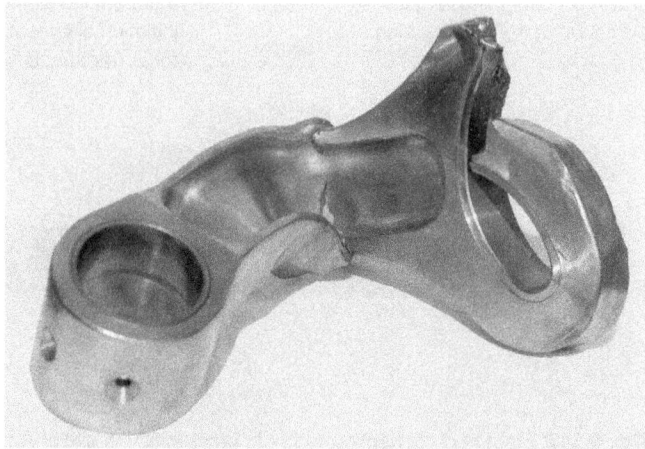

Fig. 6.150 Connecting rod failure as a result of hydraulic lock.

- Oval deformation of the bearing bore, with associated bearing seizing
- Insufficient journal bearing insert pretension, resulting in fretting corrosion—a source of fatigue failures; see Figs. 6.145 and 6.146

6.2.3.4 Failures caused by engine operation

Often, connecting rod failures develop as a result of operational irregularities and from damage to other engine components. Severe piston seizing is always associated with connecting rod damage. But even far less spectacular damage may result in connecting rod damage, e.g., leaking charge air intercoolers, cylinder seals, corrosion holes in cylinder sleeves, etc., all of which result in the cylinder filling with coolant. Similarly, fuel or lubricating oil collecting in a cylinder can lead to the feared phenomenon of hydraulic lock (Fig. 6.150).

Example: The charge air intercooler of a diesel engine developed a leak. With the engine shut down, water trickled through the intake manifold and into the nearest cylinder with open intake valves. The piston happened to come to rest below top dead center, on its downstroke. When the starter next turned the engine over, it did not immediately try to move the piston against the water in the cylinder. During the continued downward stroke, and then the upward stroke, several of the engine's fifteen other cylinders fired, so that great force was exerted by this one piston against the incompressible slug of water in the cylinder. The results were severe crank train and crankcase damage.

Hydraulic lock has also become the subject of a court decision. In accordance with German motor vehicle insurance law (specifically, AKB § 12 Section 1 IIe), on May 31, 1989, the Hamm appellate court delivered the following verdict [6-10]:

> "1. Engine damage as a result of a so-called hydraulic lock (entry into the combustion chamber of water which had been splashed up by driving through a puddle, or by another vehicle moving in the opposite direction) is considered to be an accident.
> "2. The case in question is considered to be an uninsured operational failure . . ."

The case was described as follows: ". . . at this point, a vehicle approached from the opposite direction, throwing a wave of water onto his own car. From this moment onward, the car juddered throughout the remainder of the journey, furthermore, it trailed a cloud of blue exhaust . . . partial engine disassembly revealed a bent connecting rod, which necessitated engine replacement at a cost of 8062.69 Deutschmarks . . ."

As this case shows, connecting rod failures result in severe related damage, and often lead to total destruction of the engine.

6.2.4 Crankshafts

The crankshaft is a system of serially arranged, phase-shifted eccentricities, in which the individual element—the crank throw—consists of the crankpin, the two crank cheeks, and half of each adjoining main bearing journal (Fig. 6.151). Depending on configuration, the crank cheeks may include one or two cast-in, forged-on, or bolted-on counterweights.

Crankshafts transform the forces introduced by the connecting rods into torque, "collect" the torque of individual cylinders, and "deliver" it to the flange at the output end of the crankshaft. Furthermore, crank train forces are passed to the crankcase through the crankshaft, and the crankshaft ensures a supply of oil to the crankpins. Depending on

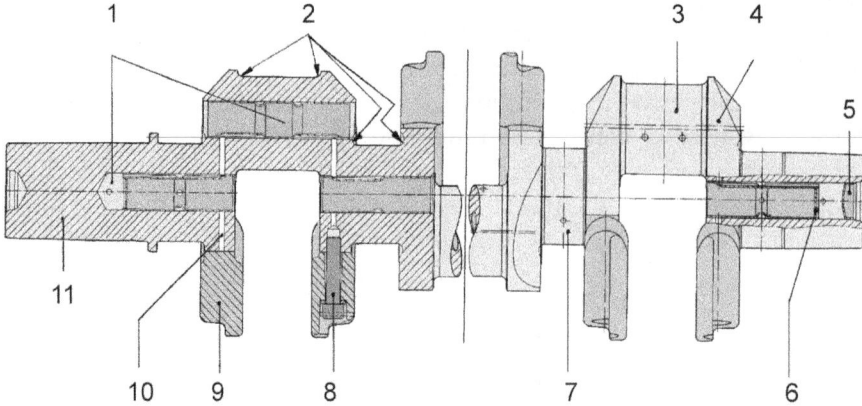

Fig. 6.151 Crankshaft nomenclature: 1, Relief bores; 2, Fillets; 3, Crankpin; 4, Crank cheek; 5, 6, Sealing plugs; 7, Main bearing journal; 8, Counterweight bolt; 9, Counterweight; 10, Oil passage; 11, Output end.

size, crankshafts can be die forged as a single unit, or crank throws may be forged individually. Crankshafts for motor vehicle engines, even for medium-sized engines as used in railway locomotives, may be cast. Connecting rod forces act upon the crankpins. These forces can be resolved into a force component tangential to the circle described by the motion of the crankpin, the tangential force F_T (or, alternatively, rotational force), and in a force component directed at the crankshaft axis, the radial force F_R. The tangential or rotational force (Fig. 6.152), combined with the crank radius, results in the torque. The radial force does not contribute to the power developed by the engine, and is therefore also described as a "blind" force.

The crankshaft is subjected to loads imposed by the tangential and radial forces. Specifically, the loads are as follows:

- Torsion (twisting):
  - As a result of the useful torque, which is the result of rated power and rated speed, and which is the sum of all torques acting on all crank throws.
  - As a result of torque pulsations caused by variation in tangential force (rotational force). Rotational forces of individual cylinders are added according to their ignition phasing, resulting in torque pulsations becoming smoother as one approaches the clutch (or coupling) end of the crankshaft. For crankshaft loading, however, the largest rotational force variation at individual crank throws is the determining factor.

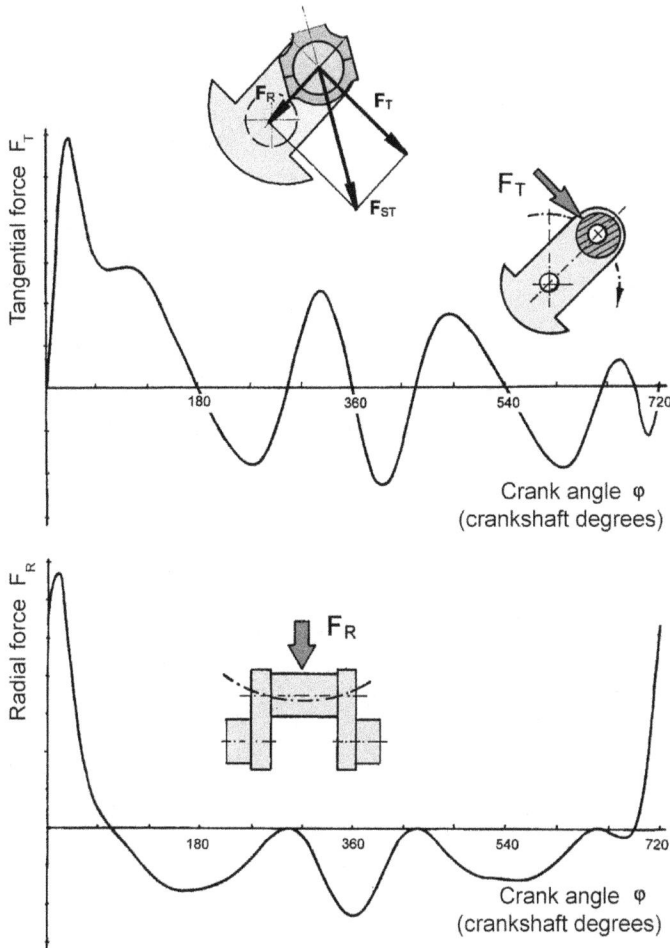

Fig. 6.152 Time history of tangential and radial forces over one working cycle.

- As a result of torsional vibrations, which lead to additional torque in the crankshaft. Torsional vibration moments may amount to a multiple of the other torques.

• Bending as a result of radial forces. In addition, the crank cheeks are bent by torque imposed by the crankpins.

In addition to loads resulting from their function, i.e., taking up and transferring forces and supporting the pistons and connecting rods, and their own intrinsic behavior (torsional vibrations), crankshafts are subjected to loading from:

• Reaction of the crankshaft to the powerplant installation, e.g., if the vehicle frame or vessel hull deforms the crankcase, and therefore the crankshaft, and from

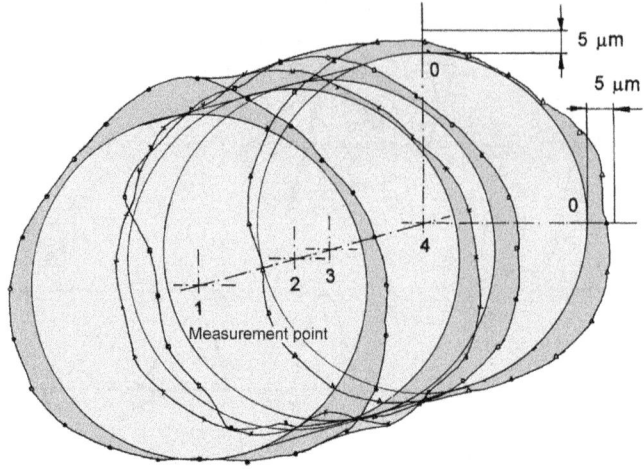

Fig. 6.153 Crankpin shape determination.

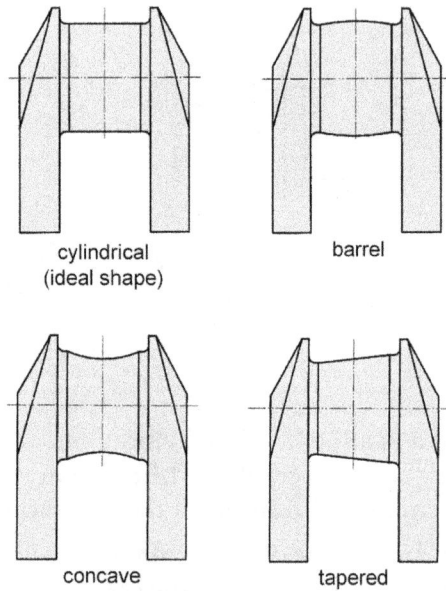

cylindrical
(ideal shape)

barrel

concave

tapered

The macroscopic shape of crankpins is
composed of various geometric components.

Fig. 6.154 Crankpin shape deviations.

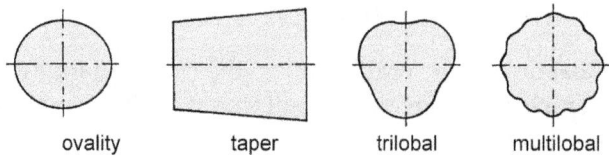

ovality

taper

trilobal

multilobal

- Torques imposed by auxiliary devices (pumps, camshaft drive, etc.)

Crankshaft function and durability demand close manufacturing tolerances and restricted dimensional and shape tolerances (Fig. 6.153), in terms of

- Deviation from an ideal cylinder
    - Taper
    - Out-of-round
- Runout and wobble
- Nonparallelism of crankpins to crankshaft axis
- Mutual eccentricity of main bearings

The macroscopic shape of crankpins—as well as other shafts—is composed of various geometries imposed by the manufacturing techniques employed. Consequently, there are various possibilities for shape deviations of main bearings and crankpins from the ideal (Fig. 6.154).

Because of the nonstationary nature of tangential (rotational) force and radial force, the crankshaft is subjected to dynamic loads. Force transfer in the crank throws leads to shape-related peak stresses, which can greatly multiply bending and torsional loads. The critical elements of a crank throw are the fillets in crankpin and main bearing journals; the nature of the imposed stresses—compression or tension—varies with loading (Figs. 6.155, 6.156, and 6.157).

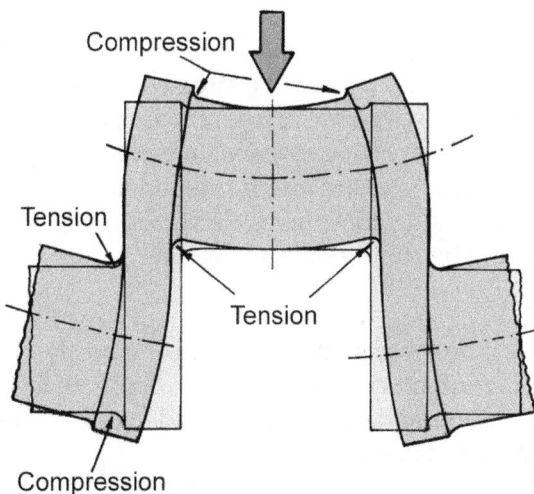

Fig. 6.155 Crank throw deformation as a result of gas loads.

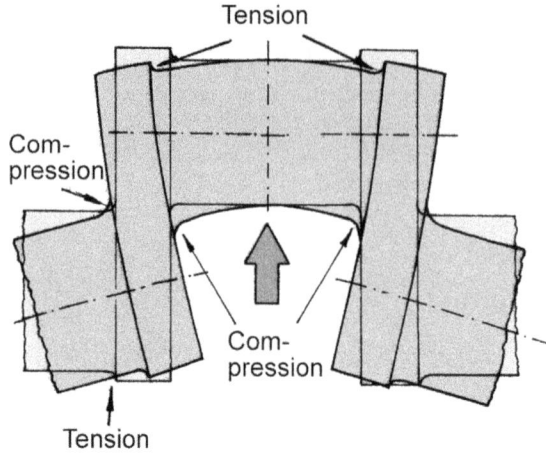

Fig. 6.156 Crank throw deformation as a result of inertia loads.

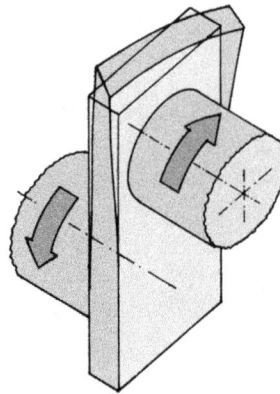

Fig. 6.157 Crank throw deformation as a result of torsional loads.

In hardening the crankpin running surfaces, the fillets are also deliberately hardened. This induces compressive stresses in the fillets. In service, operational tensile stresses cannot present a danger until they have offset these built-in compressive stresses. In view of long-term crankshaft durability, oil passages are also hardened, as deeply as possible. Another means of imposing compressive stresses on fillets is to apply mechanical hardening by rolling or shot peening.

6.2.4.1 Causes of failure

Like all engine components, crankshafts undergo complex, costly development, testing, and manufacturing processes to eliminate failures caused by layout or design. Material defects may include nonmetallic inclusions such as slag and oxide layers.

6.2.4.2 Crankshaft failures

Primary crankshaft failures consist of

- Bending fatigue failures, usually originating at highly-stressed fillets (Fig. 6.158)
- Torsion fatigue failures (Figs. 6.159 and 6.160)
    - These usually run at a 45° angle and often originate in oil passages (Fig. 6.161).
    Example: After several hundred hours of operation, a diesel engine experienced a connecting rod bearing failure. Upon disassembly, a crack was detected in the crankshaft, extending through the entire crankpin. Examination showed that the failure was primarily caused by a torsional fatigue fracture in the crankpin (Figs. 6.162 and 6.163), starting from the bottom-side fillet. The composition of the crankshaft material, and its mechanical properties, were not implicated. The crankshaft fatigue failure originated from a nonmetallic inclusion (macroscopic slag), with the origin about 4 mm below the crankpin surface. Such inclusions are very rare and had never before been encountered in this engine type. The running surface of the connecting rod bearing was scraped out by the sharp corners of the fracture surfaces. The ensuing bearing failure then additionally damaged the crankpin surface through material transfer and thermal stress cracking.
    - Torsion fractures may also run parallel to the axis, exhibiting a "burst" appearance, or as combinations of cracks propagating in various directions.
    Example: During disassembly of a diesel engine, the crankshaft was found to have multiple pronounced cracks on one crankpin. These cracks originated in a small area of the crankpin running surface, toward the adjacent crank cheeks. The cracks were torsional fatigue cracks originating at locations A1 and A2.

Fig. 6.158 Bending fatigue failure.

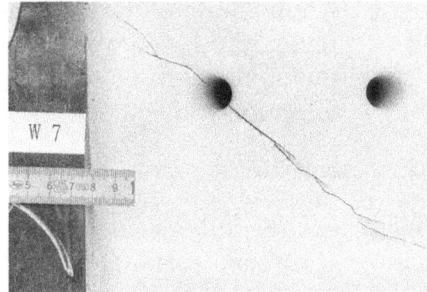

Fig. 6.159 Torsion fatigue failure, originating from an oil passage.

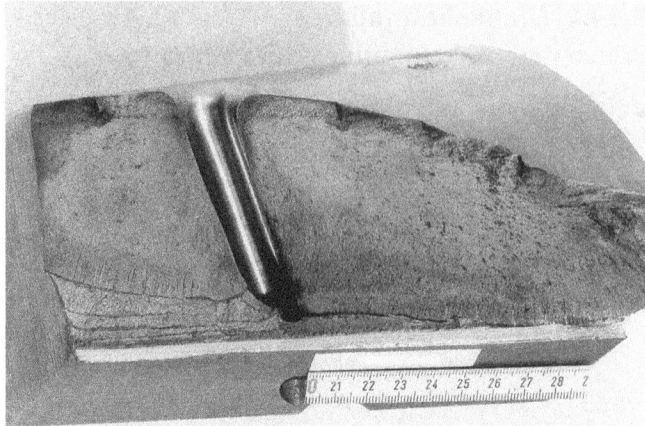

Fig. 6.160 Fracture surface of the torsion fatigue failure in the previous figure, with a large fatigue surface and small residual fracture surface.

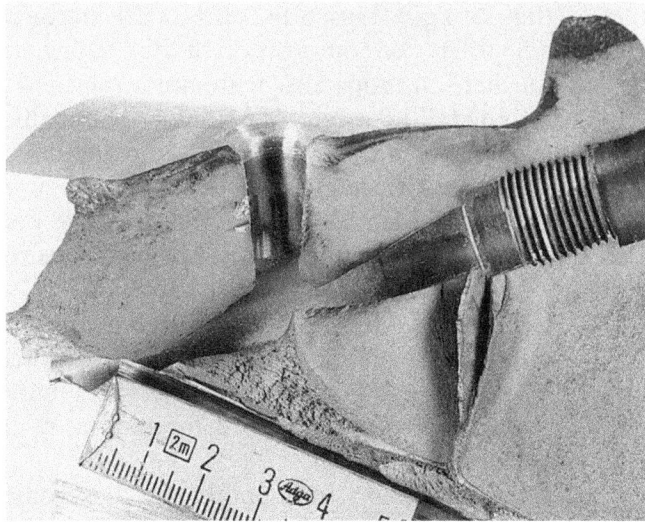

Fig. 6.161 Torsion fatigue failure through a crankpin. The fracture running through the plane of the oil passage is typical.

Hardness values and hardness distribution on the broken crankpin met specifications. In view of the long service life of 16,000 operating hours, material defects could be ruled out. Examination of the associated torsional vibration damper revealed poor damper action. The cause of this failure could therefore be traced to elevated stress levels due to torsional vibration spikes (see Figs. 6.164, 6.165, and 6.166).

Most crankshaft failures are secondary failures initiated by:

Bearing failures: fully developed bearing seizing in crankshafts with hardened running surfaces (only large diesel engines use unhardened crankshaft running surfaces) cause hot cracking. Crankshafts with minor hot cracking can be overhauled if regrinding to an approved

Fig. 6.162 Torsion fatigue failure.

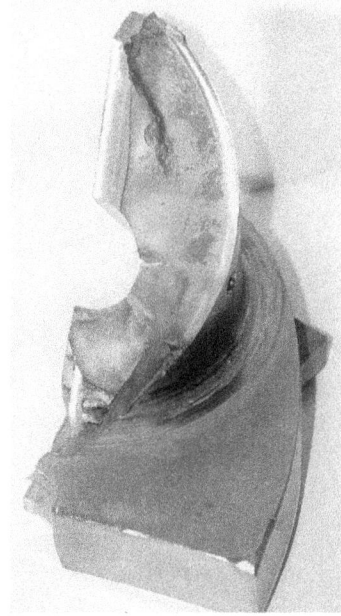

Fig. 6.163 Torsion fatigue failure fracture surface.

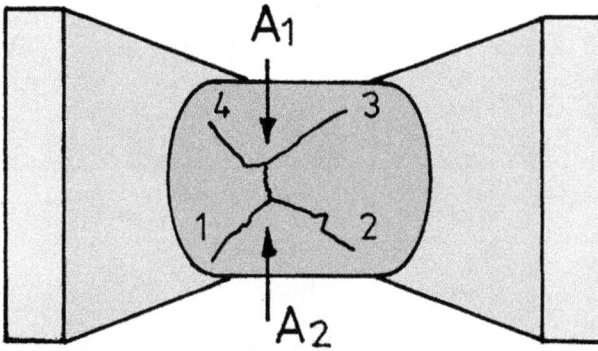

Fig. 6.164 Crankshaft torsion failure (perspective view; crank cheeks and crankpin).

undersize will remove the cracks. Hot cracks (Figs. 6.167 and 6.168) run parallel to the crankshaft axis, although transverse cracks may occasionally arise from these longitudinal cracks.

Hot cracks as extensive as those shown here generally extend all the way through the hardened layer, and initiate crankshaft fatigue failures.

Bearing failure (Fig. 6.169) not only runs the risk of causing crankshaft hot cracking, but also

Fig. 6.165 Torsional failure (fracture surface).

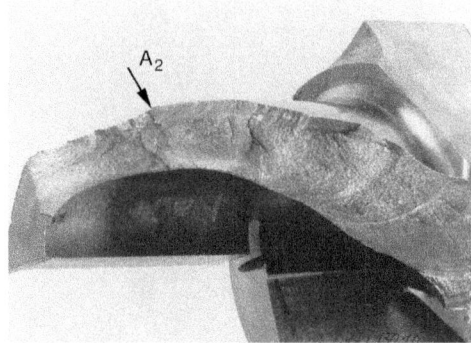

Fig. 6.166 Torsional failure (fracture surface).

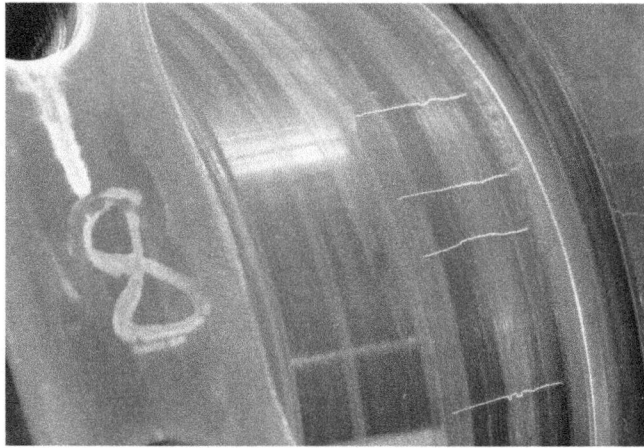

Fig. 6.167 Hot cracking of a main bearing journal.

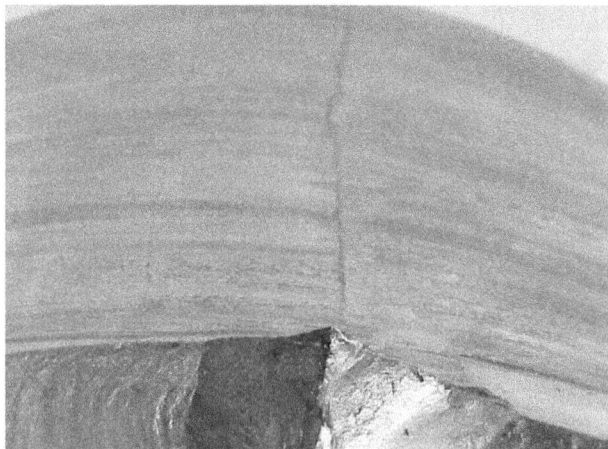

Fig. 6.168 Bending fatigue failure, originating at a hot crack.

- Mechanical failure (notch effect) in the crankshaft fillets, which are always highly stressed. This may lead to bending or torsional failures.

- In the case of bearing failures (Figs. 6.170 through 6.176)—regardless of type—bearing clearance invariably increases, and the resulting crankshaft deformation may cause loading to exceed critical levels, to the point of crankshaft failure.

- An inoperative or improperly functioning torsional vibration damper increases torsional loading beyond what the crankshaft can withstand (see Figs. 6.164 through 6.166).

Operationally, unacceptable loads may be imposed on the crankshaft as a result of

- Engine overspeeding on downgrades (commercial vehicles), or if the driver misses a shift, or in the event of

- Hydraulic lock (fluid entry into the combustion chamber: water, oil, or fuel), especially after an extended period of nonoperation

The success of any subsequent engine overhaul is strongly dependent on the repair facility's care and diligence. Crankshaft damage as a result of improper machining and treatment are becoming increasingly rare. Instead, with respect to repair facilities, possible sources or initiators of eventual failures include:

- Defective crankshaft regrinding: with excessive feed and/or insufficient cooling, grinding cracks may appear in crankpin running surfaces. Such cracks characteristically exhibit a finely branched appearance.

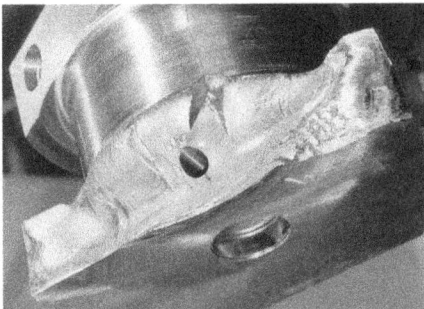

Fig. 6.169 Bending fatigue failure, originating from hot cracks resulting from bearing failure.

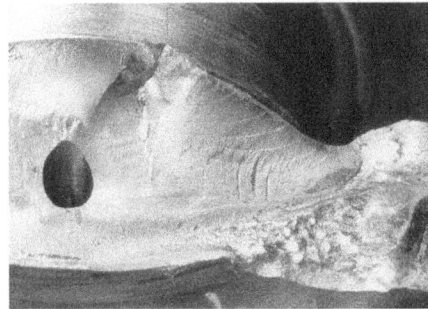

Fig. 6.170 This detail view of the foregoing bending fatigue failure clearly shows the "rest marks" ("beach marks"), as well as the roughly textured surface of the residual (overloading) fracture.

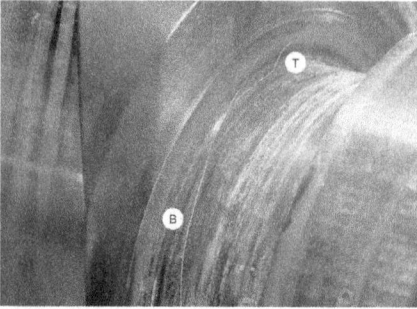

Fig. 6.171 Seized main bearing journal. A bending crack (B) and a torsion crack (T) may be seen in the crank cheek fillet.

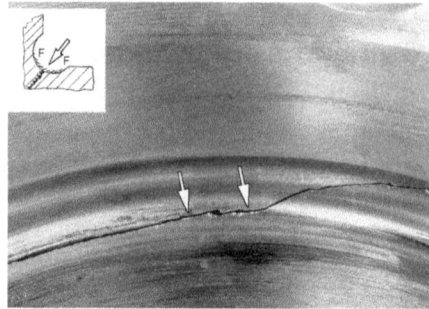

Fig. 6.172 Fracture orientation in a fillet damaged by seizing (F). The arrows show the fracture origin.

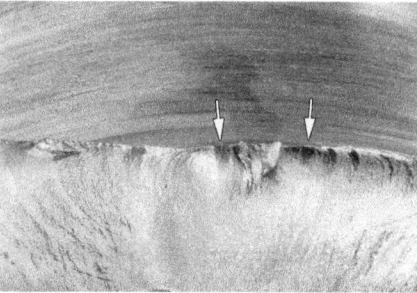

Fig. 6.173 Fracture surface in a fillet; fracture origin marked by arrows.

Fig. 6.174 Fracture surface of a crank cheek with a very small residual fracture.

- Crankshafts that are ground down too far, or whose fillets are ground incorrectly, plant the seeds for later bearing failure, and all of the associated consequences.
- Rough spots as a result of subsequent manual reworking of roundness deviations.
- Crankshaft imbalance due to counterweights switched during reassembly.
- Crankshaft out of plumb due to incorrect bearings.
- Switched main bearing caps.
- Incorrect, unsuitable flywheel or torsional vibration damper.
- Incorrect tightening torques for main bearing cap bolts or nuts.
- Impact damage and notches as a result of improper transport or storage.

Normally, crankshaft counterweights do not exhibit any damage. However, relative movement between a counterweight and its mounting surface on the crank cheek, or its attachment bolts, can result in fretting corrosion, which may lead to fatigue failure, Figs. 6.177 and 6.178.

6.2.5 Cranktrain bearings

To transfer motion, forces, and moments, rotating machine parts must by supported by a lubricating film acting against stationary or likewise moving machine parts. The lubricating film is therefore no less a "machine element" than pistons, connecting rods, or crankshaft. Along with force transfer, lubricating films are also tasked with reducing friction between moving machine components. Furthermore, the lubricant serves to remove heat and wear particles from the bearing. Bearing function depends on many factors; therefore, in connection

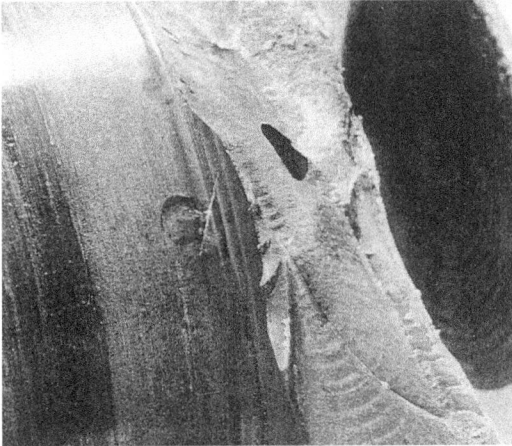

Fig. 6.175 Torsion cracking alongside the main fracture surface of a main bearing journal.

Fig. 6.176 Heat-related damage as a result of crankshaft material softening.

Fig. 6.177 Fractured counterweight, originating from fretting corrosion at the bolt head mating surface.

Fig. 6.178 Fretting corrosion under the bolt head of the counterweight in the previous figure.

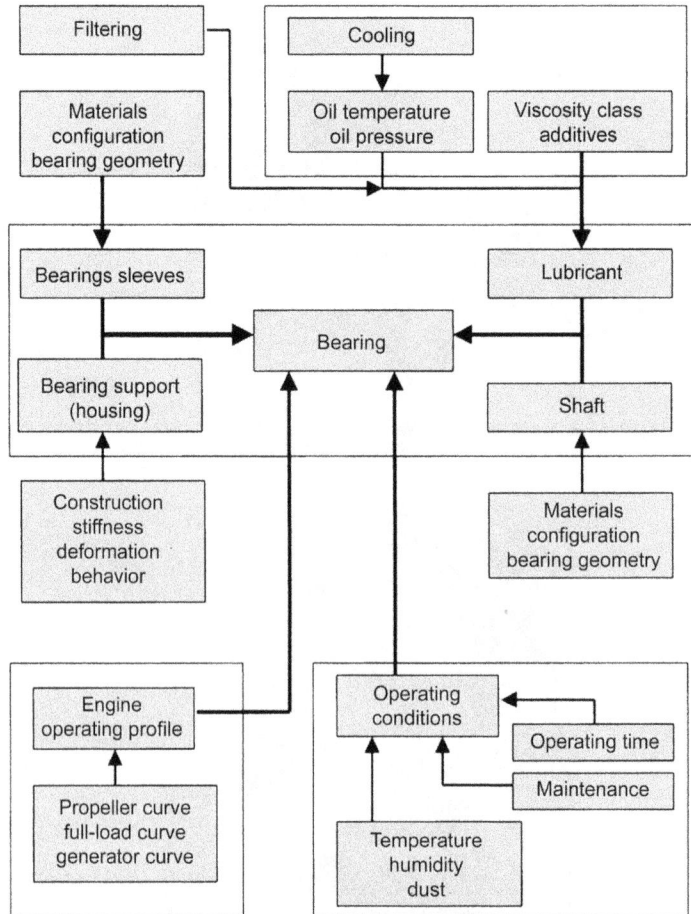

Fig. 6.179 Bearing complex.

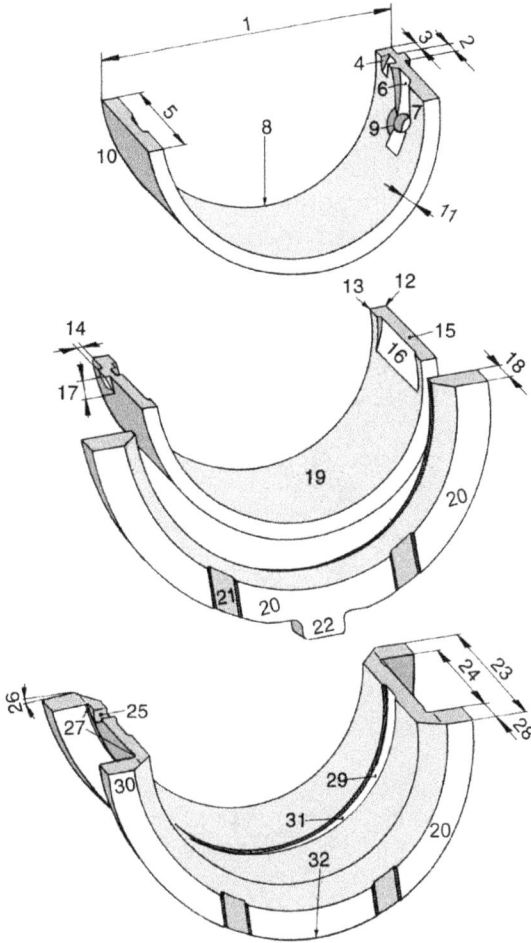

Fig. 6.180 Sleeve bearing terminology: 1, Free spread diameter; 2, Lug width; 3, Lug offset; 4, Lug notch; 5, Bearing length; 6, Partial annular oil groove; 7, Crush relief; 8, Bearing inside diameter; 9, Squirt hole with countersink; 10, Bearing backing; 11, Bearing wall thickness; 12, Outside corner; 13, Inside corner; 14, Lug projection; 15, Parting line; 16; Spreader groove; 17, Lug length; 18, Thrust washer thickness; 19, Running surface; 20, Thrust face; 21, Oil groove; 22, Locating tab; 23, Overall length; 24, Straddle; 25, Locating hole; 26, Parting line relief; 27, Relief; 28, Flange thickness; 29, Oil hole; 30, Flange face thrust relief; 31, Annular oil groove; 32, Flange diameter. Source: Miba

with bearings, one speaks of a bearing *complex* (Fig. 6.179). This is represented as a multibranched chain of cause and effect, in which irregularities in one or several links can compromise bearing function to the point of causing severe damage or even failure of the entire engine.

Depending on the type of motion, friction, and force transmission, distinction is made between rolling element and journal or plain bearings. Plain (journal) bearings have come to dominate cranktrain bearing applications (wrist pin, connecting rod, and crankshaft), with few exceptions. Plain bearings (Fig. 6.180) are force fit, have small radial and larger axial dimensions, are noiseless, and are especially suited to impact loads encountered in engine cranktrains.

Plain bearings in engines take the form of thin-walled hollow cylinders, either as single-piece bushings (Fig. 6.181) or as split bearing shells. Oil passages and grooves serve to supply and distribute lubricant. Axial forces are taken up by flange bearings; for manufacturing reasons, straight bearings with split thrust washers are becoming increasingly popular. Bearings are located in their bores by appropriate lugs.

The concept of "bearing" has two meanings, as it can be understood to mean the entire bearing complex, consisting of bearing shells, bearing housing, shaft, and lubricant, or on the other hand to mean only the actual bearing itself; i.e., the bearing shells or bushing.

Bearing clearance is the difference between bearing inside diameter and shaft outside diameter (Fig. 6.182). If this is taken relative to the nominal bearing diameter, the result is the relative bearing clearance.

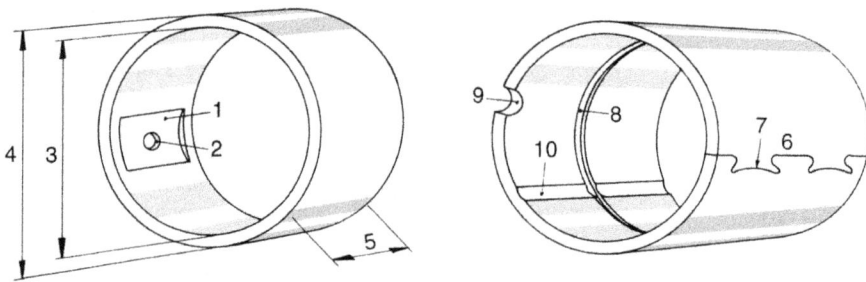

Fig. 6.181 Bearing bushings (single piece and split): 1, Oil hole; 2, Oil pocket; 3, Inside diameter; 4, Outside diameter; 5, Length; 6, Joint line; 7, Clinched seam; 8, Oil groove; 9, Locating notch; 10, Axial oil groove. Source: Miba

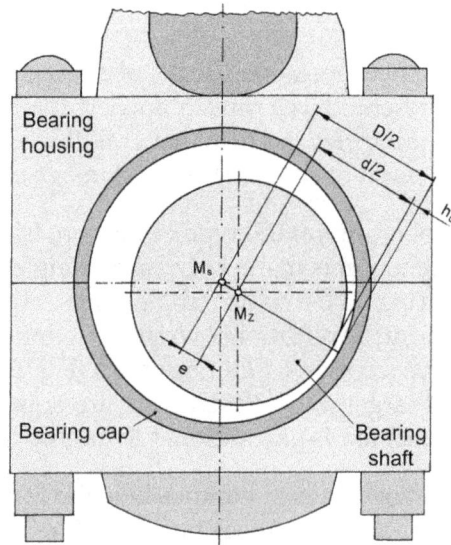

Fig. 6.182 Bearing clearance terminology: D, bearing inside diameter; d, shaft diameter; e, eccentricity; $s = D - d$ = bearing clearance; h_o, instantaneous minimum lubricating film thickness (bearing sleeve); MS, center (bearing sleeve); MZ, center (shaft); $\boxtimes = s/D$ = relative bearing clearance.

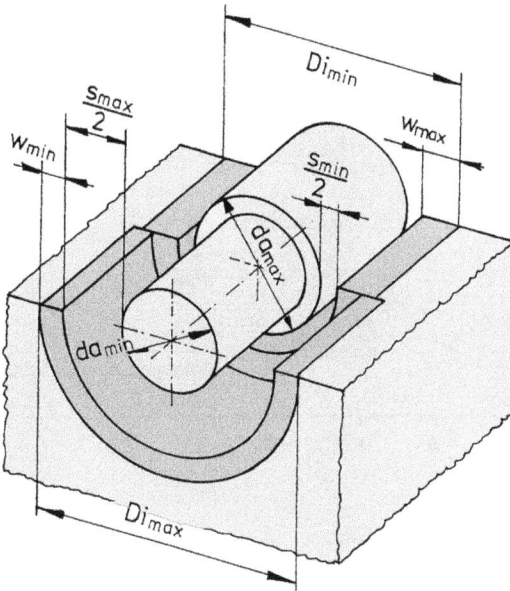

Fig. 6.183 Bearing clearance tolerances.

The (relative) bearing clearance is a vital parameter for bearing function. Because bearing shells, bearing housings, and shafts are subject to manufacturing tolerances, certain minimum clearances (Fig. 6.183) must be maintained (within an acceptable level of engineering complexity).

One can restrict bearing clearance tolerances by suitable pairing of bearing housing bore and bearing wall thickness. This requires selection and marking of bearing shells according to wall thickness. Theoretically, small clearances are advantageous, but in practice, certain minimum values must be maintained, otherwise friction losses will increase dramatically and—a more serious problem—the crankshaft will wedge itself into the tight bearings. Relative clearances for engine bearings are in the range of 0.8 to 1.6% (0.6 to 2.0%, see Fig. 6.184).

For the bearing to perform its functions, it must be firmly located (interference fit) in its housing bore. Even tiny relative motion will result in fretting corrosion, which can initiate fatigue fractures of the bearing shell and/or housing. In the event of bearing shells rotating in their housing bores, oil passages will be blocked, resulting in oil starvation and seized bearings.

This interference fit between bearing and housing is achieved by installing oversized bearings; i.e., the outside diameter of the bearing shells is larger, by a precise amount, than the receiving bore in the

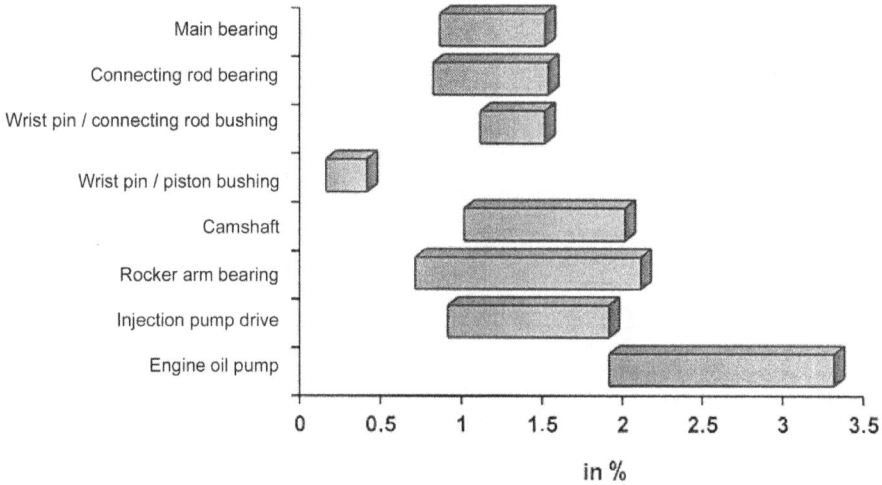

Fig. 6.184 Radial bearing clearances in a diesel engine.

bearing housing (Fig. 6.185). During bearing installation, two things happen:

- The housing bore expands, but not uniformly.
- Wherever the housing is stiff—in the connecting rod column direction for wrist pin bearings, toward the cylinder bank for the crankcase—the expansion is smaller (Fig. 6.186).
- Bearing shells are compressed tangentially. This tangential compression causes radial pressure that forces the bearing shells against the circumference of the housing bore.

Tangential stress in the bearing shells is limited:

- A lower limit is imposed by the radial pressure needed to prevent relative motion between bearing shell and housing.
- An upper limit is imposed by the maximum permissible plastic deformation of the bearing shell (about 0.01 % of the remaining compression in the circumferential direction).

Because the oversized diameter of bearing shells cannot be measured (due to their elastic spread), instead their so-called indicator crush dimension (excess circumference) is determined. The bearing shell is installed under light pressure in a test fixture. The height of the protruding bearing shell end is the indicator crush dimension (Fig. 6.187).

Although journal bearings are "age-old" machine elements, their actual function was not understood until relatively recently. In particular, the fact that the thin film of lubricant, with a thickness of only a few picometers, transmits the same forces as a massive connecting rod or crankshaft, was a difficult concept to grasp.

At rest, the shaft rests on the bearing shell. Bearing and shaft centers are at their geometrically greatest possible separation (eccentricity). Once the shaft begins to turn, a wedge of lubricant adhering to bearing and shaft is pulled into the narrowing gap. This builds up pressure in the lubricating film (see Fig. 6.188), in equilibrium with the force imposed. In the longitudinal direction, a parabolic pressure distribution is generated.

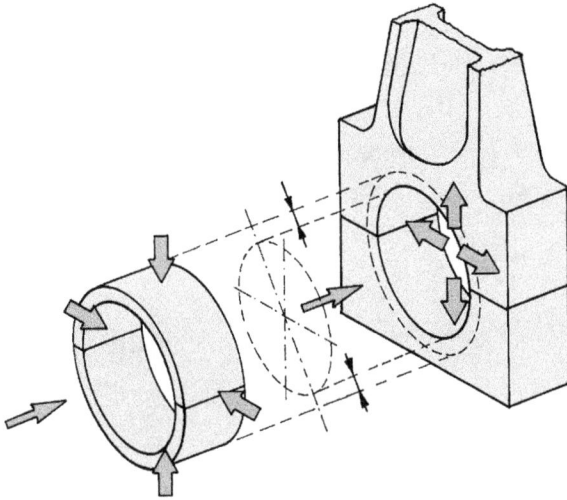

Fig. 6.185 Schematic of bearing and bearing housing press fit.

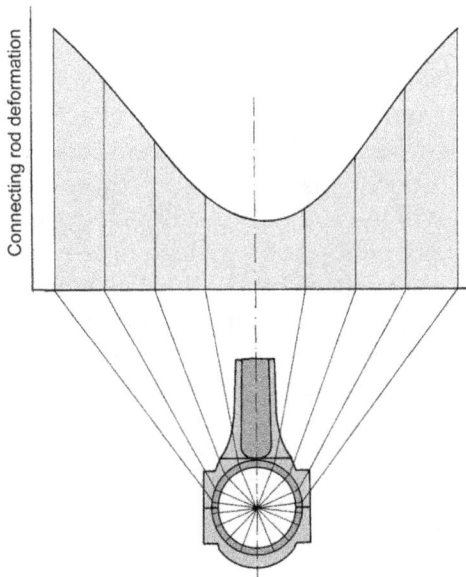

Fig. 6.186 Deformation of a connecting rod bore as a result of bearing installation. The connecting rod bore expands elastically, as a function of local stiffness.

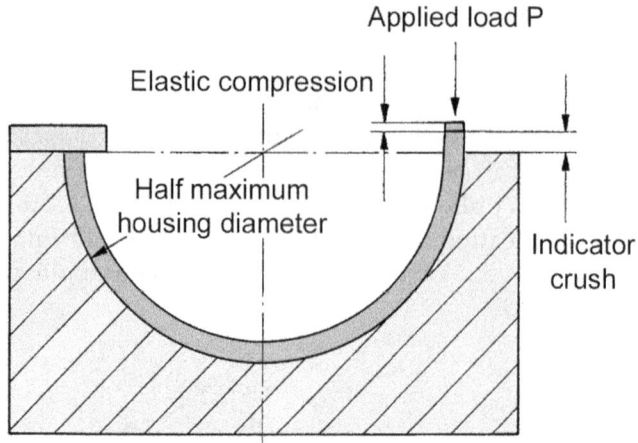

Fig. 6.187
Measurement of bearing shell in test fixture.

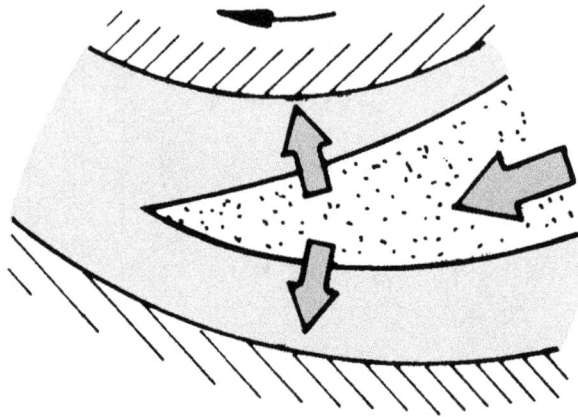

Fig. 6.188 Lubricating wedge (schematic).

The shaft center changes its position and, given constant bearing force and rotational speed, assumes a certain defined position. This is termed the rotation-induced load-carrying mechanism. However, cranktrain bearings are subjected to non-steady-state loads; i.e., bearing forces change in magnitude and direction over the course of a working cycle. The shaft therefore not only rotates (as under stationary loads), but it also shifts radially. Additional pressure builds up in the lubricant that is displaced as a result of this shift, with the magnitude of the pressure increase a function of the displacement velocity; this results in the displacement-induced load-carrying mechanism (Fig. 6.189).

The rotational and displacement load-carrying mechanisms add up to a resultant load-carrying force (Fig. 6.190).

Pressure distribution in lubricating film
rotational load carrying mechanism

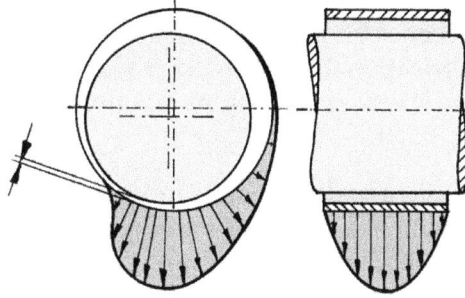

Pressure distribution in lubricating film
displacement load carrying mechanism

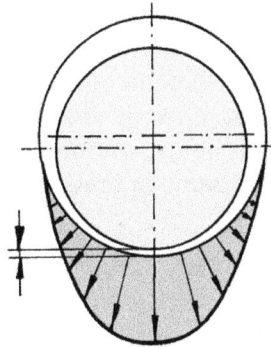

Fig. 6.189 Load-carrying
mechanism in cranktrain
bearings.

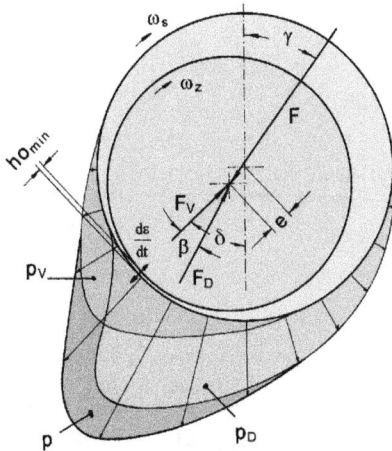

Fig. 6.190 Load-carrying mechanism for
hydrodynamic lubricant pressure in a journal
bearing with non-steady-state loading: F,
bearing force; F_D, rotational load-carrying
force; F_V, displacement load-carrying force;
p_D, lubricating film pressure as a result of
rotation; p_V, lubricating film pressure as a
result of displacement; p, resultant lubricating
film pressure; e, eccentricity; $h_{O\,min}$, minimum
lubricant clearance; $d\varepsilon/dt$, displacement speed;
ß, displacement angle; δ, angle of acting force;
ε, relative eccentricity; ω_s, angular velocity in
bearing shell; ω_z, angular velocity of shaft.

Several different friction conditions are encountered in journal bearings. At rest, the bearing is subject to static friction. Once the shaft begins to turn, this is replaced by boundary or thin-film lubrication. With rising relative speed between shaft and bearing, this in turn is replaced by mixed or semifluid lubrication, which finally transitions into hydrodynamic lubrication. These relationships are illustrated in the so-called Stribeck curve (Fig. 6.191).

Journal bearings operate in the hydrodynamic lubrication regime, even though on startup of a machine, a transition through the wear-inducing mixed lubrication regime is unavoidable. Because the bearing force changes magnitude and direction in the course of a working cycle, this force is not plotted in a Cartesian (orthogonal) coordinate system but rather in a polar diagram, as radius vectors radiating from the origin (pole), indicating force direction and magnitude (Fig. 6.192).

The coordinate system may be shaft-based or bearing-based. If bearing-related processes are of interest, a bearing-based coordinate system is chosen; if instead conditions at the shaft are of primary interest, a shaft-based system is advantageous, e.g., if lubricating film pressure in the vicinity of crankpin oil passages is being examined. Shaft- and bearing-based diagrams describe the exact same operational phenomena, but represent them differently.

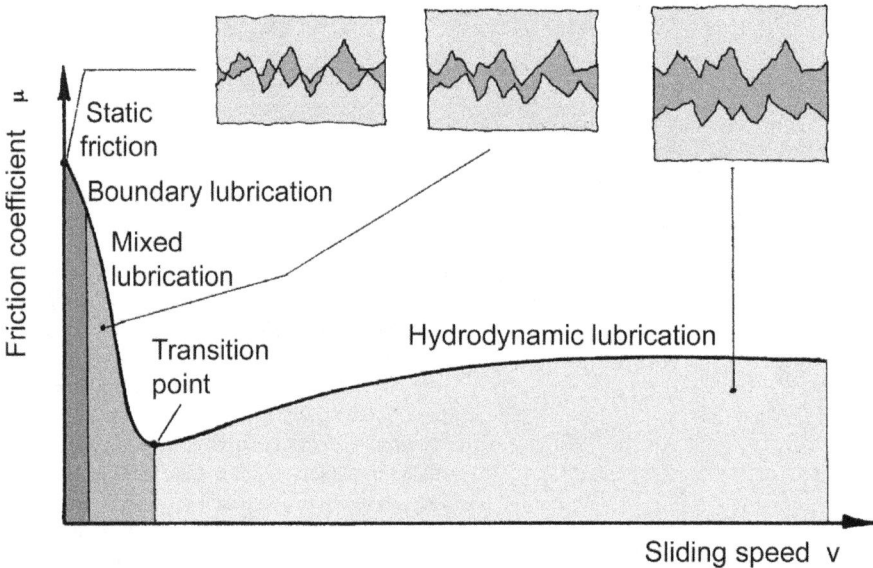

Fig. 6.191 Stribeck curve: friction coefficient as a function of sliding speed.

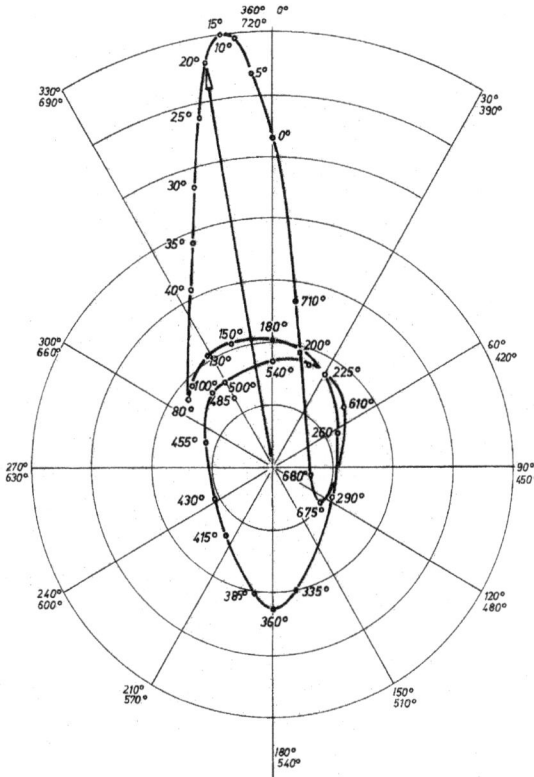

Fig. 6.192 Polar diagram of crankpin bearing force of a four-stroke diesel engine.

The shaft wanders in the bearing in response to bearing forces. The path of the shaft center is termed the shaft displacement path. The distance between shaft and bearing bore centers (the eccentricity) is continuously changing, and with it the position of the smallest clearance between shaft and bore, the minimum lubricant clearance $h_{O\,min}$. This is on the order of a few thousandths of a millimeter (µm, microns), and is therefore in the range of the roughness peaks of shaft and bearing (Fig. 6.193). Such tiny clearances are necessary to maintain high lubricating film pressures (in excess of 1000 bar) to counteract the forces acting on the bearing. The small gap acts as a labyrinth seal, preventing the lubricating film pressure from reaching equilibrium with atmospheric pressure in the crankcase. Magnitude and orientation of the minimum lubricant clearance $h_{O\,min}$, in other words the shaft displacement path, determine the operational behavior of a journal bearing.

Evaluation criteria for shaft displacement diagrams are:

- Absolute dimension of the minimum lubricant clearance $h_{O\,min}$
- Magnitude of the regimes of small lubricant clearances

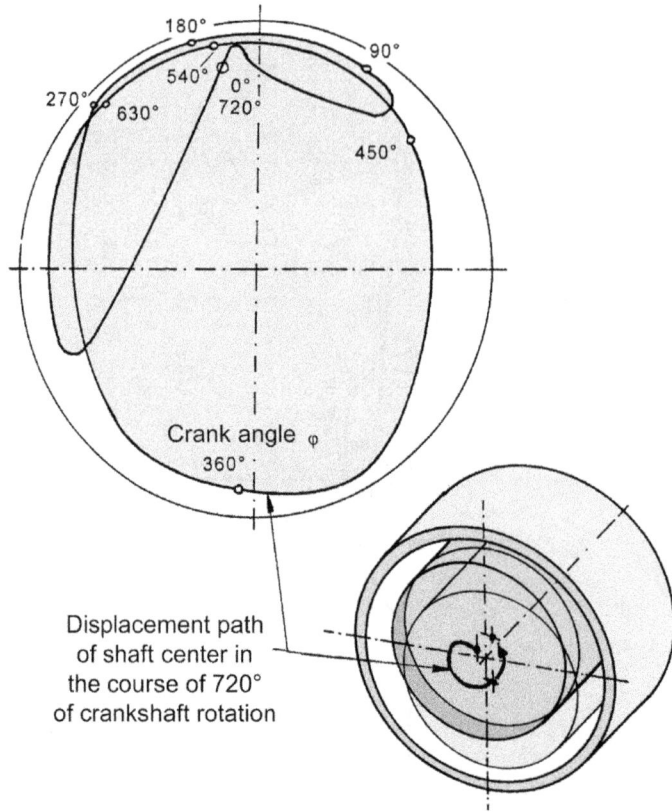

Fig. 6.193 Shaft displacement path (bearing-based coordinates) for the crankpin bearing of a four-stroke diesel engine.

Displacement path of shaft center in the course of 720° of crankshaft rotation

- Magnitude of path velocity in the direction of rotation, and opposite to rotational direction
- Possible sudden directional reversal away from the bearing shell (represents danger of cavitation)

The operational behavior of a bearing is determined by bearing force F, bearing dimensions (diameter d and width b), bearing clearance ψ, dynamic viscosity η, and relative angular velocity ω. These parameters may be expressed as a dimensionless quantity, the Sommerfeld number, which describes similitude of bearings; given identical Sommerfeld numbers and identical geometric conditions, two bearings will exhibit the same hydrodynamic behavior, with identical size and location of the minimum lubricant clearance.

$$So = \frac{F \cdot \psi^2}{b \cdot d \cdot \eta \omega}$$

With the aid of the Sommerfeld number, the effects of various parameters on bearing behavior can be examined. Bearing force and the square of bearing clearance are in the numerator, while bearing dimensions, lubricant viscosity, and relative angular velocity are in the denominator. It should be remembered that the parameters which make up the Sommerfeld number are not mutually independent. For example, large bearing clearances may be unfavorable in terms of load-carrying capacity, but permit higher oil flow rates, which result in lower bearing temperatures and therefore higher oil viscosity. This in turn benefits load-carrying capacity.

Bearing force F is the resultant of gas and inertia forces, and therefore is dependent on the engine load condition. Changing the engine operating mode (propeller curve, generator curve, full-load curve) also changes the proportion of gas and inertia forces relative to one another, and therefore also the bearing behavior. Because operating on the generator curve represents operating at a constant speed, the inertia force component remains invariant; in this mode, lubricating film thickness at part load is hardly any greater than at full load. This explains the different behavior of bearings in engines of the same type, operating in different applications (Fig. 6.194).

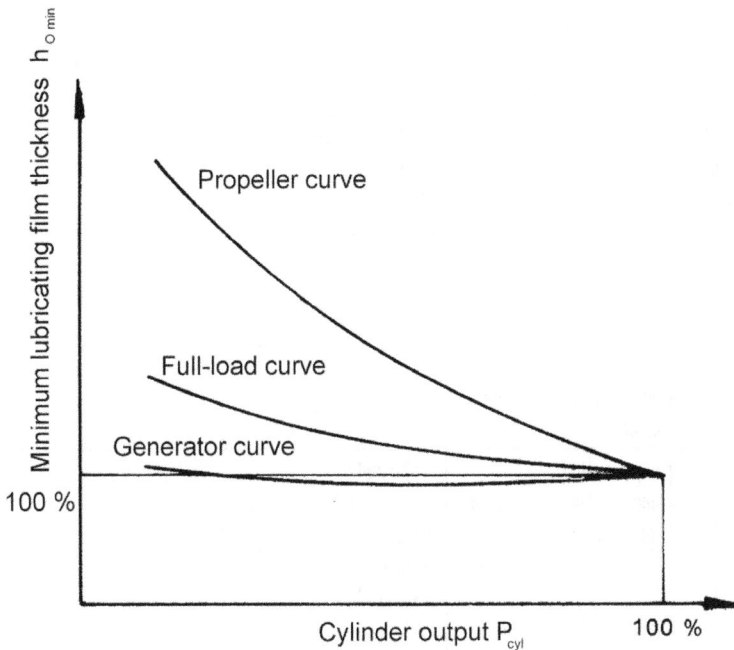

Fig. 6.194 Effect of engine operating mode on minimum lubricating film thickness (schematic).

The demands on bearing materials for highly loaded cranktrain bearings include:

- Good sliding and boundary lubrication properties
- Good wear properties
- Durability
- Ductility
- Embeddabilty of foreign bodies
- Corrosion resistance

Because these cannot be satisfied by a single material, the tasks are divided among two or more layers of different materials, to form so-called composite bearings (see Table 6.6). Modern cranktrain bearings are universally highly loadable trimetal bearings, consisting of a steel backing, the lining, a barrier layer, the overlay, and a layer of flashing. (The barrier and flashing layers aren't counted, hence the "trimetal" designation; see Fig. 6.195).

Table 6.6
Cranktrain Bearing Materials

Layer designation(s)	Function	Material	Layer thickness
Backing	Support forces, transmit forces to housing	C10	1 – 15 mm, depending on bearing size
Lining, bearing metal, bearing alloy, intermediate layer	Match bearing to shaft	Lead-bronze alloys, CuPb 22 Sn 1.5 Aluminum alloys, AlSn 6 CuNi AlSn 20 Cu Aluminum-zinc alloys, AlZn 4.5 Si	0.5 – 1.2 mm
Barrier	Prevent tin diffusion from overlay to intermediate layer	Ni, NiCr	2 – 3 µm
Overlay, galvanic layer	Improve shaft run-in in bearing, and embed small foreign particles	White metal: PbSn 10 Cu 2 PbSn 12 Cu 2 PbSn 14 Cu PbSn 18 Cu 2 PbSn 10 Cu 5 SnSb 7	20 – 60 µm
Corrosion protection layer, flashing	Prevent bearing corrosion prior to installation	Sn, PbSn	3 µm

Tin flashing, 0.003 mm

Galvanically applied overlay PbSnCu
0.020 to 0.060 mm

Nickel barrier
0.002 to 0.003 mm

Cast lead-bronze lining G-CuPb22Sn
0.5 to 1.2 mm

Steel backing C 10
5 to 15 mm

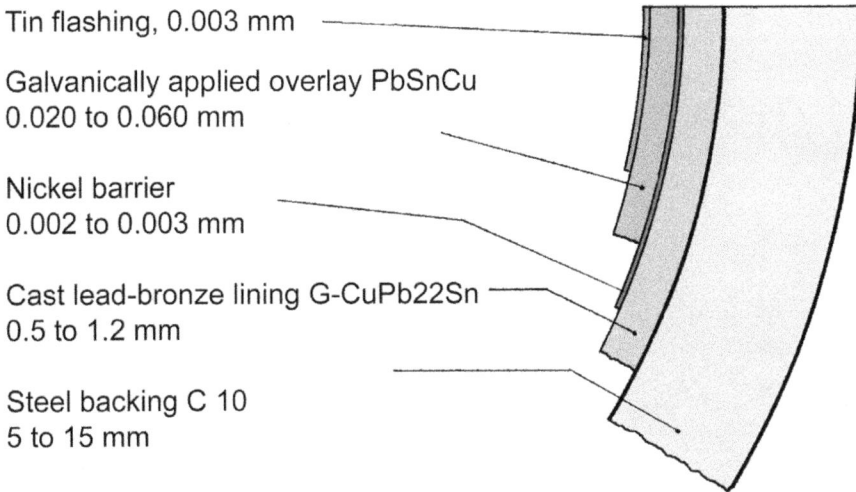

Fig. 6.195 Composition of a trimetal bearing.

Lead-bronze bearings are made by casting a lining of bearing material onto the steel backing. This manufacturing process represents a decisive step on the road toward highly loadable cranktrain bearings. Aluminum alloys are plated or rolled onto the backing, providing added cold reinforcement of the backing shell. The barrier, overlay, and flashing are applied galvanically (plated). Detailed bearing material specifications, their composition, and properties may be found in bearing suppliers' literature.

Engine power increases by means of supercharging leads to "fatter" p-V diagrams, and fuel economy improvements to higher peak pressures. These factors cause bearing loads to rise, because the duration of the period of thin film lubrication occupies a larger portion of the total cycle. Experience has shown that this causes bearings to approach, and exceed, their load limits:

- Minimum lubricant film thickness of 0.002 mm (2 μm)
- Maximum lubricant film pressure 2500 bar
- Circumferential speed 1 m/s
- Average specific thermal load 0.15 W/mm²
- Bearing temperature 140°C

For these reasons, new bearings were developed, using new materials, configurations, and manufacturing processes.

Overlay (~75% of surface area)

Aluminum alloy
(~25% of surface area)

Nickel barrier
(max 5% of surface area)

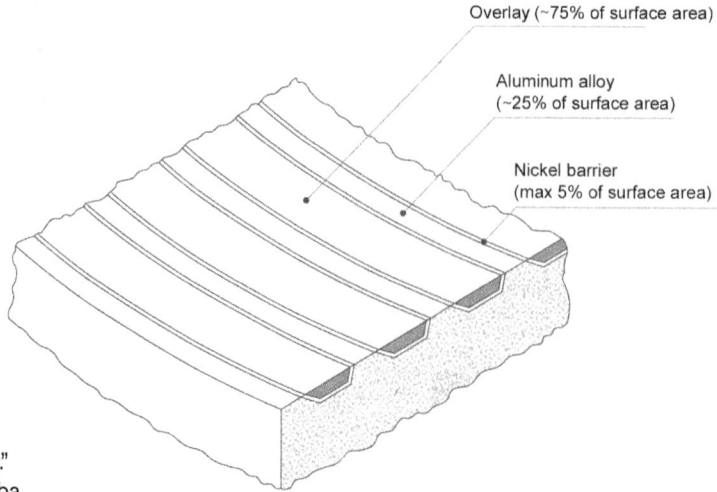

Fig. 6.196
"Rillenlager."
Source: Miba

- "Rillenlager" (from the German word for "grooved bearings," Miba,
  Fig. 6.196)
  The bearing has an intermediate layer with circumferential grooves
  (groove depth 0.025 mm). Lands in the intermediate layer provide
  lateral support for the overlay without sacrificing embeddability in
  the circumferential direction.

- Sputtered bearings (PVD, or physical vapor deposition, bearings),
  Fig. 6.197.
  An intermediate layer of AlZn 4.5 is plated onto the steel backing.
  The AlSn 20 overlay (without a nickel barrier) is sputtered; that is, a
  fine crystalline coating is applied by cathode atomization, onto the
  intermediate layer. Another option is to sputter a 0.030 mm thick
  AlSn 20 layer directly onto the backing.

Fluid friction within the lubricant converts mechanical work into heat,
which must be removed. Permissible bearing temperatures depend on
engine size and type. For motor vehicle engines, the temperature of
oil exiting the bearing may reach 150°C, for diesel engines of the size
installed in railway locomotives, about 100°C, and for large marine
engines, well under 100°C.

Oil is supplied by positive displacement pumps; in the case of direct-
drive pumps, flow rate increases linearly with engine speed. However,
this is not the case for bearing oil demand. At idle, a minimum supply
of oil must be assured, while excess oil must be diverted at rated engine
speed. Recently, so-called two-stage or sequential pumps have been

used, with one stage supplying oil in the low speed regime, and the other stage activated for the high speed range.

Because oil supply is critical for the life of an engine, it must be monitored. In general, oil pump back pressure serves as a monitoring parameter (Fig. 6.198). This is not without its problems, as not the physically relevant parameter (oil volume flow rate) but rather a dependent parameter, pump back pressure, is pressed into service for monitoring purposes. For one, pump pressure increases as the square of flow velocity (which represents volume flow rate). Second, pressure depends on flow restrictions. If an oil passage is constricted or plugged, flow resistance rises; even though less oil is flowing through the line, pump pressure does not drop. If restrictions are reduced as a result of increased bearing clearances, more oil flows through the bearing, but pressure drops and erroneously signals "inadequate oil supply."

6.2.5.1 Bearing damage

The function of a journal bearing, to transmit forces and reduce friction, is subject to so many influences that frequent appearance of bearing damage should come as no surprise. Regardless of how damage was initiated or propagated, in the majority of cases, damage results from a breakdown of the lubricating film. One should be aware that the function of a bearing depends on a film of oil whose thickness often amounts to only a few thousandths of a millimeter. It is easy to imagine that this is extremely sensitive to all manner of anomalies. The scope of bearing problems becomes clear when one examines the extensive failure catalogs published by bearing and engine manufacturers, and insurance organizations [6-11, 6-12, 6-13, 6-14, 6-15, 6-16], special literature covering journal bearing damage [6-17, 6-18], indeed the degree of detail described by the specialist literature [6-19], above all, the fact that there is even a standard for bearing damage: DIN 31661, *Plain Bearings; Terms; Characteristics and Causes of Changes and Damage.*

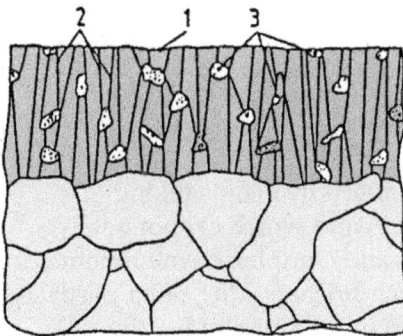

Fig. 6.197 Schematic representation of a sputtered AlSn layer: 1, AlSn20 layer structure, consisting of small, ~ 3 μm columnar structure, perpendicular to substrate; 2, with tin inclusions; 3, after Miba patent. Source: Miba

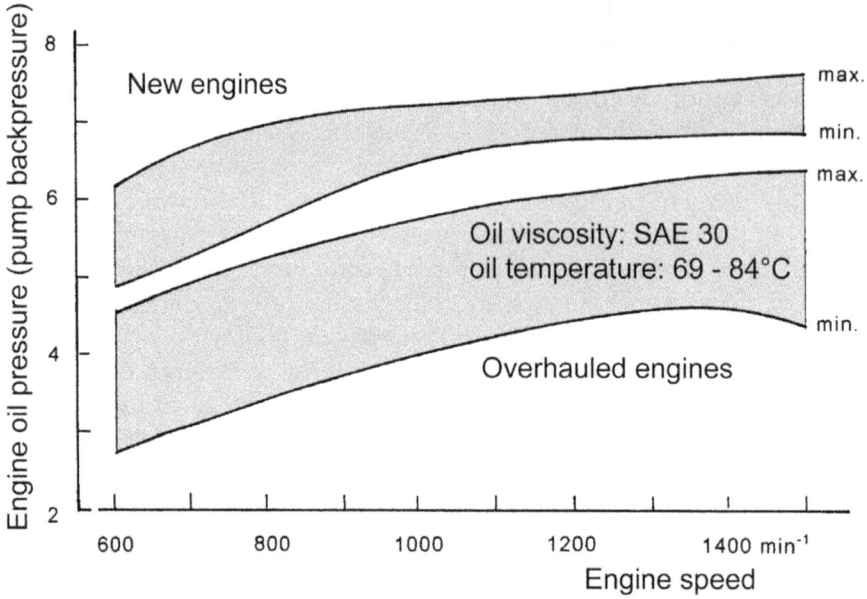

Fig. 6.198 Engine oil pressures for new and overhauled engines.

A listing of journal bearing damage from the above references highlights the following points:

- Damage to bearing material and overlay:
  - Dirt
  - Wear
  - Fatigue
  - Cavitation
  - Erosion
  - Corrosion
  - Electrical conduction
  - Bond failure

- Backing and bearing support damage
  - Fretting corrosion
  - Fatigue fracture
  - Assembly defects

Many instances of bearing damage are caused by events taking place well outside the actual bearing itself, and which cannot be eliminated despite the most painstaking and complex development and manufacturing processes. Particulates in lubricant—dirt, wear particles, carbon particles, dust, fibers, etc.—are often the cause of bearing damage.

Assembly errors, distortion of the bearing bores, bearing alley, crankshaft, shape and location anomalies of cranktrain components, as well as inadequate lubrication also play their parts. An overview of failure causes provided by an American diesel engine manufacturer attributes their various contributions as follows:

- Dirt-related failures, 45%
- Lubricant and related failures, 25%
- Installation and operating errors, 15%
- Bearing surface reactions, 15%

6.2.5.2 Wear

Bearings are subjected to wear, as are any machine parts in relative motion. Bearings react to irregularities of any sort; these may occur singly or in combinations:

- Overloading
- Compromised bearing geometry as a result of manufacturing or assembly errors
- Unsuitable or aged lubricant
- Dirt and foreign bodies in lubricant
- Lack of lubricant

Depending on the cause(s), indications of wear appear at specific locations in the bearing, characteristic of the causative agent. Wear progression, and therefore the extent of wear, is of decisive importance. Break-in wear assures matching of the two mating surfaces to one another. When starting and stopping the engine, bearings pass through the mixed lubrication regime, so that over time, visible material removal takes place. Consequently, bearings, like pistons, exhibit wear patterns (Fig. 6.199).

Normal wear primarily occurs as surface-wide material removal in the most heavily loaded zone. For main bearings, this is generally the lower bearing shell (Fig. 6.200). For connecting rod (crankpin) bearings, this is the upper (for diesels) or lower (for Otto-cycle engines), Fig. 6.201. First the break-in layer (the tin flashing) is carried away, then wear of the overlay begins. The lining wear zone is a dull gray color, while wear in the overlay has a silvery color. If a "healthy" bearing is left undisturbed, i.e., it is not removed (for example, in the course of piston removal), even bearings with extensive, prominent wear zones can achieve long service lives.

Fig. 6.199 Normal wear patterns of well broken-in bearings.

Fig. 6.200 Wear patterns of long-running main bearing shells from a supercharged diesel engine.

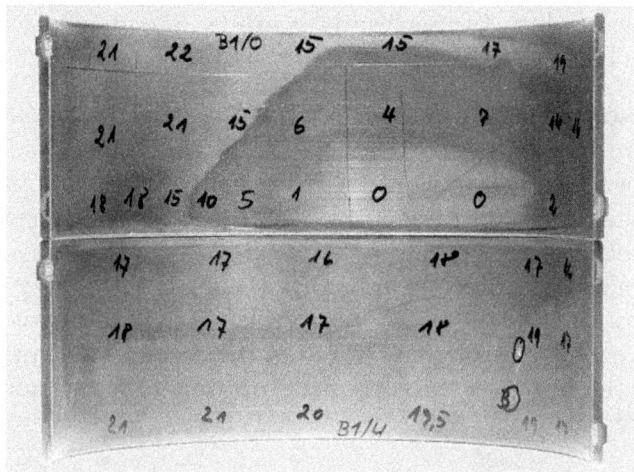

Fig. 6.201 One-sided wear pattern of a long-running connecting rod bearing. The numbers represent the remaining overlay thickness in μm.

Fig. 6.202 Light, locally limited scuff marks in a diesel engine main bearing.

Locally limited zones of visibly greater wear are termed *scuff marks* (Fig. 6.202). These may for example be the result of partial contact as a result of abnormalities in cranktrain geometry: out-of-plumb bearing bores, twisted connecting rods, or deformed crankpins.

Over the course of the engine's operating life, bearing wear increases, until the risk of scoring and the resulting seizing becomes incalculable. Reinstallation specifications indicate permissible wear limits for the overlay, with which a bearing may still be re-used (Fig. 6.203).

If the aforementioned scuff marks grow, they may develop into *scoring marks* (Fig. 6.204).

Bearings may still "recover" from scoring marks if their cause is removed, or if scoring only extends to the overlay. If scoring reaches deeper, into the lining, then scoring is a precursor to a *seized bearing*. In the event of severe overloading—regardless of its cause—the bearing first operates in the mixed lubrication regime, then in boundary lubrication; temperature rises, lubricant vaporizes locally, there is local welding of bearing to journal surface, followed by tearing. This process is self-amplifying and usually operates exponentially. The power required for seizing is easily provided by the engine, which allows seizing to progress to the point of complete bearing destruction—which makes determination of the actual cause of failure after the fact difficult at best, and often impossible (Fig. 6.205).

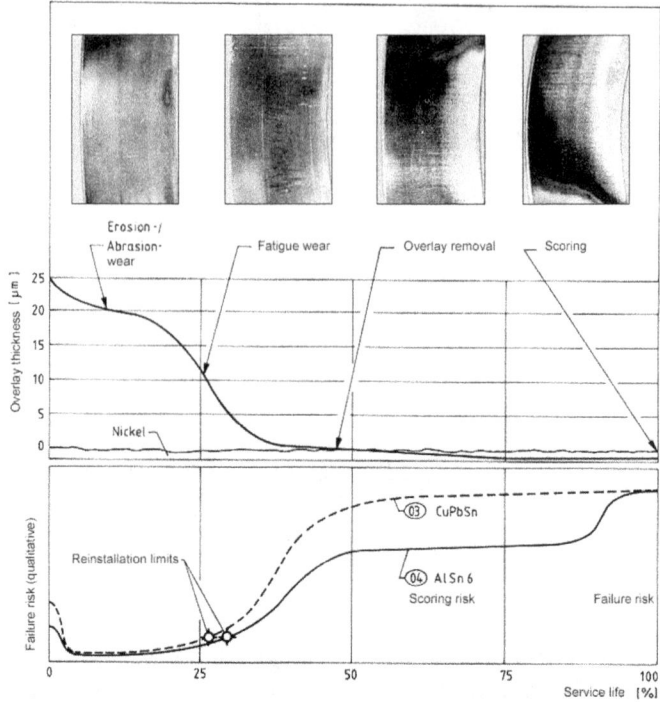

Fig. 6.203 Bearing wear over the course of usable service life. Source: Miba

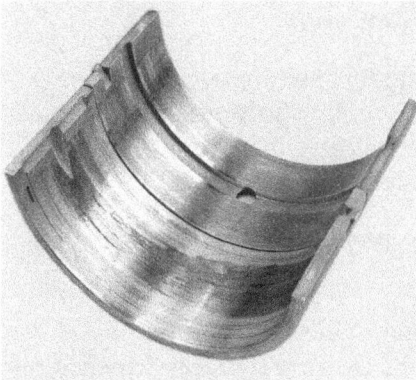

Fig. 6.204 Scuffed bearings.

Fig. 6.205 Severe seizing, completely destroyed bearings.

Contrary to appearances, the bearings in Figs. 6.206 and 6.207 were not originally flanged bearings, but rather normal straight bearing shells, which were physically rolled out in the course of bearing failure. The impressions of crankshaft fillets are clearly visible.

The extent and appearance of bearing failures can be quite remarkable, such as this case, which resulted from oil starvation. Because it was not correctly installed, a crankshaft relief bore plug had loosened. As a result, oil was flung out directly through the open bore, starving two adjacent connecting rod bearings on one crankpin (V engine, two connecting rods per crankpin) of oil. Bearing seizure developed rapidly and completely; one of the bearing shells was severely rolled out (see Fig. 6.207).

The heat of friction in the seized bearings melted the lining (AlSn 6) and allowed it to flow into the crankpin oil passages (Fig. 6.208), where it solidified once the engine was shut down and cooled (Fig. 6.208).

6.2.5.3 Contamination

Foreign bodies in lubricating oil (colloquially known simply as "dirt") represent a universal problem for combustion engine technology. This is especially true for cranktrain bearings, with their exceedingly thin oil films. Dirt has many sources, and accompanies an engine "from cradle to grave."

Initial dirt is understood to mean particles originating from engine manufacturing and assembly: foundry core sand, machining chips, weld splatter, grinding dust, scale, paint, and the omnipresent lint from cotton waste. Oil passages must be 100% inspectable for cleanliness. If they are cast into the crankcase, inspection access can only be achieved by means of machining operations; otherwise, there is no way to ensure that any remaining foundry sand (adhering to the casting skin) does not shake loose during engine operation and enter the oil stream.

Fig. 6.206 Total failure of a bearing; what could have caused this?

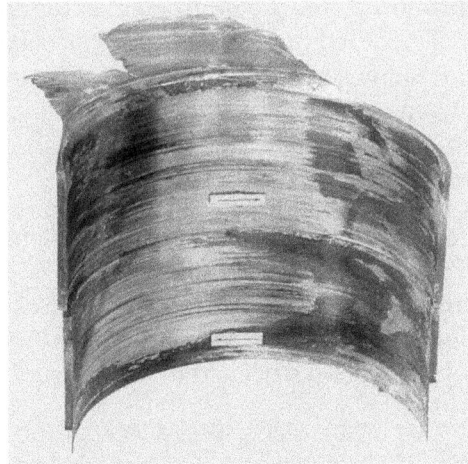

Fig. 6.207 Advanced bearing failure with squeezed-out bearing shell.

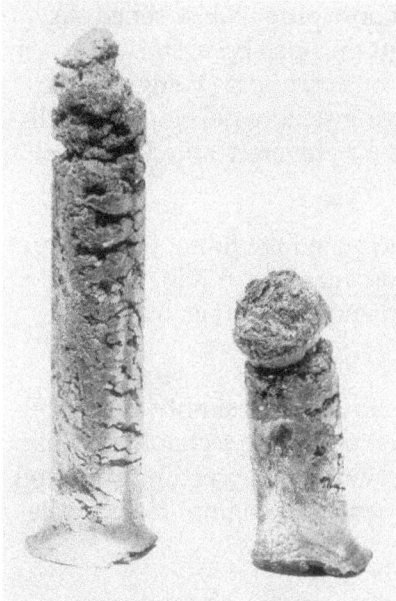

Fig. 6.208 Bearing metal forced into oil passages in the course of a bearing failure.

Fig. 6.209 Bearing metal forced into oil passage in the course of a bearing failure, still attached to the bearing itself.

Engine operation produces soot, carbon, lacquer-like substances, but also metallic particles and other wear products. Abnormalities in engine operation increase these contaminants as a result of defective injector nozzles and poor injection patterns, long idling periods, operation at low temperatures, water entry into the cylinder, and oil dilution. Inadequate filtration of air and lubricating oil, postponement of scheduled maintenance, but also extreme ambient conditions (desert operations) promote oil contamination by two mechanisms: simple increase of dirt entrained in lubricant, plus the resulting increase in wear to all moving parts of an engine, with the corresponding fallout of wear particles.

Foreign particles are carried to bearings by the lubricating oil stream. If they are small enough, they can be embedded in the overlay. Such embedding is usually recognizable as dark points, surrounded by a lighter ring. When they are pressed in, particles throw up a crater, which is soon flattened by the journal (Fig. 6.210).

If the particle size exceeds the overlay thickness of about 0.020 mm, then the particles will be carried along by the oil flow, dragging furrows as they go, until finally they are partially or fully embedded in the overlay and lining, or they are flushed out of the bearing in areas of greater oil film thickness.

If only a few gouges are present, the bearing will be unaffected; however, more extreme scoring will degrade bearing function. The oil film will be compromised, the bearing will begin to operate with local mixed lubrication, there will be scuffing, and then finally seizing will occur.

In general, main bearings are more susceptible to damage from dirt (depending on oil passage configuration), because main bearings are exposed to all particles entrained in the oil flow; connecting rod bearings, on the other hand, are primarily exposed only to smaller particles, because larger particles are centrifugally trapped in the crankpin bores and form deposits in the "stagnant" areas of the crankpin bores, which are often necessitated by design (Fig. 6.211).

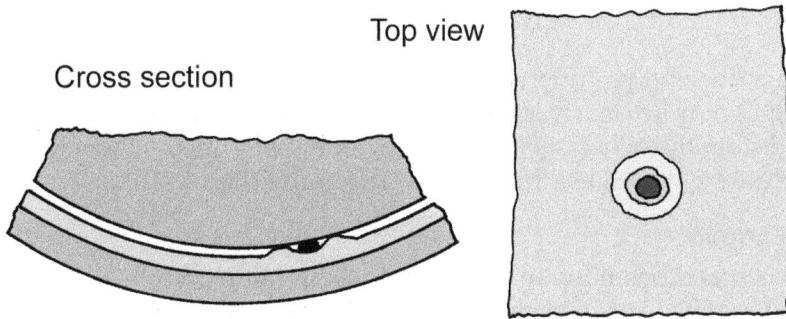

Fig. 6.210 Origin of foreign-body scoring as a result of embedded bodies in the bearing overlay (schematic). A characteristic crater ring is thrown up by embedding of such particles. This is to some extent worked down by the journal, so that when seen from above, it appears as a bright, circular ring.

Fig. 6.211 Oil sludge deposits in a diesel engine crankpin relief bore, after several hundred hours of operation.

- *Scarred bearing surface as a result of embedded small particles*
  Example: During examination of the bearings of a diesel engine, an apparent crack was observed in the lining of one bearing. Ultrasonic cleaning removed oil residue from the "crack," which could then be seen to actually be an impression, caused by a foreign body (Figs. 6.212 and 6.213).

- *Scoring originating from cavitation damage below the bearing parting line*
  Bearings may simultaneously be victims and perpetrators of their own damage, for example, when cavitation loosens particles that then flow downstream and plow furrows in the bearing overlay and lining (Fig. 6.214).

- *Circumferential scoring caused by large particles originating in a nearby seized bearing*
  Particles worn off a seized bearing also impact adjacent bearings, often to the point of total destruction (Fig. 6.215).

- *Failure of main bearings adjacent to a cylinder that has suffered piston seizing*
  Bearings are always imperiled by wear particles generated by piston seizing. It may be clearly seen that metallic wear particles have been carried from the oil passage into the lower bearing shell, where they have destroyed both bearing overlay and lining (Fig. 6.216).

6.2.5.4 Fatigue

Time-variant and positionally shifting lubricating film pressure imposes dynamic loads on bearing materials. Lubricating film pressure causes alternating tangential compressive and tensile stresses in the bearing metal, with stress levels largely determined by the lubricant

Fig. 6.212 Embedding in the overlay.

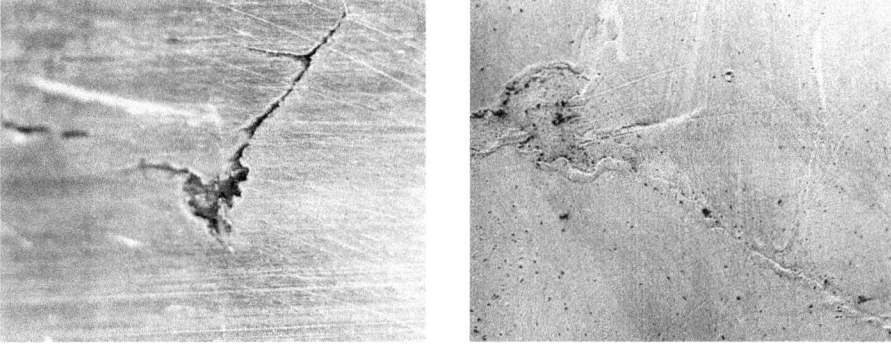

Fig. 6.213 Apparently cracked bearing lining is actually the impression of a foreign body.

Fig. 6.214 Scoring, originating from cavitation at the bearing parting line.

pressure gradient. This also explains why in general, connecting rod bearings will withstand higher loads than main bearings (Fig. 6.217)—a phenomenon that was recognized as early as the 19th century in steam engine cranktrains. Connecting rod bearings are primarily loaded by the displacement load-carrying mechanism (with its small increase in lubricating film pressure); main bearings on the other hand are dominated by the rotational load-carrying mechanism, with its one-sided, steep pressure rise.

Bearing materials fatigue as a function of load levels and frequency. Fatigue fractures will form if the fatigue strength of the bearing material is exceeded. As a consequence of tangential stresses, these appear as axial cracks. These imperil the loaded zone of a journal bearing in two ways: first, as a result of mixed lubrication in areas of minimum

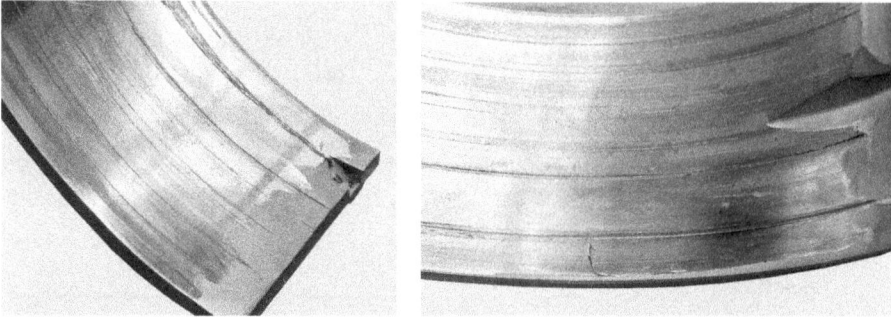

Fig. 6.215 Bearing scoring as a result of chips and wear from a seized bearing.

Fig. 6.216 Bearing damage as a result of extreme exposure to wear particles from a seized piston.

Fig. 6.217 Schematic comparison of connecting rod and main bearing loading. Connecting rod bearing: mainly displacement load-carrying mechanism, with gradual pressure rise. Main bearing: mainly rotational load-carrying mechanism, with steeper pressure rise.

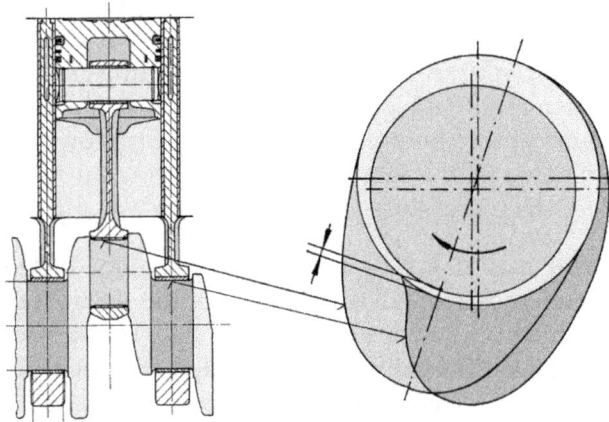

lubricant film thickness, and additionally, by the aforementioned bearing material fatigue. The onset of mixed lubrication causes bearing temperatures to rise, which in turn reduces material strength:

- Cracks in the overlay form a very fine branching network (crazing), which resemble the tracks of bark beetles under tree bark (Fig. 6.218).
- In the bearing lining, the cracks propagate radially until they reach the area of bonding with the steel backing, where the higher material strength in this area deflects them in a circumferential direction. If several cracks join, pieces of lining break out, giving a "cobblestone effect" (see Figs. 6.219, 6.220, 6.221, and 6.222).

6.2.5.5 Cavitation

If local pressure in the lubricating film drops below its vapor pressure, vapor bubbles will form. When these reach areas of higher lubricating film pressure, they collapse (implode; near walls, asymmetric pressure causes a funnel-shaped collapse), imposing locally high dynamic loading on the bearing material. The material fatigues, begins to crumble, and falls away. The bearings often appear, in the words of one experimental engineer, "as if rats have been chewing on them." Or put another way, highly localized forces cause the formation of craterlike, distinctly separated depressions in the bearing material (Fig. 6.223).

Cavitation in engine bearings is promoted by oil dilution and high oil temperatures. Even though the physical cause is always the same, cavitation in engine bearings—with their individual geometries, flow and pressure conditions, mechanical rigidity of shaft and bearing housing, as well as the shaft deflection path—may be initiated at different locations and with different manifestations (Fig. 6.224).

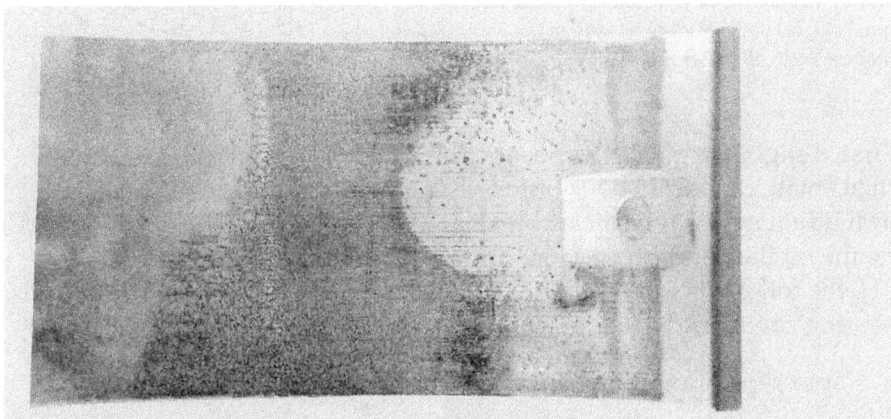

Fig. 6.218 Bearing overlay fatigue failure. Source: Miba

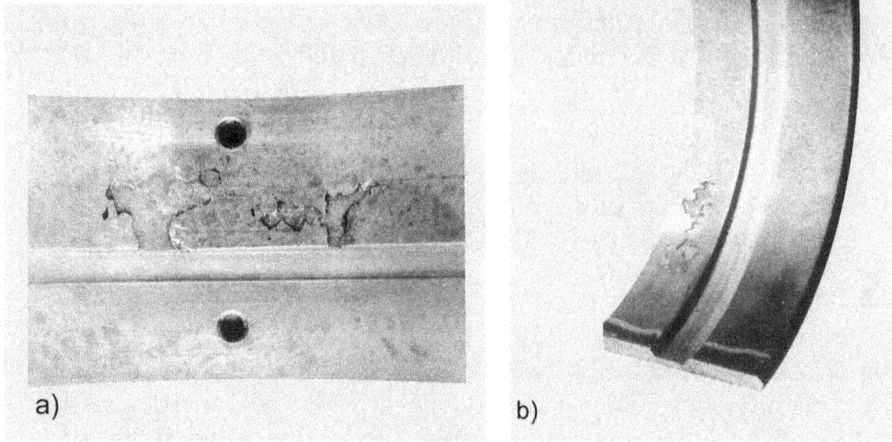

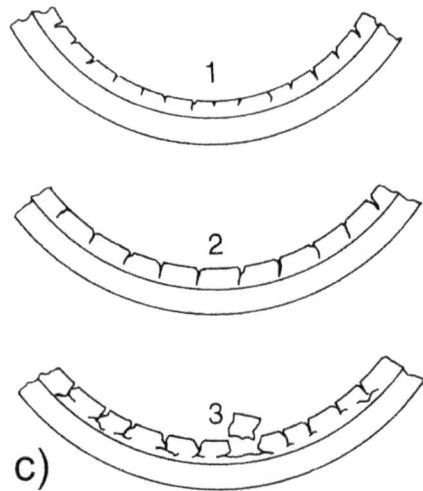

Fig. 6.219 Journal bearing overlay fatigue failure: (a) near the oil passage; (b) advanced breakout; (c) the individual phases of the "cobblestone effect." Source: Miba

First, a slight roughening appears, recognizable by a matte gray color, then small, randomly distributed, pitlike pores appear, and finally, individual material particles break out of the bearing. These are carried by the oil flow and cause more- or less-deep gouges. According to [6-21], the following effects influence the course of diesel engine cranktrain bearing cavitation:

- Rapid separation of the journal from the bearing shell
- Oil squeezed out at oil grooves

- Oil flow separation at the edges of oil grooves and pockets
- Oil pressure pulsations in the journal oil passage

Furthermore, modern terminology [see also 6-11, 6-19] draws distinction between

- *Suction cavitation* (Fig. 6.225, pressure relief cavitation) Because of its appearance, this is also known as "lancet cavitation" (Fig. 6.226). This is caused by rapid movement of the journal away from the bearing shell. The lubricant is not able to follow this rapid movement, and pressure drops below the lubricant vapor pressure. As the journal again approaches the bearing shell, the tiny vapor bubbles implode. Because of light loading and low static pressure,

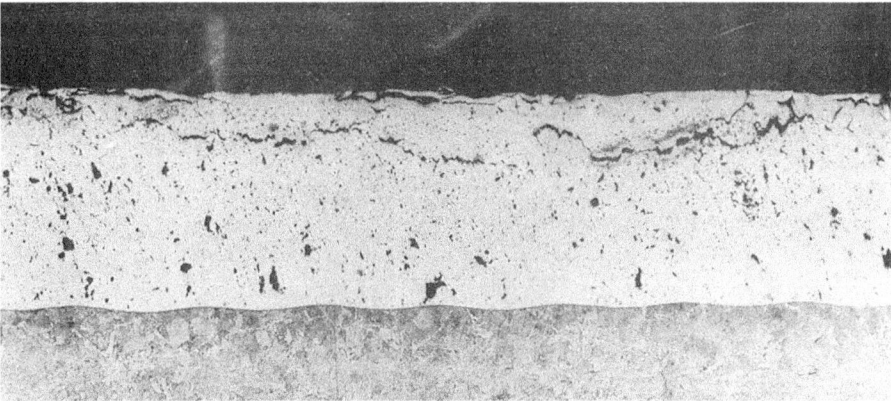

Fig. 6.220 Radial metallographic section (about 50:1 magnification) through a bearing lining. Cracks have formed below the overlay. Source: Daimler

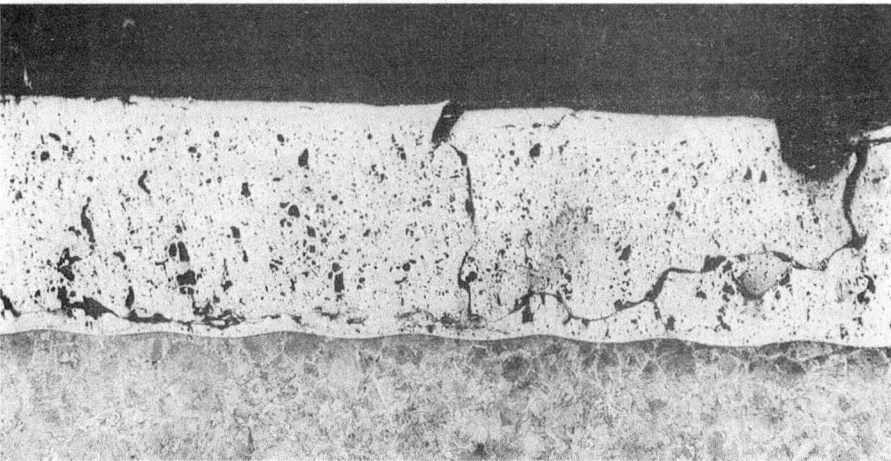

Fig. 6.221 Radial cracks are deflected in the bonding zone, and propagate circumferentially. Two lead-bronze fragments have already broken away. Source: Daimler

Fig. 6.222 Radial metallographic section (about 50:1 magnification) through the fatigue failure section of the bearing. The broken-out fragment, under continuous cyclic loading, forms hammered-out, smooth fracture surfaces. Source: Daimler

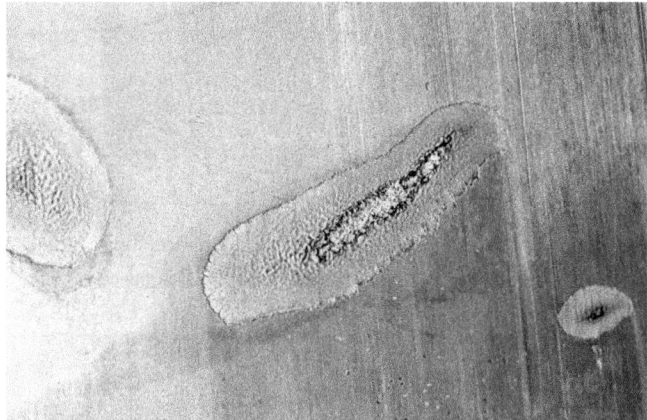

Fig. 6.223 Bearing shell cavitation.

this phenomenon is mainly observed in upper bearing shells. When oil flows into the bearing, it is distributed by an oil groove (if present) in the direction of journal rotation as well as in the direction opposite to rotation. As it exits the oil groove, oil that had been flowing opposite to the direction of rotation suddenly undergoes an abrupt change in flow direction, which leads to a drop in local pressure and formation of vapor bubbles.

- *Exit cavitation*
  Exit cavitation occurs at any location where sudden changes in cross section—these could be oil passages, pockets, grooves, but also

the bearing parting line area—cause changes in flow velocity and pressure, flow separation, and turbulence. It can also occur where oil is squeezed out of the bearing, or into the oil grooves, as the journal approaches the bearing shell. As the journal separates from the shell, the oil is no longer able to follow this movement, and pressure is reduced (Fig. 6.227).

- *Flow cavitation*
  Unfavorable oil flow, as may often be encountered at oil grooves and oil entry passages, may result in "mushroom shaped" washouts just past the oil groove runout, as well as at the edges of the oil groove. If the bearing housing is significantly deformed by operating forces, as is often the case with connecting rod bearings, cavitation may occur at the bearing parting lines (Fig. 6.228).

- *Throttling cavitation* (see Fig. 6.229, shock cavitation)
  To improve oil flow to the lower bearing shell, the upper shell has a circumferential oil groove. To prevent any disturbance to the oil

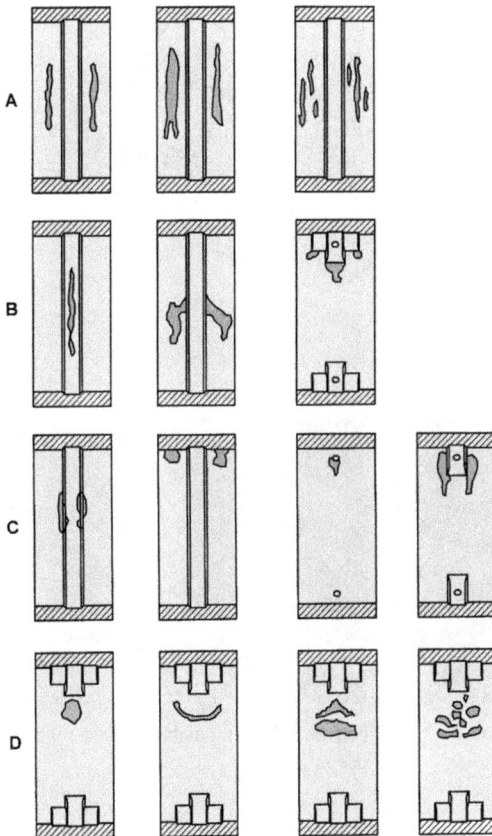

Fig. 6.224 Schematic of cranktrain bearing cavitation morphology: A, suction cavitation; B, exit cavitation; C, flow cavitation; D, throttling cavitation.

film in the most heavily loaded portion of the bearing, the lower shell itself has no groove, except for a partial groove, a "runout" of the upper groove, on the entry side of the bearing (Fig. 6.230). The connecting rod bearing is supplied by a system of oil passages in the main bearing journals, crankshaft cheeks, and crankpins. Oil flows out of the oil supply groove and into the main bearing bore. As the main bearing journal oil passage rotates past the end of the oil groove, oil flow is momentarily throttled and pressure drops. At the same time, the other end of the oil passage enters the opposite end of the oil groove, and is again well supplied with oil. Oil flow reverses, forcing oil, with entrained vapor bubbles, into the bearing. The bubbles implode just below the oil groove runout (see Figs. 6.230 and 6.231), causing kidney- or sickle-shaped cavitation patterns.

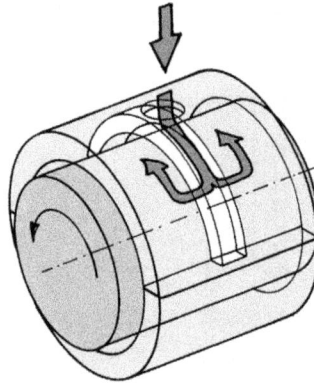

Fig. 6.225 Formation of suction cavitation as a result of oil flow reversal.

Fig. 6.226 Visual appearance of suction cavitation ("lancet cavitation").

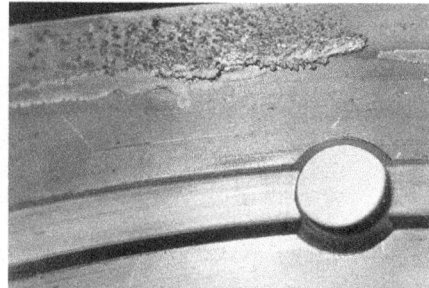

Fig. 6.227 Exit cavitation on a main bearing upper shell.

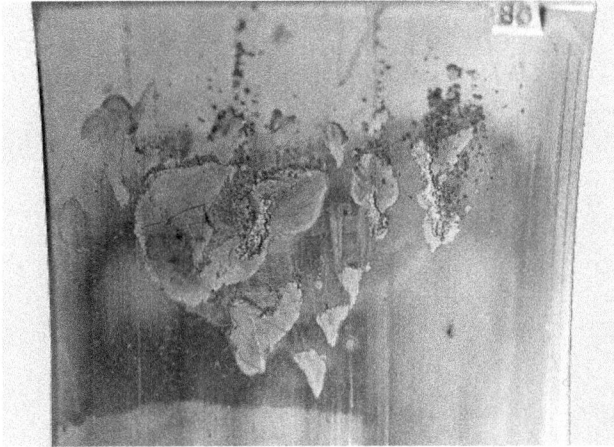

Fig. 6.228 Flow cavitation near the parting line of a connecting rod bearing.

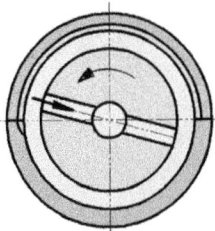

The oil passage in the main bearing journal is supplied with oil from the oil groove in the upper bearing shell.

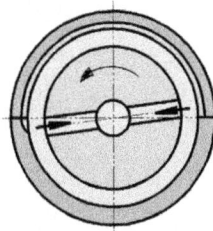

The oil passage turns past the end of the oil groove; oil flow is interrupted. Pressure drops, vapor bubbles form.

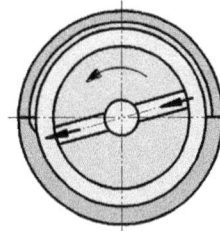

The oil passage is now entirely supplied by the oil groove on the opposite side of the upper bearing shell. Oil flow changes direction. Vapor bubbles are carried below the oil groove runout to an area of higher pressure, where they implode and cause cavitation!

Fig. 6.229 Schematic of throttling cavitation.

6.2.5.6 Erosion

Erosion is abrasive removal of surface material due to the "sandblasting" effect of very small particles (wear particles, combustion particles, dirt) carried by the lubricating oil, in which the particles are so small that they do not cause any scoring. Erosion usually occurs as a result of cavitation, because particles loosened by cavitation are carried away by the oil stream and then act as an erosive medium. Erosion traces follow the flow direction, in circumferential as well as axial flows of oil exiting the bearing (Fig. 6.232).

Fig. 6.230 Throttling cavitation at the oil groove runout in a lower bearing shell.

Fig. 6.231 Sickle-shaped cavitation of a connecting rod bearing at the lower bearing shell oil groove runout (sectioned view). The resulting scoring clearly shows the flow direction.

6.2.5.7 Corrosion

Corrosion is caused by organic and/or inorganic acids in lubricating oil. These arise from oil additives or are formed as a result of chemical changes to the oil (oil aging). Combustion of fuels containing sulfur, especially heavy fuel oil, produces sulfur dioxide, which enters the lubricating oil along with other blowby gases. Individually, causes of corrosion are

- Use of non-approved, aggressive oil additives
- Lubricating oil contamination by alkalis (e.g., antifreeze) or acids
- Extreme extension of oil service intervals (oil aging)

- Aggressive combustion products, especially in heavy-oil operation
- Excessive water content in lubricating oil (water contamination)

Bearing corrosion is characterized by a rough, porous, or satin wear surface (usually with dark discoloration) or by removal of the overlay, with transition zones of various colors. On occasion, there may be complete removal of the overlay, followed by attack on the lead-bronze lining. In particular, the lead components of the overlay and lining are removed by corrosion processes (Fig. 6.233).

6.2.5.8 Electrical arcing

If electrical arcing takes place in a bearing, the result will be highly localized welding, which is again broken apart by rotation of the journal inside the bearing (Fig. 6.234).

In theory, the lubricating film in a bearing should act as an insulator. However, the insulating ability of thin lubricating films may be so low that voltages of as little as 1 V may cause current to flow. The resulting electrical arcing melts tiny particles out of the bearing material; these are carried away by the oil flow. If the causes of such arcing are persistent, the result will be many tiny "arc craters," giving the bearing surface a sandblasted appearance. The bearing journal will also be damaged. If the bearing operates in the mixed lubrication regime, in other words during engine startup and shutdown or temporary overloading, bearing metal may be deposited on the roughened journal, which amplifies the wear rate of such electrical erosion—possibly to the point of bearing scoring, and eventually seizing. Causes of such arcing may include [6-22]:

- Generator inductive ripple voltage, caused by an asymmetric field around the armature. This ripple voltage builds up along the entire

Fig. 6.232 Erosion at an oil passage; erosion in the flow direction at the edge of the passage. Source: Federal-Mogul

length of the armature and attempts to close the electrical circuit through the front bearing, the engine bedplate or foundation, and the rear bearing. If the machinery is mounted in several bearings, current will pass through those bearings with the lowest resistance.

- Stray welding current.
- External voltages as a result of insulation failure, current leakage, or electrostatic charge separation.

Example: The main bearings of a natural gas engine used to drive a 285 kW AC generator were found to be damaged. The lower bearing shells showed pronounced signs of mixed-lubrication wear, which exposed the lead-bronze bearing lining. The transition to the remaining overlay was peppered with tiny, macroscopically barely recognizable craters (Fig. 6.235).

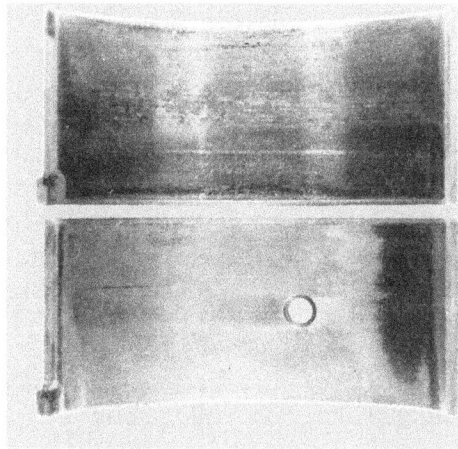

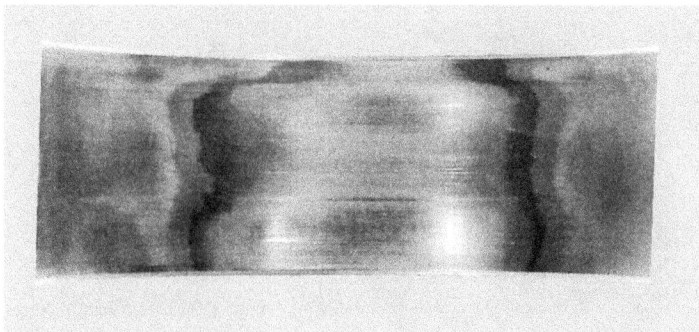

Fig. 6.233 Corroded bearing shells. Source: Miba

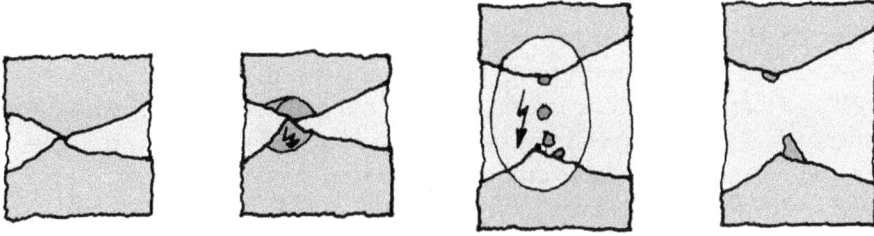

Fig. 6.234 Bearing damage as a result of electrical arcing (schematic).

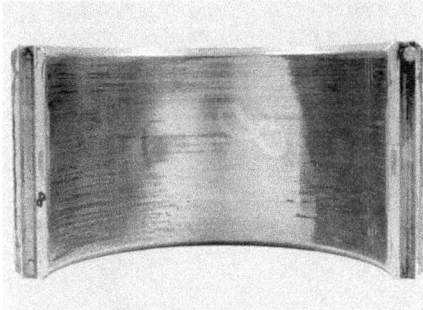

Fig. 6.235 Lower bearing shell damaged by electrical arcing. Source: Allianz

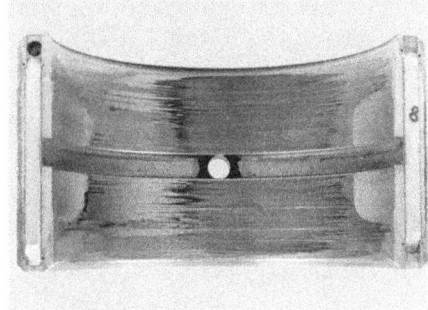

Fig. 6.236 Upper bearing shell damaged by electrical arcing. Source: Allianz

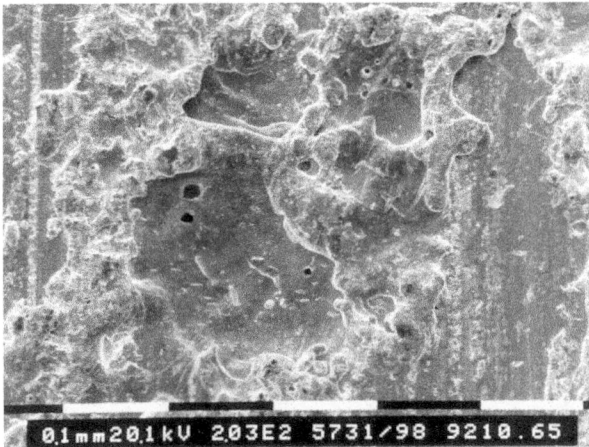

0.1mm20.1kV 2Ø3E2 5731/98 9210.65

Fig. 6.237 Reflection electron micrograph of bearing structure, showing local melting caused by electrical arcing. Source: Allianz

The upper bearing shells (with oil groove) showed even larger damaged areas, with readily visible rough surfaces, resembling sandblasting, caused by electrical arcing (Fig. 6.236).

In this area, arc-induced melting structures on the bearing metal are recognizable in a reflection electron micrograph (Fig. 6.237).

6.2.5.9 Manufacturing defects

In the bearing manufacturing process, defects may occur in the bond between the overlay and lining, between the lining and steel back, as well as separation and voids. At present, the manufacturing process for journal bearings is so well understood, and carried out with such care, that such defects may be excluded in any bearings made by reputable manufacturers. (The examples shown here, in Figs. 6.238, 6.239 and 6.240, are not contemporary.)

The section through a bearing shell in the region of a fully preserved ternary overlay shows bonding failure between the lead-bronze and nickel barrier layers (Figs. 6.241, 6.242, and 6.243).

Example: after 8000 operating hours, a marine diesel engine suffered a connecting rod bearing failure. Disassembly of the bearing showed breakouts of the bearing lining. Initially, dynamic overloading of the bearing was suspected (Fig. 6.244).

Fig. 6.238 Peeling bearing overlay.

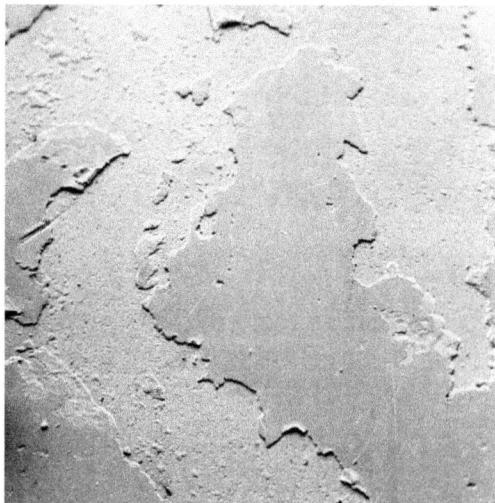

Fig. 6.239 Peeling, magnified ca. 20:1.

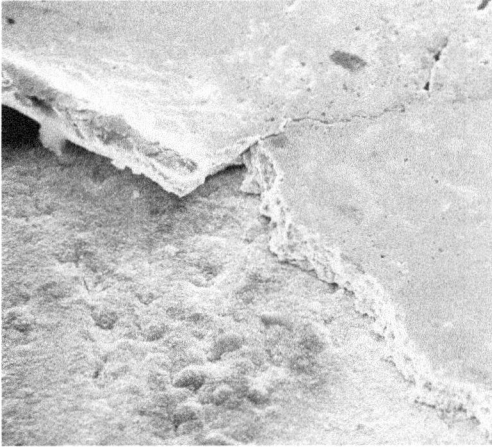

Fig. 6.240 Peeling, magnified ca. 500:1.

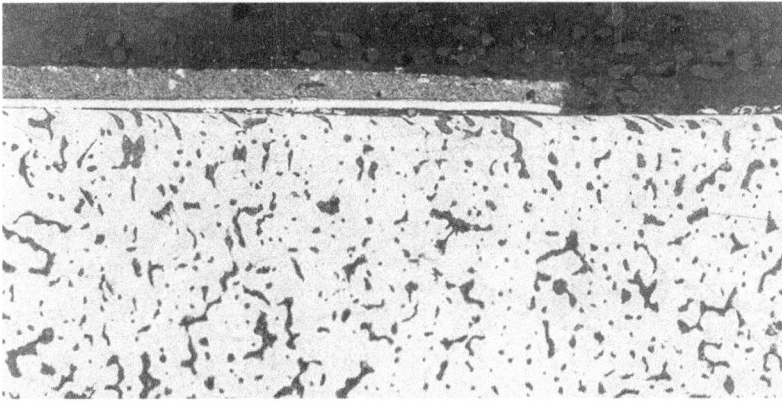

Fig. 6.241 Bond failure between lead-bronze lining and nickel barrier layer (ca. 200:1).

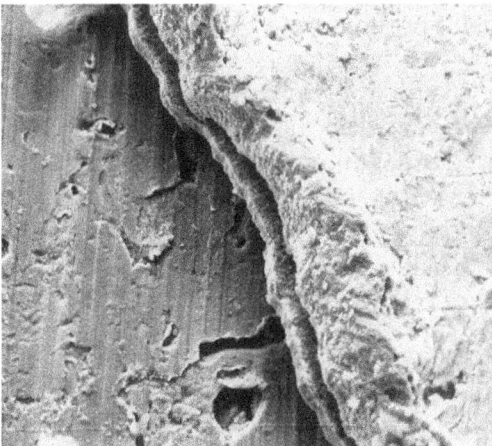

Fig. 6.242 Reflection electron micrograph (ca. 600:1) showing bond failure between overlay, nickel barrier, and lead-bronze lining.

Fig. 6.243 Broken-out inlay of a flange bearing.

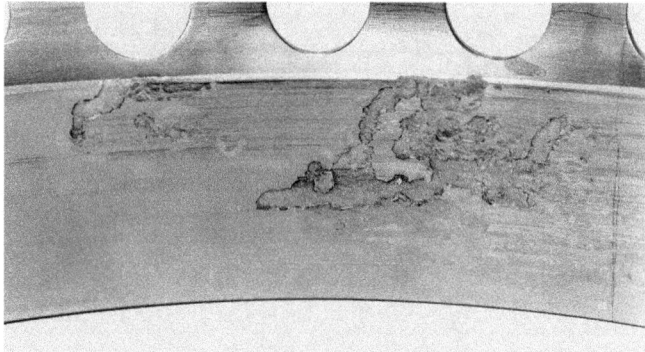

Fig. 6.244 Bearing with broken-out inlay.

To determine whether the cause was overloading or a material defect, metallographic sections were made. These showed that the breakouts were caused by porous spots in the bearing lining (Fig. 6.245).

When subjected to high operating stresses in the lower bearing shell, these caused cracking, followed by breakouts. Examination of a broken-out section with a reflection electron microscope (Fig. 6.246) showed that the porous structure was caused by the aluminum phase of the bearing lining forming a spheroidal microstructure, and that the tin phase was absent at these sites. The actual cause of this manufacturing defect could not be determined.

"Gray market" bearings, i.e., those made by unauthorized ("back-alley") manufacturers, pose a problem, as they are offered as cheap replacement parts. The resulting difficulties and costs may be considerable; engine manufacturers adamantly warn against installation of such "attractively priced" bearings (Fig. 6.247).

6.2.5.10 Anomalies in bearing geometry

Deviations in crankshaft form and orientation, errors in alignment, wobble, runout, out-of-round journals (tapered, oval, barrel-shaped), and rough journals (often as a result of manual rework) have a major effect on bearing operating behavior. If the bearing shape deviates from the (ideal) cylindrical form, the lubricating film may be disrupted to the point where several tiny wedges of lubricant are created. In the areas where these bear the load, they must build up higher pressures to counteract the forces acting on the bearing. This reduces the lubricating film thickness to critically small values; the bearing operates partly in the mixed lubrication regime, shown in Fig. 6.248.

The other lubricant wedges build up pressures that counteract not the bearing load, but rather the pressure of the out-of-round, jammed bearing shell. In extreme cases, the bearing seizes; less serious cases show increased wear and fatigue, shown in Fig. 6.249.

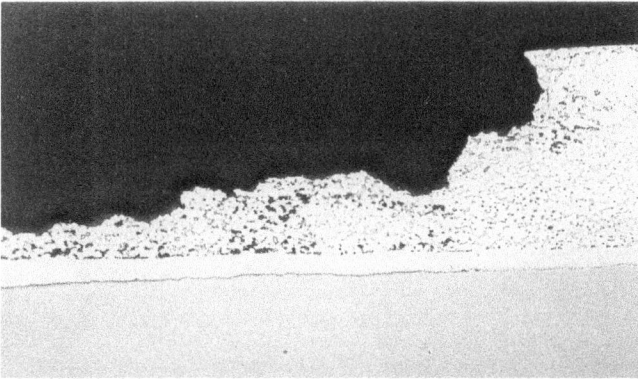

Fig. 6.245 Metallographic section of the damage in Fig. 6.244.

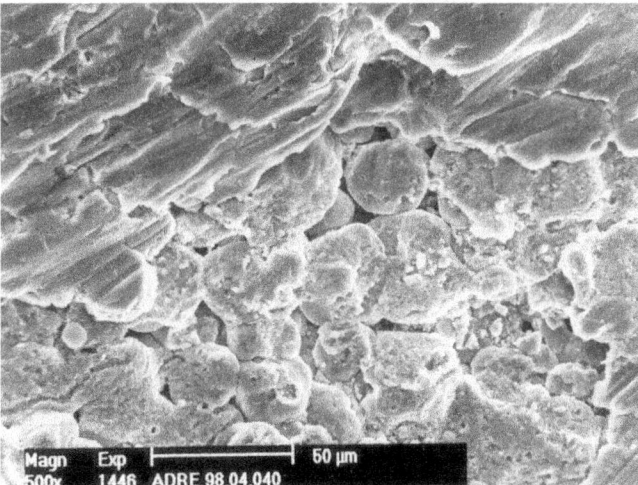

Fig. 6.246 Reflection electron micrograph of a broken-out area in a connecting rod bearing.

A common cause of problems is improper bolt torqueing during assembly. Excessive bolt tension results in horizontal oval deformation, while insufficient bolt tension results in vertical oval deformation of bearing bores. Bearing shells may be improperly inserted in the bores, resulting in crushed locating lugs. Particles trapped between the bearing housing bore and bearing shell are impressed into the bearing shell wear pattern. Insufficient interference between the bearing shells results in inadequate pretension and loose fit; bearing shells and

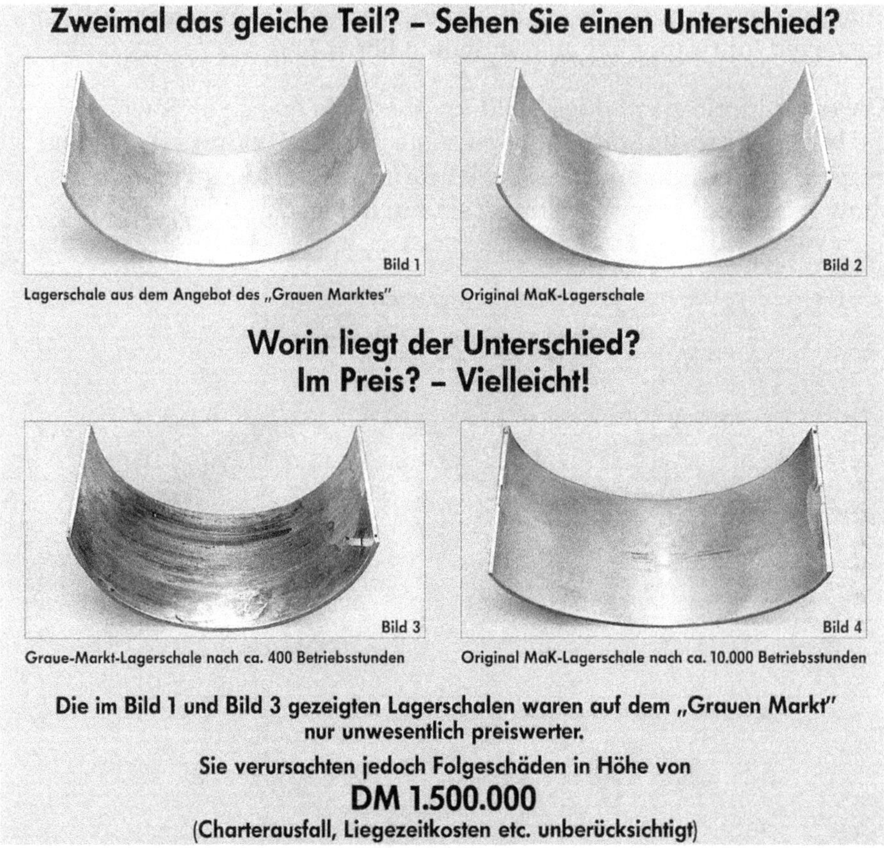

Zweimal das gleiche Teil? – Sehen Sie einen Unterschied?

Bild 1
Lagerschale aus dem Angebot des „Grauen Marktes"

Bild 2
Original MaK-Lagerschale

Worin liegt der Unterschied?
Im Preis? – Vielleicht!

Bild 3
Graue-Markt-Lagerschale nach ca. 400 Betriebsstunden

Bild 4
Original MaK-Lagerschale nach ca. 10.000 Betriebsstunden

Die im Bild 1 und Bild 3 gezeigten Lagerschalen waren auf dem „Grauen Markt" nur unwesentlich preiswerter.

Sie verursachten jedoch Folgeschäden in Höhe von

DM 1.500.000

(Charterausfall, Liegezeitkosten etc. unberücksichtigt)

Fig. 6.247 Information provided by an engine manufacturer regarding the risks inherent in gray-market bearings.
The same part? Can you spot the difference?
Bild 1: Bearing shell from a "gray market" supplier.
Bild 2: Genuine MaK bearing shell.
What's the difference? Price? Perhaps . . .
Bild 3: Gray market bearing shell after about 400 operating hours.
Bild 4: Genuine MaK bearing shell after about 10,000 operating hours.
The gray market bearing shells shown in Figs. 1 and 3 were only slightly cheaper, but they caused damage in the amount of 1.5 million Deutschmarks (not counting loss of charter, demurrage, etc.). Source: Caterpillar

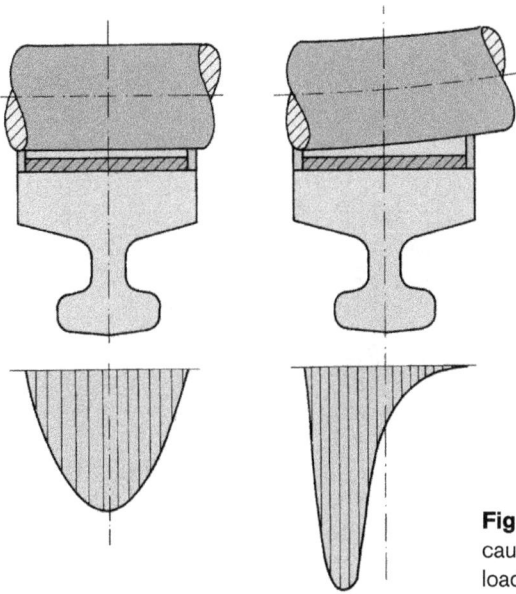

Fig. 6.248 Pressure increase caused by asymmetric bearing load due to canted shaft.

housing experience relative motion, resulting in fretting corrosion. Oil passages may even be blocked. Such damage may be recognized by signs of heavy contact marks, severe localized wear, and sometimes by cracks in the bearing overlay.

6.2.5.11 Edge wear
Single-ended (Fig. 6.250)
Same-ended
As a result of axial displacement of the bearing shells, the journal cuts into one side of the bearing, leaving a small, unworn strip of bearing at the edge of the bearing shell.

In the case of tapered (conical) journals and/or bearing housings, misaligned or twisted bearing bores, or with severely deformed (bent) crankshaft, the bearings show pronounced one-sided contact. Journal angularity (non-orthogonality) causes wobble, which in turn causes the bearing clearance to enlarge to one side. Lubricating film pressure cannot be maintained on the widened side, and bearing load must be taken by correspondingly higher pressure in the remaining bearing area with tighter clearances. This results in one-sided overloading—edge contact. The extent of this zone of heavy wear, circumferentially as well as axially, depends on the severity of the underlying cause of edge contact (Figs. 6.251, 6.252, and 6.253).

Fig. 6.249 Damaged main bearing shells as a result of out-of-round beyond tolerances, as well as misalignment of an adjoining bearing.

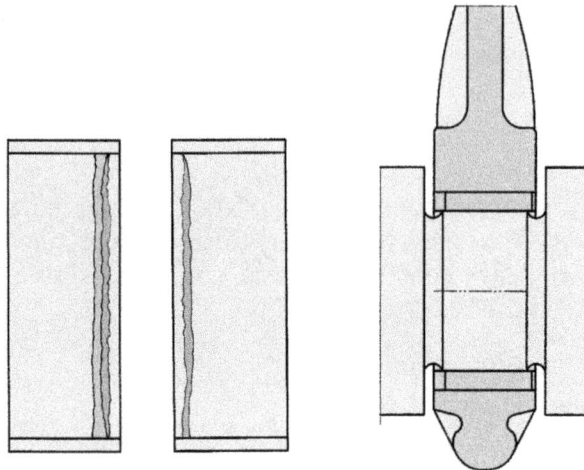

Fig. 6.250 Single-ended edge wear caused by axial shift of bearing shells.

Single-ended, narrow worn strip, at the same location for both upper and lower bearing shells, caused by axial shift of bearing shells

An extreme case of same-ended edge contact is shown by the motorcycle engine bearing of Figs. 6.254 and 6.255. One end of the bearing surface is virtually undamaged, but the other, by contrast, has been completely destroyed.

Alternating wear
If the journal is canted, the bearing shells will each show contact on one end only, but upper and lower shells will contact on opposite ends. Other causes of both bearing edges showing contact include improper

Single-ended, narrow worn strip on the same side for both upper and lower bearing shells

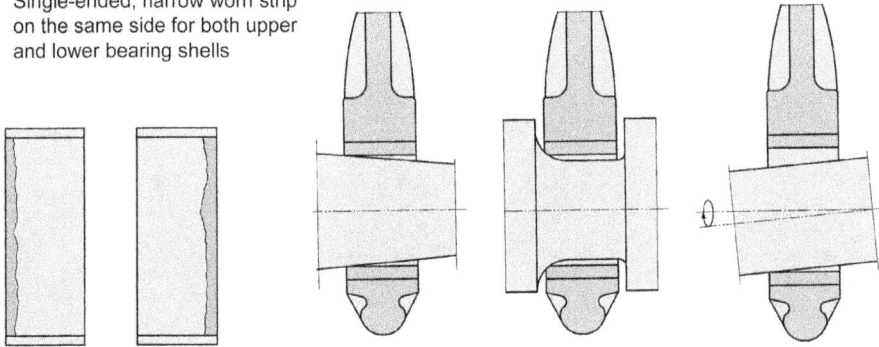

Fig. 6.251 Causes of single-ended, same-ended edge contact.

Fig. 6.252 Narrow circumferential zone of edge contact. Source: Miba

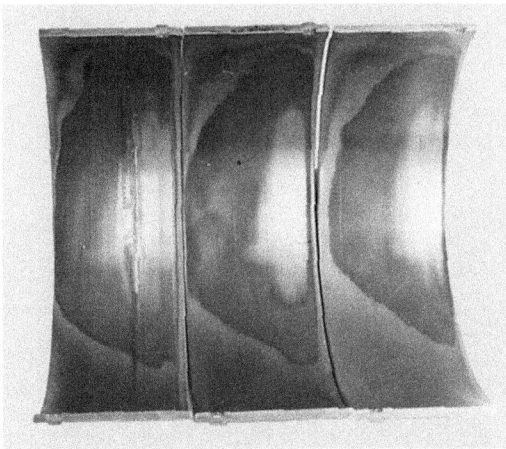

Fig. 6.253 Pronounced single-ended edge contact (connecting rod bearing shells) caused by non-orthogonal bearing bores.

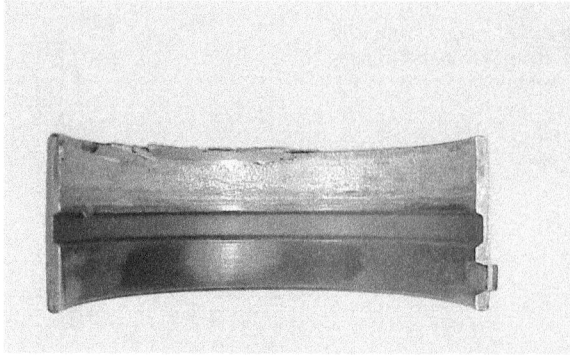

Fig. 6.254 Single-ended contact in a crankshaft bearing (motorcycle engine).

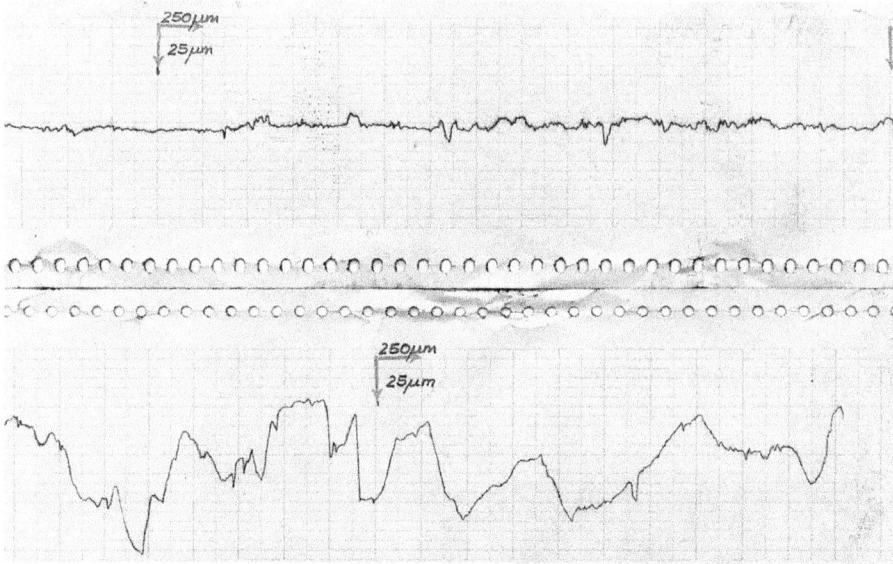

Fig. 6.255 Profilometer trace of the bearing in Fig. 6.254: "good" and "bad" sides.

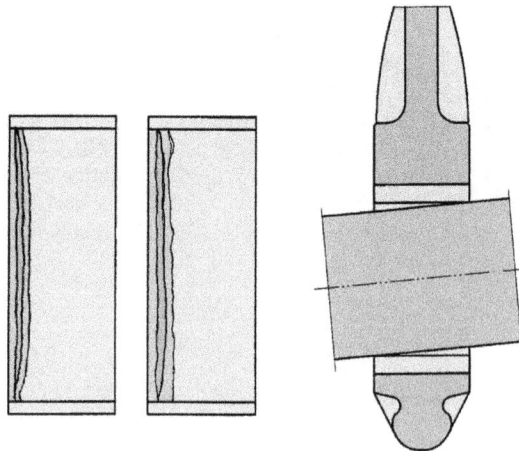

Fig. 6.256 Cause of single-ended, alternating edge contact.

fillet radii, wobbling connecting rods, or excessive axial bearing clearance. The bearings shells show edge contact (Figs. 6.256 and 6.257).

Two-ended

Tapered journals or bearing bores, as well as two-sided contact against the journal fillets, cause both ends of the bearing shells to show heavy wear; see Figs. 6.258 and 6.259.

Contact marks in center of bearing

If the bearing has a barrel-shaped contour, either through barrel-shaped bearing shells, housing bore, or journal, the bearing will exhibit a broad contact stripe along its centerline; the bearing material will "smear," and "bark beetle" cracks may be observed in the lining (Figs. 6.260 and 6.261).

Narrow, wear-free zones at the bearing edges

If, as a result of a manufacturing error, the hardened crankshaft running surface is narrower than the bearing shells, the journal will burrow into the bearing or bearing shell (Figs. 6.262, 6.263, and 6.264).

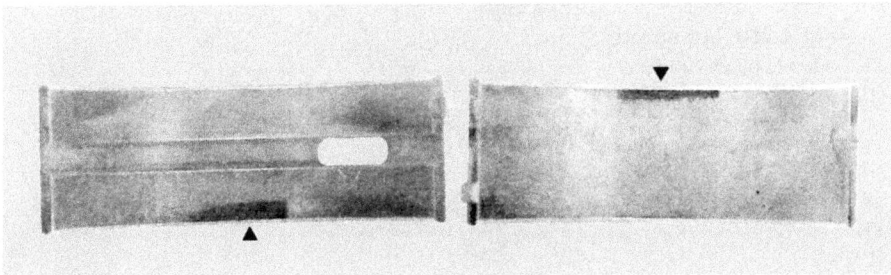

Fig. 6.257 Single-ended, alternating edge contact. Source: Miba

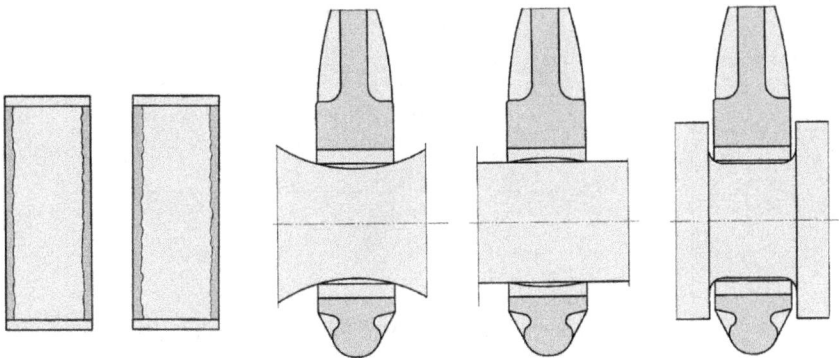

Fig. 6.258 Causes of two-ended edge contact.

6.2.5.12 Comb wear

If the underside of a journal does not wear in the area of the bearing oil groove, a ridge will appear in the center of the journal. This results in increased bearing wear, in the form of a central stripe. Such wear may also be caused by improper radiusing of the journal oil passage (Figs. 6.265 and 6.266).

6.2.5.13 Surface wear

Offset bearing bore halves (bearing cap offset)

If bearing cap and bearing housing are offset relative to each other, the bearing geometry will inevitably be incorrect. The bearing shell that protrudes in the direction of rotation will scrape oil away from the

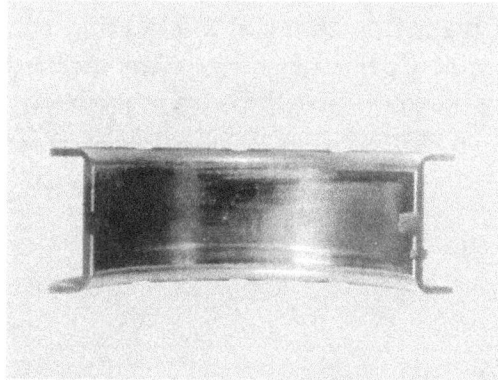

Fig. 6.259 Two-ended edge contact. Source: Miba

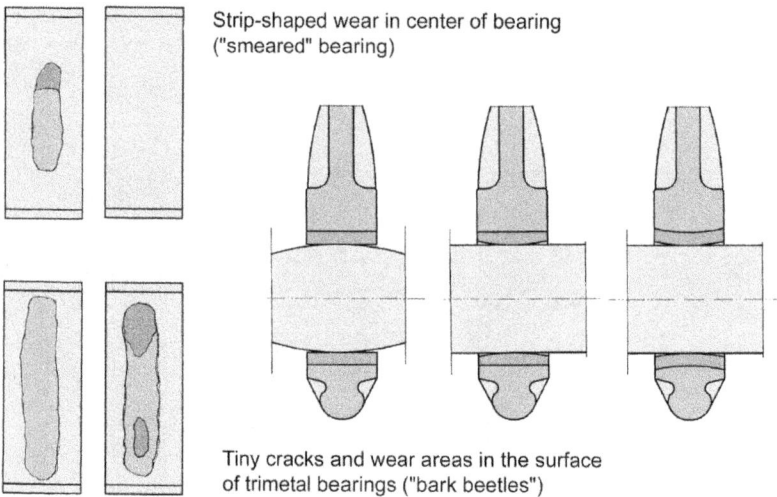

Strip-shaped wear in center of bearing ("smeared" bearing)

Tiny cracks and wear areas in the surface of trimetal bearings ("bark beetles")

Fig. 6.260 Causes for central contact.

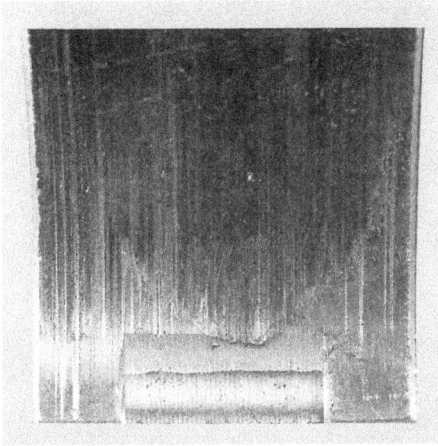

Fig. 6.261 Bearing with central contact wear. Source: Miba

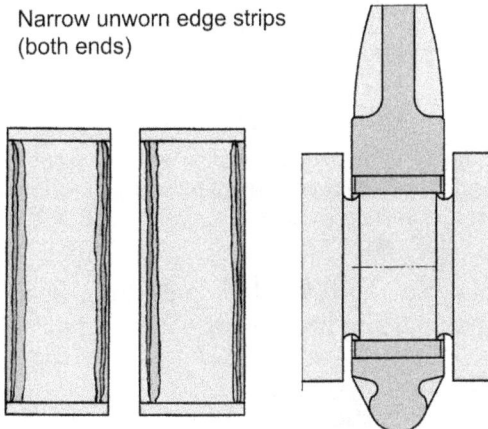

Narrow unworn edge strips (both ends)

Fig. 6.262 Cause of two-ended wear-free zones: Burrowing of journal into bearing or bearing shell.

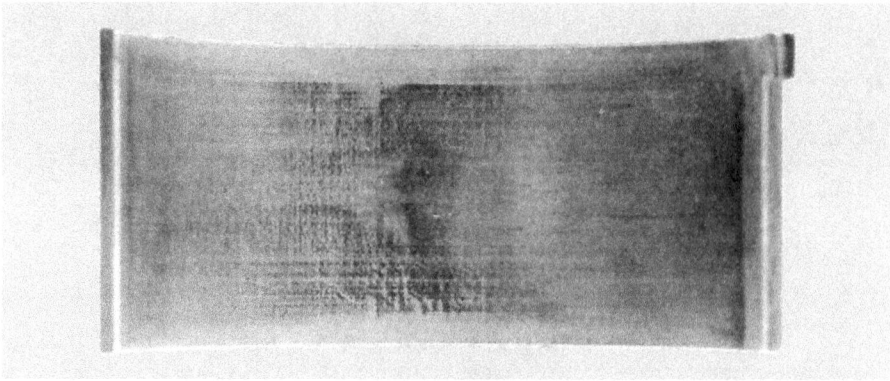

Fig. 6.263 Bearing with unworn edge strips. Source: Miba

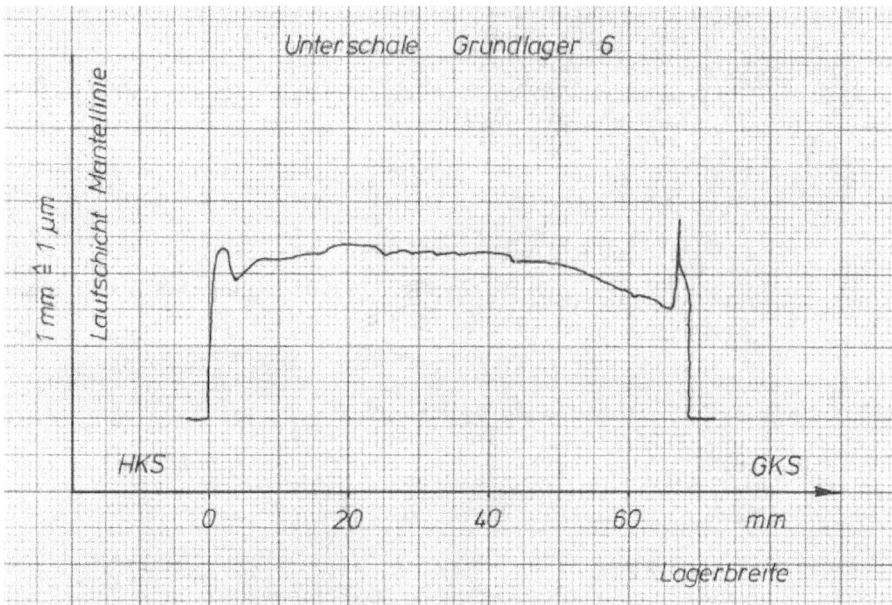

Fig. 6.264 Profilometer trace of the lower bearing shell with severe burrowing by journal:"Unterschale Grundlager 6" = lower shell of #6 bearing; "Laufschicht Mantellinie" = overlay profile; "Lagerbreite" = bearing width.

journal. The result is wear, and possibly contact marks at diagonally opposed bearing shell parting lines (Figs. 6.267 and 6.268).

Out-of-round bearing housings

If bearing shell bores are vertically oval, either as a result of incorrect bolt tension, or high connecting rod inertia loading, the bearing will exhibit heavy wear in the parting line areas, possibly with contact marks or seizing. In cases of less extreme ovality, the bearing may operate for some time, but "bark beetle" cracks may form in the overlay (Figs. 6.269, 6.270, 6.271, and 6.272).

Horizontal oval bearing shell bores result from excessive bolt tension (incorrect tightening), extreme "setting" of the connecting rod or housing parting surfaces, or extreme compressive loading of the connecting rod. In this situation, added force is imposed on the bearing shell crowns (Figs. 6.273 and 6.274).

Example: After 8000 operating hours, an engine experienced connecting rod bearing failure. The serrations between connecting rod and cap had set, causing severe horizontal oval deformation of the connecting rod bores (Fig. 6.275). Upper and lower bearing shells exhibited pronounced wear patterns. Two connecting rods, however, had been operating for only 500 hours; their bearing bores were still round, as clearly indicated by their bearing shells (second and third from the top, on the right).

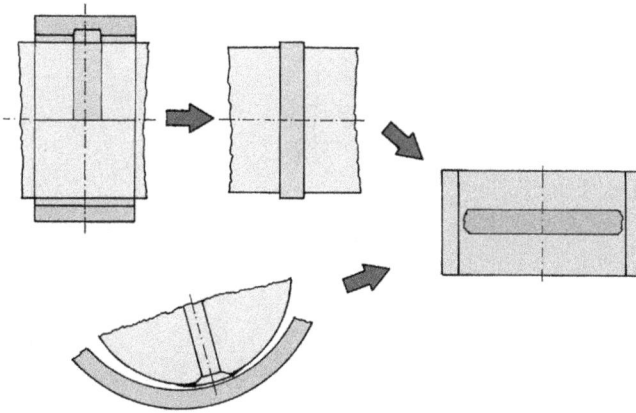

Fig. 6.265 Origin of comb wear (schematic).

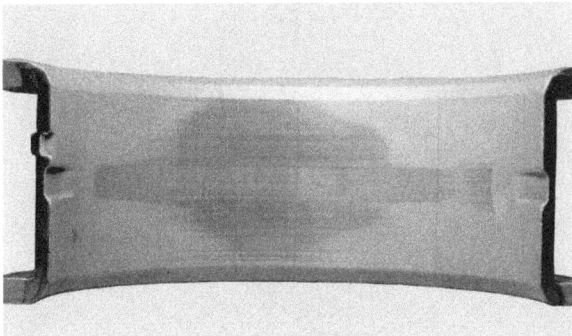

Fig. 6.266 Comb wear with added pronounced wear pattern. Source: Zollern BHW

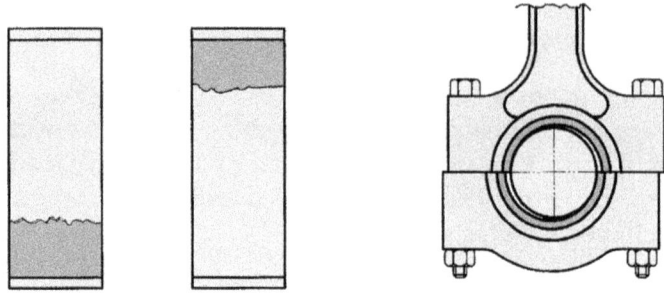

Fig. 6.267 Bearing cap wear (schematic). One-sided wear at opposite parting lines caused by bearing shell offset.

Fig. 6.268 Bearing with scuffing in offset, opposite parting line areas. Source: Miba

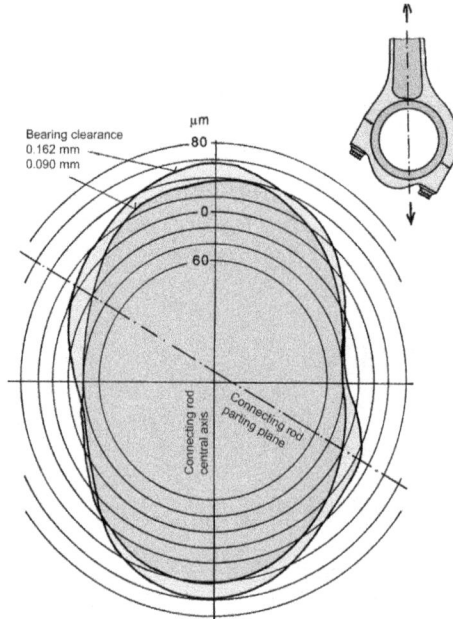

Fig. 6.269 Oval deformation of a connecting rod bearing bore subjected to inertia forces (four-stroke diesel engine, ca. 165 mm cylinder bore).

6.2.5.14 Insufficient bearing shell interference
Fretting corrosion of bearing back

If bearing shells do not have adequate interference in their bore due to insufficient crush or because of insufficient bolt tension, the bearing shells will "work" in their housing bores, causing fretting corrosion on their backs, at their parting lines, and in their housing bores. In its lightest form, fretting is recognizable by dark, slightly roughened surfaces; more advanced fretting typically shows a scarred surface, and, finally, cold-welded areas (Figs. 6.276, 6.277 and 6.278).

Example: After more than 2000 operating hours, bearing-specific wear particles were detected in the oil filter of a marine engine. In the course of scheduled routine inspection, the crank train was disassembled. It was discovered that one main bearing had seized. Large areas of the cap side bearing back showed fretting and axial scoring. The bearing bore in the cap showed the same phenomenon. Formation of fretting corrosion on the bearing back was a secondary effect caused by loss of pretension and free spread diameter of the bearing shells, caused by bearing seizing (Figs. 6.279 and 6.280).

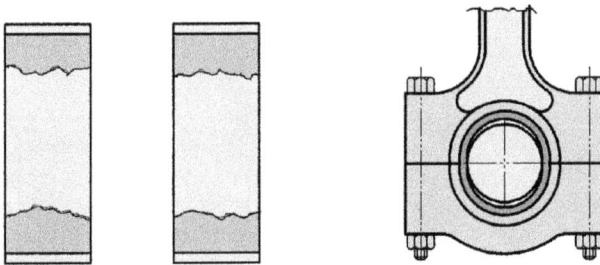

Fig. 6.270 Wear pattern for vertical oval deformation of bearing bore (schematic). Two-sided wear at the parting lines.

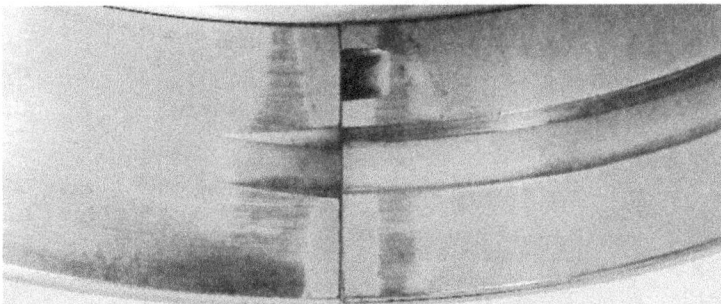

Fig. 6.271 Contact marks and hard wear pattern near the bearing shell parting line, caused by vertical oval deformation of bearing bore. Source: Miba

Bearing shell fatigue fracture

Fretting corrosion may result in fatigue fracture of bearing shells as well as connecting rods (Fig. 6.281).

Installation errors

* *Covered oil passages*

  Incorrect installation of bearing shells can result in serious damage if oil passages are covered. Oil starvation rapidly leads to seized

Fig. 6.272 Out-of-round bearing with heavy wear pattern below the parting line, and light "bark beetle" marks at edge of shell.

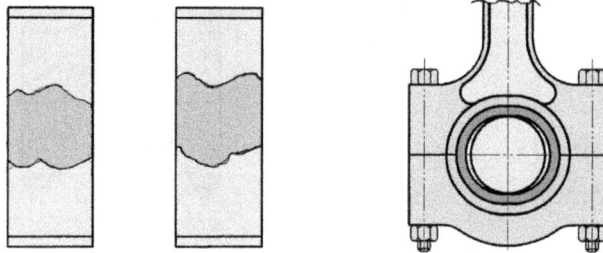

Fig. 6.273 Wear pattern for horizontal oval deformation of bearing bore (schematic). Large area of wear, in direction of load.

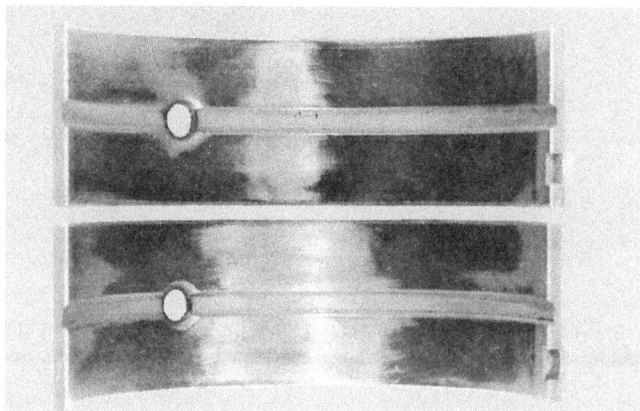

Fig. 6.274 Contact marks and heavy wear pattern in bearing shell crown area, caused by horizontal oval deformation of bearing bore. Source: Miba

Fig. 6.275 Wear pattern of connecting rod bearings with damage caused by horizontal oval deformation.

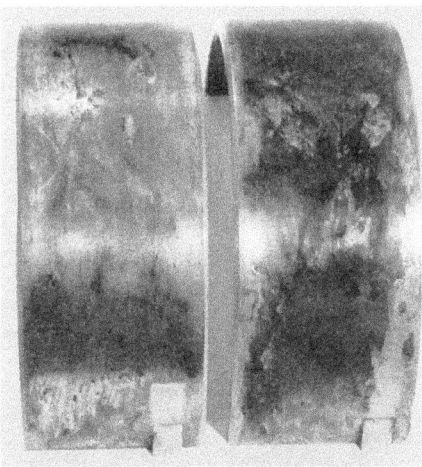

Fig. 6.276 Fretting corrosion on bearing backs caused by insufficient interference in bearing bore. Source: Miba

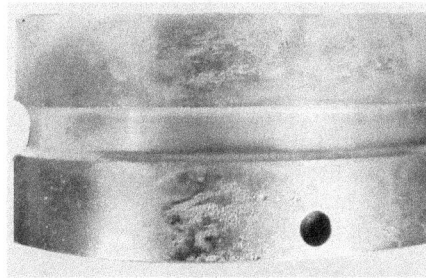

Fig. 6.277 Advanced fretting corrosion on bearing back. Source: Miba

bearings. The imprint of the blocked oil passage is usually visible on the bearing back (Fig. 6.282).

- *Incorrect lug seating*
  If a bearing shell is incorrectly installed so that the lug is not properly seated in the corresponding notch in the bearing bore, the lug will be damaged, and the crushed lug will force the bearing inward, causing scuffing (Fig. 6.283).

Fig. 6.278 Fretting corrosion. Top: at bearing shell parting line; bottom: on bearing back.

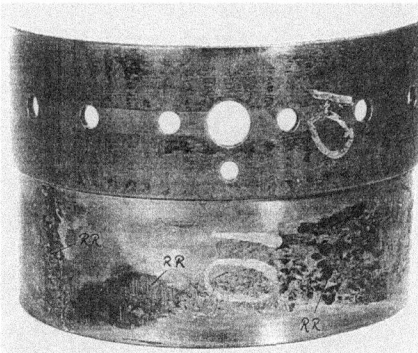

Fig. 6.279 Fretting corrosion damage (marked "RR") and axial scoring on back of cap-side bearing shell.

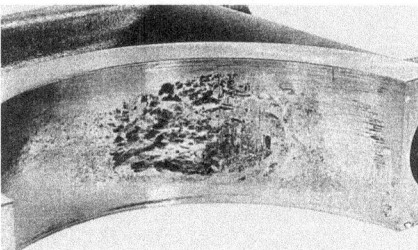

Fig. 6.280 Bearing cap with same type of fretting corrosion and surface damage as on back of bearing shell.

Fig. 6.281 Fractured bearing shell as a result of fretting corrosion on bearing back. Source: Miba

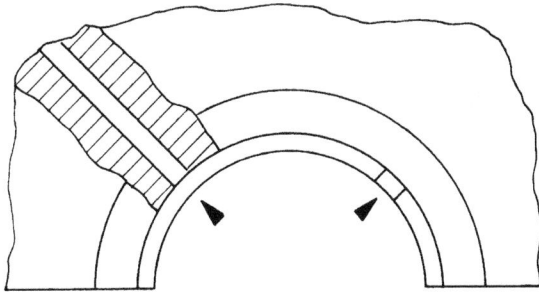

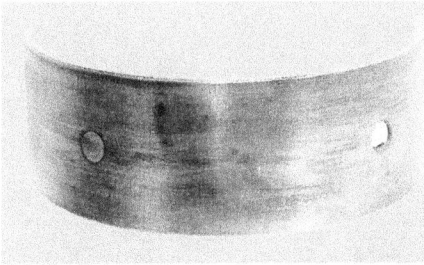

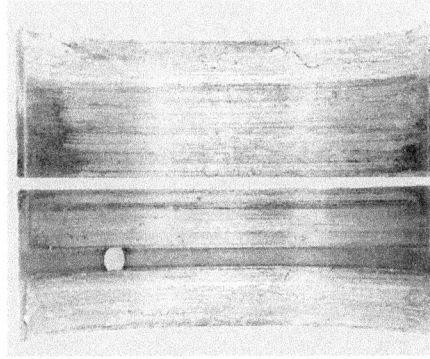

Fig. 6.282 Oil starvation caused by improperly installed bearing shell and blocked oil passage. Source: Miba

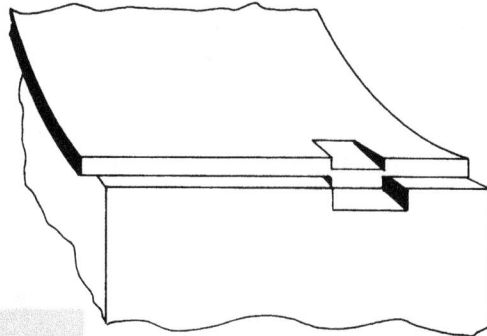

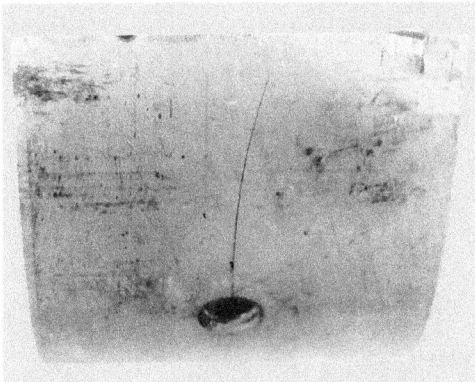

Fig. 6.283 Heavy wear and scuffing as a result of damaged locating lug (assembly error). Source: Miba

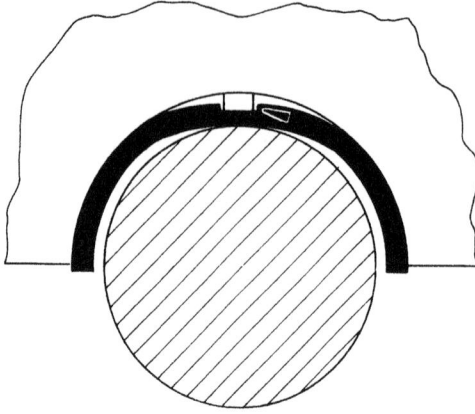

Fig. 6.284 Heavy wear in deformed area caused by improperly installed locating pin. Source: Miba

- *Locating pin interference*
  If the locating pin is too long or extends too far, for example due to dirt in the locating pin bore, the bearing shell will be forced inward by the locating pin, with the same consequences: increased wear, scuffing, and possibly seizing. The unsupported area of the back is clearly visible (Fig. 6.284).

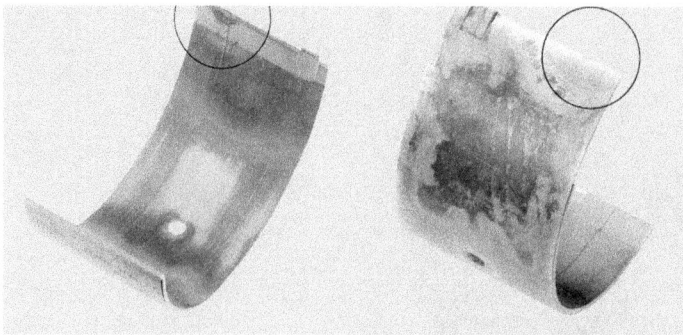

Fig. 6.285 Heavy wear caused by contaminants between bearing back and bearing housing bore. Source: MAHLE Engine Systems UK

- *Dirt on the bearing back*
  Dirt between bearing back and bearing housing bore prevents proper seating of the bearing shell, and, again, forces the shell inward, with the consequences of increased wear and fretting corrosion of the bearing back (Fig. 6.285).

6.2.6 Engine oil

Engine oil is a crank train component. As already described, lubricating films, only a few thousandths of a millimeter thick, are required to withstand the same forces as pistons, connecting rods, and crankshaft—themselves generously dimensioned components made of high-grade heat-treated steels. No wonder, then, that the slightest impairment of this lubricating film can result in operational malfunctions and engine damage. Furthermore, oil serves an important function during engine assembly, in that it assures low and—equally important—well-defined friction conditions during bolt tightening, press fitting of components, etc.

Oil fulfills the following tasks:

- Power transmission:
  functions as a machine element

- Lubrication:
  reduces friction and wear

- Microsealing:
  in principle, sliding components can only be sealed by means of a lubricating film

- Impact and vibration damping

- Noise reduction

- Cooling:
  heat transfer from thermally loaded components

- Cleaning:
  removal of all manner of particles, especially wear particles; engine housekeeping

- Rendering harmless any undesirable products

The significance of oil for the operation, operational assurance, and satisfactory service life of engines is indicated by the emphatic requirements by engine manufacturers with respect to oil types to be utilized in their products. Targeted development of special oil concepts is carried out in association with new engine designs. Yet despite the importance of oil for engine function, operation, and longevity, engine manufacturers have only limited influence on whether engine operators actually adhere to their requirements for engine operating fluids. Unsuitable or aged ("exhausted") oils are the source of many types and examples of engine damage.

The ability of lubricants to transmit forces ultimately rests on their internal friction acting to resist deformation. The governing oil property for this behavior is its viscosity. This is highly dependent on temperature and (to a lesser extent) on pressure. Oil viscosity behavior is described by its viscosity index. The operational behavior of engine oils is strongly dependent on engine operating modalities, specifically engine temperature—and it is precisely this parameter that is subject to extreme variation in motor vehicle applications.

One peculiarity of engine oils is that their engine-relevant properties cannot be described by unique physical quantities; there is simply no physical property for "lubricating ability." Physical properties such as viscosity or density are important properties, but these alone are inadequate for engine operation. Over time, the properties required of engine oils came to be recognized. These properties are determined by means of special tests in specific test and production engines, under precisely defined conditions. But this alone is not sufficient. Oils are subjected to road tests and fleet trials; ultimately, they must acquit themselves in actual service. Once again, the engine itself has the last word. Engine oil properties are described by so-called *classifications*, while *specifications* indicate the requirements to which an engine oil is subjected. Engine oils must exhibit a multitude of properties, summarized in Table 6.7.

Engine oils are stressed by the following:

- Power concentration
  - Rising power levels
  - Rising engine speeds
  - More compact engine designs
- Frequent engine load and speed changes
- Low oil fill volumes
- Extended service and oil change intervals
- Degradation in fuel quality, especially for general-purpose diesel engines

Superimposed on this general development trend are more-demanding operating conditions. What is regarded as normal and what is regarded as more-demanding depends entirely on the type of engine and its characteristic operating conditions:

- Motor vehicles are subjected to operation across a broad range of ambient temperatures. The arctic extremes of a Scandinavian, Russian, or North American winter must be withstood as well as the sweltering heat of deserts and tropical climates. But even Central European operations expose motor vehicles to unfavorable operating conditions. Frequent starts and driving in low gears cause fuel condensation against cold cylinder walls, along with condensed combustion gases. Cold starts hamper oil supply to the bearings. Frequent and sudden shifts from low to high load (acceleration) stress the engine oil. For motor vehicle engines, primary short-range operation in stop-and-go conditions represent an added difficulty, as does operation in a dusty atmosphere.

  Offsetting these factors, motor vehicles in most industrialized countries operate with fuels of comparatively good quality. The diesel fuel sold in Central and Western Europe may be regarded as the "champagne" of diesel fuels.

- Medium and large engines are almost exclusively operated within specified temperature limits; in low ambient temperature conditions, the engines must be pre-heated. A preluber ensures that the bearings are well supplied with oil even before the engine is turned over. On the other hand, such engines are supplied with fuels of mediocre to

Table 6.7
Engine Oil Properties

Mechanical properties	Chemical properties	Qualitative properties	Economic properties
Load carrying capacity	Low temperature viscosity	Bearing durability	Usable in various engine types
Friction reduction	Shear stability	No objectionable odors	Manufacturing economy
Sealing capacity	Neutralizing capacity	Compatibility with humans and environment	
Wear protection	Dispersion capacity	Consistent quality	
Adhesion	Low sensitivity of viscosity to temperature changes	Longevity	
	Resistance to water and coolant		
	Compatibility with metals and paints		
	Compatibility with elastomers		
	Low volatility		

low quality, in extreme cases with heavy fuel oil containing up to 5% sulfur. Long periods of idling are especially disadvantageous for large engines.

To ensure that motor oils are able to perform their duties under such conditions, their properties are improved by means of appropriate additives. Motor oils consist of a base oil (mineral, synthetic, or a blend of both) and 10 to 20% of additives. These additives improve oil properties, suppress undesirable properties, and impart properties that are not present in the native oil. These additives perform the following functions [6-23]:

- Antioxidants retard oil aging.
- Detergents and dispersants encapsulate solid and liquid contaminant particles and keep them in suspension, and prevent their coagulation and deposition on engine parts; they also neutralize acidic combustion and aging products.
- Viscosity improvers improve startup behavior and lubrication at low temperatures, ensure lubrication and microsealing at high temperatures, reduce oil consumption, and reduce wear.
- Pour point depressants ensure that oil is able to flow at low temperatures.
- Extreme pressure additives increase load-carrying capacity and reduce wear at high loads.
- Anti-foaming agents change the surface tension of oil to suppress foaming.
- Anticorrosion additives prevent rust and corrosion by reacting with metals and by forming a protective film.

Crankcase oil is exposed to blowby combustion gases, and is finely atomized by the rotating crank train. Because of the large surface-to-volume ratio of small droplets, air is given easier access to the oil. Such close contact with air promotes oxidation. As gas content of oil increases, its load-carrying and heat-transfer ability decreases. As it is exposed to heat, air, metallic wear, and combustion products, oil undergoes changes; it "ages." This also reduces reaction temperatures. Low temperatures also promote oil aging, because the associated larger piston clearances allow more combustion products to blow past the pistons and rings and into the crankcase. The chemical and physical properties of the oil are degraded, and sludge deposits form on engine components (Fig. 6.286).

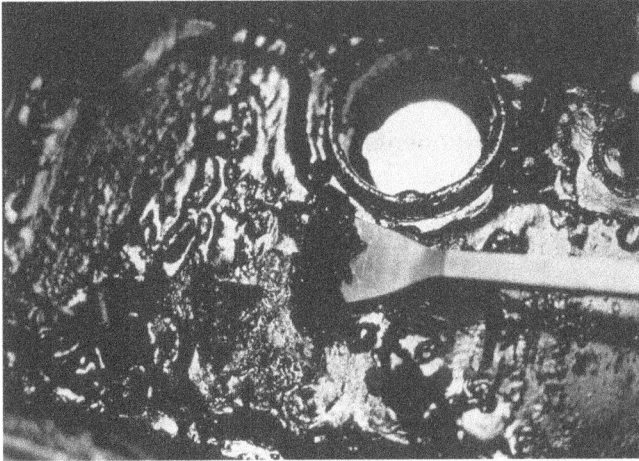

Fig. 6.286 Oil sludge deposits (shown here inside valve cover); thickness and consistency are such that they had to be scraped off with a spatula. Source: Klaver

In the case of diesel engines, soot is the primary agent acting to degrade engine oil. The concept of "oil aging" encompasses various physical and chemical processes:

- *Sludge formation*
  Oxidation and nitration products, and water (originating from the combustion process as well as condensate) form a sludge-like emulsion. This process is reinforced by foreign particles and acids. Copper and iron metallic wear particles catalytically promote this process and lower the reaction temperature for oil sludge formation. Oxidation products act as acids and promote corrosion of engine components. This oxidation is an irreversible process, but can be reduced by oil detergent action. Distinction is made between
  - Cold sludge
    Cold sludge forms primarily at low ambient temperatures in stop-and-go operation. It has been determined that reducing coolant temperature from 80 to 40°C increases sludge formation by a factor of 25.
  - Hot sludge
    Hot sludge consists of oil-insoluble combustion products that enter the crankcase with blowby gases.

According to [6-24], *engine sludge* is defined as ". . . deposits consisting primarily of oil and combustion products, which do not run off but which may be removed by wiping . . ."

Sludge formation is generally undesirable because it may clog oil pump strainers and oil passages. "Black sludge" consists primarily of oil-insoluble combustion products that are not removed by oil changes (Fig. 6.287).

- *Vaporization losses*
  Vaporization of low-boiling components thickens oil, increases its viscosity as well as oil consumption. These losses depend on oil type, viscosity, and temperature levels.

- *Viscosity increase*
  Oxidation, vaporization, and addition of foreign materials (contaminants from combustion air and combustion products) thicken oil and increase its viscosity. In diesel engines, a certain proportion of combustion soot in oil is unavoidable; 1 to 3% is regarded as normal. Nevertheless, just 1% of soot content raises oil viscosity by one SAE grade.

- *Viscosity reduction*
  Mechanical stresses on oil—shear, i.e., mutual displacement of oil layers—and fuel entering the crankcase both act to lower oil viscosity. As such, 2% fuel content in lubricating oil is regarded as normal, but 4% lowers viscosity by one SAE grade, and 10% is regarded as "excessive." Fuel entrained in oil eventually evaporates, but in motor vehicle engines, oil only reaches its operating temperature after 15 to 20 km (8 to 12 miles). Tests have shown that after two hours of driving, only 80 to 85% of the fuel has evaporated [see also 6-23]. In World War II, fuel was added to engine oil by the German air force and army to improve cold-start behavior of aircraft, tank, and vehicle engines ("Rechlin Cold Start Method").

- *Acidification*
  Oxidation processes form organic, oil-insoluble acids in engine

Fig. 6.287 Oil pickup strainer completely clogged by black sludge. The consequences are not difficult to imagine. Source: Klaver

oil. Combustion of sulfur-containing fuels produces sulfuric and sulfurous acids. These processes are promoted by frequent engine cooling in winter stop-and-go operation. These acids attack the cylinder walls and lead to corrosion of bearing materials.

The described oil alterations promote the formation of varnish, resinous, and sludgelike deposits which—wherever they are subjected to high combustion chamber temperatures—harden to form "carbon" deposits. On crank train and combustion chamber components, they hinder motion of pistons, piston rings, and valves, often resulting in the phenomena of piston rings riding on ring groove deposits, stuck piston rings, seized pistons, and stuck valves described elsewhere. In particular, the phenomenon of stuck piston rings has repeatedly proven to be a serious problem: cold-stuck rings compromise cold startability and increase pollutant emissions; they are a preliminary stage of hot-stuck rings, which ultimately cause engine failure. Varnish deposits on pistons reduce heat transfer, which causes piston temperatures to rise and promotes ring sticking.

The engine oil aging process begins slowly, but is accelerated by the catalytic action of wear particles, especially copper and iron. Today, wear rates have been reduced, thanks to improved materials and manufacturing methods. As operating time builds up, oil additives are altered and consumed. The oil's capacity to carry contaminants decreases, and viscosity increases. Because modern engines use less oil, occasional replenishment with fresh oil is no longer sufficient to make up for lost additive properties.

The concept of oil consumption encompasses two different processes: actual oil consumption and oil loss.

- *Oil loss*
  Is represented by oil leaving the engine. For this, there are many opportunities, specifically at all fixed or movable joints on the exterior engine surface: the joint between the crankcase and oil pan, camshaft drive box and cylinder heads, as well as cylinder heads and valve covers, oil filter and oil cooler flanges, etc. Leaking oil drain plugs contribute to oil loss, as do the crankshaft seals—depending on engine design, at the clutch flange only, or also at the opposite end of the crankshaft.

  Outward leaks are recognizable by oil puddles under the vehicle, especially in the case of older engines. Marine and stationary engines are preferentially given light-colored paintwork to make oil leaks more obvious. Engines in construction equipment are often covered by a thick crust of oil and dust.

- *Oil consumption*
  This is understood to mean the physical consumption of oil,
  primarily through combustion, but also through evaporation.
  Internal leaks may be explained by unsuitable or worn piston ring
  packages, worn piston ring grooves, incorrect cylinder honing or
  "bore polishing" of the upper ends of cylinder walls, and excessive
  valve stem clearances. Until very recently, the products of one
  well-known automobile manufacturer could be identified at great
  distances by the cloud of blue oil smoke that was emitted whenever
  the driver lifted off the throttle during shifts. With the throttle closed,
  engine vacuum pulled oil into the combustion chamber, where it
  burned. Evaporative losses also affect oil consumption. Engine oil
  consumption can only be given in general terms. For passenger cars,
  0.1 to 0.25 liters per 1000 km is considered "normal"; for commercial
  vehicles, 1 to 3 liters/1000 km; and for general-purpose engines,
  0.5 to 2 g/kWh. Absolute oil consumption increases with increased
  engine speed, while specific consumption decreases (Fig. 6.288).

Lubricating oil failure may be due to several, often related causes:

- *Use of unsuitable oil*
  This "deadly sin" is to be avoided at all costs; engine manufacturers'

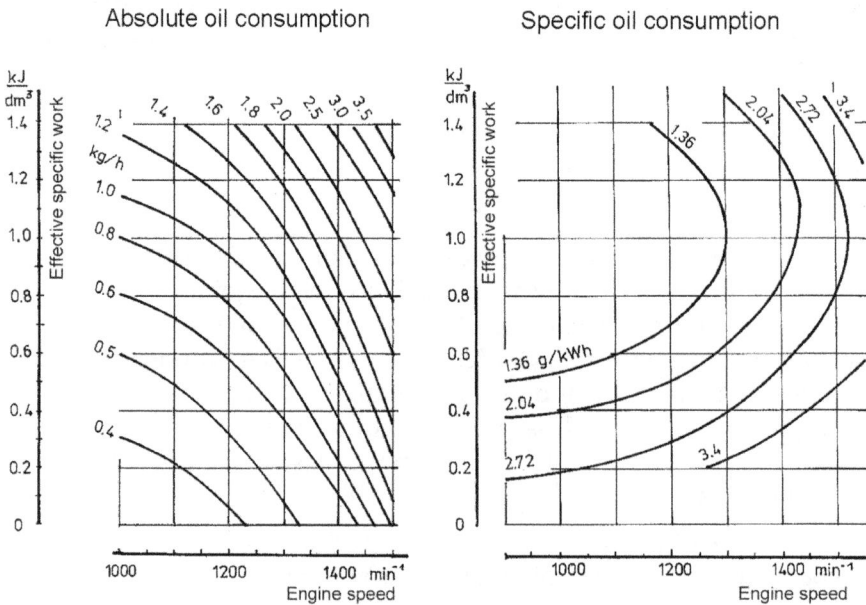

Fig. 6.288 Absolute and specific oil consumption maps of an older design, twelve-cylinder
prechamber railway locomotive diesel (190 mm bore, 230 mm stroke, 1030 kW at 1500 rpm).
SAE 40 oil, oil temperature upstream of engine 80 to 85°C.

oil recommendations should be followed without exception. Failure to do so may result in incalculable damage.

- *Loss of relevant lubricating oil properties as a result of excessively extended oil change intervals*
  Time and again, attempts are made to extend oil change intervals by means of add-on bypass filtration. Comparative tests, however, have shown that (given the state of the art and the conditions at hand), oil changes and thickens in the course of its permissible operating life. Driving comparisons over distances of 60,000 km (36,000 miles) have shown that 13.5 kg of savings in fresh oil were bought at the cost of 250 liters of increased fuel consumption (as a result of increased friction) [6-25].

  Oil change intervals are established by engine manufacturers or operators to reflect specific applications and operating conditions, either in terms of distance covered, in operating hours, or in liters of fuel used. Factors that shorten oil intervals for general purpose applications include, for example, extreme climatic conditions, frequent starts, frequent and extended idling periods or periods of light-load engine operation, or high fuel sulfur content.

- *Contamination as a result of*
  - mixture formation and combustion problems,
  - coolant and fuel system leaks,
  - dust, dirt, and wear
      all lead to excessive oil contamination by unburned fuel, soot, acids, and water.

Individual engine manufacturers, in specifying their test methods, provide analytic limits for oil properties, which include the following properties (among others):

- Overall contamination
- Viscosity loss as a result of fuel dilution
- Viscosity increase as a result of aging and contamination
- Flash point, water content
- Total Base Number

The following tests are conducted within the framework of operational monitoring:

- Determination of dispersability (spot test)
- Determination of diesel fuel fraction in oil
- Determination of water in oil

In association with any engine damage, initial examination of the engine's lubricating oil is in order:

- What is the oil's appearance (color, consistency)? What is the appearance of deposits on engine components?

- How does the oil smell—sour, pungent, burnt, like fuel, etc.?

- Can water be detected in the oil?

- How viscous is the oil?

- Is there any unusual discoloration of engine components?

If warranted, this would be followed by more extensive laboratory tests.

6.3 Crankcase and ancillary components

6.3.1 Crankcase

The crankcase is the central component of the engine, carrying and containing all engine functional groups and connecting components. It represents the system boundary of the engine, in that it, and conjoined components, separate the engine from the external environment, and prevent working gases, coolant, and lubricant from escaping, and dust, moisture, and contaminants from entering.

In principle, the crankcase or engine block ("cylinder crankcase") consists of intermediate walls (webs), the upper deck, longitudinal walls, and end walls (faces). The intermediate walls incorporate bearings for the crankshaft and in some cases camshaft(s). Medium and large engines have access holes in the longitudinal walls, with removable covers, lids, or even doors, which permit access to the crank train for inspection, maintenance, and repair purposes; some of these lids are configured as explosion relief doors or valves. In high-speed (i.e., smaller) engines, water jackets are cast into the engine block, as are oil and coolant passages, and possibly also charge air intake passages. At the bottom, the crankcase is closed by the bedplate or oilpan. To extend crankcase operating life, provisions are made for remachining, for example, at the cylinder liner flange mounting faces, cylinder sleeve spigots, and crankshaft bearing bores. Crankcase designs are differentiated by:

- Number and location of crankcase parting lines

- Crankshaft bearing configuration ("hanging" or supported from below)

- Method of joining individual crankcase components

- Material and fabrication method (cast or welded)

In special cases, the crankcase also serves as the load-bearing structure for the machine that it powers (e.g., tractors; see Fig. 6.289).

These basic requirements must be satisfied along with various related conditions:

- Good utilization of the available installation space, with the lowest possible component mass
- Adequate structural stiffness with regard to bearing bore and cylinder spigot deformation, as well as sealing of ancillary components such as oilpan or cylinder heads

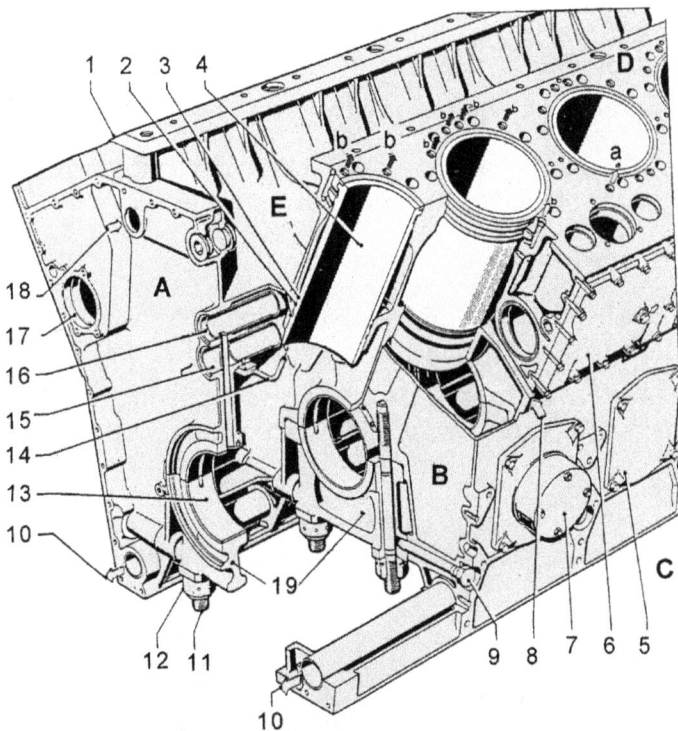

Fig. 6.289 General-purpose diesel engine crankcase (V configuration): construction and terminology: A, Crankcase end wall (face); B, Crankcase intermediate wall (web); C, Crankcase longitudinal wall; D, Crankcase top deck; E, Crankcase valley. 1, Crankcase; 2, 3, sealing rings; 4, cylinder liner; 5, crank train inspection port; 6, camshaft inspection port; 7, explosion relief valve; 8, from coolant passage; 9, transverse anchor bolt; 10, to engine oil pump; 11, main bearing stud; 12, main bearing nut; 13, main bearing; 14, piston oil cooling spray nozzle; 15, main piston oil cooling passage; 16, main cranktrain oil passage; 17, camshaft bearing; 18, piston cooling oil from gear chest; 19, main bearing caps. a, Engine oil to cylinder head; b, engine coolant to cylinder head. Source: MTU

- Substantial integration of engine accessories
- Good accessibility to individual functional groups for maintenance and repair operations

The crankcase is subjected to forces and moments acting within and upon the engine, and transmits these to the engine mounts. In addition, the crankcase is subjected to external forces:

- Forces from engine accessories
- Radial and axial forces from the driven machinery (reaction forces, axial thrust)
- Engine mounting forces (e.g., from deformation of a marine vessel's hull)
- Installation forces
- Forces resulting from thermal expansion

In designing a crankcase, along with function and concept, primary consideration must be given to forces and moments.

Cylinder gas pressure acts upon

- The cylinder head, which exerts force upon the crankcase intermediate walls by means of the cylinder head bolts, as well as
- The crankshaft main bearing caps by means of pistons and crank train. The bearing caps in turn act upon the crankcase intermediate walls (Fig. 6.290).

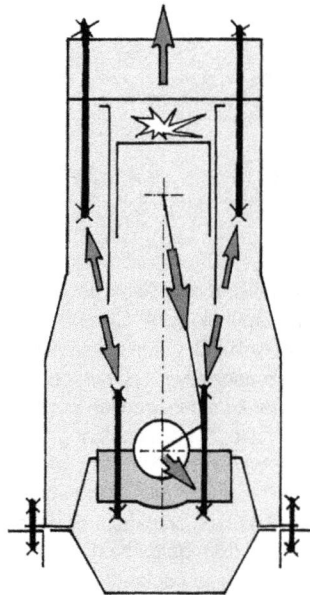

Fig. 6.290 Force flow in crankcase intermediate wall (schematic).

This closes the flow of forces. The crankcase wall is loaded in tension (and dynamically, at that). The cylinder head bolts (four, six, or eight per cylinder, depending on engine size) are arranged around the cylinder(s), permitting bolt forces in the vicinity of the crankcase walls to be transferred directly into the crankcase. Bolt forces near the crankshaft plane are transferred to the intermediate walls by means of special design features such as tension bands, ribs, or belts (Fig. 6.291).

Force transfer to the crankcase intermediate wall, combined with complex cross section shapes, results in additional loads. Regardless of the fact that the crankcase is the largest and heaviest engine component, it undergoes appreciable deformation in response to operating and assembly forces—quite in keeping with the old adage that *where there are loads, there will be load paths.*

Deformation as a result of assembly and operating forces in the top cylinder spigot area affect piston action (Fig. 6.292).

Although piston clearance is matched to the cylinder liner and its deformation, differences in stiffness—for example, between inner and corner cylinders—sometimes causes difficulties. Crankcase forces may cause bearing bore deformations that approach the magnitude of the bearing clearances themselves, shown in Fig. 6.293.

6.3.2 Crankcase damage and failure
6.3.2.1 Incipient cracks, cracks, and fatigue fractures
Unfavorable loading, manufacturing defects (e.g., deviations from flatness), or assembly errors lead to incipient cracking in the fillets of cylinder liner mounting flanges, as well as fretting corrosion between cylinder barrels and mounting flanges.

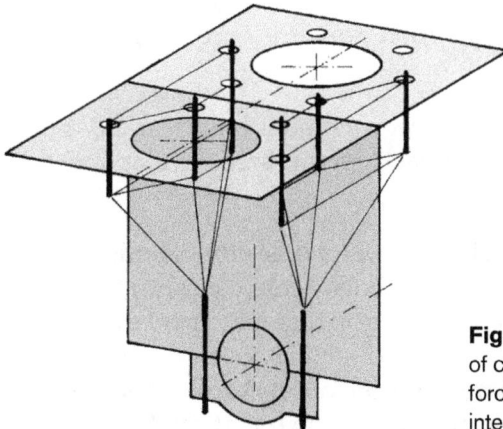

Fig. 6.291 Transfer of cylinder head bolt forces to crankcase intermediate wall.

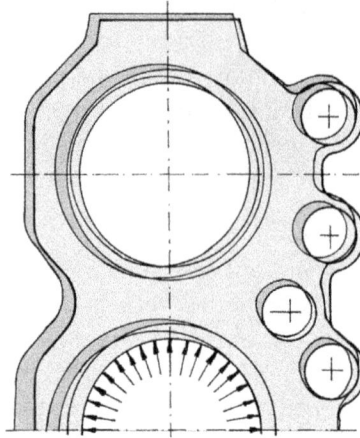

Fig. 6.292 Cylinder top deck deformation in response to cylinder ignition pressure loading.

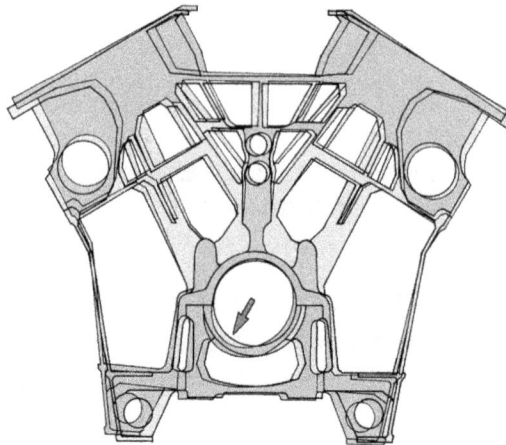

Fig. 6.293 Deformation (indicated by dark shading) of a V-engine crankcase intermediate wall in response to cylinder ignition pressure.

In the event of a defective (eccentrically installed) cylinder head gasket, the combustion chamber seal ring may break out if static preload is superimposed on dynamic ignition pressure, exceeding the permissible load.

A critical section is the area between cylinder bolt clearance holes and the cylinder liner spigot. Cracks in this area develop into fatigue fractures.

Due to their configuration, blind tapped holes for cylinder head bolts and crankshaft main bearing bolts act to concentrate stresses. Manufacturing defects and assembly errors may increase these stresses to the point of incipient cracking, which may in turn develop into cracks. Vibration of auxiliary components and accessories, transmitted through mounting flanges, may lead to incipient cracking.

Crankshaft main bearing caps with transverse loading applied by transverse anchor bolts are usually installed with interference fits. Forced installation, i.e., excessive spreading of the bearing web, may result in incipient cracking. There is also a danger that metal particles may be split off and enter the lubricating oil stream.

6.3.2.2 Fracture as a result of excessive force

Generally, failures of this sort are the consequence of other failures. The most common cause is connecting rod failure, in which the (now unguided) connecting rod penetrates the crankcase wall ("thrown rod," Figs. 6.294 and 6.295).

If coolant with insufficient antifreeze is used, low ambient temperatures may cause the coolant to freeze, thereby cracking the crankcase ("cracked block").

6.3.2.3 Wear and material removal

Cavitation and corrosion may damage the walls of coolant passages to the point of penetration. However, even less extreme pre-existing damage may, under certain conditions, act as the origin for fatigue failures. Therefore, modern midsized and large medium-speed diesel engines are designed so that coolant, contained in passages and coolant rings, no longer comes into contact with the crankcase itself.

Fig. 6.294 Passenger car engine crankcase, penetrated by a connecting rod.

Fig. 6.295 Detail of damage caused by thrown connecting rod.

Fig. 6.296 Cavitation of an upper cylinder liner spigot.

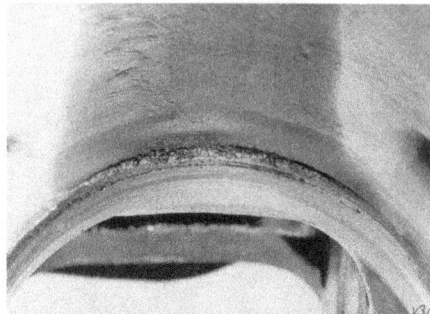

Fig. 6.297 Cavitation of a lower cylinder liner spigot.

Example: The crankcase of a medium-sized diesel engine exhibited large, deeply pitted areas in the upper and lower cylinder spigots (Fig. 6.296). The upper spigot has in part been eaten away as far as the liner flange mating surface. The water gallery itself shows severe pitting, as well as a heavy coating of rust. Severe cavitation resulted in water penetration through the lower cylinder liner spigot, into the lubricant (Fig. 6.297), as evidenced by pronounced rust scarring of the crankshaft cheeks. The excessive water content of the lubricating oil resulted in a seized bearing. Furthermore, the cylinder heads were thermally overloaded as a result of the reduced coolant level, which resulted in severe loss of material in the cylinder heads and pistons. The pronounced cavitation damage was traceable to use of an unsuitable coolant, also indicated by severe rust deposits throughout the cooling circuit. Material loss at the upper cylinder liner spigot resulted in increased clearance.

On motor vehicles, oilpan damage is usually the result of impacts, for example if the oilpan contacts the ground. This danger does not arise for general-purpose engines, e.g., railway or marine engines. Damage to their oilpans is usually caused by improper welding, vibration (excited by the engine), or oil mist explosions.

6.3.3 Cylinders, cylinder liners, and cylinder jackets

Colloquially, the concept of "cylinder" has several meanings; it is applied to cylinder crankcases, cylinder blocks, or cylinder liners. A more precise definition according to DIN ISO 7967 is in order. According to this standard, a cylinder consists of a component containing a single piston, with or without a cylinder liner and/or cylinder head. The cylinder jacket surrounds the cylinder, contains coolant, and is attached to the crankcase or frame. The sleeve inserted in the cylinder jacket, containing the running surface for the piston, is termed the cylinder liner.

In small engines (passenger car engines), cylinders are cast into the crankcase and machined in place (Fig. 6.298). Larger engines have individual dry or wet cylinder liners, as this permits use of the most suitable materials for individual functions. Liners are considered wear items and are replaceable. Cylinder jackets serve to locate and support the liners, and provide passages for coolant and lubricant as well as mounting points for engine accessories.

Cylinder liners perform the following functions:

- Limit the working space (combustion chamber)
- Guide the crank train
- Transfer forces to the crankcase
- Transfer heat to the coolant
- Assist in gas transfer (in two-stroke engines)

The basic form of cylinder liners is that of a circular cylinder. Dry liners are made with or without flanges, and are inserted into cylinder bores in the engine block. Wet liners, i.e., those whose outer walls are directly exposed to coolant, have a flange at their upper end, which is

Fig. 6.298 Cylinders with and without upper deck plate ("closed deck" vs. "open deck" engine blocks). Source: KS

axially supported directly by the cylinder deck or by screwing into the engine block and against a mating flange. The liner is free to expand downward. Radially, liners are located by the upper and lower spigots in the engine block (Fig. 6.299).

To prevent corrosion, the outer wall of the cylinder liner is provided with a layer of corrosion protection. Engine type and size, as well as operating conditions, determine cylinder liner configuration, material, wall thickness, flange design, and cooling.

Air-cooled engines have individual cylinders, configured as

- Shrink-fit aluminum barrels over cast-iron liners
- Aluminum barrels with cast-in gray iron liners, joined to the barrel by
    - Intermetallic bond between cast iron liner and aluminum barrel ("Al-Fin" process), or as
    - Rough-cast cylinder liner

In view of their function, cylinder liners must seal outward as well as inward:

- Gas pressure inside the cylinder is maintained by the seal between cylinder liner and cylinder head (cylinder head gasket) as well as the sealing function of piston and piston rings.

- The upper end of the coolant gallery around the cylinder liners must be sealed by the liner flange seated against the cylinder flange, and the lower end toward the crankcase sealed by one, two, or three O-rings set in the liner, whose function is based on their deformation. If these O-rings are not sufficiently deformed, they cannot seal properly; excessive deformation can also compromise sealing. One problem is posed by the fact that after installation, the condition of the sealing rings can no longer be monitored. Therefore, a relief groove is often machined into the liner between the upper and lower sealing rings. A matching relief passage is located in the crankcase or block. If a sealing ring above the relief groove fails, coolant is collected by the relief groove and is able to escape through the relief passage, where it can be observed.

Cylinder liners are subjected to mechanical, thermal, tribological, and chemical loads.

- Tensioning the cylinder head bolts imposes large forces on the liner flange. The cylinder spigot exhibits nonuniform radial deformation in the cylinder head bolt area, especially if cylinder head bolts are not tightened in the proper sequence and with the correct torque.

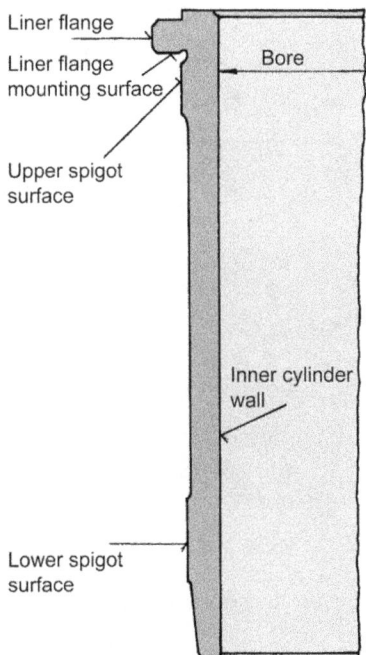

Liner flange

Liner flange mounting surface

Bore

Upper spigot surface

Inner cylinder wall

Lower spigot surface

Fig. 6.299 Cylinder liner terminology.

This leads to constriction of the cylinder liner. The shape of cylinders machined directly into the crankcase is also influenced by installation of engine accessories (Fig. 6.300).

For passenger car crankcases without a top deck plate; i.e., with free-standing cylinders (so-called "open deck" designs), cylinder shape is less susceptible to bolt tension than is the case with "closed deck" designs (Fig. 6.301).

- Gas pressure can literally "inflate" the cylinder liner; it deforms in time with the engine working cycle.
- Oscillating piston normal forces excite cylinder liner vibrations.
- High gas temperatures, steep temperature gradients, and temperature changes due to load changes give rise to quasistatic thermal stresses and thermal deformation (Fig. 6.302).

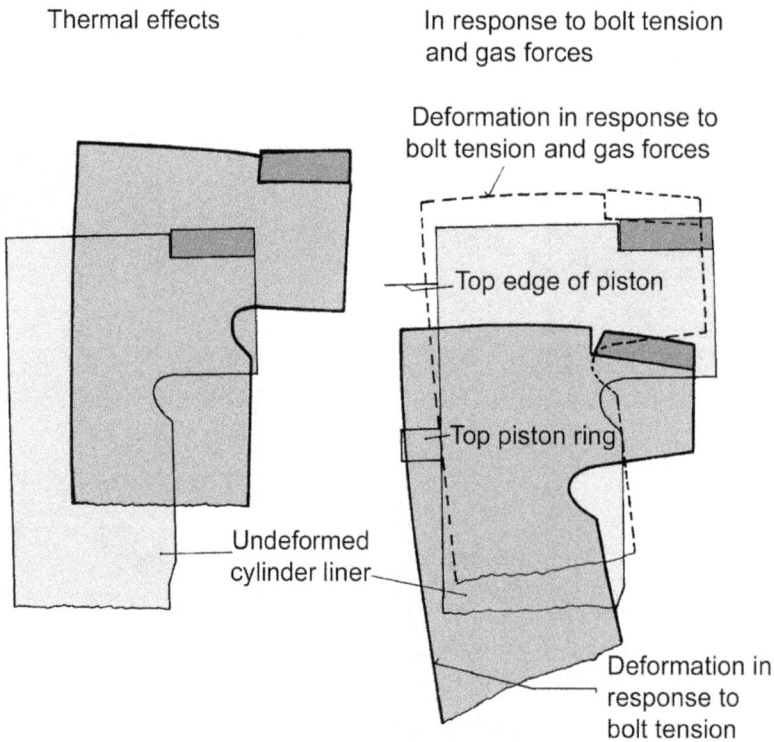

Fig. 6.300 Cylinder liner deformation. Effect of bolt tension on cylinder shape.

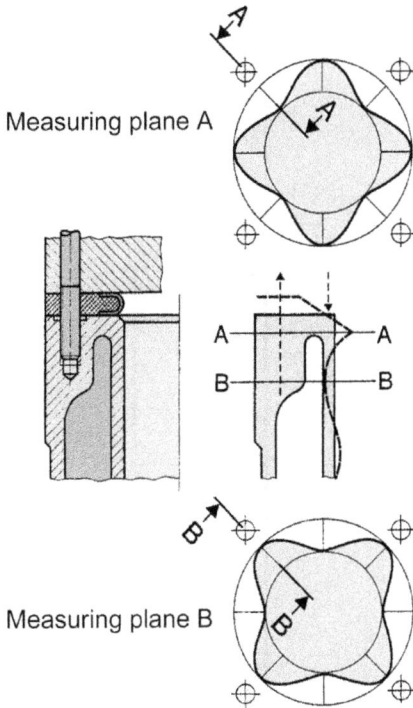

Measuring plane A

Measuring plane B

Fig. 6.301 Cylinder deformation of a closed deck crankcase. Conventional design showing typical elastic cylinder deformation caused by attachment of cylinder head. Source: Federal-Mogul

- Additional loads result from dissimilar cylinder and cylinder head temperatures, which try to equalize through the cylinder gasket plane.

- Tribologically, cylinder walls are subject to loads imposed by mixed lubrication between the piston ring and cylinder liner; mechanical wear takes place.

The properties required for cylinder liner function are achieved by material selection and manufacturing processes.

- **Material and microstructure:**
  Diesel engine cylinder liners employ a cast gray iron alloy with pearlitic microstructure and a fine, hard phosphide network (providing a support lattice function), specifically,
    - Cr alloy GGL (standard material; cast iron with lamellar graphite),
    - CrNi alloy GGL (greater resistance to wear and mechanical loading), and
    - High carbon CrMo alloy GGL (very good wear characteristics).

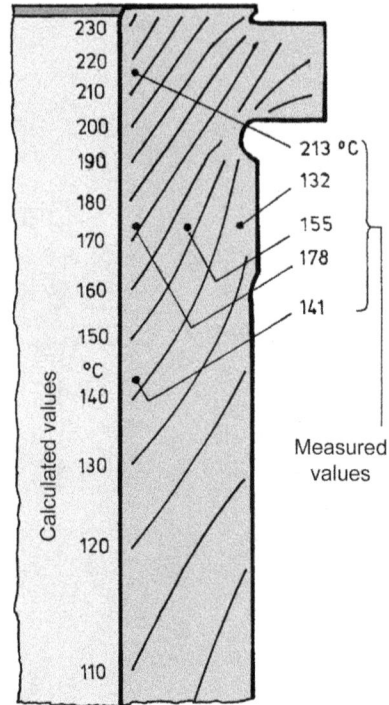

Fig. 6.302 Temperature distribution in a cylinder liner. Four-stroke diesel engine, bore ca. 165 mm, power output 1000 kW, speed 2200 rpm.

Note: Compound castings consisting of lamellar cast iron (good wear properties) and modular cast iron (high strength) permit taking advantage of the properties of different materials.

Passenger car engines with aluminum-alloy engine blocks employ either armored inner cylinder walls or coated pistons running directly on aluminum cylinder walls.

- **Factors determining liner shape:**
  Macrogeometric shape of liners is achieved by the manufacturing process and installation, crankcase design (stiffness), type and arrangement of cylinder head bolts, cylinder and crankcase wall thickness.

Additionally, the tribological properties of wear and running properties are improved by:

- Hardening (induction hardening, nitriding, laser hardening).

- Phosphating (coating): phosphate crystals on the cylinder surface improve oil adhesion and thereby sealing function, which helps to prevent piston ring burning.

- Manufacturing processes to achieve specific surface finish. The microgeometry of the cylinder running surface significantly determines the running behavior of piston rings and piston; this is why cylinder surfaces are honed (Fig. 6.303).

With increasing power output, engine cylinder walls are susceptible to formation of a so-called metal jacket, which results if cutting pressure of the honing tool crushes the desirable honing marks and covers up lamellar graphite structures (Fig. 6.304).

Cylinder walls are therefore subjected to a special honing process, so-called plateau honing (Fig. 6.305). Peaks are cut down to produce smoother plateaus, separated by deep valleys. It is necessary to achieve a uniform honing finish, with well-defined crosshatch pattern, without torn or folded material. Such a honing pattern is achieved by finish

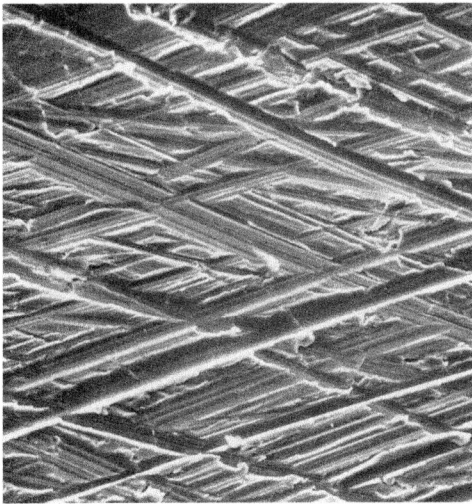

Fig. 6.303 Normal honing structure.

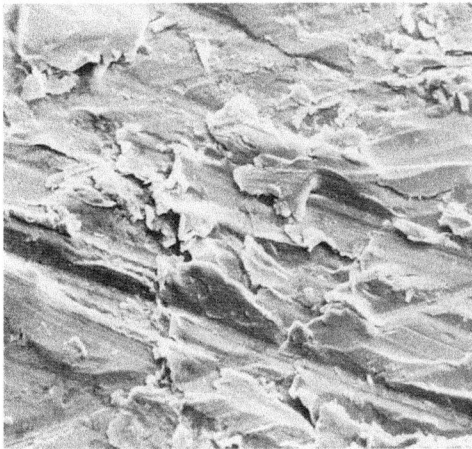

Fig. 6.304 Honing structure with "metal jacket" formation.

honing, which does not reach the deepest valleys created by the pre-honing process (Fig. 6.306).

The crosshatch angle should be between 30 and 60° (relative to the cylinder axis). The plateau structure is evaluated according to the so-called Abbott-Firestone curve (Fig. 6.307) using the criteria of roughness, percent surface contact, and average depth of "outliers."

Honing quality is of decisive importance for the wear behavior of the cylinder/piston/piston ring functional group. However, good honing is a complicated process from a manufacturing point of view, and requires appropriate knowledge and experience. Gray-market cylinder liners, so-called pirate copies, represent an appreciable risk for engine operators.

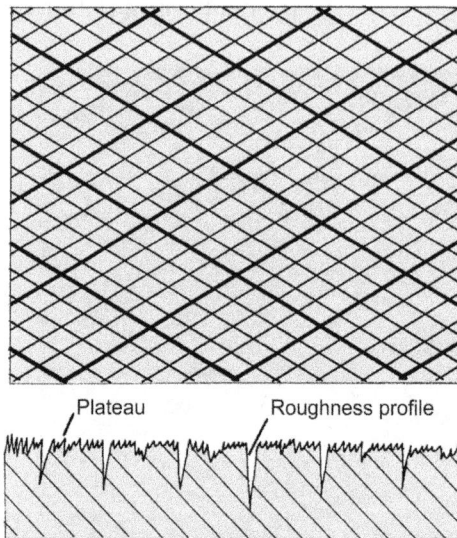

Fig. 6.305 Plateau honing, schematic.

Fig. 6.306 Plateau honing (reflection electron micrograph). Source: MTU

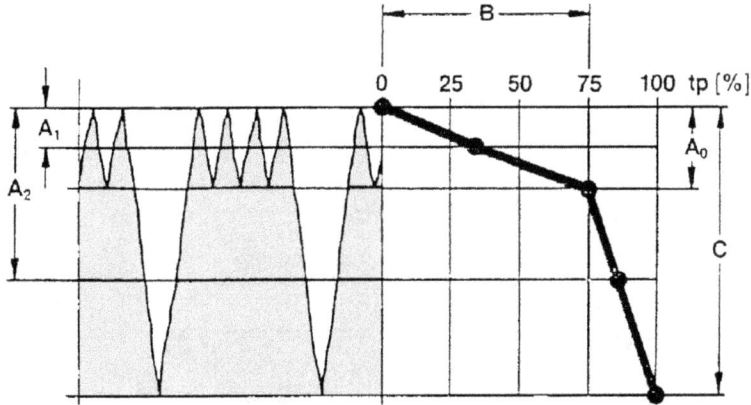

Fig. 6.307 Evaluation of honing structure using Abbot-Firestone curve. A_0, roughness of base structure; A_1, A_2, valley depths; B, percent surface contact of load-carrying structure; C, base structure.

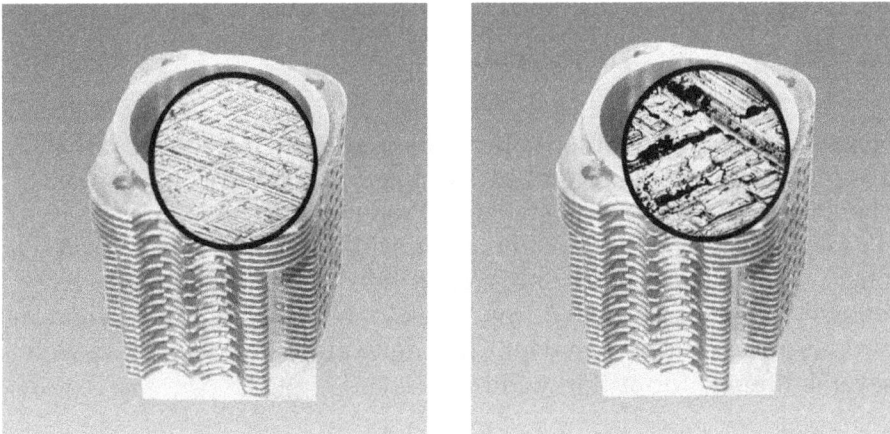

Fig. 6.308 Honing structures of air-cooled cylinder barrels. Original manufacturer replacement (left), and pirated copy (right). Source: DEUTZ

Comparison of honing structures in an original replacement part and a counterfeit is shown in Fig. 6.308 [6-26].

6.3.4 Cylinder damage

6.3.4.1 Wear

Cylinder wear is unavoidable. Deciding factors are wear rate and the specific affected regions of the cylinder wall. Cylinder wear is driven by several different factors (Fig. 6.309).

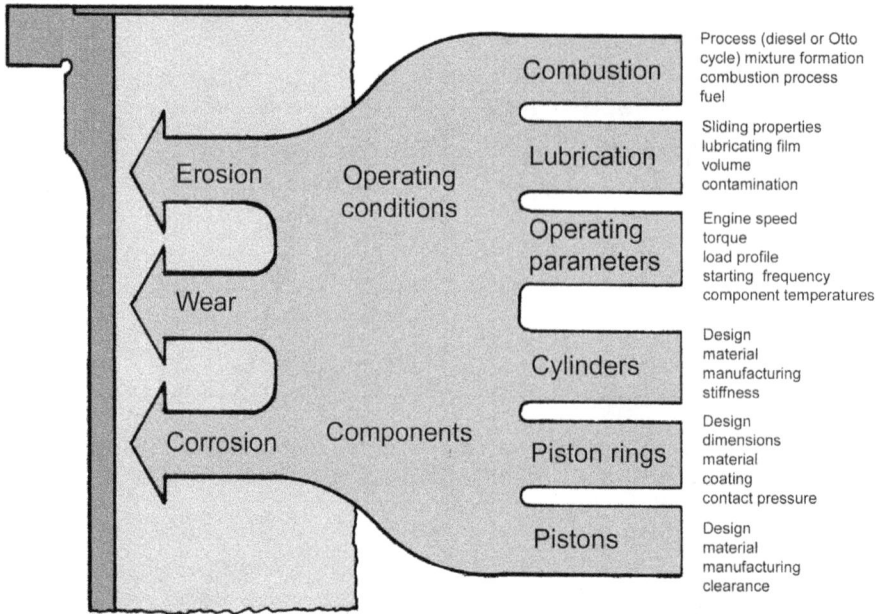

Fig. 6.309 Factors affecting cylinder wear.

Because operating conditions often have a greater effect on wear than materials, microstructure, and surface finish, the bandwidth of wear phenomena is correspondingly large. Proper engine break-in is decisive. It is desirable to achieve rapid break-in, while avoiding burned piston rings and cylinder wall scoring. Wear products (iron oxide) act as an abrasive. Cylinder shape (i.e., roundness or out-of-round, and cylinder outline) affect running behavior. Cylinder wall wear manifests itself in several different forms (Fig. 6.310).

6.3.4.2 Ring reversal (top dead center) wear

In the area where the top piston ring reverses direction, the hydrodynamic lubricating film collapses. At ignition top dead center, combustion gas acts like a cutting torch. In addition, gas pressure forces the top ring to operate in the wear-inducing mixed lubrication regime. This is aggravated by a corrosion component if combustion gas temperature drops below the condensation point of SO_2, allowing the formation of sulfurous acid. Ring reversal wear is intensified by increasing gas pressures and poor fuel quality (heavy oil); see Fig. 6.311.

6.3.4.3 Adhesive wear

The materials of members in sliding contact form microscopic "contact bridges," which are subsequently separated, tearing particles out of the cylinder wall in the process (Fig. 6.312).

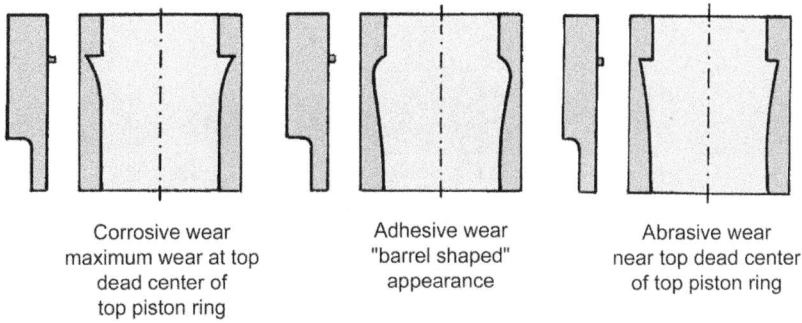

| Corrosive wear
maximum wear at top
dead center of
top piston ring | Adhesive wear
"barrel shaped"
appearance | Abrasive wear
near top dead center
of top piston ring |

Fig. 6.310 Wear of medium and large diesel engines consists of several different components (schematic of wear progress after Aeberli and Lustgarten).

Fig. 6.311 Corrosive wear.

Fig. 6.312 Adhesive wear.

6.3.4.4 Abrasive wear

Wear particles in fuel, especially in low-quality fuel, can reach the cylinder walls and act as abrasive compound (Fig. 6.313).

If carbon particles from the combustion chamber polish the cylinder wall in the upper third of the piston stroke, bright areas appear, spreading from the minor thrust side and extending over large areas of the cylinder wall (Figs. 6.314 and 6.315).

One particular characteristic of abrasive wear is *bore polishing* (Figs. 6.316, 6.317, and 6.318). This is understood to mean smoothing wear by

Fig. 6.313 Abrasive wear.

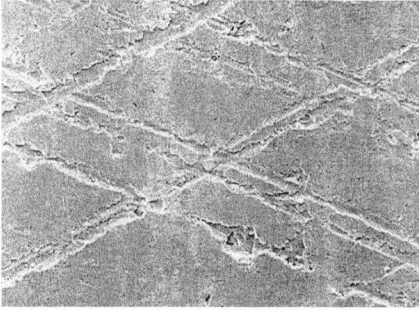

Fig. 6.314 Cylinder bore polishing: appearance of smoothened, worn honing plateaus with deformation burrs pushed into honing valleys. Source: MAHLE

Fig. 6.315 Cylinder bore appearance outside the polished zone. Source: MAHLE

the grinding action of carbon deposits on the piston crown (fire land). Carbon deposits in the piston ring groove roots cause the piston rings to "ride up," which also promotes bore polishing. The phenomenon derives its name from the resulting brightly polished cylinder surfaces. Bore polishing leads to increased blowby and increased oil consumption, and ultimately leads to ring and piston seizing due to local oil starvation.

Initially, mitigation was attempted by means of increased fire land clearance. Ultimately, an effective remedy was found in the form of a flame ring (aka fire ring, carbon cutting ring, anti-polish ring, shown in Fig. 6.319). The flame ring consists of a loose cast iron ring inserted in a corresponding channel at the top of the cylinder liner, which extends slightly beyond the cylinder wall. It reduces piston clearance and so makes it more difficult for carbon deposits to form on the piston fire land. Bore polishing may also be addressed from a lubrication perspective, by selecting "super high performance diesel" (SHPD) oils or "long distance oils" (LDO) [6-27].

6.3.4.5 Scoring

Axial scoring may be caused by seized pistons and rings, fuel flooding, foreign particles carried by combustion, as well as wear particles.

6.3.4.6 Seizing

Obviously, seized pistons and rings always affect the cylinders or cylinder liners (Fig. 6.320). These often represent consequential damage. Although the original causes may be rooted in cylinders or liners, in most cases they may be traced back to piston rings and pistons, or on interactions between these sliding members.

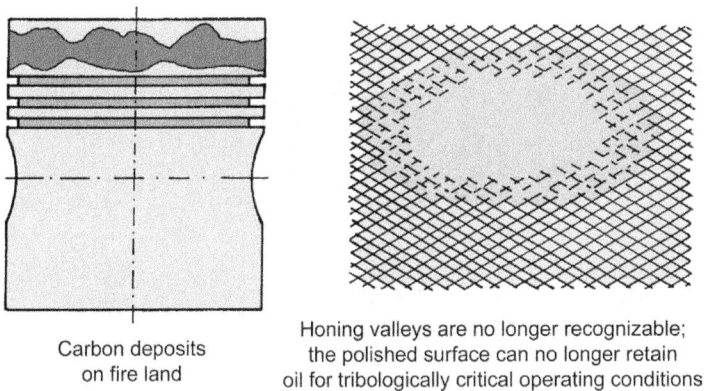

Carbon deposits on fire land

Honing valleys are no longer recognizable; the polished surface can no longer retain oil for tribologically critical operating conditions

Fig. 6.316 Bore polishing (schematic). Uneven polishing wear of cylinder running surface caused by grinding effect of piston fire land carbon deposits. Source: MAHLE

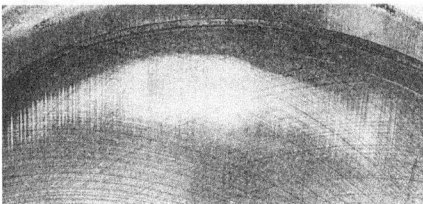

Fig. 6.317 Pronounced, extensive bore polishing.

Fig. 6.318 Pronounced, limited-area bore polishing.

Example a: During a cross-country trip, a tour bus with a rear-mounted engine suffered a burst coolant hose. The resulting loss of coolant allowed engine temperatures to rise to the point where all pistons seized, more or less solidly, in the upper cylinder liner area (Figs. 6.321 and 6.322). The bus driver only noticed a problem as the vehicle lost power and came to a stop. This failure could easily have been avoided if the customary visual coolant temperature gauge had been supplemented by an acoustic warning upon reaching the maximum permissible coolant temperature. A driver cannot continuously monitor engine operating parameters such as coolant temperature or engine oil pressure; he is far too occupied with traffic situations. The ultimate

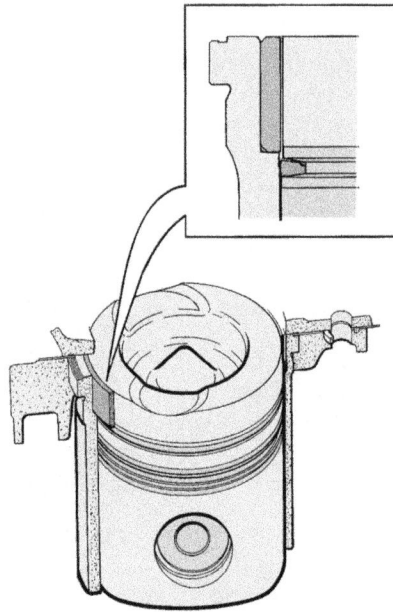

Fig. 6.319 Cylinder liner with flame ring. Source: Scania

Fig. 6.320 Streak-like seizing marks as a result of piston ring seizing.

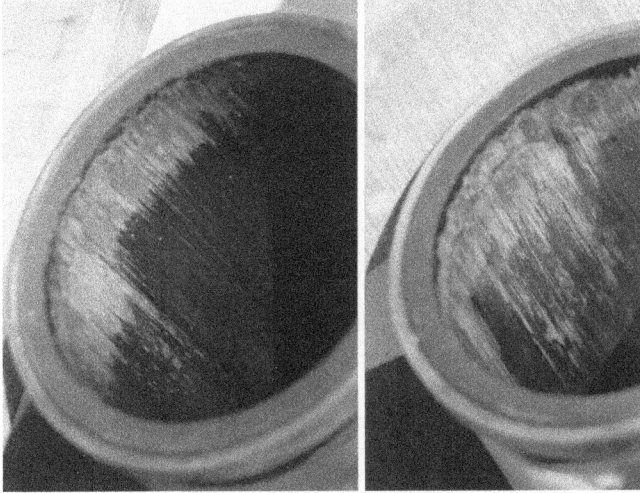

Fig. 6.321 Seizing marks in cylinder liners as a result of engine overheating.

Fig. 6.322 Piston firmly seized in cylinder as a result of engine overheating.

cause of this avoidable failure is a design fault on the part of the vehicle manufacturer.

Example b: In the course of operation, the cylinder barrels of an air-cooled engine became clogged with dirt, largely filling the spaces between the cooling fins. This reduced heat transfer, resulting in overheating and severe piston seizing (Fig. 6.323).

6.3.5 Cavitation

Within a tightly limited wall area on the coolant side of cylinder barrels, opposite the major thrust side of the piston and cylinder, hard, alternating piston contact may result in cavitation. The piston impact excites vibrations in the cylinder liner; the liner deforms, and the coolant is no longer able to follow this deformation. Locally, pressure

Fig. 6.323 Piston seizing in an air-cooled engine as a result of overheating. Top: Removed cylinder showing seizing marks; bottom: destroyed piston with seizing marks and local melting.

drops below the coolant vapor pressure. As the cylinder wall bounces back, the vapor bubbles implode, with the expected consequences. If the coolant contains enough free oxygen for oxide formation, the effect of cavitation is amplified by corrosion (Figs. 6.324 and 6.325).

The following measures may be used to combat cylinder cavitation:

- Increased coolant system pressure
- Coolant de-gassing (coolant preparation)
- Piston design
    - Reduced piston clearance
    - Controlled-expansion pistons. In one case, application of controlled-expansion pistons finally conquered a cavitation problem in one railway locomotive engine design.
    - Piston offset toward minor thrust side (see Fig. 6.7). This is understood to mean that the wrist pin axis does not intersect the cylinder axis, but rather is offset several millimeters toward the major thrust side. Because diesel engines normally have wrist pins offset toward the minor thrust side to reduce carbon deposits, the advantages and disadvantages of these contrary measures must be evaluated on a case-by-case basis.

6.3.5.1 Crack cavitation

To damp the transmission of coolant pressure fluctuations to the liner sealing rings, so-called damping grooves are employed. However, these in turn promote the formation of crack cavitation. This arises because cylinder liner vibrations force coolant out of the crack rapidly enough for the coolant to drop below its vapor pressure.

6.3.5.2 Cracks and fractures

Cracks and the resulting fractures most commonly occur in the flange area. They may originate from an entire series of causes (Figs. 6.326, 6.327, and 6.328):

- Incorrect cylinder head gasket.
- Improperly shaped or dimensioned liner flange or crankcase seat.

Fig. 6.324 Cylinder liners showing cavitation.

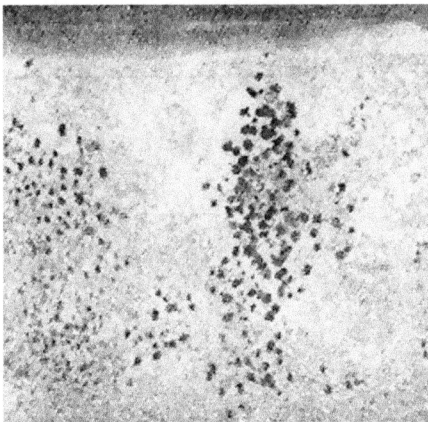

Fig. 6.325 Detail view of cavitation damage to a cylinder liner.

- Improper installation: inadequately or excessively tightened cylinder head bolts. In the case of insufficient tightening, the liner flange and its seat in the crankcase are able to move relative to each other. The result is fretting corrosion, which leads to cracks in the flange as well as the crankcase seat.

- Uneven mating of flange and seat as a result of corrosion, fretting corrosion, foreign particles, coolant scale deposits.

- Fillet fracture caused by corrosive attack.

Example: After long service, longitudinal and transverse cracks between the flange and upper cylinder spigot were discovered in the cylinder liner of a marine engine, along with displaced longitudinal cracks. The cracks extended to the combustion chamber, leading to gas leaks and exhaust gas bubbling in the coolant recovery tank. Corrosive loss of material resulted in kerfs that led to fatigue failures (fatigue crack

Fig. 6.326 Cracked cylinder liner; crack at thinnest cross section of liner, with two slightly offset transverse cracks.

Fig. 6.327 Cleaned cylinder liner with cracks broken open.

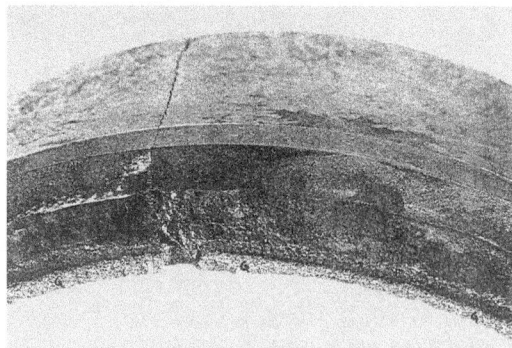

Fig. 6.328 Detail view of cylinder liner crack.

corrosion). Presumably, crack propagation was accelerated by corrosion. This corrosive attack was caused by faulty coolant preparation.

6.3.6 Cylinder heads

The cylinder head forms the boundary of the working volume, carries intake and exhaust valves, gas ports, injector nozzle and prechamber or swirl chamber (if used), in large engines also the safety and decompression valves as well as the air start valve. The valve actuation mechanism and valve rotators are installed on the cylinder head.

Cylinder heads (Fig. 6.329) are boxlike structures consisting of the compression surface at their lower extremity, the upper deck, and side walls. Gas ports, valve guides and ports for fuel injector, spark plug, prechamber or swirl chamber, with intermediate walls and support ribs, brace the compression surface and stiffen the structure. Coolant flow through the cylinder head is controlled by the intermediate webs, the

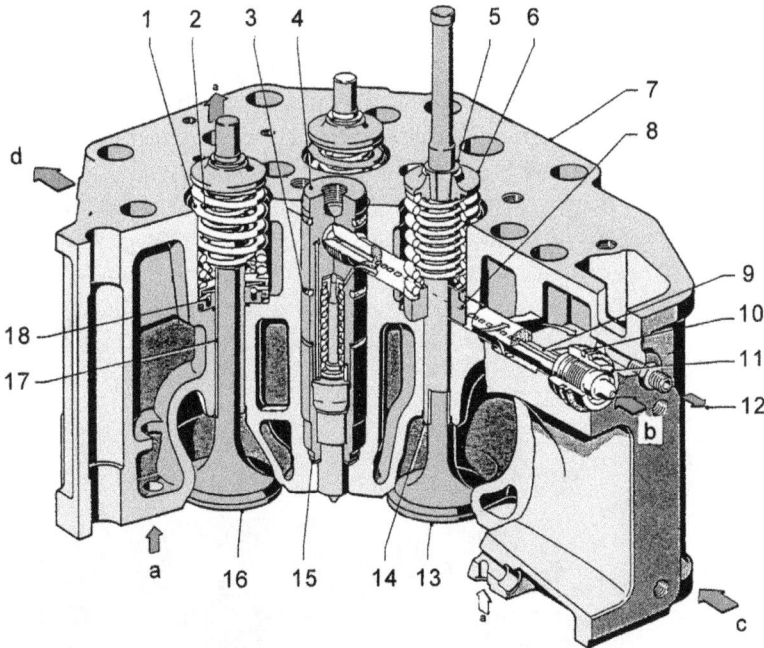

Fig. 6.329 Schematic of diesel engine cylinder head assembly and nomenclature: 1, Outer valve spring; 2, Inner valve spring; 3, Sealing ring; 4, Fuel injector; 5, Valve keeper; 6, Upper valve spring retainer; 7, Cylinder head; 8, Lower valve spring retainer; 9, Supply tube; 10, Sealing ring; 11, Hollow nut; 12, Fuel leakoff port; 13, Intake valve; 14, Intake valve guide; 15, Sealing ring; 16, Exhaust valve; 17, Exhaust valve guide; 18, Valve rotator. a ,Engine coolant; b, Fuel; c, Air; f, Exhaust. Source: MTU

bracing between ports and gas passages, and openings through these braces and webs.

In smaller engines, cylinder heads of an entire cylinder bank are cast as a single unit. Especially in the case of passenger car engines (Fig. 6.330), this simplifies camshaft mounting ("overhead camshaft") within the cylinder head. In larger engines, a cylinder head covers two or three cylinders. Individual cylinder heads are used on engines with cylinder bores of about 120 mm and above. Individual heads have advantages in terms of manufacturing and installation (modular construction), and are stiffer, which again has advantages in force transmission and sealing.

Design differences may be found in the number and location of valves. Passenger car and small commercial vehicle engines have two, three, four, or even five valves per cylinder; larger engines have four; special designs have as many as six valves (Fig. 6.331).

Each valve can control a gas flow through a single passage, or a single passage may supply two or more valves. A crossflow configuration offers flow advantages, but locating both intake and exhaust ports on the same side of the engine (counterflow or "single port face" head) has advantages for turbocharging, thanks to shorter passages to the turbocharger. Intake manifolds and passages may be configured to induce swirl, or may be designed as swirl-free, low-restriction passages for better filling. Cylinder heads are subjected to high thermal (low cycle) and mechanical (high cycle) loading. They are made of lamellar or nodular gray cast iron, or in special cases of cast steel. Passenger car engines usually employ aluminum cylinder heads. Valve seats are inserted to prevent valve recession into the relatively soft heads.

Fig. 6.330 Aluminum alloy cylinder head for a four-cylinder passenger car engine. Source: KS

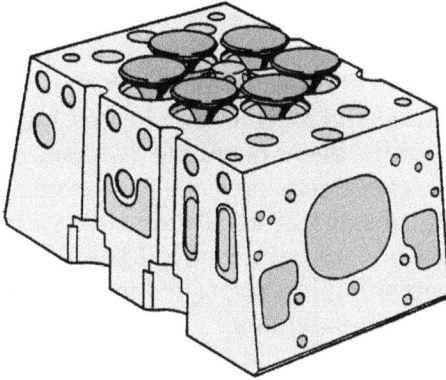

Fig. 6.331 Six-valve cylinder head of a high-performance diesel engine (cylinder bore: 185 mm). Source: MTU

Cylinder heads are simultaneously subjected to high thermal as well as mechanical loads, which leads to conflicting goals in material selection and dimensioning; see Table 6.8. A thin-walled combustion chamber face provides good thermal conductivity, but high gas forces must be countered by a thicker face. To control deformation despite high gas pressures, low-speed two-stroke and medium-speed four-stroke engines employ drilled cooling passages in the valve seat and web areas. Coolant flow in the injector nozzle and exhaust valve areas is positively directed to reduce seat temperatures. Injector nozzles are mounted either in a thin-walled sleeve or directly in a dedicated port in the cylinder head.

Intake valve seats are inductively hardened; exhaust valve seats are hardfaced. In medium and large engines, special valve seat rings are used to prevent recession due to carbon deposits, wear, and corrosion. Depending on size, configuration, and design, valve seats may be cooled (Fig. 6.332). In particular, exhaust valves, subjected as they are to severe thermal loading and hot corrosion, are mounted in valve cages to permit rapid exchange during maintenance.

Table 6.8
Diesel Engine Cylinder Head Temperatures

	Medium-speed engines °C	High-speed engines, high-performance engines °C	High-speed engines, commercial vehicle engines °C
Combustion chamber face (water side)	210 – 250 240 – 260	120 – 150 130 – 140	
Valve web	260 – 300	280 – 290	200 – 350

The gas passages, bolt holes and ports for injector nozzle or spark plug, and various reinforcing and coolant deflector ribs all serve to create a cylinder head structure of varying stiffness, which is well able to withstand normal thermal and mechanical stresses, but which is sensitive to cracking in the event of engine overloading or combustion anomalies. Incorrect bolt tension leads to uneven cylinder head preloading and warping, unsatisfactory sealing and cylinder head gasket failure. In the case of non-maintenance-free cylinder head gaskets, the cylinder head bolts should be retorqued in accordance with the engine manufacturer's recommendations.

6.3.7 Cylinder head damage

6.3.7.1 Valve seat wear

The causes of excessive valve seat wear may include:

- Incorrect valve clearance (valve lash)

- Carbon deposits and/or wear particles

- High exhaust gas temperatures as a result of engine overloading, improper mixture formation, or combustion anomalies

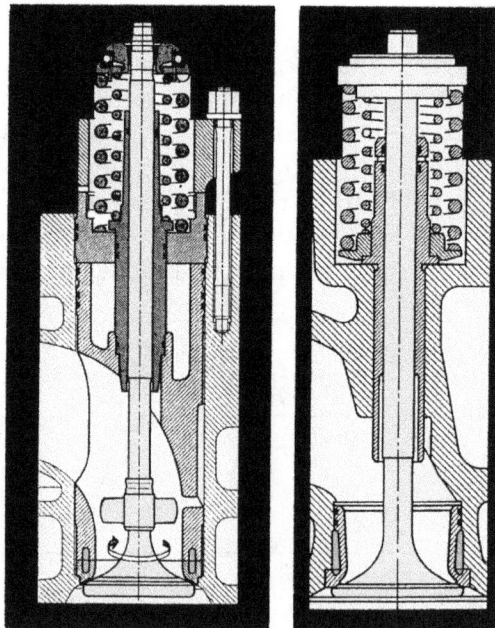

Fig. 6.332 Various configurations of valve seat cooling as found in large medium-speed four-stroke diesel engines. Left: valve cage with cooled valve seats (MAN L 58/64); right: cooled valve seat (MAN L 25/30). Source: MAN

6.3.7.2 Cracks and fractures

Thermal and mechanical overloading result in cracking at critical locations, which differs between individual engine types. Particularly affected are the webs between intake and exhaust valves. These represent low-cycle failures that arise in response to temperature changes caused by engine load changes. Superimposed on these are high-cycle loads that amplify crack formation and propagation (Fig. 6.333).

Origination of such cracks is initiated or promoted by coolant-side scale deposits, which impede heat transfer and lead to local overheating. On the combustion side of the head, iron oxide scale provides an indication of elevated temperatures resulting from water-side insulating deposits. Unsuitable or improperly prepared coolant therefore becomes a medium- to long-term cause of severe cylinder head damage. Also, poor injector patterns with associated uncontrolled combustion and elevated ignition pressures have been identified as a cause of cylinder head damage.

Example: Cracks were detected in the cylinder head of a four-valve gas-fueled Otto cycle engine, between the spark plug port and the intake and exhaust valve seats (see Figs. 6.334 and 6.335). Destructive testing of the damaged component yielded no clues to material or casting defects; the material, GG 20, met specifications. There were no coolant-side deposits that might have explained the cracks as a result of compromised cooling. The cracks, which extended to the water gallery, may be regarded as fatigue cracks caused by locally high

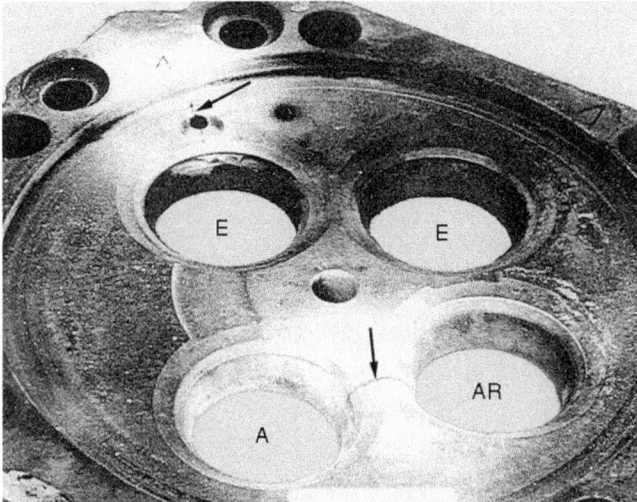

Fig. 6.333 Cracked webs between exhaust valves.

thermomechanical dynamic overloading. Burnt lubricating oil additives in the combustion chamber, recognizable as gray deposits, indicate excessively high combustion temperatures. These might be caused by incorrect ignition timing (early ignition) or combustion knock (the engine was not equipped with knock sensors).

Cracked webs may also be caused by material inhomogeneities and casting flaws (shrink holes), as shown in Figs. 6.337 and 6.338.

Cracks originating from valve seats are often caused by broken valve seat inserts. Microscopic movement of the broken parts against mating components results in fretting corrosion, which serves as the crack origination site. Damaged valve seat inserts hamper valve rotation and may therefore also cause valve damage. Furthermore, there is a danger that parts of a broken valve seat may drop into the combustion chamber. Severe cylinder head damage may also be caused by stuck valves (Fig. 6.339).

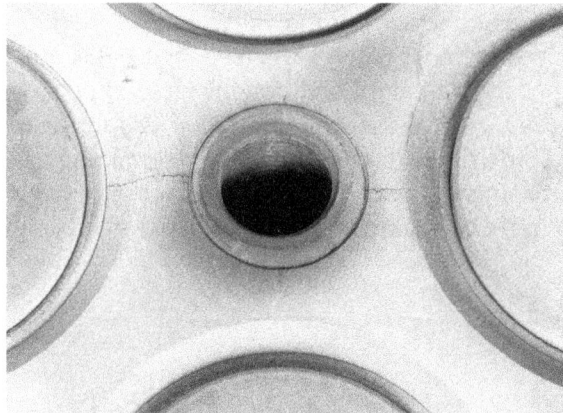

Fig. 6.334 Cracked cylinder head of an Otto-cycle gas engine. Source: Allianz

Fig. 6.335 Fracture surface of the crack in Fig. 6.334. Source: Allianz

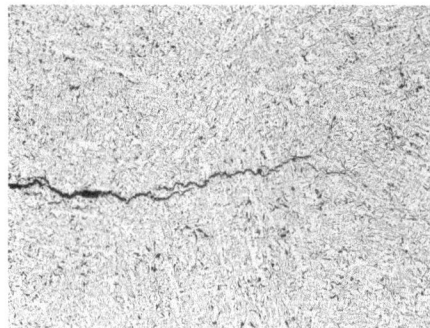

Fig. 6.336 Enlargement of the crack shown in Fig. 6.335. Source: Allianz

Fig. 6.337 Cylinder head of a motor vehicle diesel engine showing shrink hole in web area.

Fig. 6.338 Cracked web as a result of a shrink hole.

Fig. 6.339 Cylinder head damaged by stuck valve and subsequently dropped valve head.

Example: Carbon deposits between valve guide and valve stem led to seizing of the stem and guide, and eventually to a stuck valve. The valve head broke off ("dropped valve"), was trapped between piston and cylinder head, where it was beaten into both piston and cylinder head. The combustion chamber face at all four valve sites was hammered through to the water gallery (Fig. 6.340). The valve seats were hammered and in part broken away. Portions of the valve guides were broken away, as well as portions of the casting contours of the valve guide bores.

6.3.7.3 Erosion and corrosion

High flow velocities may be found in coolant passages where they enter and exit the cylinder head. Fine particles entrained in coolant may cause erosive material removal, especially if flow direction changes drastically. The combustion face of the cylinder head may experience

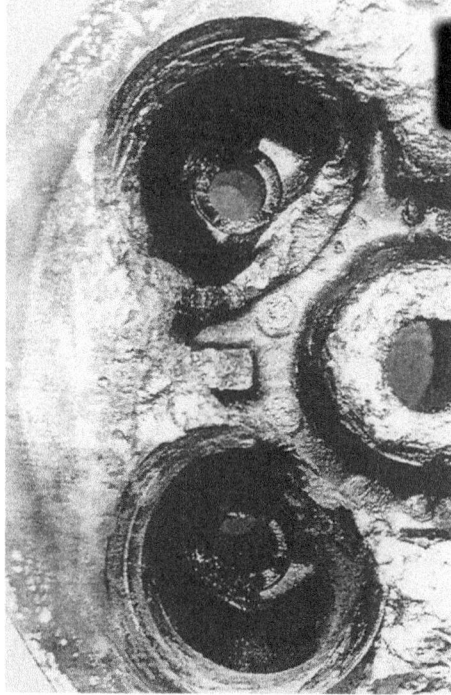

Fig. 6.340 Punctured cylinder head face with hammered valve seats (portion of previous illustration).

corrosive material removal if (small amounts of) coolant enters the combustion chamber (e.g., as a result of penetrated cylinder liner flange seats in the crankcase), especially in the event of long periods of nonoperation (Fig. 6.341).

6.3.7.4 Cylinder head face distortion

Cylinder heads may be distorted as a result of excessive temperatures, above all if multiple cylinders are covered by a common cylinder head. The heads can no longer be sealed properly. Causes may include:

- Lack of coolant
- Restricted coolant flow (clogged passages or hoses)
- Insufficient coolant pump flow volume
- Failure of the coolant thermostat

6.3.7.5 Valve guide wear

Manifestations of wear are dependent on operating conditions. In the case of exhaust valve guides, trumpet-shaped expansion as a result of abrasive wear caused by corrosive material removal may be observed. Exiting combustion gases carry sulfur dioxide and water vapor into the gap between valve stem and valve guide. Once temperature drops

Fig. 6.341 Corroded cylinder head face.

below approximately 140°C (284°F), sulfurous acid is formed, which attacks the valve guide and valve stem ("cold corrosion"). Possible causes include:

- Unsuitable fuels (high sulfur content) combined with unsuitable or over-aged lubricating oil
- Shape abnormalities
- Excessive clearances and/or
- Excessive valve rotation speed caused by scatter in rotocap function

This may be remedied by:

- Improved corrosion and wear resistance of the valve guide material
- Increased temperature to prevent wet corrosion
- Improved gas sealing
- Targeted lubricating oil supply

6.3.7.6 Cylinder head gasket leaks
The cylinder head gasket is unable to seal properly if

- Foreign particles are trapped between the gasket and its sealing surfaces
- The engine overheats

- The cylinder head bolts are torqued incorrectly. In practice, this is usually found to be the case with non-maintenance-free gaskets, if the cylinder head bolts are not retorqued at the operating interval specified by the engine manufacturer.

The consequences of such cylinder head leaks include:

- Blow-through of gases to the coolant

- Entry of coolant to the cylinder, which may lead to hydraulic lock

6.3.7.7 Cylinder head blow-through at spark plug thread, glow plug thread, or injector mount

Such damage is the result of assembly faults; incorrect installation of the spark plug thread in the cylinder head permits blow-through of combustion gases.

6.4 Valve train

The valve train is tasked with controlling gas exchange within an engine by means of timely opening and closing of intake and exhaust ports. A camshaft, driven from the crankshaft by gears, timing belts, or chains, turns at half crankshaft speed (four-stroke engines) or full crankshaft speed (two-stroke engines). The camshaft drive may be taken off the crankshaft on the clutch (engine output) end, which has advantages in terms of torsional vibrations, but this may impose layout penalties. Therefore, in most cases, camshaft drive is taken off the end opposite the output flange. Timing chains and timing belts stretch over the course of their operating lives; this is compensated by spring loaded or hydraulic tensioners. In commercial vehicles and medium to large engines, camshaft and engine auxiliaries (pumps, etc.) are driven by gears (gear train). The camshaft actuates mushroom-shaped reciprocating valves directly, or by means of finger followers (Figs. 6.342 and 6.343). This configuration is primarily found in passenger car engines, as well as high-speed, high-performance diesel engines. In diesel engines, the valves are universally actuated by pushrods, rocker arms, or finger followers.

The functional principle of a valve train (Fig. 6.344) is based on conversion of rotary motion to reciprocating motion by means of an eccentric cam. Such eccentric cams are mounted on a camshaft. As long as the valve or its actuating mechanism (finger follower, pushrod, etc.) rests on the cam base circle, the valve remains closed. Once the base circle transitions to the eccentric portion of the cam (the cam lobe), the actuating mechanism begins to lift, but the valve remains closed against its seat until clearances are taken up. (For the valves to

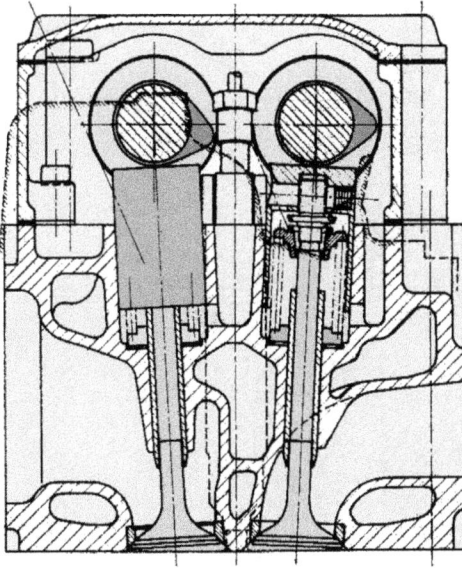

Fig. 6.342 Overhead cam valve train.

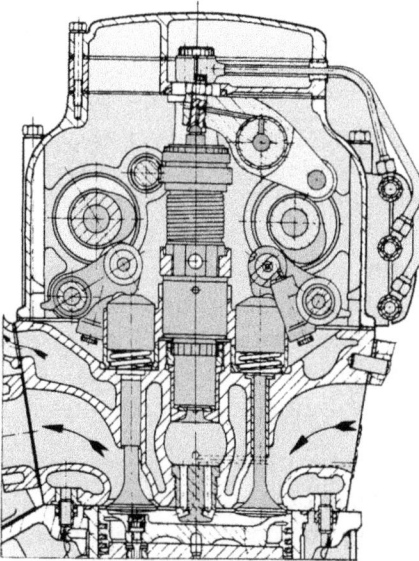

Fig. 6.343 Overhead cam valve train (example: a high-speed, high-performance diesel engine by MTU Friedrichshafen). The exhaust camshaft actuates three exhaust valves per cylinder by means of finger followers, as well as the unit injector. Source: MTU

close tightly, sufficient clearance must exist between the valve and its actuating mechanism to ensure that thermal expansion of the valve does not lift it from its seat). Once valve clearance ("valve lash") has been taken up, forced contact is once again established between the elements of the actuating mechanism; the valve quickly accelerates, up to its maximum. If left to their own inertia, the actuating components

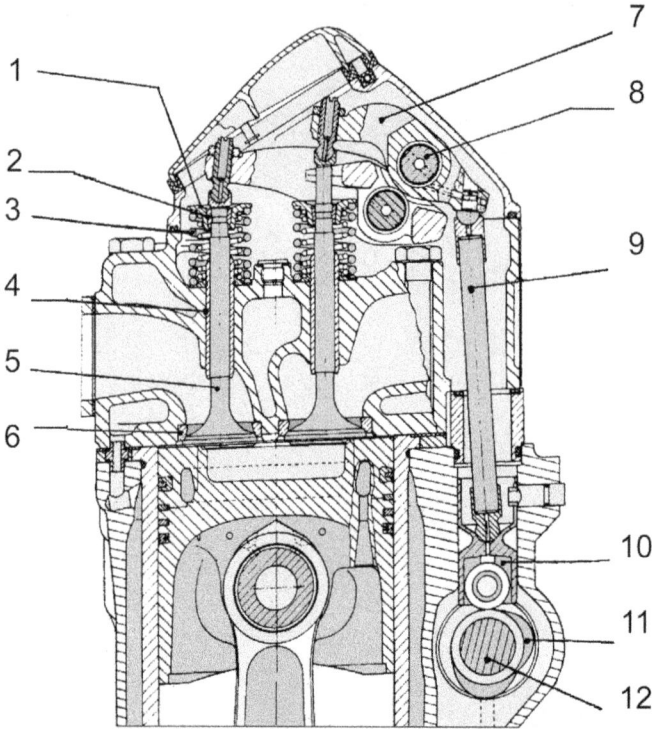

Fig. 6.344 Valve train configuration and terminology: 1, Valve spring retainer; 2, Keepers; 3, Valve spring; 4, Valve guide; 5, Valve; 6, Valve seat; 7, Rocker arm; 8, Rocker arm shaft; 9, Pushrod; 10, Roller tappet; 11, Cam; 12, Camshaft. Source: MTU

would continue to move at this maximum speed. However, the valve is intended to move slowly, to permit large port openings to remain open for a comparatively long time. Therefore, deceleration built into the cam profile must be forced upon the valve by means of valve spring(s). After the valve has opened completely, the valve spring again accelerates it toward the valve seat.

Valve motion is in part controlled by force, in part by shapes of the involved components. The valves should open and close as rapidly as possible, first to open ports quickly and second to prevent bleeding of gases (in particular for exhaust valves). The valve timing diagram specifies the opening range of the valves; valve motion is controlled by the cam profile.

The valve actuation system is a spring and mass system which, given periodic excitation, is prone to unintended oscillation. In view of its oscillatory properties, the valve actuation mechanism must be sufficiently rigid, but in view of inertia effects its components should be as light as possible. The camshafts of vehicle engines may be hollow cast or built up; i.e., the cams may be pressed onto the camshafts.

If valve actuating components are deformed, or in the event of oscillation, valve movement is no longer defined by the cam profile, but rather opens with improper timing (Fig. 6.345).

Valve train oscillations are excited by

- The cam profile, in particular cams subjected to "jerk."
- Manufacturing defects, e.g., an eccentric base circle.
- Camshaft torsional vibration.
- Nonuniform cam drive.
- Bending vibration of camshaft and/or pushrods. The valve actuation mechanism not only undergoes oscillation in the direction of

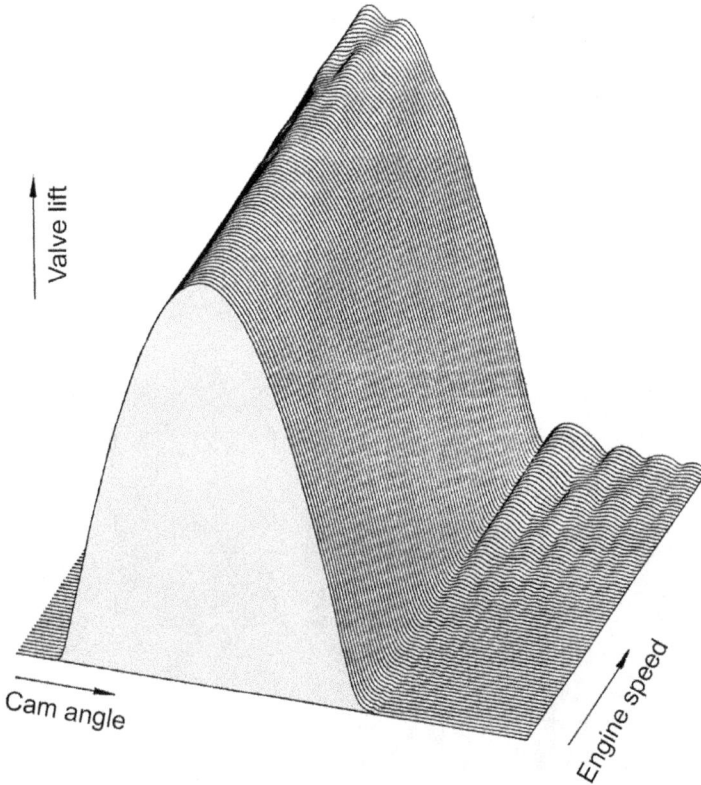

Fig. 6.345 Dynamic (i.e., actual) reciprocating motion of diesel engine valves, with engine load represented by a propeller curve. With increasing engine speed and consequently increasing load (in keeping with propeller behavior), the actual lift curve begins to deviate significantly from the theoretical curve. Source: MTU

force (longitudinal vibrations) but also in bending. These bending oscillations are excited by friction in the spherical ends and seats of pushrods and rocker arms.

6.4.1 Valve springs

In general, the function of springs is to take up forces while undergoing relatively large changes in shape; in this way, they are able to store mechanical energy. They serve to balance energy, force, and distance.

Valve springs, with few exceptions, consist of cylindrical coil compression springs made of circular section wire (Fig. 6.346). They are expected to fulfill the following tasks:

- During the "rest period," the valve must be held closed against its seat.

- Positive engagement (form fit) between the actuating components and the cam must be ensured.

- The valve, accelerated by the ramp and flank of the cam profile, must be decelerated to zero velocity at maximum lift.

- The valve must again be accelerated toward its seat.

Valve springs undergo axial compression, which imposes dynamic torsional loads on the spring material. Because of the curvature of the spring wire, the inner diameter of the spring windings experiences greater shear forces than the outer diameter. The long-term fatigue resistance of valve springs depends on material (type of material,

Fig. 6.346 Valve springs. Source: SCHERDEL

microstructure, heat treatment), but above all on their surface finish. In fatigue testing, the acceptable number of load cycles is determined for a given load.

Valve spring failures

Subjected to high dynamic loads, valve springs are extraordinarily sensitive to material and manufacturing faults of all types. In particular, surface damage—even the tiniest flaws—are the cause of many a broken valve spring, as shown in Fig. 6.347 [6-28]. These may arise

- As a result of the manufacturing process: shrink holes, slag inclusions, foreign bodies, drawing cracks, faulty surface treatment

- In storage: as a result of mechanical damage and/or corrosion

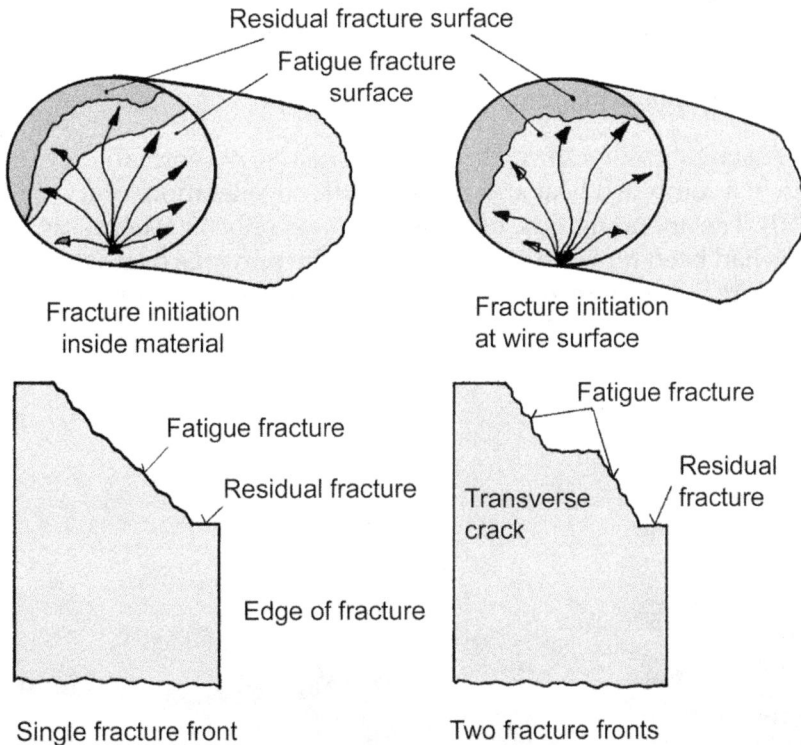

Fig. 6.347 Fatigue failure of springs: failure mode and location (after Pomp and Hempel).
1. Fatigue and transverse fractures occur in one or more windings of a single spring. In the first case, the fracture proceeds at an angle of about 45°, in the second nearly perpendicular to the wire axis. With increasing oscillating loads, the fracture angle is shifted toward angles under 45°.
2. In smooth-edged fatigue fractures, the fracture proceeds along a fracture front. If the fracture shows a steplike discontinuity traceable to a tiny surface defect running parallel or perpendicular to the wire axis, the crack will propagate along two or more fracture fronts. (Source: Hempel, in Konstruktion, Vol. V (1953), No. 10).

- During assembly and repair operations
- During engine operation

The combination of apparently miniscule inhomogeneities or pre-existing mechanical damage, with highly dynamic loading, leads, after longer or shorter operating time, to fatigue failure.

Example: In a diesel engine, the outer valve spring of one exhaust valve broke after about 10,000 km. The fracture was located at roughly the midpoint of the spring (Fig. 6.348). The fracture surface indicated a rapidly progressing torsional fatigue failure. This fracture originated in a pointlike kerf (K) on the inside diameter of the coil (Fig. 6.349).

Given the fatigue strength of the spring steel in question, an added dynamic component as a result of the spring's own oscillation will impose very high loads, making failure as a result of such surface defects unavoidable. However, a flawless surface provides an adequate margin of safety against failure.

Example: An engine valve spring broke in the course of (normal) operation of a combined heat and power (CHP, cogeneration) unit (Fig. 6.350). The operating time of this spring was only 3800 hours, as all springs had been replaced in the course of the previous overhaul. Fatigue failure was recognized as the cause of this spring failure.

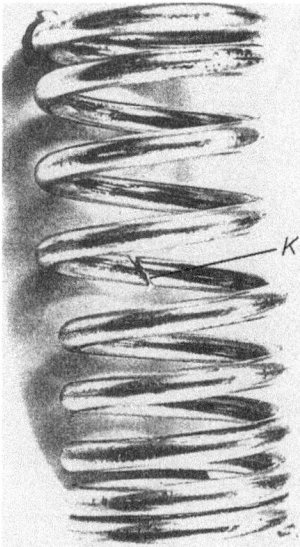

Fig. 6.348 Fracture point of an exhaust valve.

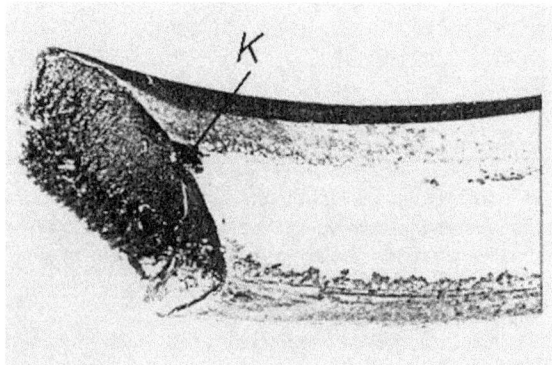

Fig. 6.349 Spring fracture origination point (notch, or kerf, K).

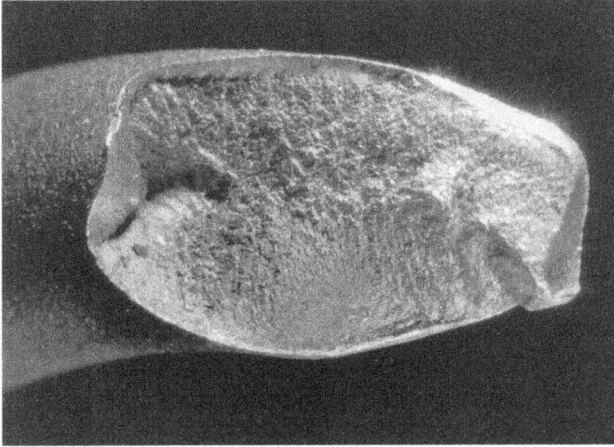

Fig. 6.350 Broken valve spring from a co-generation engine. Source: Allianz

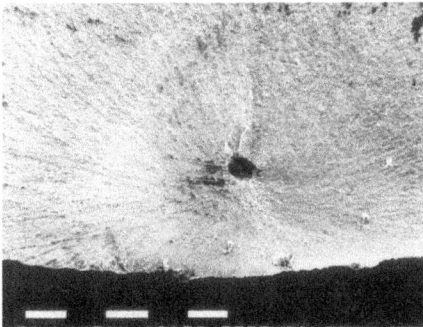

Fig. 6.351 Point of origin of the above spring fracture. Source: Allianz

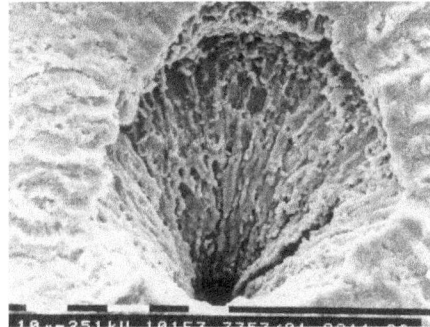

Fig. 6.352 1000X magnification of a void in the spring material. Source: Allianz

The failure origin was not located at the spring surface, where the highest torsional stresses are found, but rather 0.25 mm below the surface (Fig. 6.351). The failure initiator was a sinkhole-like void in the spring material (Fig. 6.352).

Often, the origin of fatigue failures may be found in the flat-ground spring ends. Relative motion between spring end and contact face of the mating surface promotes fretting corrosion, which initiates a fatigue failure (see Figs. 6.353, 6.354, and 6.355).

6.4.2 Valves

6.4.2.1 Introduction

Valves (Fig. 6.356) serve to completely block all gas flow; they work in the direction of flow, or opposed to flow direction. Modern combustion engines use poppet valves exclusively. A poppet valve consists of a disc-like blocking element, the valve head, whose tapered seat forms the

Fig. 6.353 Valve spring fatigue failure, caused by fretting corrosion on a ground spring end.

Fig. 6.354 Detail of the above spring fracture.

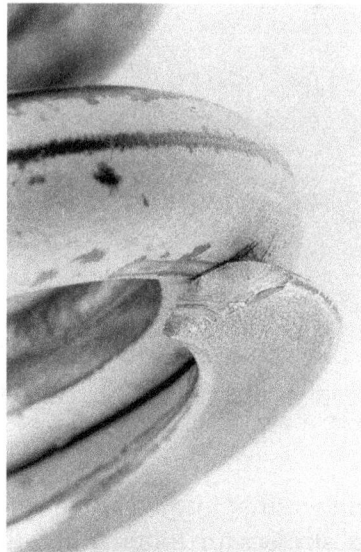

Fig. 6.355 Detail of the above spring fracture; view of fracture surface.

actual sealing surface, and the valve stem, which guides the valve motion. Each valve is actuated by a rocker arm, finger follower, tappet, or directly from the cam contacting the end of the valve stem; and by a valve spring, which is compressed between the cylinder head and the valve spring retainer. The valve spring retainer is provided with an internal taper; two externally tapered valve keepers are wedged into this recess. Ridges in the keepers fit grooves in the valve stem. If a gap is left between the two keepers, differential mating conditions will ensure that they are tightly wedged against the valve stem. If, however, the valve stem is intended to rotate, the keepers will be forced against each other (non-wedging action). As the valve stem extends into the combustion chamber, there is a danger of carbon deposit buildup that might hinder valve motion. To counteract this, the valve stem diameter is reduced to form a scraping edge that removes deposits from the valve guide as the valve opens.

Understandably, valves are unfavorably shaped in terms of flow and heat transfer considerations, as they represent significant flow obstructions, and the valve head constitutes a large heat-absorbing surface with only a narrow seat and the valve stem available for heat removal. The single, but deciding, advantage of poppet valves is

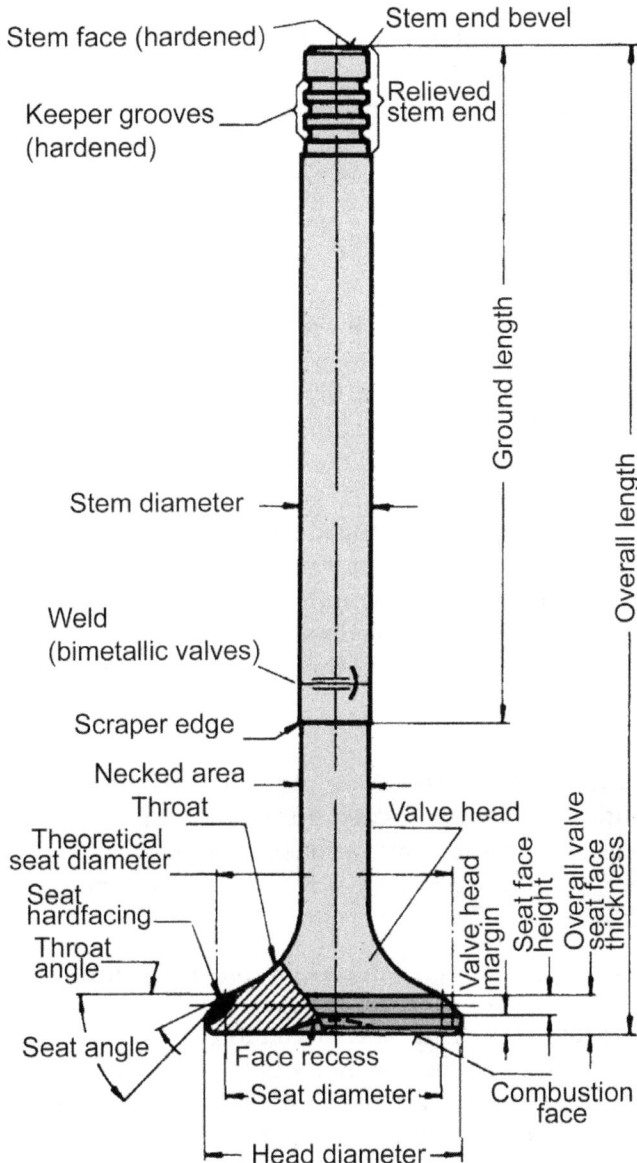

Fig. 6.356 Valve terminology.

that they are pressed against their seats with even more force as the pressure against which they are required to seal increases. This is a self-amplifying process, as the system components themselves, by virtue of their shapes and orientation, support the desired (sealing) function.

The valve head is heated during the time between begin of combustion and end of the expansion process; it is in turn cooled, half during the intake process, half during the compression stroke. The thermal behavior of the valve seat surface is strongly dependent on the conditions at hand. Heat is transferred through the valve seat and valve stem. With regard to good heat transfer, the seat surface should be as large as possible; with regard to sealing pressure, it should, however, be small. Clearance between valve stem and valve guide has to be large enough to permit low-friction valve motion, but small enough to permit good heat transfer from the valve stem. Recommended stem clearance for intake valves is 6% of stem diameter, for exhaust valves 9% of stem diameter.

Design to reduce loads on valves includes

- *More valves per cylinder*—three, four, five, even six valves (see Fig. 6.331), which is advantageous in several respects:
    - Reduction in accelerated valve mass, for reduced inertia forces (inertia forces rise as the square of engine speed)
    - Smaller heat-absorbing surfaces and therefore lower thermal loading of the valves
    - Larger port opening areas, for improved gas transfer
    - Possibility for centrally located injector nozzle or spark plug
    - Self-acting valve clearance adjustment (Fig. 6.357)
    - Valve seat cooling by means of cooled valve seat inserts or removable valve cages (in the case of medium and large diesel engines; see Fig. 6.332)
- *Valve rotators*
  To prevent deposit formation on valve seats and to achieve even temperature distribution, valves are rotated during their opening, closing, or both phases, for example, by various proprietary mechanisms ("Rotocap," "Rotomat," etc.) or rotary vanes on the valve stem (Fig. 6.358).

Valves, in particular exhaust valves, are subjected to severe loads:

- Mechanically, by gas pressure, spring, and inertia forces. In passenger car engines, valves hit their seats about 50 times per second; commercial vehicle diesels, about 17 times per second; and high-speed marine diesels, three to four times per second.

- Thermally, by combustion gases (see Table 6.9). Temperatures range from 600°C for intake valves, to 800–1000°C for exhaust valves. Gas speeds can reach 70–100 m/s and result in intensive heat transfer.

- Tribologically, by operation in the mixed lubrication regime.

- Chemically, by hot corrosion.

Table 6.9
Diesel Engine Exhaust Valve and Valve Seat Temperatures

	Low-speed engines	Medium-speed engines	High-speed engines, high- performance engines	High-speed engines, commercial vehicle engines
	°C	°C	°C	°C
Valve head: Combustion chamber side, Valve seat side	500 – 600 360 – 450	450 – 550 360 – 440	550 – 560 500 -- 550	570 – 690 240 – 350
Stem	200 – 380	180 – 240	250 – 260	260 – 300

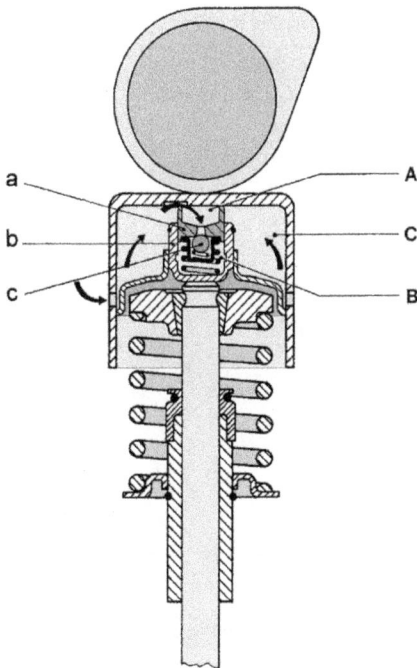

Fig. 6.357 Hydraulic valve clearance adjustment: a, Pressure pin; b, Guide sleeve; c, Ball check valve; A, Reservoir; B, Working chamber; C, Valve lifter chamber. Hydraulic valve lash adjustment provides continuous contact between valve train components. A spring-loaded ball check valve c separates two oil-filled chambers, the working chamber B and the reservoir A. When the valve is rapidly loaded by the cam or valve spring, the oil in the working chamber acts as a rigid, force-transmitting member. In an unloaded condition, oil from the engine lubrication system is able to flow through the valve lifter chamber C, the reservoir A, through the opened ball check valve c into the working chamber B. With gradual loading, such as valve temperature changes, oil is displaced through a restricting clearance between pressure pin a and guide sleeve b, in an amount representing the change in valve length. Source: Daimler

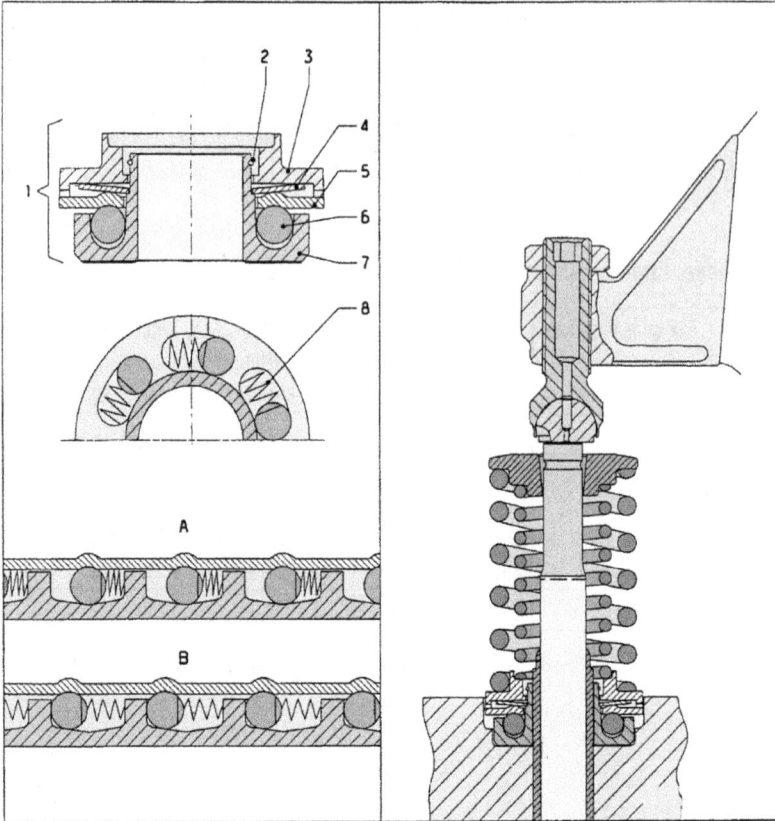

Fig. 6.358 Detail of valve rotator: 1, Valve rotator assembly; 2, Securing ring; 3, Cover; 4, Belleville spring; 5, Ball race; 6, Steel ball; 7, Housing; 8, Coil spring; A, Valve opened; B, Valve closed. Source: MTU

Valve materials therefore have to meet the following requirements:

- High strength at elevated temperatures
- Good thermal conductivity
- Good tribological properties
- High resistance to wear
- Corrosion resistance

An overview of the most important valve materials is provided by Table 6.10.

Because these requirements cannot be completely satisfied by a single material, severely loaded exhaust valves employ a composite solution,

Table 6.10
Standard Valve Materials (Source: [6-29])

Designation, after DIN 17006	Short designation	Applications
X 45 CrSi 9 3	Cr-Si steel	Intake valves, normal service; stem material for bimetallic valves
X 85 CrMo V 18 2	Cromo 193	Intake valves, heavier duty, good resistance to scale and wear
X 53 CrMnNiN 21 9	21-4 N	Intake and exhaust valves for improved hot and long-term durability, as well as corrosion resistance; standard exhaust valve with Stellite F hardfacing for passenger car engines
X 50 CrMnNiNbN 21 9	LV 21-43	Intake and exhaust valves for improved hot and long-term durability, as well as corrosion resistance; standard non-hardfaced exhaust valve for commercial vehicle engines
X 60 CrMnMoVNbN 21 10	Resis TEL	Intake and exhaust valves for improved wear resistance, improved hot and long-term durability; non-hardfaced exhaust valve for truck engines
Ni Cr 20 TiAl	Nimonic 80 A	Extreme duty exhaust valve for large heavy-fuel engines

in the form of bimetallic valves: the valve head, with a more or less long portion of the stem, is made of heat-resistant austenitic steel. The stem end is made of martensitic steel, which can be hardened and provides good sliding properties. The two components are butt-welded together. Valve stems are chrome plated for improved sliding properties. One solution employed by vehicle engines to compensate for poor thermal conductivity of austenitic steels is to use hollow valve stems, about 60% filled with metallic sodium with a melting point of 97°C. The molten sodium is shaken back and forth, and improves heat transfer from the valve head to the stem; valve head temperatures may be reduced by 80 to 120°K. Hollow valves are not employed in medium and large diesel engines; intake valves are monometallic, exhaust valves are butt-welded bimetallic valves. The stems may be salt bath nitrided or chrome plated.

As protection against high thermal, mechanical, and chemical loads, the seat surfaces of intake valves is inductively hardened, while exhaust valve seat surfaces are hardfaced with special cobalt and nickel alloys. The stem end is hardened, hardfaced, or fitted with a hardened button.

Valves are loaded in various ways:

- High cycle mechanical loading by gas pressure, with ignition frequency
- Thermally
    - High cycle thermal loading by non-steady-state heat transfer from valve head to cylinder head
    - Low cycle thermal loading during engine warmup and load changes

6.4.2.2 Valve failure

Thermal overloading

The already high gas temperatures to which valves are subjected are increased even more by

- Engine overloading, and by
- Engine operational anomalies such as
    - Lean mixture and/or incorrect ignition timing for Otto-cycle engines
    - Combustion problems as a result of defective fuel injection for diesel engines

Heat transfer from valve head to seat or stem is hampered if:

- The valve does not seal properly against its seat, and in the event of
- Excessive valve guide clearance (temperature rise of about 100°K!)

Furthermore, the valve head is heated unevenly if valve rotation is compromised. Valve blow-through is the result of faulty valve closing

- As a result of faulty valve geometry and kinematics and/or
- Deposits

The consequences of such thermal overloading are:

- Changes in microstructure
- Thermal cracking
- Valve head breakouts
- Valve head burn-through

Example: After various operating times, two identical gas-fueled Otto cycle engines in a combined heat and power generating plant exhibited excessive exhaust gas temperatures during engine operation, and loss of compression when stationary (Fig. 6.359). The cause in each case was burned exhaust valves (blowthrough) of one cylinder. As a result of improperly set valve clearance, the valves no longer rested against their seats, allowing combustion gases to leak past and heat the valves to their melting point. The exhaust gas stream carried away the molten valve material, leading to blowthrough.

Mechanical overloading

The entire valve train is subjected to high mechanical stresses. As a spring and mass system undergoing periodic excitation, the valve train has a tendency toward resonant vibration (Fig. 6.360). Lately, torsional

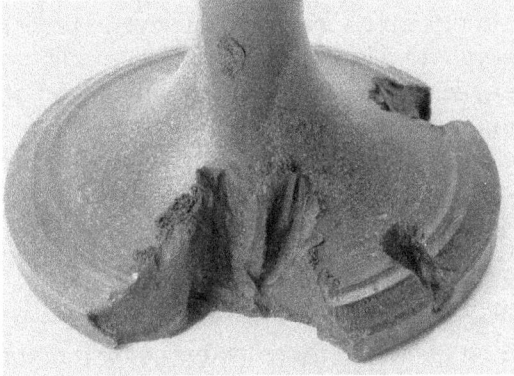

Fig. 6.359 Burned exhaust valve of a natural gas engine. Source: Allianz

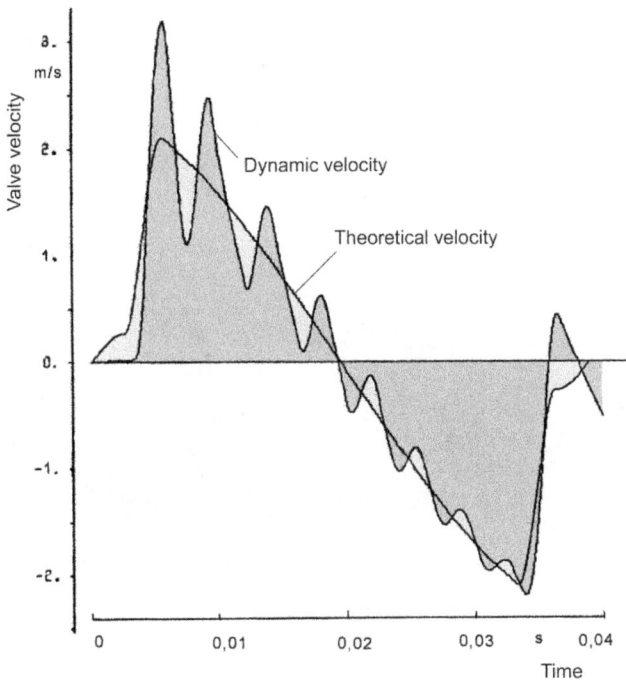

Dynamic velocity

Theoretical velocity

Fig. 6.360 Valve velocity aberrations induced by oscillations. Engine speed 1500 rpm.

oscillation of the camshaft, which also drives the injection pump, has been identified as an exciter in direct-injection passenger car diesel engines. Irregular motion overloads the valve train and results in functional faults, which impose additional thermal loads on the valves.

Associated with this phenomenon, distinction is made between

- *Valve float*
  In which the valve lifter loses contact with the cam lobe in those portions of its travel where the valve spring is intended to ensure contact.

- *Valve bounce*
  In which the valve again lifts off its seat (once or several times) after the actual closing process, and rebounds into the seat at high speed (see Fig. 6.345). This may lead to the valve head breaking off at the transition between stem and head (Fig. 6.361), in particular if there are microstructural inhomogeneities or pre-existing mechanical damage.

Under certain conditions, fracture may occur in the area of the valve keepers. In commercial vehicles, it is not uncommon for the engine to be over-revved, either because the driver has missed a shift, or in descending grades. Due to high inertia forces, valves may fracture. If the valve is not precisely guided, because the valve guide is worn or has worn conical as a result of a manufacturing defect, the valve will undergo a "wobbling" motion, with the end result that the stem is bent and the valve fractures at the head or spring retainer (Fig. 6.362.)

A valve guide that has been expanded by excessive wear may allow the valve head to scrape the inner edge of the cylinder liner while opening; the resulting bending load will lead to a fatigue fracture.

Fig. 6.361 Broken valve head originating from the valve throat. Source: TRW

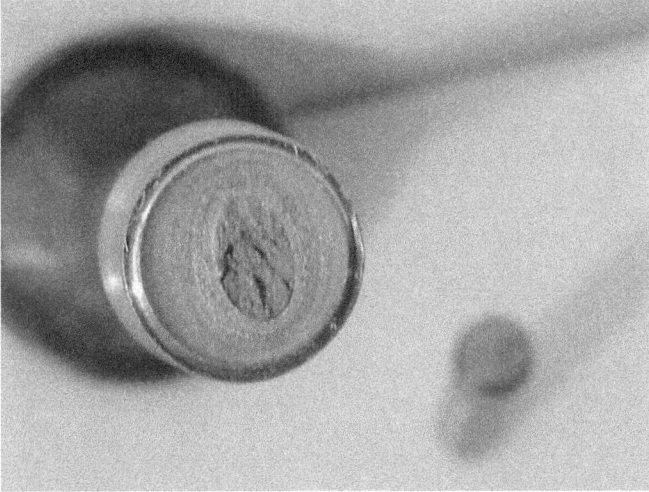

Fig. 6.362 Rotating bending fatigue fracture in valve keeper groove area.

Wear

- *Valve seat*

  Microscopic movement caused by elastic deformation of the valve head in response to high ignition pressures, or in the case of cylinder heads with a "soft" structure, give rise to fretting corrosion between the valve and valve seat. Intake valves were and are most susceptible; by contrast, exhaust valve seats are protected by thin nonmetallic layers (oil deposits, metal ash, soot particles). In diesel engines operating on heavy fuel oil, hard valve seat deposits combined with valve rotation result in scoring (Fig. 6.363).

  The same effect may be caused by excessive rotation speed, or by erratic Rotocap function.

- *Valve guide*

  Shaft seizing is caused by
    - Insufficient clearance
    - Orthogonality problems, and/or
    - Lubricant starvation.

Stuck valves

If carbon deposits build up in the valve guides (as a result of overextended oil change intervals or unsuitable oil), freedom of valve motion may be compromised to the point of seizing. Thickness measurements of such deposits show that they can completely take up all of the available valve guide clearance (Figs. 6.364 and 6.365).

Fig. 6.363 Exhaust valve seat wear of a medium-speed diesel engine after 7400 operating hours.

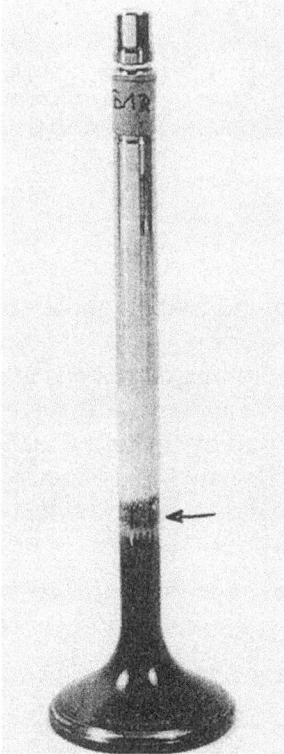

Fig. 6.364 Exhaust valve with scoring and carbon deposits on stem.

Fig. 6.365 Enlarged view showing scoring and carbon deposits.

Depending on where in its travel the valve sticks, it may experience more- or less-severe damage. If it strikes the piston, the valve will be severely deformed and possibly broken (Fig. 6.366).

Depending on the situation, this may result in the valve head breaking off without any visible deformation of the valve stem (Figs. 6.367, 6.368, and 6.369).

Fretting corrosion

Fretting corrosion at the contact surfaces between valve keepers and valve stem may initiate fatigue fractures (Figs. 6.370, 6.371, 6.372, and 6.373), causing the valve to drop into the engine. Distinction is made between

- Primary fatigue failure with a very small residual force fracture in the upper radius of the keeper halves, and
- Secondary fatigue failure in the valve spring retainer area

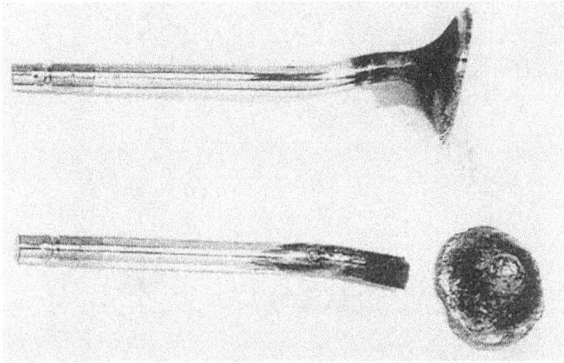

Fig. 6.366 Bent valve, and broken valve showing severely damaged valve head.

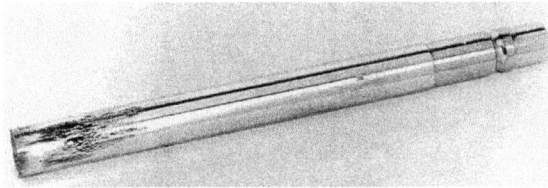

Fig. 6.367 Broken valve stem showing evidence of severe seizing.

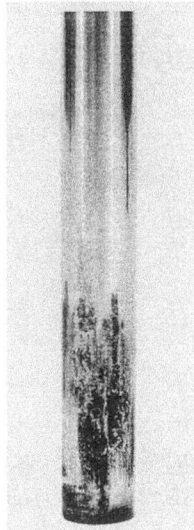

Fig. 6.368 Valve stem (fracture at bottom) with severe seizing marks and carbon deposits.

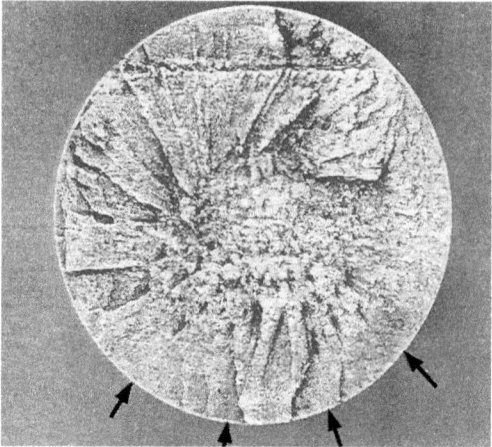

Fig. 6.369 Fracture surface of previous valve stem, showing fracture origins (arrows) at valve stem surface.

Fig. 6.370 Broken valve caused by fretting corrosion at valve keeper recesses. Left: primary fracture A. Right: resulting secondary fracture B at valve head transition. Compare to Figs. 6.372 and 6.373.

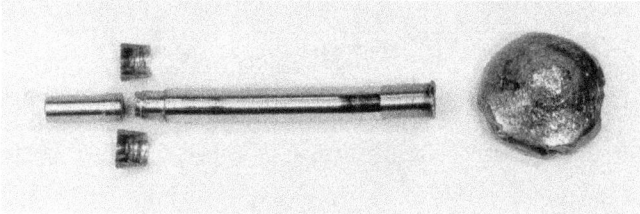

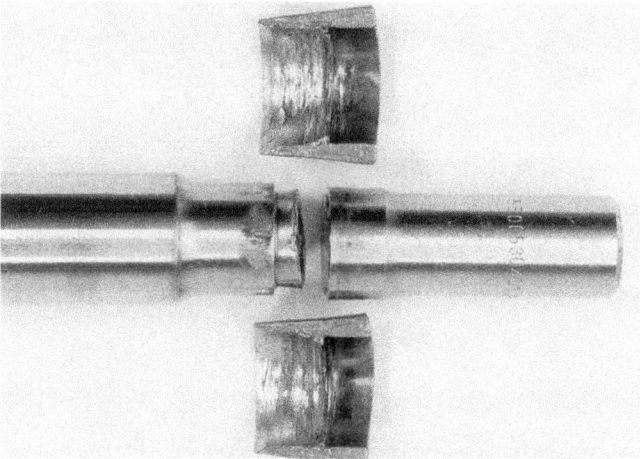

Fig. 6.371 Detail view showing fretting corrosion and wear marks on valve keeper halves and keeper grooves in stem.

Deposits

After long service, oil deposits may form on intake valves as a result of lubricating oil passed into the intake air by the turbocharger or valve guides, and landing on the outside (stem side) of the valve heads. Over time, these deposits carbonize and are firmly attached to the valves (Figs. 6.374 and 6.375).

Corrosion
If temperature drops below their dew point, sulfuric and sulfurous acids are formed, which can attack the valve stem chrome plating and penetrate to the actual valve material. Of course, the valve guides are also attacked by such corrosion. Causes may include:

- Component temperature too low
- High sulfur content in fuel
- Insufficient acid neutralization action of oil
- Excessive blowby

Long-term operation with a leaking charge air intercooler, which admits ocean water to the charge air, may also result in corrosive attack on valves.

Example: After an operating time of about 3000 hours, pinpoint corrosion was detected in the valve guides of a ship's powerplant (Fig. 6.376). Sulfur and chlorine were detected by means of qualitative X-ray microanalysis. Sulfur probably originated in the fuel, while chlorine was carried by salt air.

Combustion of heavy oil forms carbon deposits that adhere to the components making up the combustion chamber. However, carbon particles can also be carried along by the exhaust stream, to be trapped against their seats by closing exhaust valves. Carbon deposits in heavy-

Fig. 6.372 Fracture surface A.

Fig. 6.373 Severely hammered fracture surface B at valve head transition.

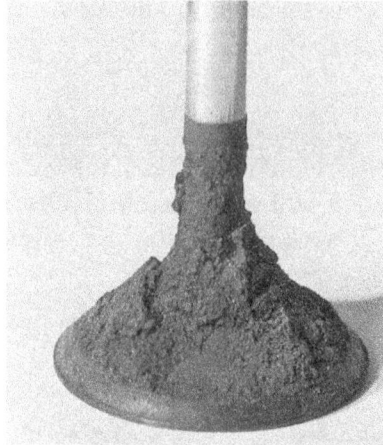

Fig. 6.374 Intake valve with carbon deposits. Source: TRW

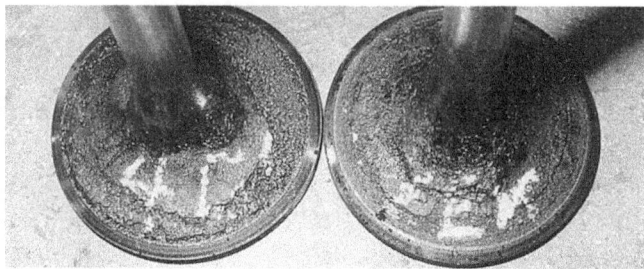

Fig. 6.375 Burned-on intake valve deposits from a medium-speed diesel engine after 7400 operating hours.

oil operation are more dangerous than those resulting from gasoil (diesel fuel) operation, because they promote hot corrosion.

Surface-wide corrosion may also be observed on the undersides (faces) of valves. This leads to material removal, thereby weakening the valve (Fig. 6.377).

Hot corrosion

Corrosion is an especially critical issue in engines operating on heavy fuel oil. Vanadium, sodium, and sulfur contained in heavy oil burn to form oxides (SO_2, SO_3, V_2O_5, and Na_2O). These oxides react with one another to form salts, which melt at relatively low temperatures (for example, sodium vanadyl vanadate, $Na_2O \cdot V_2O_4 \cdot V_2O_5$, has a melting point of 530 to 540°C). The melts can adhere to components at temperatures of 500 to 520°C, and cause surface-wide hot corrosion by dissolving protective layers of metal oxide. They also leach alloying elements from valve seat hardfacing, and additive components from lubricating oil, and admit oxygen from combustion air to valve material, which further promotes corrosion. Pressure of the closing valve squeezes out some of the molten salt deposits, while the remainder hardens and turns brittle. These hard valve seat deposits are cracked by the impact of the closing valve and flake away. This produces gaps (channels), at first only in the deposits, which permit combustion gases to leak through. The cracks and channels in the deposits grow, and act as conduits for hot combustion gases; local temperature rises, and heat transfer from valve head to seat in the cylinder head is impaired. Next, cracks and channels appear in the hardfacing. Due to the appearance of the resulting surface structure, this is sometimes called "cobblestone corrosion" (Fig. 6.378). Ultimately, the cutting torch effect of outrushing exhaust gases leads to complete destruction of the valve [6-30].

Possible remedies:

- Leaner mixture (higher λ)
- Intensive valve seat cooling to temperatures below 450°C
- Improved valve design
- Higher-grade valve materials

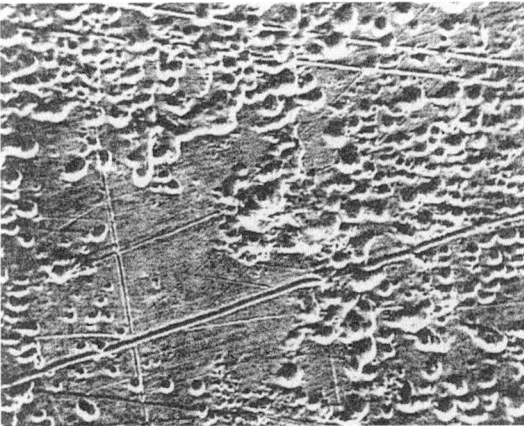

Fig. 6.376 Inside surface of a valve guide, showing pinpoint corrosion.

Fig. 6.377 Valve combustion face corrosion after 7400 operating hours.

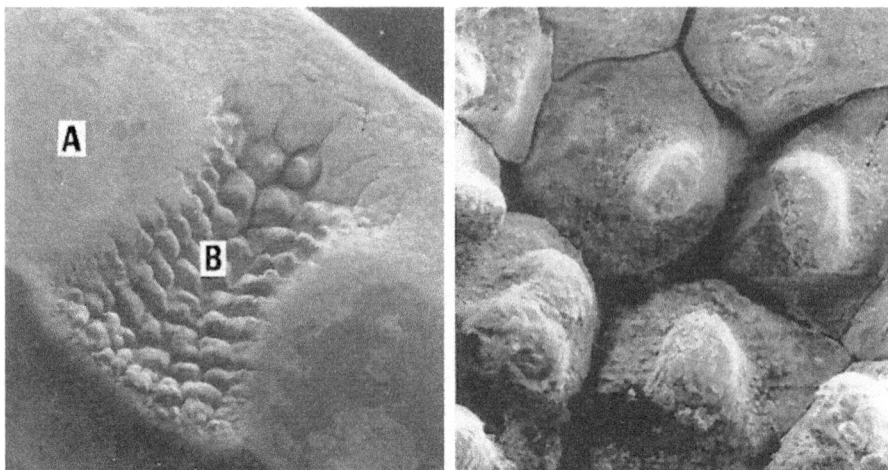

Fig. 6.378 Hot corrosion on hardfaced valve seat (A) of a marine diesel engine, and enlarged view of area B, clearly showing "cobblestone" corrosion. Source: TRW

Erosion

Tiny particles carried by the gas stream, impacting the valves at high speeds, lead to erosive wear. The protective function of hardfacing is reduced, and seat geometry is altered.

Valve rotator faults

- *Contamination*
  Valve rotator function may be compromised by contamination due to lubricating oil residue, wear, or combustion products. These deposits gum up individual parts and hinder the movement of balls in the rotator mechanism.

- *High resistance to rotation*
  Excessive friction between rocker arm and valve stem end, or from valve stem seals, increase the torque needed to rotate the valve. With extreme valve spring angularity, the cover (item 3 in Fig. 6.358) may induce considerable frictional torque in the rotator housing (item 7) itself, resulting in cessation of rotator action.

6.4.3 Camshaft and cam followers

The camshaft actuates the valve train and, depending on engine configuration, also drives the fuel injection equipment. The camshafts of commercial vehicles and medium and large engines are mounted within the crankcase; passenger car engines and high-performance diesels whose valves are actuated directly by the camshaft or by finger followers have camshafts mounted in the cylinder head or a separate upper head section.

Various forces act on the camshaft:

- Valve actuation forces
- Inertia forces
- Spring forces
- Gas forces
- Friction forces
- Fuel injection equipment forces

The camshaft is required to exert considerable force; actuating the cams of a high-speed, high-performance diesel engine (cylinder bore ca. 230 mm) may require as much as 40,000 N (4 metric tons) of force; actuating the injector cam of the individual unit injectors requires about 60,000 N (6 metric tons) [6-31]. As these forces act off-center, the camshaft is loaded in torsion, both statically and dynamically. Line contact is present between cams and lifters—either flat tappets or roller lifters—which leads to high Hertzian contact stress. In addition, lubrication conditions at this line of contact are problematical. Hertzian contact stress and lubrication are therefore the decisive factors for operational reliability and durability of cams and lifter.

Valve motion, which governs gas exchange, is determined by the cam profile. The associated valve lift diagram (which defines *travel*, *velocity*, *acceleration*, and *inertia forces*) determines the mechanical loads and resonant behavior of the entire valve train. The cam profile is composed of several sections:

- Base circle,
  on which the lifter or tappet rolls or slides while the valve is closed
- Ramps,
  which lessen the impact of valve train parts as clearances are taken up
- Cam flanks and nose,
  which open and close the valve

Because individual sections of the cam profile are defined by mathematical functions, they have different radii of curvature and therefore discontinuities at their transitions, which lead to sudden changes in acceleration; these in turn lead to imposition of high mechanical loads on the valve train. These loads may be reduced by so-called "jerk free" cams, with smooth transitions between various sections of the cam profile.

The camshaft exhibits its most severe torsional deflections at ignition top dead center, i.e., at the time of fuel injection (injection equipment). The resonant behavior of the camshaft influences resonance of the entire valve train:

- Radial oscillations disturb valve clearances (valve lash), including during the return phase of the lifter to the base circle.
- Torsional vibrations are always a factor if a lever arm is present to convert vibrations into reciprocating motion.

Cams and cam followers (flat or roller lifters, rocker arms, finger followers, bucket tappets) are subjected to Hertzian contact stresses.

High contact stresses resulting from (theoretical) line contact causes surfaces to flatten. The surfaces are hardened to prevent plastic deformation. Compressive stress builds up in a semicircular stress distribution along the line of contact. Shear stresses build up a few tenths of a millimeter below the surface; these may cause microcracking. Such cracks are promoted by material inhomogeneities and inclusions; the material fatigues, fatigue fractures run at an angle to the surface. Surface pitting takes place.

These processes are aggravated by corrosion. Lubrication is the decisive factor in the operational behavior of such highly loaded components. Whether or not the required elastohydrodynamic lubricating film can form, depends on the design of individual components, material pairings, lubricant supply, and on the lubricant itself. To reduce wear, cams are offset, cam lobes are ground with tapered contours, matched to crowned lifters to ensure lifter rotation; oil-related measures include "extreme pressure" additives.

Camshaft and lifter damage includes:

- Wear
- Fatigue
- Contact corrosion
- Gray staining (micropitting)
- Fretting corrosion
- Damage caused by excessive force
- Seized bearings

Camshaft wear

- Frinding cracks from regrinding

- Consequences of alignment problems

Example: Offset camshaft alignment resulted in point contact on the cam lobes. Valve train geometry and kinematics were compromised to the point where the exhaust valve was no longer able to open correctly. This ultimately led to piston overheating. The roller lifter was also badly worn and damaged (Figs. 6.379 and 6.380).

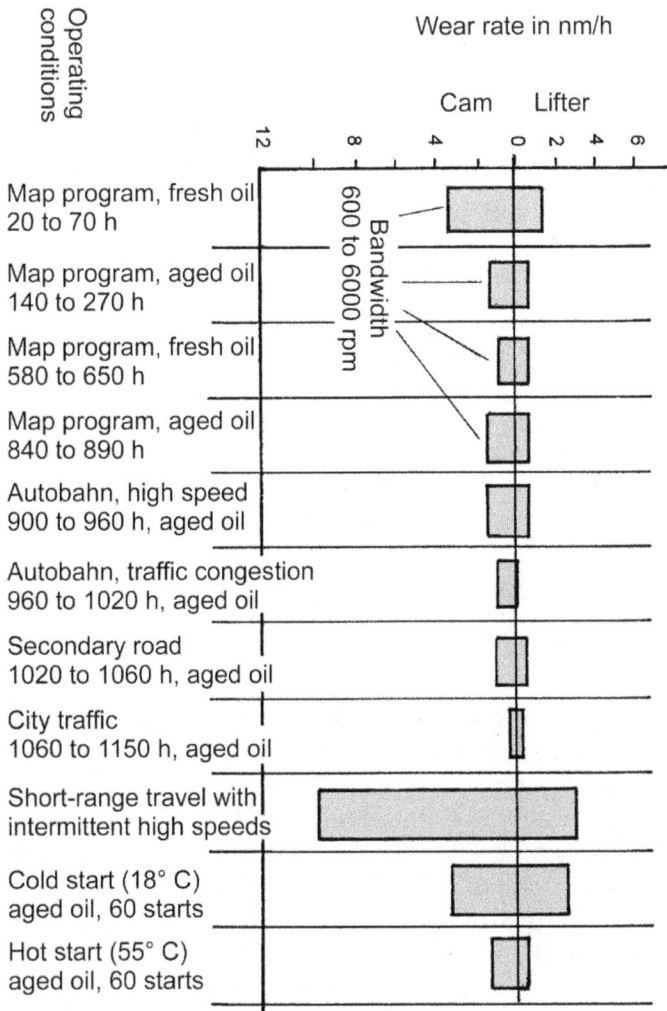

Fig. 6.379 Cam wear as a function of operating conditions. Parameters: lubricant, operating hours at time of measurement.

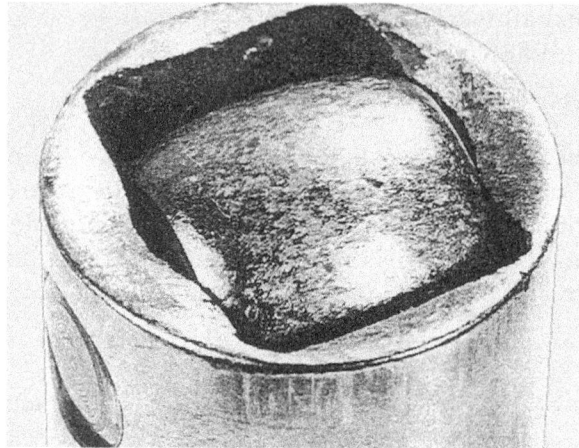

Fig. 6.380 Roller tappet wear.

Gray staining (flecking, micropitting)

Recently, the wear surfaces of cams and cam followers (lifters) have begun to show a phenomenon heretofore regarded as typical of highly stressed gears: gray staining. This is understood to represent material fatigue resulting from the combination of compressive loads, sliding contact, and mixed lubrication, causing microcracking of the surface. High mechanical loads produce plastic deformation of the surface on a microscopic scale, which leads to cracking. Corrosive attack by lubricant additives aggravates these microcracks, which grow in response to dynamic loading along a main shear stress plane below the surface, with a tendency to turn back toward the surface. Such crack fields occur in hardened gears (HRC > 55); their dull gray appearance leads to the term "gray staining." In the final stage, entire sections of material crumble, resulting in pitting.

Example: After operating times of 4000 to 13,000 hours, diesel and natural gas-fueled Otto-cycle engines of various manufacturers, employed in cogeneration plants, were discovered to have more or less severely worn cams and bucket tappets. All cases involved well-proven engine types, and the particular material pairings had previously not drawn any attention to themselves. To achieve longer operating times, each engine had been operating with a lubricating oil containing additives (based on zinc dialkyl dithiophosphate) for improved load-carrying properties. In all cases, analysis of the lubricating oil showed that the additive package exhibited signs of relatively early breakdown. This happened because the engines were continuously operated at maximum power and/or were sometimes overloaded by fluctuations in the natural gas composition. Engines operating on landfill gas are especially at risk. An added difficulty is that oils containing zinc dialkyl dithiophosphate additives have only a limited load capacity in association with gray staining. It was precisely such micropitting, typical of gray staining, that was detected on the bucket tappets of a gas engine with a worn camshaft (Fig. 6.381).

Pitting exposes carbide particles in the hard tappet material, which then act on the cam lobes like abrasive cloth (Fig. 6.382). Over time, this leads to extreme wear of the cam lobe surfaces (Fig. 6.383), which show actual wearing-in of the cam surface (Fig. 6.384).

6.4.4 Timing belts, chains, and gears

A precondition for engine functionality is correct operation of functional groups such as the valve train, lubrication, cooling, and scavenging systems. A portion of the engine's output must therefore be diverted to drive the valve train, pumps, blowers, as well as auxiliary devices. Depending on engine size and type, power transmission to these devices is accomplished by means of belts, chains, or gears. All elements that directly control engine processes or exert influence on those processes (valve train, fuel injection pump, ignition distributor drive) must operate with kinematic exactitude; i.e., they must not

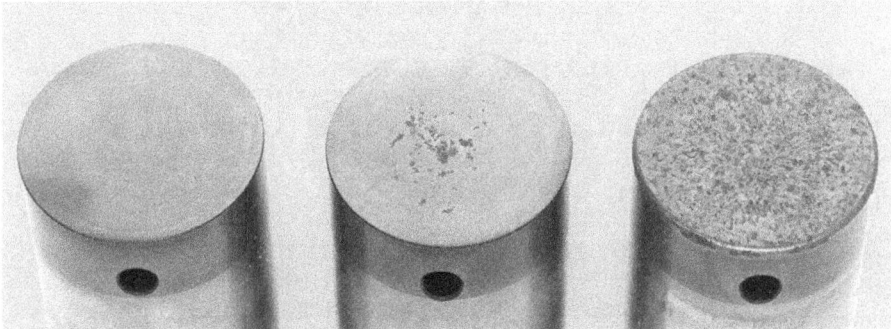

Fig. 6.381 Worn bucket tappets of a gas-fueled Otto-cycle engine. Source: Allianz

Fig. 6.382 Worn cam lobe of a gas-fueled Otto-cycle engine. Source: Allianz

exhibit any slip. Motor vehicle engines must also drive other devices: power steering pumps, air conditioning compressors, air compressors, and alternators or generators. Often, this is accomplished by belt or chain drives—roller chains, V-belts, poly V belts—to transmit torque from one shaft to another. The shafts must have parallel axes, and gears, sprockets, or pulleys must be coplanar.

The advantages of belt and chain drives include freedom in choice of shaft spacing, and the ability to span large distances between shafts—several meters in the case of large two-stroke marine diesel engines. An added advantage is the possibility of extreme redirection of the belt or chain; in this way, multiple drives may be served in a compact arrangement. Belts and chains do not demand especially precise dimensioning of their power transmission elements.

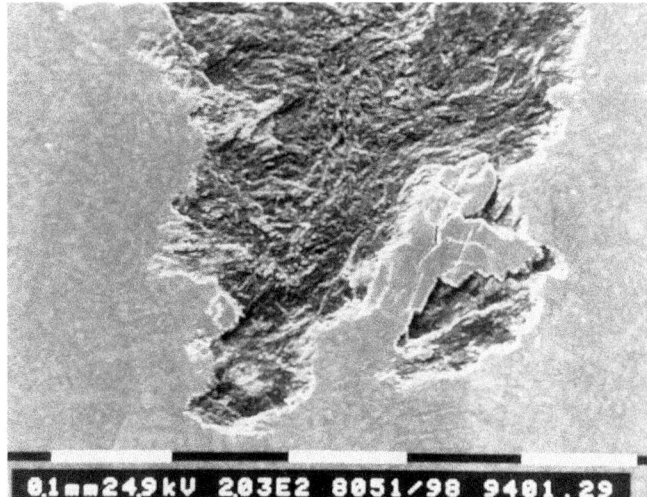

Fig. 6.383 Reflection electron micrograph (200X magnification) of the wear zone of a cam showing "gray staining." Source: Allianz

Fig. 6.384 Cam lobe of a gas-fueled Otto-cycle engine showing worn-in grooves.

6.4.4.1 V-belts

V-belts transmit torque by means of friction between belt and pulleys, with the required friction provided by belt pretension. To allow the belt to be pretensioned, it must exhibit a certain degree of elasticity. Tensioning rollers or ramps are used to compensate for minor changes in length due to varying loads, temperatures, and aging. The portions of the belt not in contact with pulleys are designated the tight (or tension) and slack sides. The higher tension in one side of the belt leads to different stresses, and therefore differential expansion of the belt. This expansion is compensated by belt creep, which is not to be confused with belt slip caused by overloading. Belt drives are loaded in tension, dependent on the tension force and belt cross section— and on wrapping around pulleys and rollers. V-belts have a roughly trapezoidal cross section and run in correspondingly shaped pulleys (Fig. 6.385).

Belt pretension and loading pulls the belt into the pulley grooves [6-32]. Recently, so-called V-ribbed belts (and various manufacturers' trademarked names such as Poly-V®, Micro-V®, Multirib®), both single-sided and two-sided, have been introduced (Fig. 6.386). The backs of these belts may also be used to transmit power, thereby driving even more accessories. They present virtually no maintenance demands.

V-belt and V-ribbed belt faults and failures [6.33]
Hardened, polished belt flanks are the result of

- Improper pretensioning, or if the
- Reinforcing member has been damaged by improper installation

Uneven wear takes place if

- The pulleys are misaligned, or
- The belt drive is subjected to extreme oscillations

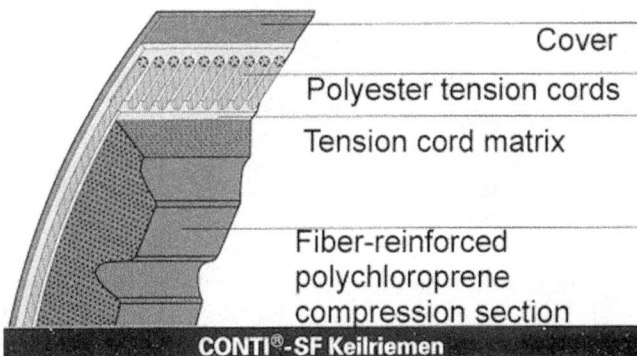

Cover
Polyester tension cords
Tension cord matrix
Fiber-reinforced polychloroprene compression section
CONTI®-SF Keilriemen

Fig. 6.385 V-belt (basic construction). Source: ContiTech

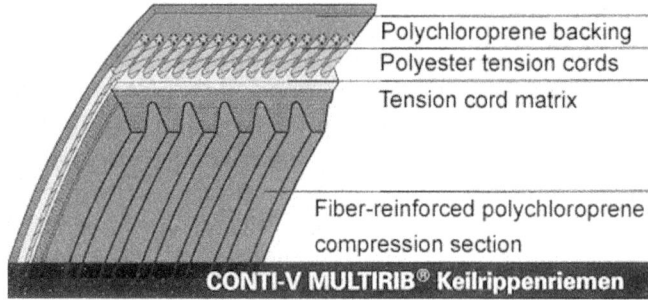

| Polychloroprene backing |
| Polyester tension cords |
| Tension cord matrix |
| Fiber-reinforced polychloroprene compression section |

CONTI-V MULTIRIB® Keilrippenriemen

Fig. 6.386 V-ribbed belt (basic construction). Source: ContiTech

Noise is the result of

- Insufficient belt pretension, or
- Excessive operating time, i.e., exceeding the belt service life

Breaks and tears of the substrate or profile are the result of

- Insufficient belt tension,
- Exceeding the belt service life, or
- Foreign objects caught between belt and pulley

Many engine manufacturers provide information on the permissible extent of belt wear ("normal wear") or at what level of wear the belt should be replaced (Fig. 6.387).

The back of the belt (for V-ribbed belts) may be damaged (Fig. 6.388) if

- The reversing pulley is defective (difficult to turn), or
- The running surface has been damaged by foreign objects

Premature belt separation (breakage) after only a short service life (Fig. 6.389) may be the result of

- Improper installation of the reinforcing member or
- Excessive belt pretension

V-belt flanks show excessive wear or turn brittle (Fig. 6.390) if

- Belt slip is excessive,
- Pulleys are misaligned, or
- Pulley grooves are washed out

Belt flanks harden and exhibit a polished appearance if

- Pulleys are misaligned,

- The reinforcing is damaged by improper installation, or
- An incorrect belt package was installed

6.4.4.2 Timing belts

Timing belts (Gilmer belts, synchronous belts) combine the elasticity of belt drives with the kinematic precision of gears. They have been widely adopted in modern passenger car engines for this reason (Fig. 6.391).

Timing belts run quieter and smoother than chains, and do not require lubrication. However, both sides must be constrained to prevent slipping off their sprockets. In cross section, timing belt teeth are trapezoidal or have a rounded profile (Figs. 6.392 and 6.393).

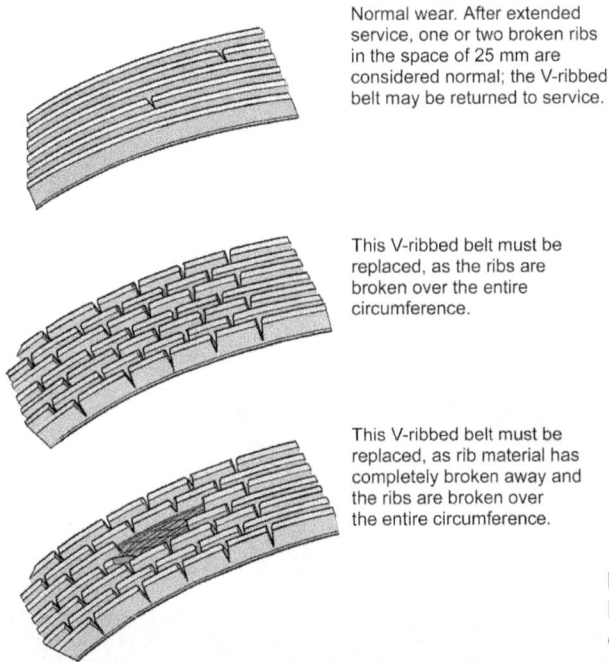

Normal wear. After extended service, one or two broken ribs in the space of 25 mm are considered normal; the V-ribbed belt may be returned to service.

This V-ribbed belt must be replaced, as the ribs are broken over the entire circumference.

This V-ribbed belt must be replaced, as rib material has completely broken away and the ribs are broken over the entire circumference.

Fig. 6.387 V-ribbed belt damage: broken-out ribs. Source: MTU

Fig. 6.388 Damaged back of a V-ribbed belt. Source: ContiTech

Timing belts require a certain level of maintenance, and their service life is limited. Therefore, engine manufacturers specify definitive timing belt replacement intervals for every engine type.

The bandwidth of these manufacturer-specified service intervals is comparatively large, between 40,000 and 120,000 km (25,000 to 75,000 miles). Still, these recommendations are often ignored, which leads to timing belt damage and its consequences: valves striking pistons, and, in worst-case scenarios, damage to connecting rod and crankshaft bearings, and to the cylinder head. As in other tension-based drive systems, proper timing belt pretension is imperative for satisfactory operation. Excessive pretension leads to "howling." On overhead cam engines, excessive belt pretension may lead to seizing of the front camshaft bearing. In situations where the camshaft of passenger car diesel engines also drives unit injectors with injection pressures of up to 2000 bar, the timing belt must be relieved of loading by suitable design. This is achieved by increasing the tooth clearance at a specific point on the timing belt sprocket. This compensates for periodic stretching of the timing belt caused by camshaft deceleration during the fuel injection

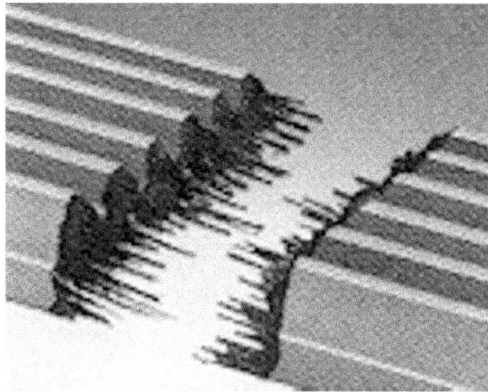

Fig. 6.389 Torn V-ribbed belt. Source: ContiTech

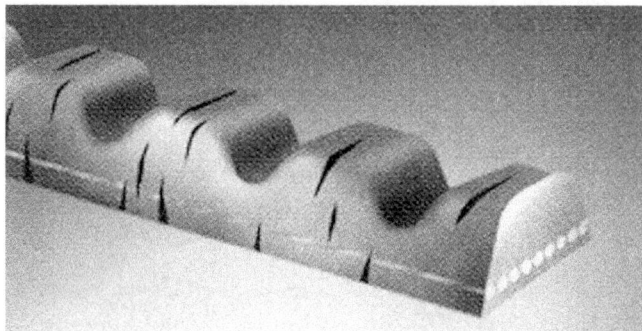

Fig. 6.390 V-belt with severely worn flanks. Source: ContiTech

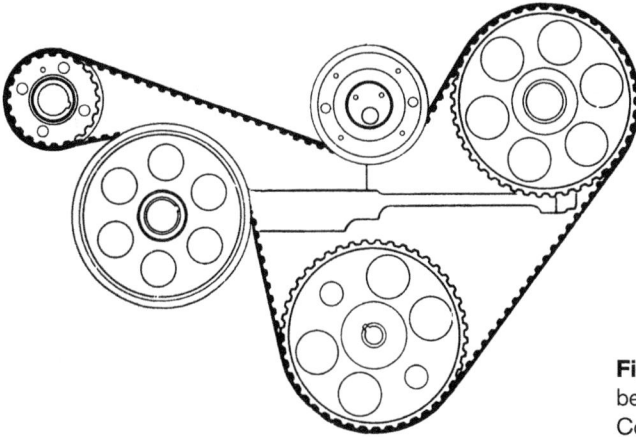

Fig. 6.391 Timing belt drive. Source: ContiTech

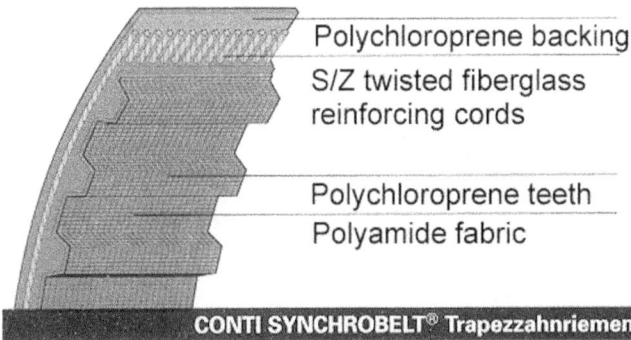

Polychloroprene backing

S/Z twisted fiberglass reinforcing cords

Polychloroprene teeth
Polyamide fabric

CONTI SYNCHROBELT® Trapezzahnriemen

Fig. 6.392 Timing belt with trapezoidal teeth. Source: ContiTech

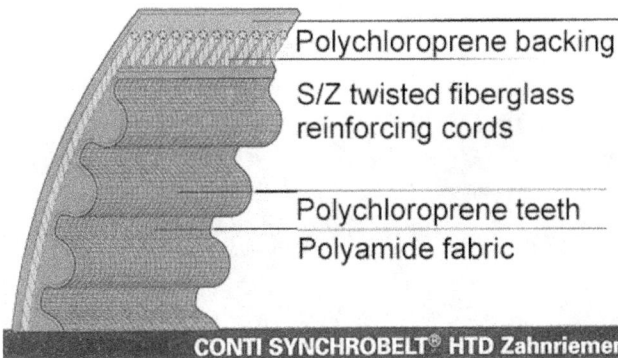

Polychloroprene backing

S/Z twisted fiberglass reinforcing cords

Polychloroprene teeth
Polyamide fabric

CONTI SYNCHROBELT® HTD Zahnriemen

Fig. 6.393 Timing belt with round tooth profile. Source: ContiTech

process, and simultaneous acceleration of the crankshaft at the begin of combustion.

Low pretension—a loose timing belt—induces belt oscillation, which may cause it to jump its sprockets. The result: valve timing and crank train motion will be improperly synchronized, with severe engine damage as the consequence.

Timing belt damage

Edge wear

Edge wear (Fig. 6.394) results when

- Timing sprocket axes are not parallel; the timing belt will run against the sprocket flanges, or
- The sprocket flange has gaps, and
- Individual components have incorrect bearing clearances

Backing worn in roots

Possible causes of worn backing (Fig. 6.395) include

- Excessive pretension and/or
- Excessive belt temperatures

Fig. 6.394 Timing belt edge wear. Source: ContiTech

Fig. 6.395 Timing belt with worn reinforcing. Source: ContiTech

Fig. 6.396 Timing belt with cracked tooth roots and sheared-off teeth. Source: ContiTech

Fig. 6.397 Teeth and reinforcing separating from backing. Source: ContiTech

Tooth root breakage and shear

These failures (Fig. 6.396) occur as a result of

- Excessive tangential force due to the torque transmitted, or incorrect belt pretension, if:
- The belt exceeds the permissible length specification, if
- Dividing (tooth spacing) errors caused backing wear, or if
- The belt drive often operates in its resonant regime, which induces severe vibration of the free spans, overloading the belt. And finally
- The effects of foreign substances, for example, oil.

In extreme cases, entire teeth and reinforcing material will separate from the backing.

Teeth and reinforcing material separated from backing (Fig. 6.397)

- As a result of engine compartment leaks, e.g., oil, coolant (antifreeze), etc.

Wear marks on the tooth side (Fig. 6.398)
are indications of

- Foreign objects in the belt drive system,
- Sprocket damage caused by foreign objects or improper assembly

Cracks in timing belt backing (Fig. 6.399)
may be the result of

- Excessively high ambient temperatures, e.g., continuous operation at 100°C and above
- Extremely low ambient temperatures, e.g., −40°C
- The effects of aggressive media such as paints, solvents, etc.

Separated (torn) timing belts (Fig. 6.400)
- May be caused by foreign objects in the belt drive system,
- Foreign media attacking the belt,
- Excessive pretension, or
- Creasing of the belt prior to installation

Fig. 6.398 Timing belt with wear marks on tooth crowns. Source: ContiTech

Fig. 6.399 Timing belt with cracked backing. Source: ContiTech

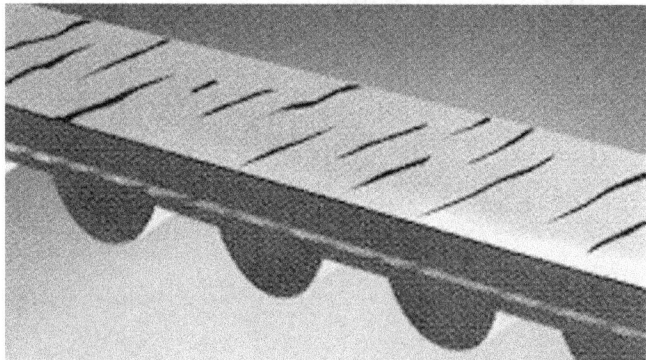

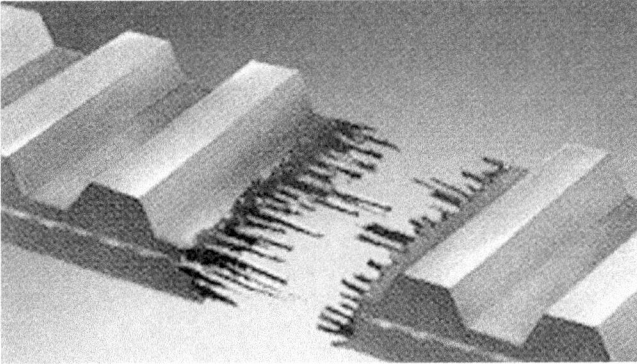

Fig. 6.400 Torn timing belt. Source: ContiTech

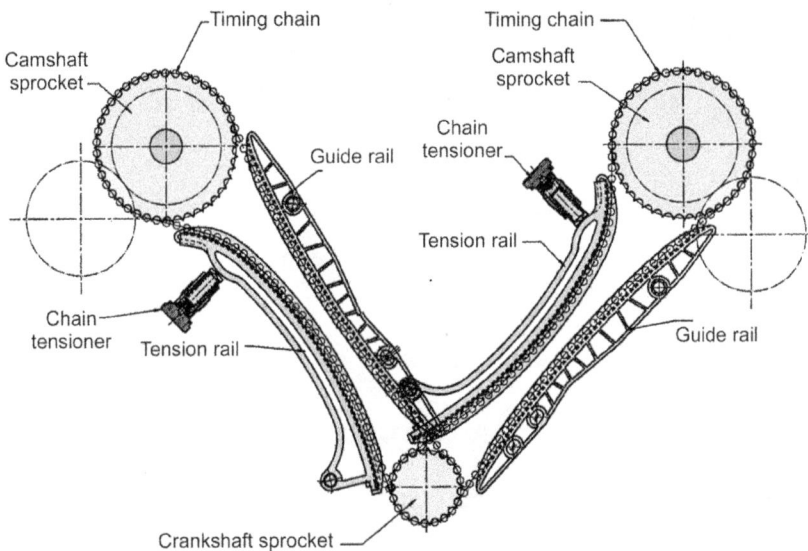

Fig. 6.401 Schematic of chain drive for a V engine with two timing chains, and one set of meshing timing gears per cylinder bank driving the exhaust camshafts. Source: iwis

6.4.4.3 Chain drives

Chain drives are tension drives whose individual link plates are joined by articulated connections. Forces or moments are transmitted by form-fitting interfaces, without any slip between chain and sprockets, which permits large forces to be transmitted with high efficiency (Fig. 6.401).

Combustion engine designs usually employ roller chains in the form of single strand ("simplex") or double strand ("duplex") chains. Triple strand ("triplex") chains are seldom used in modern engine designs

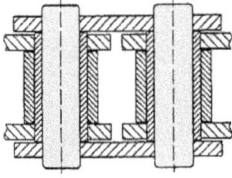

Single-row roller chain   Single-row bushing chain
(simplex chains)

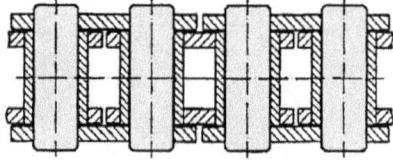

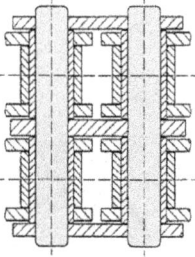

Double-row roller chain   Double-row bushing chain
(duplex chains)

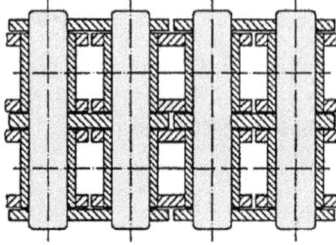

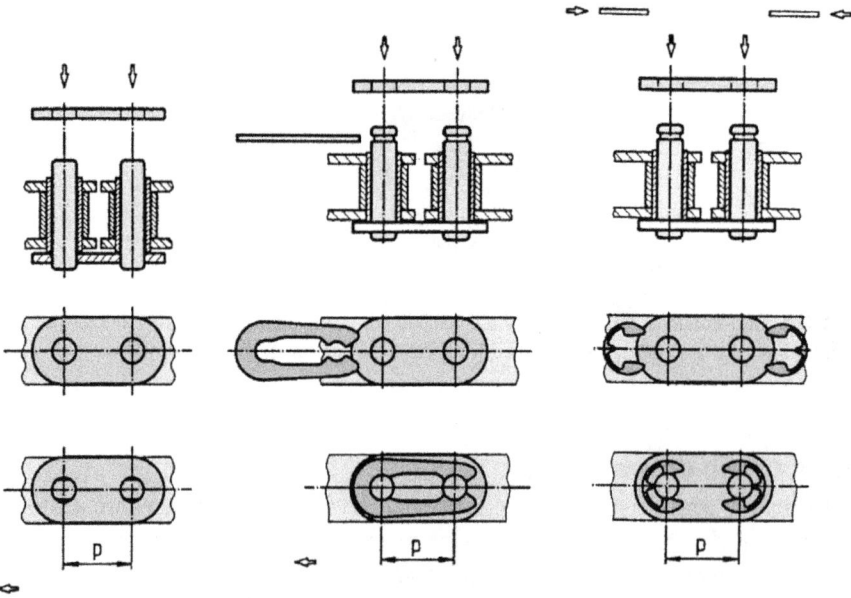

Joining and securing chain links by means of
Riveted master link   Master link with spring clip   Master link with E-clips

Fig. 6.402 Timing chain configurations. Source: iwis

(Fig. 6.402). Outer link plates are joined by pins and bushings. Rollers are free to turn on the bushings.

The timing chain functional group consists of the timing chain itself, at least two sprockets, a tensioning rail, guide rail(s), and hydraulic or mechanical tensioner(s).

Because the chain, as it wraps around its sprockets, deviates from the ideal circular shape (polygon effect), chain tension fluctuates by a certain amount. The link plates are loaded in tension, while pins and bushings are loaded in shear, and the articulated links are subjected to surface compressive forces. Chain drives have an advantage in that they operate perfectly even at high temperatures, and an oil film in the joints acts as a buffer, reducing noise. Sprocket loads are distributed through multiple chain links and sprocket teeth, providing elastic power transmission.

Like other functional groups of the modern combustion engine, chain drives undergo a complex, costly development process, and are thoroughly tested and meticulously manufactured. Chain drive failures are primarily caused by errors in assembly or maintenance (usually in the course of engine repairs), abnormal operating conditions, or serious faults and damage originating in other engine components.

Chains operate correctly only if given proper pretension:

- Excessive pretension is audible as "howling" of the drive, because the oil film between chain links is disrupted.
- Insufficient pretension tends to make chains "slap" (oscillate), and, in worst cases, may cause them to jump sprocket teeth. (Chain slap usually occurs at low to medium engine speeds; at higher speeds, centrifugal force provides outward chain tension.)

In combustion engines, chains are excited to oscillation by nonuniform rotation of the crankshaft and camshaft(s) (Fig. 6.403), which in turn disturbs valve and ignition timing. Spring loaded or hydraulic tensioning rails are used to combat this, and also for improved chain wrap around the sprockets. Modern designs no longer operate in resonant regimes, because the chain drive system has already been tested for resonance during the development process, and individual components have been designed accordingly.

Chains may lengthen (stretch) in service as a result of roller and bushing wear. Chain tensioners compensate for changes in length and tolerances; they pretension the chain and damp out oscillations.

Fig. 6.403 Roller timing chain of a large marine engine (cylinder bore 700 mm, stroke 2200 mm).

Two-wheeled vehicles may use roller chains between the crankshaft, clutch, and transmission; in passenger car engines, chains are used to drive the valve train and engine accessories, as well as balance shafts. In large two-stroke diesel engines, camshafts as well as balance shafts are chain driven. Chain drives must be lubricated.

Chain drive damage

Extreme changes in chain length (stretching) after only short service, up to and including chain breakage (Figs. 6.404, 6.405, and 6.406),

- May be detected by excessive side play; sprocket tooth imprints may be seen on rollers and bushings. Depending on the degree of stretch, valve head imprints may be seen on piston crowns, as valve timing can no longer be maintained as designed.

- Furthermore, sections of chain may be thermally discolored (so-called oil discoloration).

Such changes in chain geometry may cause the chain to jump its sprockets, which will

- Impose added loads on the chain, resulting in
  - Broken sprocket teeth as well as
  - Valve interference (impact with piston),
- Chain tensioner failure, and

Fig. 6.404a Contact marks on top and center of link, with material removal. Source: iwis

Fig. 6.404b As a result of Fig. 6.404a, the chain runs crooked. This results in diagonally offset end wear of the bushings as well as clearly visible material removal on the bushing, in the tension direction. Source: iwis

Fig. 6.404c Cocking of inner links as a result of one-sided link wear resulting from inner links contacting the sprockets, combined with surface damage to rollers. Source: iwis

Fig. 6.404d Duplex chain with contact marks and/or burrs on backs of center links. Such burrs and material deformations may result in increased wear of tension and guide rails. Source: iwis

Fig. 6.404e Roller surface damage as a result of striking the timing chain sprocket. Source: iwis

Fig. 6.405a Fatigue cracking of roller, as a result of overloading. Source: iwis

Fig. 6.405b Roller crack that has progressed to roller fracture. Source: iwis

Fig. 6.405c Further development of a roller fracture. Source: iwis

Fig. 6.406 Broken chain caused by overloading, from chain riding up on sprocket. Briefly, the chain completed several revolutions in this condition, until ultimately breaking. Source: iwis

- Damage to the shaft/flange interface
- Worn chain links
- Clogged oil passages, leading to
    - Increased wear to guide rails and tensioning rails, as well as
    - Increased noise generation because oil damping is absent

Possible causes of chain lengthening are:
- Lubrication deficiencies as a result of
    - Insufficient oil quantity
    - Unsuitable oil, or
    - Oil foaming
- Worn timing chain sprockets
- Assembly errors due to
    - Improperly installed chains or
    - Incorrect phasing of crankshaft and camshaft
- Lockup (preventing rotation) of engine accessories

Wear and surface damage
In case of severely out-of-parallel or wobbling timing sprockets, or skewed installation of tensioning and guide rails, the inner links of the timing chain will run against the sprocket teeth, resulting in

- Damaged links
- One-sided chain loading, and
- Uneven wear of chain links

In cases of more severe wear, the chain links may climb up the sprocket teeth, which imposes severe loads on the rollers, resulting in pitting or surface breakouts (shown in Figs. 6.404a to 6.404e).

Guide and tension rail failure

These failures usually occur in the center of the part, as a result of overloading or fatigue failure. These failures make themselves known through

- Excessive chain wear
- Deep chain wear marks
- Pitting
- Uneven wear of rails

The consequences of broken rails are

- Increased friction.
- Chains are forced to run crooked.
- Break-in difficulties.
- Increased noise generation.

Possible causes are

- Overloading as a result of blocked rotary motion
- Defective component
- Overloading as a result of a failed chain tensioner
- Extremely stretched chain
- Faulty assembly
- Thermal and/or mechanical overloading

Broken bushings

Possible causes are extremely high dynamic tensile loads in the chain, resulting in high impact loading as well as sprocket tooth loading during the chain break-in period, caused by

- Tensioner failure as a result of insufficient or improper oil
- Failure of tensioning components (e.g., broken springs)
- Chain stretch
- Incorrect tensioner installation
- Defective sprocket teeth
- Sprocket tooth flank wear

Figs. 6.405a through 6.405c document the development of chain damage.

Chain breakage
When subjected to unacceptable loads, chain links are damaged, ultimately causing the chain to break. Individually, the following damage may be observed:

- Bent and broken links and pins
- Elongated link holes
- Crushed and/or broken sleeves and rollers
- Severe surface damage to components

Causes may include:
- Improper assembly
  - Installation of incorrect chain
  - Misaligned sprockets
  - Chain components installed crooked, causing chain link to run off-center on sprockets
  - Mistake in closing chain link
  - Failure to observe side play tolerances
  - Re-use of damaged or severely worn chain components
- Inadequate lubrication as a result of
  - Unsuitable oil
  - Over-aged oil, or
  - Insufficient oil (oil starvation)
- Chain riding up on sprockets (e.g., as a result of tensioner failure; see Fig. 6.406)
- Seizing of crankshaft or driven shafts as a result of
  - Seized pistons, with abrupt deceleration from perhaps rated engine speed to zero in a fraction of a second, imposing deceleration forces that the chain cannot withstand. The same applies to
  - Seized camshafts, and
  - Seized accessories such as oil pumps or oil pump drives

6.4.4.4 Gears
On medium and large engines, as well as general-purpose engines, the torque needed to drive the valve train, injection pumps, water and oil pumps, as well as other engine accessories, is not transmitted by flexible tension drives but rather by gear trains (Fig. 6.407). In general, these consist of spur gears; in the past, special cases employed bevel gear ("quill shaft" or "bevel shaft") drives.

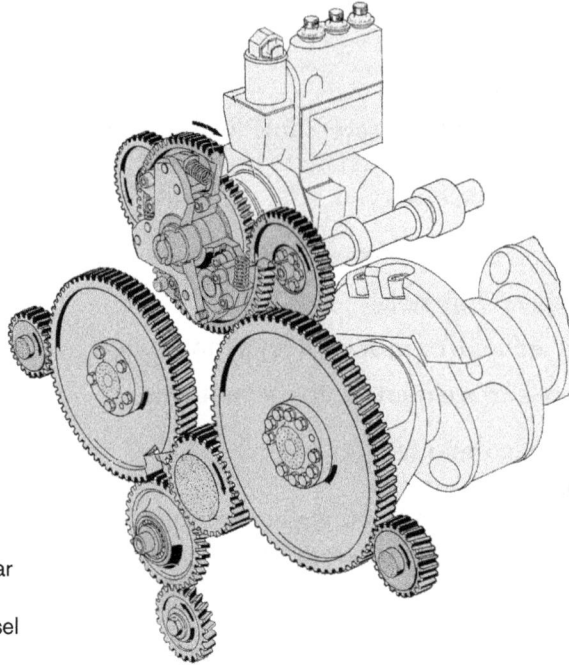

Fig. 6.407 Timing gear train of a high-speed, high-performance diesel engine. Source: MTU

Gears operate as form-fitting, mating components, precisely transmitting kinematic motion, and are suitable for high and extreme loads. Individual gears of a gear train may be made of steel, cast iron, aluminum alloy, or molded materials (e.g., Novotex: phenolic resin-impregnated, laminated fabric). For higher loads, only steel gears come into consideration. Helical spur gears are used for jerk-free and quiet transmission of forces. The relatively large clearances found in journal-bearing crankshafts, and rotating speed variations caused by cyclic loading over the course of a single working cycle, must be taken into consideration when designing a gear drive; these factors could promote gear pitting.

Gear damage [6-34]
Pitting
Subsurface material fatigue, which leads to cracking and crack propagation. Oil penetrates these cracks. If such cracks are closed as gears mesh, material particles are blasted out of the surface (Figs. 6.408 and 6.409).

Gear tooth breakage
- *Force failures (overloading)*
  Primarily occur if particles or even larger fragments are caught between meshing gears.

- *Fatigue failures*
  Result from a combination of dynamic loading and pre-existing damage from which the fatigue failure develops. At low loads, such fractures propagate slowly, with tightly spaced "beach marks" ("rest lines") and a small residual fracture surface. At higher loads, fracture occurs relatively quickly, with a large residual fracture (Figs. 6.410 and 6.411).

Galling

Galling of gears occurs primarily in areas of high Hertzian contact pressures and high sliding velocities (in the gear tooth roots and at the tips) if otherwise adequate lubrication is simply overcome by these operating conditions, or if lubrication is inadequate for other reasons, so that meshing gear teeth operate under mixed and boundary lubrication. If the galling process is initiated by just a single load spike, the galling marks may be smoothed out by subsequent normal operating conditions (Figs. 6.412, 6.413, and 6.414).

Fig. 6.408 Mild pitting at the tooth roots of a helical spur gear. Source: ZF

Fig. 6.409 Pitted surfaces with breakouts; scalloped structure in substrate. Source: ZF

Fig. 6.410 Fatigue failure at gear tooth root of a straight-cut spur gear. Source: ZF

Fig. 6.411 Tooth end fatigue failure: only a portion of the tooth, starting from the end, has broken away. Source: ZF

Fig. 6.412 Galling over the entire load-carrying width of gear teeth. Source: ZF

Fig. 6.413 Mild galling in a helical spur gear, which smoothed out again after returning to "normal" operating conditions. Source: ZF

Wear

Wear is a question of load and operating time. If gear backlash increases over the gears' operating life, this will become noticeable as a harsh noise.

Micropitting (flecking, gray staining, microspalling)

Gears are subjected to high mechanical loads. Operation under conditions of partial mixed lubrication, even temporarily, will result in a form of material fatigue. Due to its visual appearance, this is sometimes known as "gray staining": extremely tiny particles are broken out, and

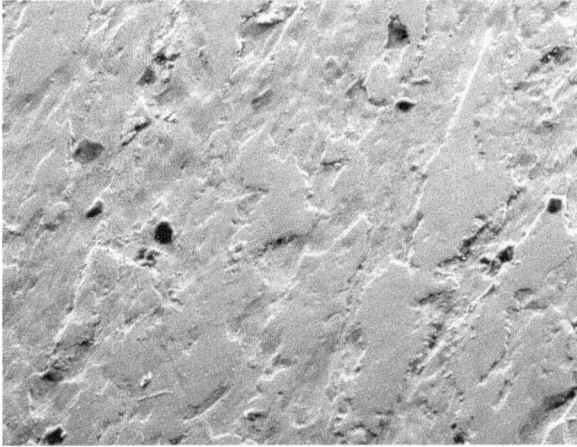

Fig. 6.414 Reflection electron micrograph detail of galling. The surface flow phenomena caused by pressure, friction, and temperature are clearly visible. Source: ZF

these zones expand to cover larger areas. Although not individually recognizable to the unaided eye, these tiny cracks present an overall appearance of gray flecks or spots (Figs. 6.415, 6.416, and 6.417).

6.5 Fuel injection and ignition systems

6.5.1 Diesel engine mixture formation and combustion

The fundamental characteristic of diesel engines is autoignition of fuel mixed with highly compressed (and therefore hot) air. The diesel engine thermodynamic process is described as a "mixed" process; heat, in the form of fuel whose energy content is released by the combustion process, is introduced in part at constant volume (isochoric portion) and in part at constant pressure (isobaric portion). The constant volume portion takes place near piston top dead center. Constant volume heat addition therefore represents infinitely rapid combustion, which of course is not possible in a real process. Real thermodynamic cycles proceed somewhat differently. In simplified form, diesel combustion may be divided into four stages (Fig. 6.418).

Ignition delay
Ignition delay is the time between the begin of fuel injection and the onset of combustion. Fuel is injected, warms up, vaporizes, and mixes with air. This constitutes mixture formation. Ignition delay may increase in the case of a cold engine, ignition-resistant fuels, or other unfavorable circumstances. During the ignition delay period, mixture temperature and pressure increase only as a result of compression.

Uncontrolled combustion with steep pressure rise
Combustion begins, throughout the combustion chamber. There is a steep rise in gas pressure and temperature. This greatly improves conditions for mixture formation, which in turn accelerates combustion.

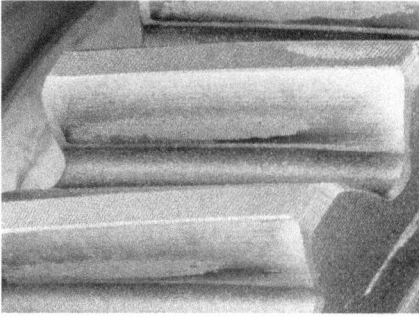

Fig. 6.415 Gray staining of a helical spur gear. Source: ZF

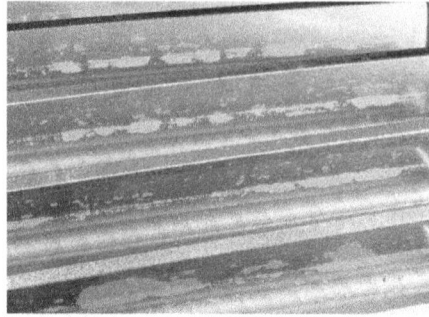

Fig. 6.416 Gray staining of a helical spur gear, with stains appearing primarily in tooth root areas. Source: ZF

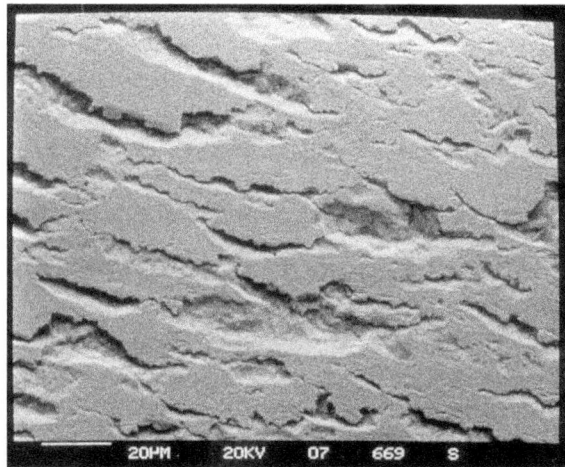

Fig. 6.417 Reflection electron micrograph detail of gray stained zone (magnification ca. 1000X). Source: ZF

Fuel injection continues during this phase. Thermodynamically, this represents heat addition at constant volume. The greater the ignition delay, the more pronounced the period of uncontrolled combustion; this explains the diesel rattle typically encountered with cold engines. As soon as the engine has warmed up, ignition delay is reduced, and the combustion process becomes "softer."

Controlled combustion
Combustion continues at an extremely high rate. Fuel injection ends. As the piston again moves downward, pressure rise due to combustion is largely offset by the increasing volume in the cylinder; thermodynamically, this represents heat addition at constant pressure. By contrast, temperature reaches its peak.

Afterburning
The rate of combustion decreases, combustion gas expands, cylinder pressure drops, as does temperature, to a lesser degree; thermodynamically, this represents heat addition at constant temperature.

Diesel engine mixture formation and combustion are closely related, and intertwined; to some extent, they take place simultaneously and are mutually supporting (Fig. 6.419).

The main problem for diesel combustion is the short period of time available. Fuel can only be injected near the end of the compression stroke (35 to 5° before top dead center); on the other hand, combustion must end at the proper time. Diesel fuel is more viscous than gasoline or other Otto-cycle fuels, and its individual components (fractions) have high boiling points, factors that also contribute to difficult mixture formation. In order to achieve good mixture formation in the short period of time available, diesel engines must operate with excess air. At full load, the relative air/fuel ratio (excess air factor) λ is in the range of 1.3 to 2.0. This is possible because the diesel mixture is ignitable across a wide range of fuel/air ratios. Diesel engine power output can be controlled by means of the injected fuel quantity, i.e., without throttling, a further reason for the good part-throttle behavior

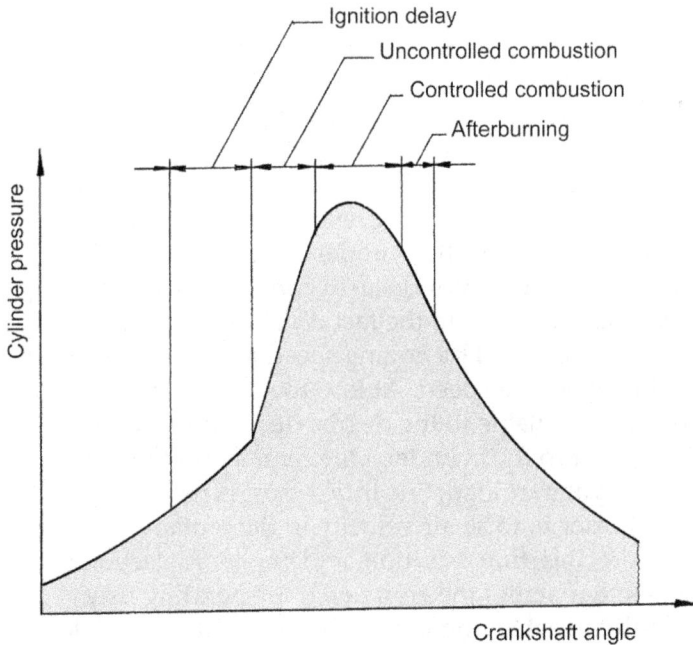

Fig. 6.418 Individual phases of diesel combustion.

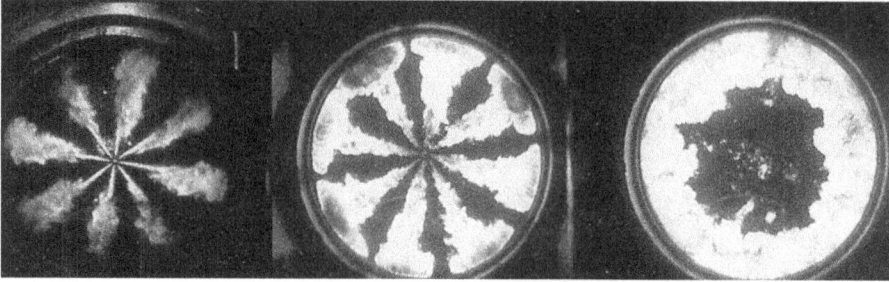

Fig. 6.419 Diesel engine mixture formation and combustion. Source: Daimler

of diesel engines. However, excess air means that more air needs to be inducted in the cylinder than is actually needed for combustion. The engine displacement is not fully utilized, and therefore diesel engines have lower specific output (horsepower per cylinder volume) than a comparable Otto-cycle engine, but in return have better pollutant performance. The quality of combustion, and therefore mixture formation, may be discerned from

- Power output
- Fuel consumption
- Mechanical and thermal loads
- Pollutant emissions and noise

These properties should be optimized for all operating conditions—full load, part load, and idle, i.e., across the entire engine operating map. This is achieved by the fuel injection system. Distribution of fuel within the cylinder and its mixing with air by means of the fuel injection nozzle is supported by very specific air motion. Through appropriate shaping of intake passages ("swirl ports"), swirl is imparted to the intake air; that is, it rotates around the cylinder axis. This motion continues even after the intake valves close. In conventional fuel injection systems, the difficulty lies in the fact that both fuel injection pressure and duration are defined by engine speed and load, but swirl is primarily a function of engine speed. At low speeds, greater swirl is needed, but this is not available to the degree desired due to the low speed of air flowing into the cylinder. One remedy (in the case of four-valve heads) is that at part load, one intake port is blocked off by a throttle butterfly, in order to raise air velocity in the remaining port. Swirl, injection pressure, injection duration, and begin of injection must be matched to one another across the entire engine operating map. Because of the difficulty in achieving the "right" amount of swirl, low-swirl combustion processes are more desirable. These are achieved with high injection pressures, more nozzle holes with smaller diameters, and a wide, flat piston bowl (Fig. 6.420).

The time history of combustion (the "heat release") is termed the burn rate. Simplified, the plot of burn rate over time has a triangular shape. It is determined by the fuel injection rate, and this, in turn, is determined by the fuel delivery rate (time history of fuel delivery by the injection pump). We therefore have the following functional chain (Fig. 6.421):

Fig. 6.420 Injection spray pattern of a ten-hole nozzle in a medium-speed four-stroke diesel engine. Source: Caterpillar

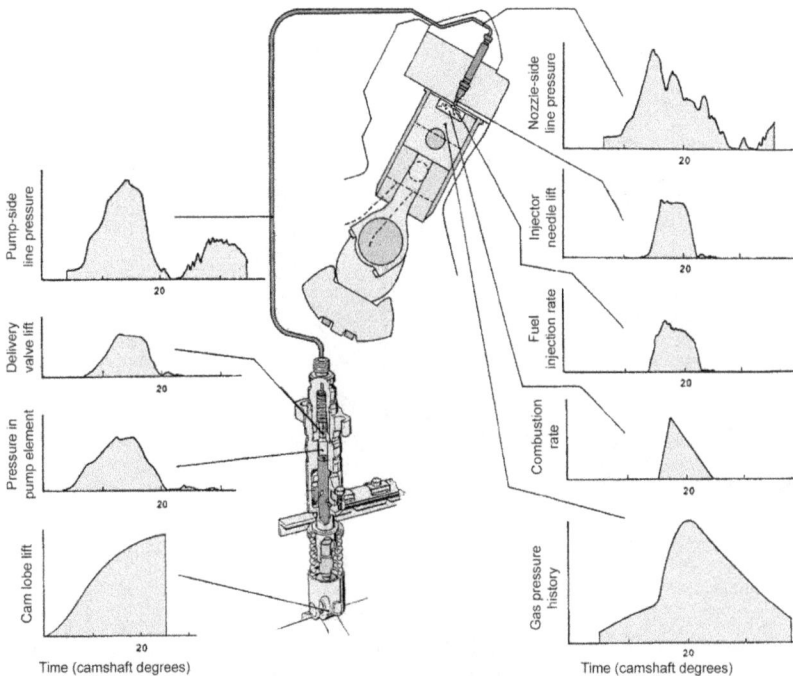

Fig. 6.421 Functional chain of engine power development, from injection pump cam lift to gas pressure history in engine cylinder.

Fuel delivery → fuel injection → combustion → pressure rise in cylinder → conversion of gas pressure into torque by crank train.

Conventional mixture formation systems (Fig. 6.422) consist of

- A fuel injection pump to pressurize and meter fuel
- Fuel injection lines to supply fuel to injectors
- Injector nozzles ("injectors") to distribute fuel within the combustion chamber, as well as
- Accessories: feed pump, filter, and preheating equipment

Decisive factors for engine behavior are injection begin, rate, and duration; these are controlled by the fuel injection pump. Now, diesel engine mixture formation throws up a number of obstacles:

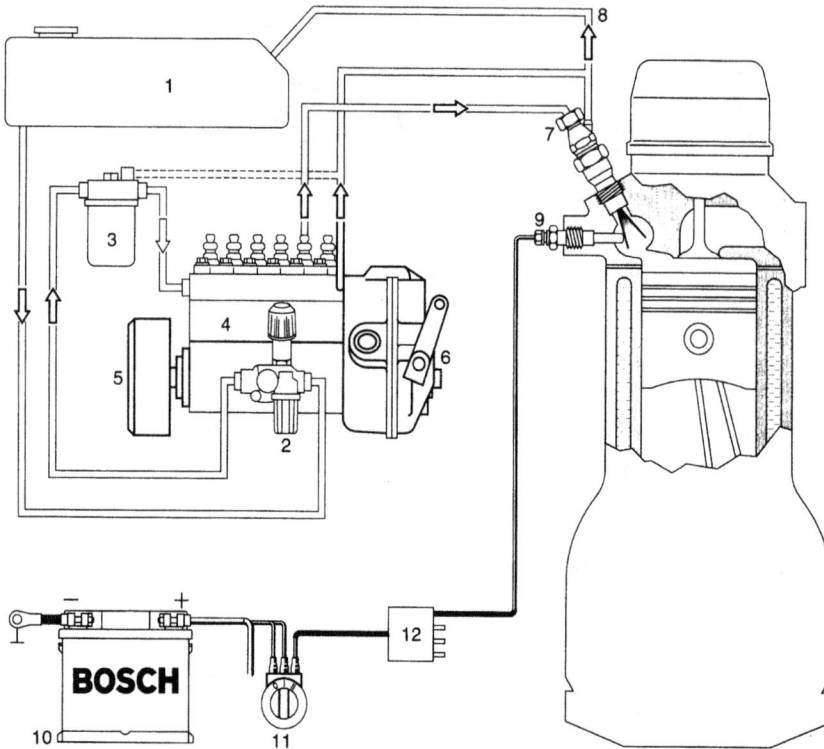

Fig. 6.422 Fuel injection system with mechanically governed "PE" model inline injection pump: 1, Fuel tank; 2, Feed pump (lift pump, supply pump); 3, Fuel filter; 4, Inline fuel injection pump; 5, Timing device; 6, Governor; 7, Nozzle holder with nozzle; 8, Fuel return line; 9, Glow plug; 10, Battery; 11, Glow plug/starter switch; 12, Glow control unit. Source: VDI

- Very small fuel quantities must be metered with great precision. An example will illustrate this: a four-cylinder diesel engine develops 75 kW at 2400 rpm, and has a specific fuel consumption of 0.200 kg/kWh. In other words, it uses 15 kg of fuel per hour. In this time, 288,000 injection events take place. Each injection event represents 0.052 g of injected fuel, or 65 mm^3 at a density of 0.8 g/cm^3. For comparison, a raindrop has a volume of about 30 mm^3.

- These already tiny quantities are reduced even more at part throttle.

- The individual cylinders of an engine must be given the same fuel quantity, within a narrow tolerance band.

- Injection timing must be controlled very precisely.

Ideally, fuel injection would take place synchronously with pressure rise and injection pump delivery; one could meter injection directly by means of the injection pump. In reality, conditions are more difficult, because

- Fuel is not incompressible, as is usually assumed for fluids.

- The fuel injection pump operates discontinuously (with periodic interruptions), which creates pressure waves.

- Although these pressure waves travel at the speed of sound in their medium, this is nevertheless a finite propagation speed; and

- The fuel injection lines expand elastically in response to the pressure waves.

The consequences are that
- Fuel injection lags behind fuel delivery (injection delay), and
- The fuel injection rate deviates from fuel delivery rate.

Furthermore, less fuel is injected than is delivered, because the pump must "offer" the injector more fuel than it needs. The difference between delivered and injected fuel flows back, unpressurized, to the fuel tank, as overflow (or leakoff) return. The overflow lubricates the injector needle and serves to cool the injector nozzle.

The factors determining injection delay are:

- The length of injection lines (the longer the line, the longer the propagation time for the pressure wave)

- Engine speed, because with rising speed, the available time (in milliseconds) for injection becomes shorter

Injection line delay may be compensated by

- Equal-length injection lines from pump to individual cylinders; this explains the often artfully bent injection lines found on diesel engines (Fig. 6.423)

- Engine speed-dependent delay introduced by the automatic timing advance unit that alters the timing of the injection pump camshaft relative to the engine crankshaft

For mixture formation, the following injection system parameters are of primary importance:

Begin of delivery and begin of injection

Begin of delivery and begin of injection are specified in crankshaft degrees relative to top dead center, e.g., 15° BTDC or 5° ATDC ("before top dead center" and "after top dead center," respectively). As injection in cylinders of four-stroke engines takes place only on every second crankshaft rotation, the injection pump camshaft turns at half crankshaft speed. One degree of (pump) cam angle represents two crankshaft degrees. Combustion should take place near top dead center (Fig. 6.424). Too early begin of combustion leads to a steep pressure rise, places unduly high mechanical loads on engine components, and is acoustically objectionable; late injection begin causes combustion to extend far into the expansion stroke, to the detriment of engine efficiency and particulate emissions.

With respect to performance, consumption, and exhaust emissions, there is no optimum begin of injection. Early injection leads to high ignition pressures and oxides of nitrogen emissions, but also to low fuel consumption; late injection, by contrast, raises HC and particulate emissions, as well as fuel consumption. A reasonable compromise between these contrary demands requires exceedingly tight tolerances for begin of injection, on the order of ±1° crank angle; at 2000 rpm, this

Fig. 6.423 Fuel injection lines on a commercial vehicle diesel engine.

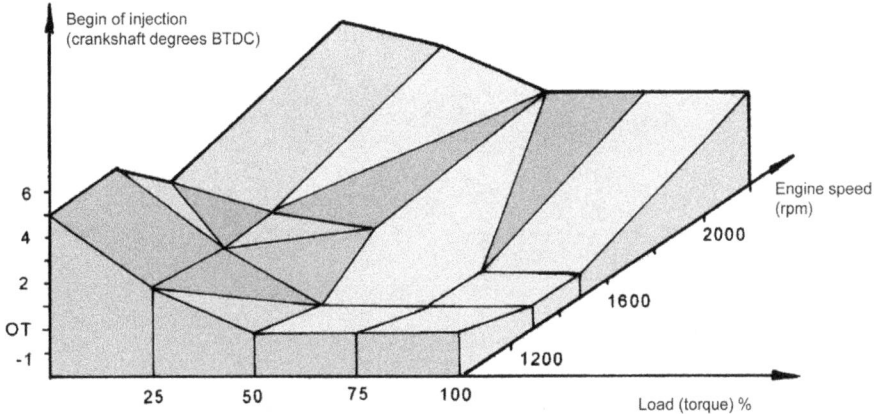

Fig. 6.424 Combustion optimization (begin of injection map). Source: MAN

represents all of 0.08 milliseconds. For noise and emissions reasons, part-load operation often employs a different begin of injection than full load. The injection timing map shows injection begin as a function of load and engine speed.

Engine speed-dependent modification of injection timing is accomplished by the automatic timing advance unit, while load-dependent modification is performed by the pump plunger helix.

Injection duration and injection rate
The injection pump delivery rate diagram usually exhibits a trapezoidal shape and is dependent upon pump geometry. The injection pump cam profile determines the delivery rate; the delivered quantity is the result of the delivery duration. Because of the aforementioned elasticity of fuel and injection lines, injection begin is not only displaced with respect to delivery begin, but also the injection rate is distorted compared to delivery rate; i.e., the shape of the delivery rate diagram at the pump does not match the injection rate diagram at the injector. The fuel injection system must therefore not only ensure exact injection timing and total injection quantity, but also distribute the injected quantity (for example, 0.25 mm^3/crankshaft degree/stroke) uniformly across 15 to 20 crankshaft degrees, and distribute and vaporize this quantity evenly throughout the combustion chamber. Injection duration for a given fuel quantity is a function of injection pressure and nozzle hole cross sections.

Injection pressure
Fuel must be introduced to the cylinder in a very short period of time, and atomized finely enough to form an ignitable mixture with air. The

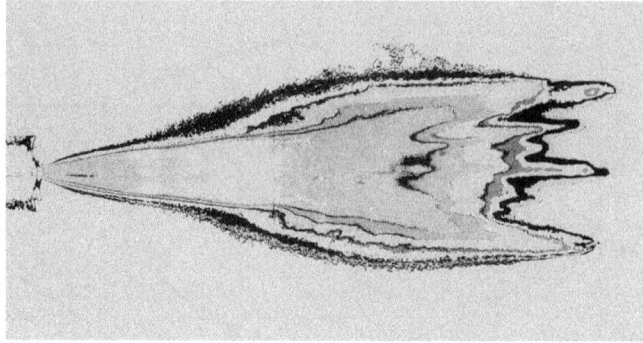

Fig. 6.425 Equidensity image of an injection spray. Source: VDI

injection spray (Fig. 6.425) must, on the one hand, penetrate far enough to encompass as large a volume of the cylinder as possible, yet it should not impinge on the cylinder wall where it would wash away lubricating oil.

Good fuel distribution within the cylinder can be achieved by injecting at high pressures through small holes, moreover, through multiple holes. This causes fuel to mix well with air in the cylinder. Indirect-injection engines may employ injection pressures of 400 bar, while direct-injection engines may use up to 1000 bar, with development tending toward even higher pressures.

Various diesel combustion concepts require various injection durations; direct injection engines need about 25 to 30 crankshaft degrees at rated speed, while indirect-injection engines require 35 to 40 crankshaft degrees of duration. At an injection pump speed of 2000 rpm, injection duration of 30 crankshaft degrees (representing 15 camshaft degrees) implies an injection duration of 1.25 milliseconds.

For low fuel consumption and soot emissions, injection duration must be established as a function of operating point on the engine map, and matched to the begin of injection. At the beginning of the injection process, only a small amount of fuel should be injected, so that not too much fuel is processed during the ignition delay period, which would cause excessively rapid combustion. This limited early fuel injection serves to limit the pressure gradient dp/dt. On the other hand, the injector should close quickly and securely, to prevent any secondary injection. In secondary injection, the injector nozzle opens again briefly after its initial closing, and delivers fuel, which, however, cannot be mixed properly and which extends combustion into the expansion phase. Secondary injection may also lead to formation of cone-shaped carbon deposits at the injection nozzle holes, which will compromise injection spray formation.

The demands for mixture formation, as given by the engine map, must ultimately be realized by the fuel injection pump. The injection pump map must satisfy the engine map in all its operating modes—starting, idling, part load, and full load, with additional boundary conditions for

- Smoke limit
- Cylinder pressure
- Exhaust gas temperature
- Maximum engine speed

Starting

Combustion engines are not capable of self-starting; they must be brought up to starting speed by external forces. Small engines are started mechanically by kick starting or by means of a starter pull cord; motor vehicles are started by electric starter motors, and medium and large engines are started by compressed air. Actual engine starting consists of igniting mixture and accelerating the crank train to the point where it becomes self-sustaining, in the process overcoming the so-called breakaway torque. Diesel engines must be brought up to ignition speed, to provide sufficiently high compression temperature to ignite the mixture (about 500–700°C). Engine starting must be assured even at low temperatures. This is opposed by the following:

- With a cold engine, breakaway torque and turning resistance are higher because the cold lubricating oil is more viscous.
- Starter battery capacity diminishes as temperature is lowered.
- Compressed air in the combustion chamber is subject to significant leakage losses, because of large piston clearances and the necessary lubricating film to ensure sealing between piston rings and cylinder walls is insufficiently developed.
- Compressed air transfers a large part of its heat to the cold combustion chamber surfaces.

Indirect injection engines (those with auxiliary combustion chambers—prechamber, swirl chamber) suffer especially high heat losses, because

- Air forced into the auxiliary chamber is cooled by throttling, and
- The auxiliary chamber experiences intensive cooling because of its large surface-area-to-volume ratio.

To ensure starting even at low temperatures, the following measures are applied:

Fuel matching
Heated fuel filters or direct fuel heating prevents fuel blockage as a result of paraffin in diesel fuel separating in the form of wax crystals.

The ability of diesel fuel to flow at low temperatures may be improved by adding ordinary gasoline. As a guideline, 10 to 30% (by volume) of gasoline may be added, depending on temperature. Engine manufacturers' recommendations must be followed in any case. Winter-grade diesel fuels available at service stations are typically formulated to remain fluid at expected minimum local temperatures.

Starting assistance
Starting of direct-injection diesel engines at low temperatures may be assisted by prewarming intake air by means of flame glow plugs. Indirect-injection engines are less willing to start, and therefore employ a glow plug in the prechamber to assist in starting. Glow plugs with a heating duration of only a few seconds permit rapid starting. Direct-injection passenger car engines are equipped with pencil-type glow plugs, not only to improve starting, but also to shorten the warmup phase, with its higher exhaust emissions.

Injection system matching
An engine needs appreciably more fuel for starting, both to compensate for high condensation and leakage losses, as well as to build up torque in the speed run-up phase. A further measure is advancing injection timing to compensate for ignition delay. The correct begin of injection must be maintained within narrow limits. If fuel is injected too early, it condenses against cold cylinder walls and washes away the lubricating film. If it is injected too late, ignition occurs in the expansion stroke, with corresponding loss of performance.

Idling and part load operation
Immediately after starting, the engine will not have reached its operating temperature; ignition delay is long and the engine "knocks." HC emissions are high; at compression temperatures of less than 250°C, this is visible as white smoke, at higher temperatures as blue smoke. With an idle fuel injection rate of only 5 to 7 mm^3 per injection stroke (motor vehicle engines), it is obvious that fuel must be metered very precisely; just 0.5 mm^3 amounts to 10% of the total injected volume. Engines are designed to deliver their rated power; losses increase during operation at other load points. Although diesel engines provide considerably better part-load efficiency than Otto-cycle engines, the

injection system must be carefully tuned for precisely this operating regime, because many engines spend most of their lives at part load.

Full load

"Full load" is understood to mean the maximum torque that the engine can develop, given various boundary conditions. Theoretically, the full load curve of normally aspirated engines is independent of engine speed. Reality, however, is quite different. At low engine speeds, the actual developed torque is (primarily) decreased by heat losses and cylinder leakage; at the upper end of the engine speed range, torque is (primarily) decreased by flow and fiction losses. The fuel delivery curve of the injection pump must satisfy the engine's demand for fuel, which works out quite nicely, because both the pump and engine are piston machines and therefore have somewhat similar characteristics. But because cylinder filling at high speeds and loads is reduced, the injection pump supplies too much fuel under these conditions. This is countered by speed-dependent adjustment. In addition, atmospheric pressure, and, for supercharged engines, boost pressure, must be taken into consideration in fuel metering. A further adjustment (restricting injected fuel quantity) is applied as engine speed decreases, to counteract exhaust smoke.

In the event of sudden unloading, diesel engines will rev freely; i.e., the crank train will be accelerated by the full engine torque so quickly that the driver or operator would have no chance to shut the engine down quickly enough. This might happen if a drivetrain element breaks, or, in marine applications, if the ship's screw is lifted clear of the water in heavy seas. To prevent such self-destructive engine behavior, diesel engines are equipped with governors that automatically adjust the fuel rack to idle by mechanical, hydraulic-mechanical, or electrical/ electronic means once a certain speed has been attained.

6.5.2 Fuel injection systems

Diesel fuel injection systems consist of several components that fulfill various functions, in part based on different operating principles.

The actual injection system serves to

- Generate high pressures
- Meter fuel quantity as a function of engine demand (idle/part load/ full load)
- Inject fuel into the engine

A precondition is that the injection system be supplied with fuel by a fuel feed pump that transfers fuel from the fuel tank to the injection pump, with a certain specific fuel feed pressure. To prevent mechanical damage and operational faults, a fine-mesh filter is installed in the fuel supply line. With increasing pressures found in modern fuel injection systems, adequate fuel filtration becomes ever more important. Depending on application, a preheater may be installed to ensure fuel flow even at low temperatures. Individual fuel injection systems (Fig. 6.426) differ in how and where their subsidiary functions of pressure generation, metering, and injection are achieved.

The most important identifying characteristics are:

- Combined pressure generation and injection (conventional systems) or decoupled systems (common rail systems)
- Pressure generation and fuel metering in the injection pump or in an injector
- Engine cylinders supplied by a single unit injector per cylinder or jointly by a common plunger (distributor pumps)
- Physical separation of pump and nozzle (pump—line—nozzle) or incorporation into a single entity (unit injector)

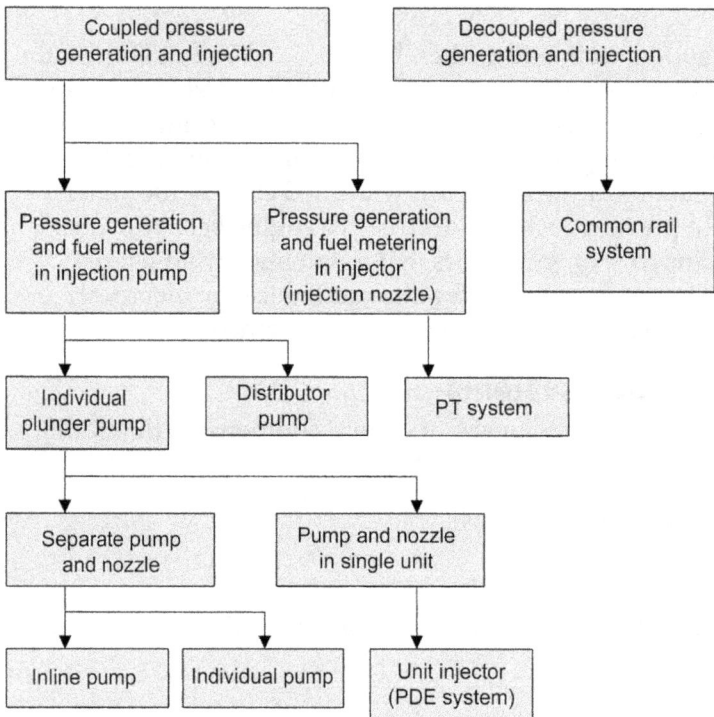

Fig. 6.426 Fuel injection systems.

- Inline injection pumps incorporating multiple plungers and barrels, driven by a common injection pump camshaft, or individual pumps driven by the engine camshaft

Individual-plunger pump
In this configuration, every engine cylinder is supplied by its own dedicated pump (plunger and barrel) unit. Pressure is generated by a cam-driven plunger, operating on an overflow principle with port and helix metering. The helical grooves are machined into the upper portion of the pump plunger; depending on rotary position, these permit a portion of the fuel to flow back to the fuel gallery. Rotating a control sleeve around the plunger changes the relationship between the helix and the inlet port, allowing the fuel delivery to be modulated continuously from zero to full load. Begin of delivery and end of delivery are mechanically established by the position and shape of the control helix relative to the inlet port. A flat-topped helix provides a constant begin of delivery (regardless of engine load); a flat lower end of the helix provides constant end of delivery. In marine engines, engine speed and torque, effectively a function of injected fuel quantity, are related by the propeller law, which is taken into account by angled upper and lower control helixes (Fig. 6.427).

Once pressure in the plunger and barrel assembly has reached the correct value for fuel delivery, a spring-loaded delivery valve opens.

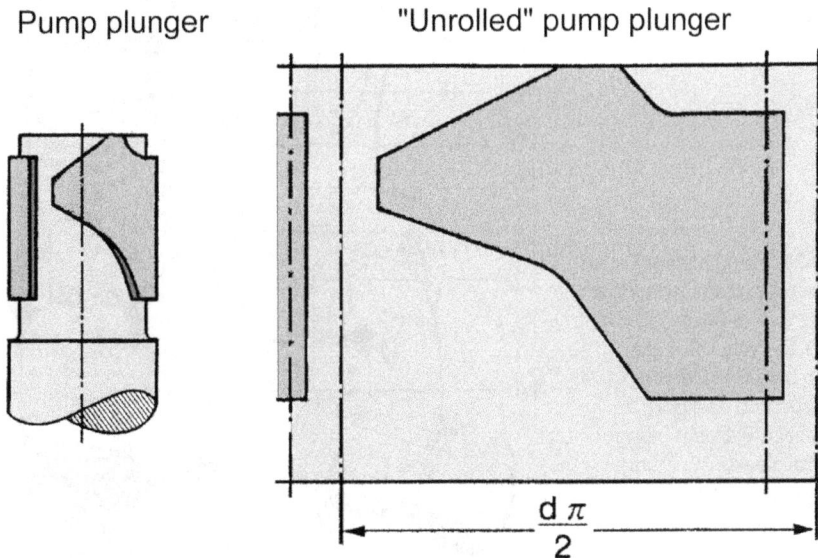

Pump plunger   "Unrolled" pump plunger

$$\frac{d\,\pi}{2}$$

Fig. 6.427 Pump plunger helixes. Source: MAN

The pump continues to supply fuel until the plunger helix again restores a connection to the inlet port and fuel gallery, at which point delivery pressure collapses. The spring force on the delivery valve predominates, and seals the high-pressure fuel delivery line from the pump chamber. This increases the volume in the high-pressure fuel delivery line, and pressure drops abruptly to prevent secondary injection. Delivery is completed (Fig. 6.428).

Pump elements (plungers and barrels) feeding individual engine cylinders are combined in multicylinder injection pumps (Fig. 6.429).

Larger engines are fitted with separate single-cylinder plunger pumps that are driven by the engine camshaft(s) (Fig. 6.430).

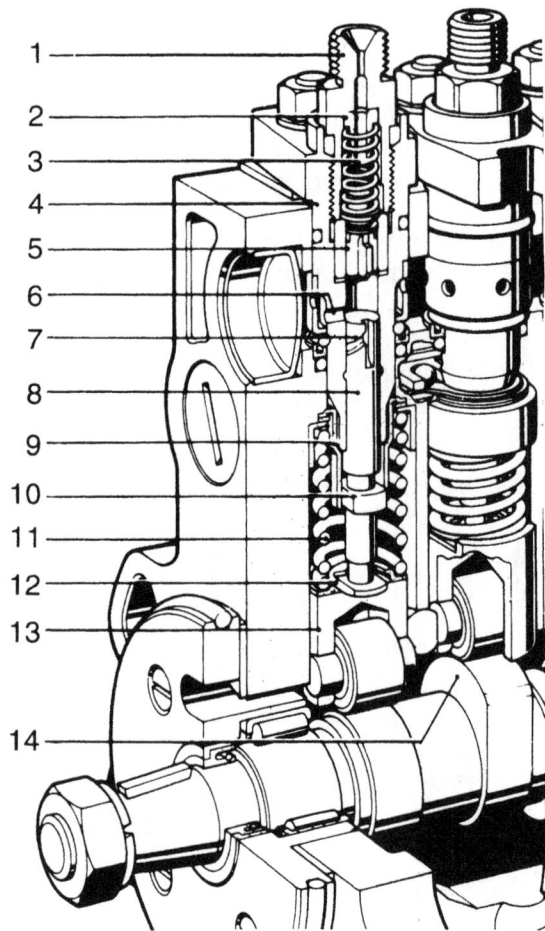

Fig. 6.428 Pump element of an inline injection pump, with drive: 1, Delivery valve holder; 2, Filler piece; 3, Delivery valve spring; 4, Pump barrel; 5, Delivery valve; 6, Inlet port and spill port; 7, Control helix; 8, Pump plunger; 9, Control sleeve; 10, Plunger control arm; 11, Plunger return spring; 12, Spring seat; 13, Roller tappet; 14, Cam. Source: VDI

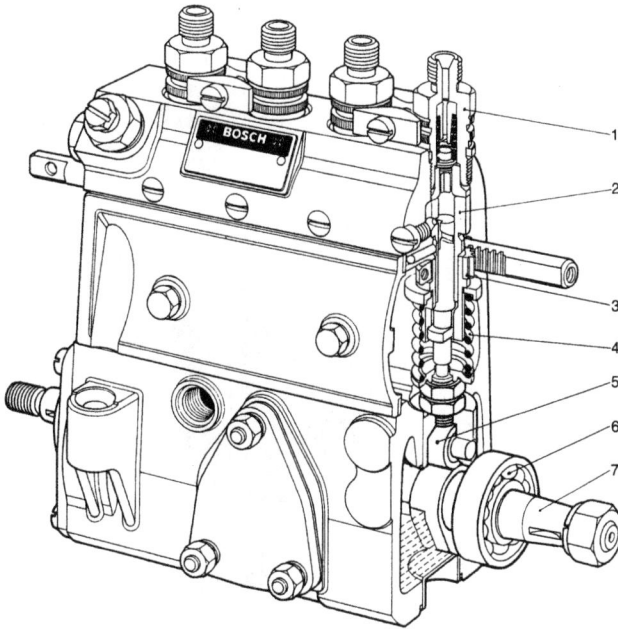

Fig. 6.429 Bosch inline fuel injection pump. Source: VDI

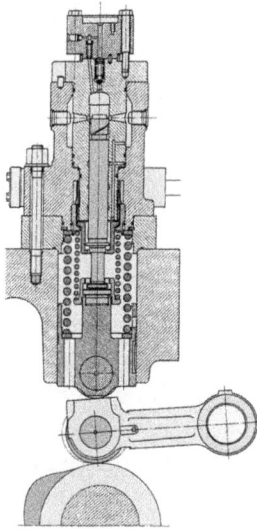

Fig. 6.430 Single-cylinder injection pump. Source: MAN

Unit injectors

Unit injectors, or PDE injectors (from the German "Pumpe-Düse Einheit—pump/nozzle unit) (Fig. 6.431) consist of an injection pump and injector nozzle combined into a single unit. General Motors has employed large numbers of unit injectors for two-stroke diesel engines; in Germany, Maybach equipped four-stroke diesel engines for railway locomotive and marine service with unit injectors. More recently, unit

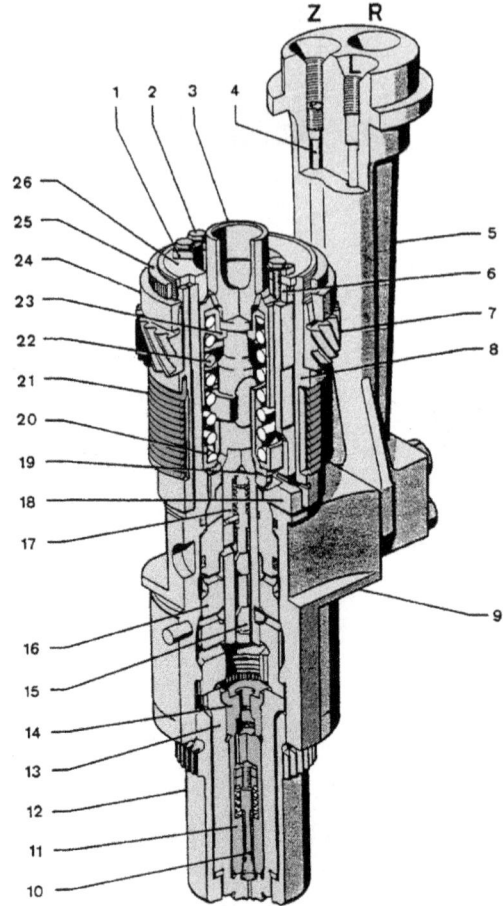

Fig. 6.431 Unit injector (MTU/ L'Orange type): 1, Plate; 2, Screw; 3, Lifter; 4, Filter screen; 5, Connector element; 6, Key; 7, Helical gear; 8, Control sleeve; 9, Housing; 10, Valve head; 11, Valve seat; 12, Threaded barrel; 13, Nozzle cap; 14, Pressure valve; 15, Suction valve; 16, Pump piston guide; 17, Pump piston; 18, Guide sleeve; 19, Dog; 20, Spring plate; 21, Torsion spring; 22, Compression spring; 23, Plunger base plate; 24, Clamping ring; 25, Threaded collar; 26, Adjusting ring. Z, Fuel supply; R, Fuel return; L, Fuel leakoff. Source: MTU

injectors have found their way into commercial vehicle and passenger car diesels. Due to elimination of the high-pressure injection lines, these systems are hydraulically "stiffer," which allows lower pump pressures than those found in conventional pump/line/nozzle systems; on the other hand, these systems permit achieving extreme pressures (2000 bar). The problem of injection line cavitation is eliminated, as is line delay (time delay of injection begin as a result of line length).

Distributor pump

The distributor pump is a structural combination of high-pressure pump and distributor, mechanical speed governor, hydraulic timing device, supply pump, and shutoff device (Fig. 6.432). It permits exact metering of even the tiniest fuel quantities, and is therefore especially well suited for smaller engines, i.e., passenger cars, light commercial vehicles, tractors, etc.

The function of distributor pumps is based on superposition of plunger stroke to generate pressure, and plunger rotary motion to distribute fuel to individual cylinders. In axial piston pumps, a cam disk converts rotary motion of the pump driveshaft into a combined rotary/reciprocating motion of the distributor plunger. This accomplishes supply port opening and closing, pressure generation, delivery port opening, and end of delivery by opening the spill port in the distributor plunger by means of a control sleeve.

In the radial piston pump, input shaft and distributor shaft are separate. A drive plate, carrying radially arranged pistons, turns within a cam ring with an appropriate number of lobes. These lobes actuate the pistons by means of rollers. Radial piston pumps are designed for larger fuel delivery quantities and higher pressures (Fig. 6.433) [6-35].

In the previously described fuel injection systems, begin of delivery is controlled by fuel pressure buildup in the pump element pressure chamber; delivery quantity is controlled by the duration over which high pressure is maintained in the chamber. Delivery is cut off by establishing a connection between the pressure chamber and the fuel

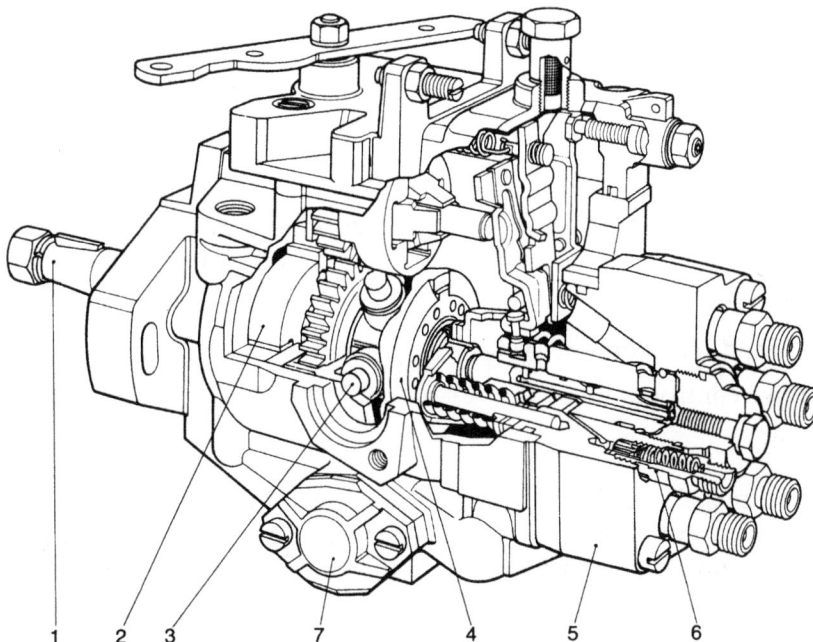

Fig. 6.432 Distributor injection pump (axial piston pump, Bosch): 1, Drive shaft; 2, Feed pump; 3, Roller holder; 4, Cam disk; 5, Distributor body with plunger and control sleeve; 6, Delivery valve holder with compression spring and delivery valve. Source: VDI

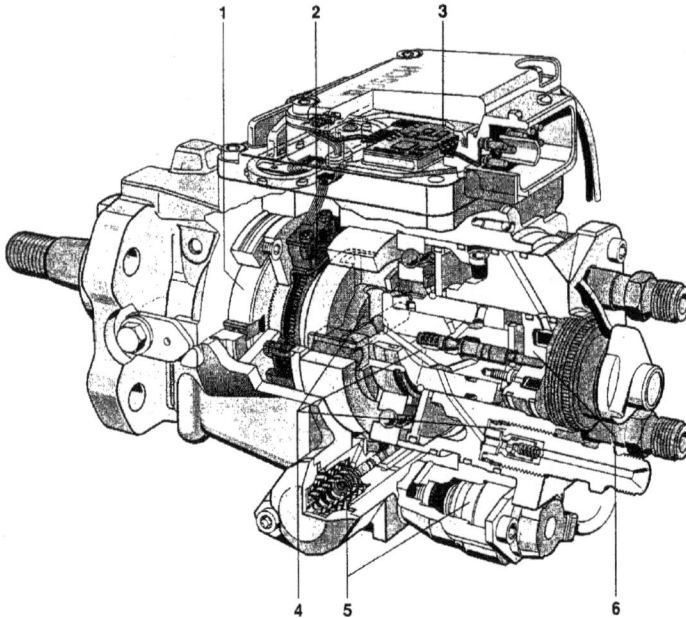

Fig. 6.433 Distributor injection pump (radial piston pump, Bosch): 1, Vane type supply pump with pressure control valve; 2, Angle of rotation sensor; 3, Pump electronic control unit (ECU); 4, Radial piston high pressure pump with distributor shaft and delivery valve; 5, Timing device and pulsed timing solenoid; 6, High-pressure solenoid. Source: BOSCH

gallery; in effect, a hydraulic short circuit. In the single-plunger pump, this is achieved by the plunger helix, and in the axial piston pump by a control sleeve. At present, the opportunities offered by electronic controls are finding increased application in diesel fuel injection equipment. Electronic diesel controls consist of:

- Sensors and setpoint devices
- Electronic control unit
- Actuator

In inline pumps, the fuel rack is moved by a linear magnet controlled by an electronic control unit. Axial piston distributor pumps use a solenoid to rotate the cam ring. In a further development stage, fuel metering is accomplished by high-pressure solenoids.

Development of dependable, reliable high-pressure solenoids has permitted electronic control of begin and duration (and, therefore, quantity) of fuel delivery or injection. Electronic injection systems can now be precisely tailored to the demands of individual engine map operating points, while incorporating additional parameters. Because development of electronically controlled diesel fuel injection is a rapidly evolving field, reference should be made to the current specialist literature.

Pressure generation in the injector nozzle

If one regards the unit injector as a "simplification" of the conventional pump/line/nozzle system, this path can be taken one step further in that the nozzle itself is used to generate pressure and meter fuel. Such systems have been developed in the United States by the Cummins Engine Co. (Fig. 6.434). The injector plunger (colloquially, the "needle") is actuated, against spring force, by means of a rocker arm, pushrod, and a separate cam on the engine camshaft. Metering for any given load condition is accomplished by storing the appropriate amount of fuel in a system of throttling and metering reservoirs. Fuel supply pressure to the injector is modulated by a fuel feed pump as a function of the desired engine output.

Recently, the Cummins injection system has also been converted to electronic operation. Fuel metering and begin of injection are controlled by a solenoid valve (the Cummins "CELECT" system).

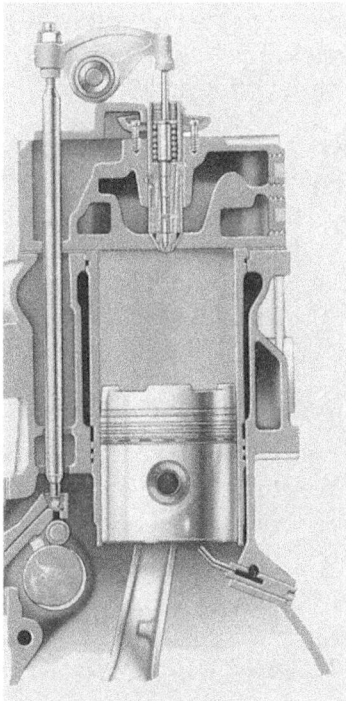

Fig. 6.434 Cummins PT injection system. Source: Cummins

Common rail injection systems: decoupling pressure generation and injection

In conventional fuel injection systems, injection is coupled to pressure generation (regardless of whether this is accomplished in the pump or the injector itself), and therefore coupled to crankshaft rotation; both the injection pump as well as the injectors are driven indirectly by the pump or directly by the engine camshaft, and therefore, directly or indirectly, are driven by the crankshaft. To match fuel injection to engine demands in all operating regimes, suitable measures such as appropriately shaped plunger helixes, timing advance units, etc. are needed. Furthermore, conventional systems have a disadvantage in that injection pressure increases with engine speed and injected quantity; larger quantities are injected at higher pressure than smaller quantities. Peak pressure is more than twice as high as average injection pressure.

In the common rail principle, injection takes place independently of pressure generation. Injection pressure is therefore freely selectable, and the injection rate diagram may be shaped as desired. Common rail systems are built on a modular principle (Fig. 6.435), consisting of a high-pressure radial piston pump, the common rail itself, solenoid-actuated injectors, and an electronic control unit. The high-pressure pump generates pressure in the fuel rail, independently of the actual injection system. Fuel is individually metered to cylinders by means of electronically controlled injectors. Injected quantity, injection pressure, and injection timing are freely selectable, as functions of pressure in the high-pressure rail and the opening duration of the solenoid valve. The advantage of such a system is that pre- and/or post-injection are easily achievable.

Pre-injection serves to precondition, i.e., improve, mixture formation conditions by reducing the ignition delay period as well as the combustion peak pressure ("softer" combustion). Post injection serves to reduce NOx emissions; after-injected fuel is not burned but rather only vaporized. Downstream, in an NOx catalytic converter, it serves as a reducing medium for oxides of nitrogen. Engine design advantages of common rail injection include:

- Greater freedom in designing the injection system.

- The high-pressure pump no longer needs to be driven by timing gears, chains, or belts, or from the camshaft; this unloads these components, and they may be redimensioned accordingly. Furthermore, the camshaft will be less subject to torsional vibration.

- Greater design freedom for shaping the cylinder head, e.g., compared to a unit injector system.

- Reduced noise emissions.

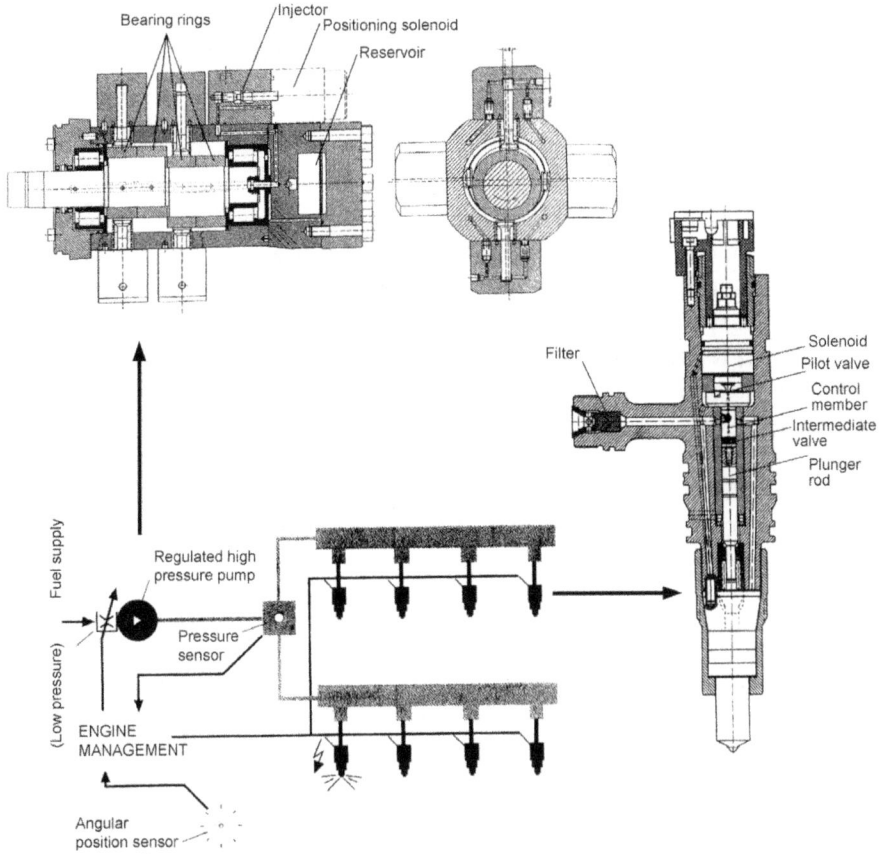

Fig. 6.435 Common rail injection system (L'Orange type). Source: L'Orange

Injection nozzles and injectors

The conventional injection nozzle (nozzle holder assembly with nozzle, Fig. 6.436) fulfills several tasks:

- Injection
- Fuel preparation for mixture formation
- Shaping the fuel injection rate
- Sealing the fuel system against combustion chamber gases

Nozzle opening by actuation of the needle is accomplished

- Hydraulically, by fuel pressure acting on a shoulder of the needle
- Mechanically by means of cam, pushrod, and rocker arm (Fig. 6.347), or, most recently
- Electrically, by activating a solenoid valve (Fig. 6.438)

The needle is returned to its seat by a compression spring. Direct-injection engines employ multi-hole nozzles; certain combustion processes may employ single-hole nozzles. Indirect injection engines use pintle nozzles.

6.5.3 Fuel injection system damage

In many cases, fuel injection system damage is caused by inadequate fuel filtration, either by filter deficiencies or by fuels of extremely poor quality, as are often encountered in maritime service. In the former Warsaw Pact nations, fuel quality still leaves much to be desired; in particular, its high water content remains a problem. Thanks to close tolerances of only a few microns (µm), the fuel injection system is extremely sensitive to dirt and fuel contamination. Suitable fine filtration is intended to remove contaminants from the system, yet time and again, dirt manages to find its way into injection systems. The ensuing wear enlarges tight clearances between sliding components of

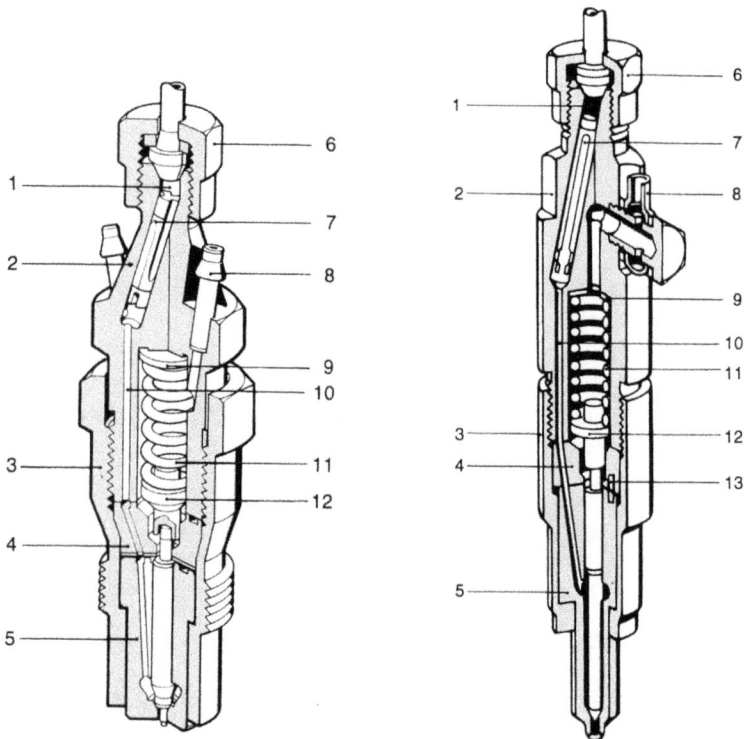

Fig. 6.436 Bosch-type nozzle holder designs. Left: with throttling pintle valve; right: with multi-hole nozzle. 1, Inlet; 2, Nozzle holder body; 3, Nozzle retaining nut; 4, Intermediate element; 5, Injection nozzle; 6, Union nut with fuel injection line; 7,Edge filter; 8, Leakoff connection; 9, Pressure adjusting shims; 10, Pressure channel; 11, Compression spring; 12, Pressure pin; 13, Locating pins. Source: VDI

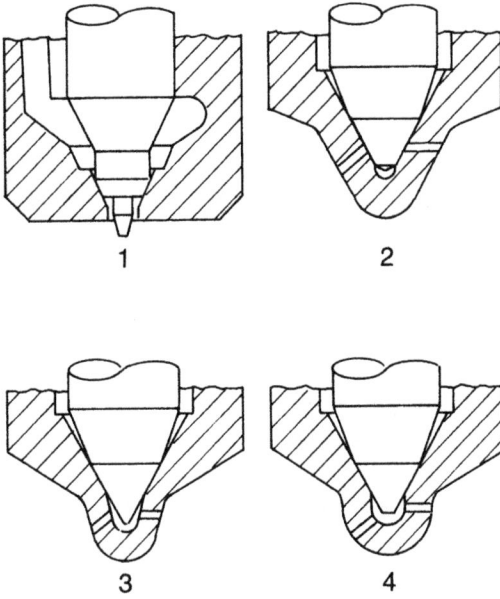

Fig. 6.437 Nozzle configurations:
1, Throttling pintle nozzle; 2, Seat-hole nozzle; 3, Hole-type nozzle with conical sac hole; 4, Hole-type nozzle with cylindrical sac hole. Source: VDI

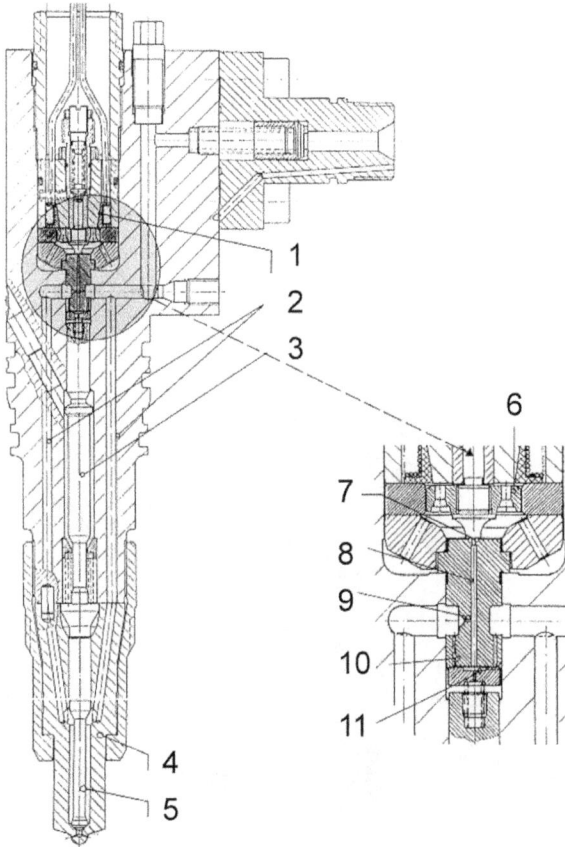

Fig. 6.438 Solenoid-actuated injector (L'Orange): 1, Solenoid; 2, Fuel supply; 3, Plunger rod; 4, Injector nozzle; 5, Injector needle; 6, Solenoid armature; 7, Fuel return restrictor; 8, Control member; 9, Fuel supply restrictor; 10, Fuel supply passages; 11, Intermediate valve. Source: L'Orange

pumps and nozzles. The resulting leakage losses reduce fuel injection quantity and delay pressure buildup, causing delayed injection. Experimentally, it has been possible to quantitatively demonstrate the relationship between injection pump plunger clearances and engine torque or power [6-36]. Because they are directly exposed to hot combustion gases and their corrosive components, injector nozzles often exhibit signs of damage [6-37]. An additional damage complex is cavitation, often associated with fluid flow processes.

6.5.3.1 Injection pumps
Plunger pumps
Cavitation

Even before actual damage occurs, cavitation may make its presence known by a drop in delivery volume. Cavitation is promoted by

- Deviation of the plunger helix geometry from design, and by
- Operation (even briefly) with water-contaminated fuel

Typical cavitation phenomena range from light cavitation "shadows" to severe pitting

- Between the upper and lower plunger helixes and in the
- Pump housing (low-pressure side, at the edge of the spill port)

Consequences of cavitation and erosion (Fig. 6.439) include wear, cracks, and fatigue fractures. Remedies may include design changes and/or increased system pressure.

Fig. 6.439 Cavitation/ erosion wear to an injection pump plunger after several thousand operating hours.

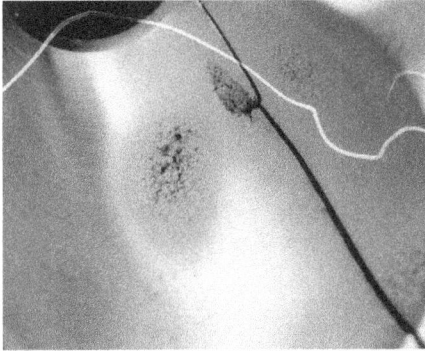

Fig. 6.440 Development of cavitation damage in spill port.

Fig. 6.441 Cavitation/erosion areas in spill port passages.

Fig. 6.442 Fatigue failure surface of a pump barrel.

The approximately 2-mm-deep cavitation crater in the spill port of this pump element (Figs. 6.440 and 6.441) initiated a slowly propagating fatigue fracture, with nearly perfectly "classical" beach marks (Fig. 6.442). The walls of the inlet passage (top, in illustration) as well as the plunger bore have suffered corrosive attack.

Corrosion
Water, especially seawater, intrusion into the fuel system will lead to corrosion. The corrosion products will cause mechanical damage.

Seizing
Compromised lubrication, caused by dirt contamination or water in the fuel, or by other agents, may lead to plunger seizing (Fig. 6.443).

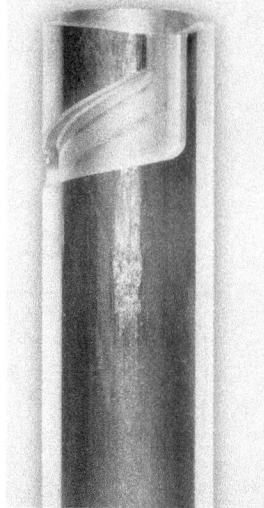

Fig. 6.443 Seized plunger.

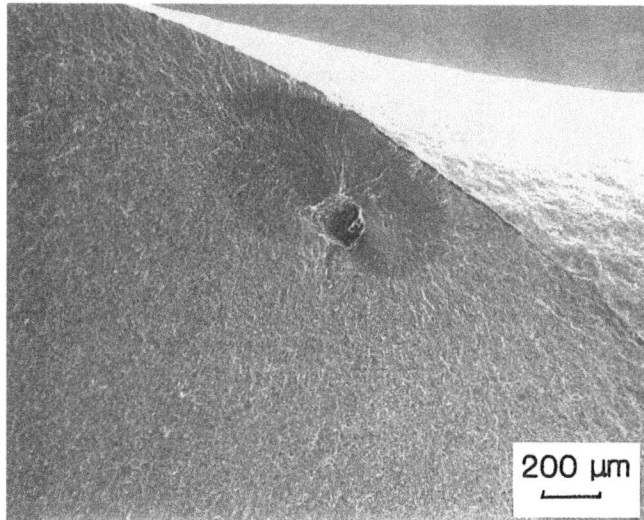

Fig. 6.444 Nonmetallic inclusion in an injection pump spring. Source: SCHERDEL

200 μm

Injector springs

The same types of failures encountered in engine valve springs are also seen in the springs of fuel injection systems—fatigue failures originating from the flat-ground spring ends, or as a result of material inhomogeneities (Figs. 6.444 and 6.445).

Example: Fatigue failure of a fuel injection pump spring was caused by a nonmetallic inclusion, of size no less than 100 μm. Energy-dispersive X-ray diffraction analysis showed that this inhomogeneity, very untypical of this spring material, consisted of difficult-to-melt zirconium oxide, which had not been removed with slag during the manufacturing process.

Injection pump cams

- Dirt impressions and dirt scoring (Fig. 6.446).

- Wobble wear (Fig. 6.447).

- Mechanical damage:
  Improper installation after overhaul may result in mechanical damage, as can extreme plunger seizing, which will also affect other pump components (Figs. 6.448 and 6.449).

Defects on specific high-pressure components of the injection pump include:

- Fretting corrosion to nozzle holder, and

- Fretting corrosion to roller follower pins (Fig. 6.450)

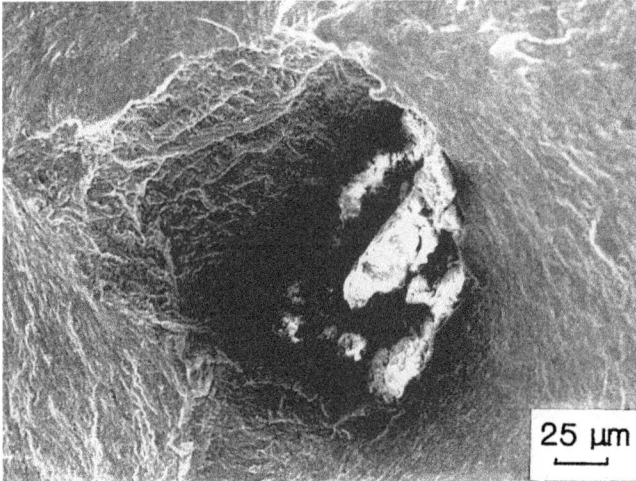

25 µm

Fig. 6.445 Point of origin for fatigue failure of an injection pump spring. Source: SCHERDEL

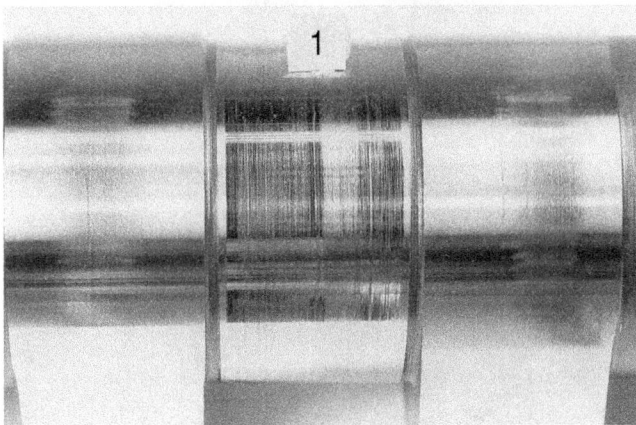

1

Fig. 6.446 Dirt scoring of a pump cam (1).

6.5.3.2 Injector nozzles

Damage caused by dirt

Inadequate fuel filtration naturally has consequences for injector nozzles. Dirt may originate in the fuel itself, or, as is often the case after engine overhauls, metal shavings or slag from poorly cleaned fuel lines may cause scoring and seizing of injector needles.

Wear

Wear inevitably accompanies engine operation. This is only amplified by dirt carried by fuel (Figs. 6.451, 6.452, and 6.453).

Fuel-borne dirt, in particular sand, has a devastating effect on injector nozzles.

Example: Wear on the pressure pin (pressure plate) increases needle lift, which in turn raises the impact speed of the injector needle against its

Fig. 6.447 Typical wobble wear patterns in "rising" cam ramps (2, 3).

Fig. 6.448 Severely deformed and damaged "rising" cam ramps.

Fig. 6.449 Cracked, broken control sleeve.

Fig. 6.450 Roller follower pins with fretting-like damage pattern.

seat. This leads to higher loads, which may ultimately cause fatigue fractures and the breakage of the nozzle tip (Fig. 6.454).

In one case, pressure pin wear was so great that this was effectively punched through. In addition, sand had been hammered into the needle seat, preventing proper sealing and causing fuel dribbling. The resulting carbon deposits clogged the nozzle holes (Fig. 6.455).

This condition was "remedied" in the course of repair by the vehicle operator, who "cleaned" the nozzles with abrasive cloth, a fact that was subsequently obvious by traces of Al_2O_3 (grinding carborundum) packed into the nozzle holes (Fig. 6.456).

Damage of this sort underscores not only the need for faultless fuel filtration upstream of the injection pump, but also for measures to prevent entry of dirt (sand) while refueling, or through the fuel tank vent. This is especially important for vehicles whose engines are

Fig. 6.451 Normal appearance of an injector needle seat surface, with impressions and embedded foreign matter in the actual seat area.

Fig. 6.452 A sectioned nozzle reveals the mating needle seat surface. Two pronounced impressions are visible, matching those on the mating needle.

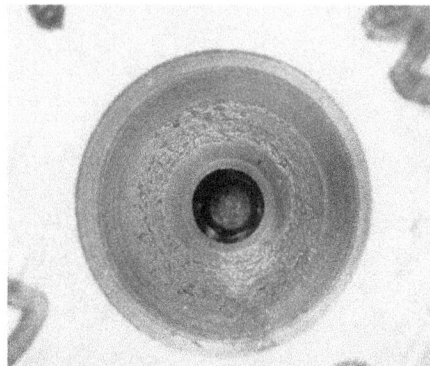

Fig. 6.453 Severely worn needle seat area in nozzle, with recognizable "break through." This nozzle no longer seals properly.

required to work under extremely dusty conditions, e.g., construction equipment or vehicles in arid climates.

Cavitation

Cavitation leads to loss of material and material fatigue along the walls and inlet passages of injector nozzles and nozzle holders (Fig. 6.457).

Corrosion

Use of high-sulfur fuels, as is common in maritime service, leads to condensation of SO_2, which has an extraordinarily corrosive action. In medium and large engines, this tendency is aggravated in cooled injector nozzles. According to injection equipment supplier Bosch, the average injector nozzle operating temperature should not fall below 110°C (230°F; see Fig. 6.458).

Excessive operating temperatures
In the course of performance increases by means of supercharging, injector nozzles, like other components, are subjected to higher temperatures. Temperatures in excess of 200°C (390°F), recognizable by the appearance of "tempering colors" on the nozzles, result in a decrease in needle seat hardness, which shortens nozzle life. An additional effect of high temperatures is fuel cracking inside the nozzle. Wear increases, and nozzle opening cross sections are reduced by deposits (Fig. 6.459).

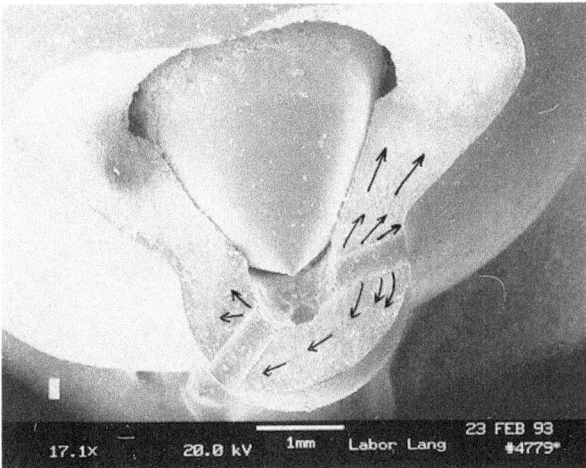

Fig. 6.454 Fatigue failure of an injector tip (injector needle shown in closed position). The fatigue fracture propagation directions are indicated by arrows.

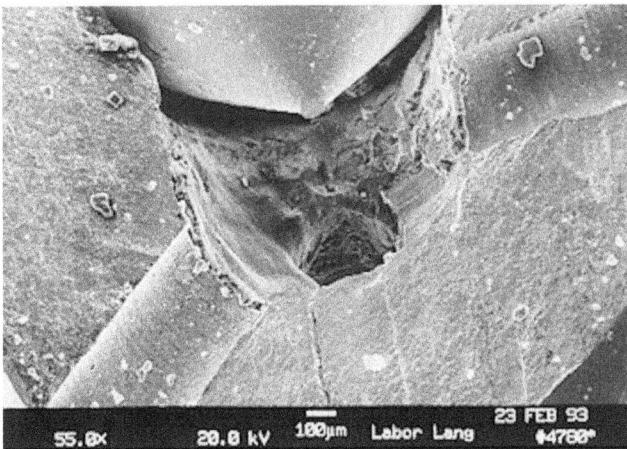

Fig. 6.455 The area around the needle tip shows carbon deposits.

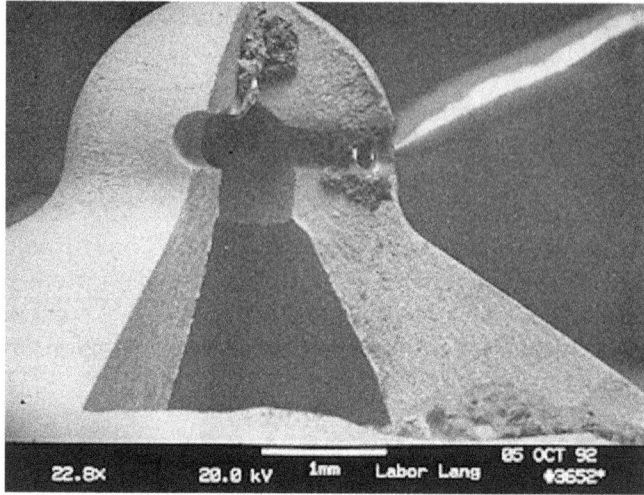

Fig. 6.456 The nozzle holes are clogged by grinding compound.

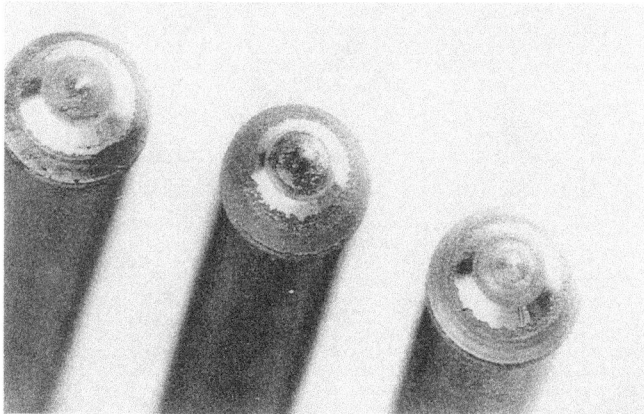

Fig. 6.457 Injector needles with severe cavitation wear above the seat surface.

Uncapping

In cases of extreme abrasive wear, the lands between individual nozzle holes are weakened to the point where they are mechanically overloaded and crack. The nozzle tip breaks off (the nozzle is "uncapped"), partially or completely. Causes may include:

- Poor fuel filtration, or fuel with a high abrasive particle content
- Deposit formation on the injector needle guide, due to poor fuel quality
- Insufficient nozzle opening pressure
- Cold corrosion

Deposits on the needle seat within the nozzle impair motion of the needle; combustion gases may enter the nozzle and destroy the seat.

Soot-blackened needle points indicate nozzles that no longer seal completely (Fig. 6.460).

The resulting uncontrolled combustion leads to overheating and destruction of the nozzle. The same failure sequence occurs if nozzle opening pressure has dropped too low, due to poor maintenance and/or excessive wear to the nozzle holder pressure pin. Cold corrosion, especially in lightly stressed engines, may also lead to destruction of nozzles, in particular if maintenance intervals are exceeded.

Example: Fuel containing seawater led to deposit formation in the nozzle, which hampered movement of the injector needle. Injection and combustion processes became increasingly uncontrolled, and the nozzle overheated and was destroyed (Figs. 6.461, 6.462, 6.463, and 6.464).

Injector nozzle compression springs
Injector nozzle compression springs are subject to the same failure modes as those of injection pumps.

Example: Scanning electron microscopy of the compression springs of several injector nozzles from a marine diesel engine with an operating time of ca. 2500 hours showed that spring fractures were caused by notching as a consequence of corrosion pitting. The gap at the coil-bound end of the spring helped to promote the failure (crack corrosion effect; see Fig. 6.465). Because energy-dispersive X-ray diffraction analysis of the corrosion products did not show any seawater-specific elements such as chlorine, the corrosive attack was probably caused by fuel-borne water. This was confirmed by fretting corrosion in the

Fig. 6.458 Corroded injector nozzle.

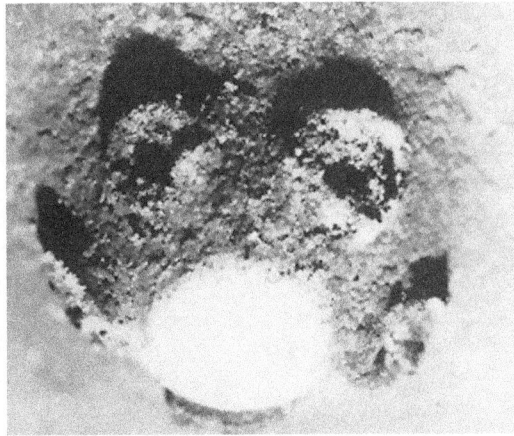

Fig. 6.459 Carbon deposits on an injector nozzle.

high-pressure section (Figs. 6.466 and 6.467), which, from experience, is aggravated by water-contaminated fuel.

6.5.3.3 Fuel injection lines

To prevent secondary injection, attention is given to ensuring rapid pressure release in the pump pressure chamber by means of a delivery valve incorporating a constant retraction volume piston. However, this sudden pressure relief promotes cavitation in the injection lines. It is difficult to find effective measures against this cavitation, because a small retraction volume combined with a large fuel delivery quantity is just as unfavorable as a large retraction volume combined with smaller fuel delivery quantity. Remedies were tried with constant-volume retraction valves incorporating return flow restrictors, yet cavitation continued to appear (Fig. 6.468).

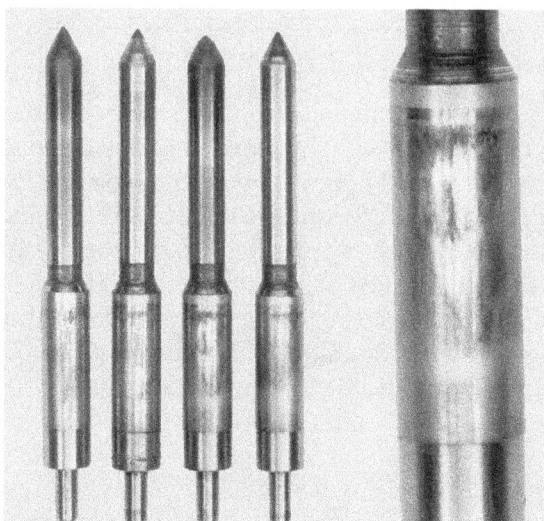

Fig. 6.460 Signs of seizing and binding, as well as deposit formation, on injector needles, caused by substandard fuel.

Fig. 6.461 Severe burnoff around injector holes and on needle tip.

Fig. 6.462 Nozzle tip burned away as far as the nozzle holes; needle tip burned back to seat area.

Fig. 6.463 Completely "uncapped" nozzle tip.

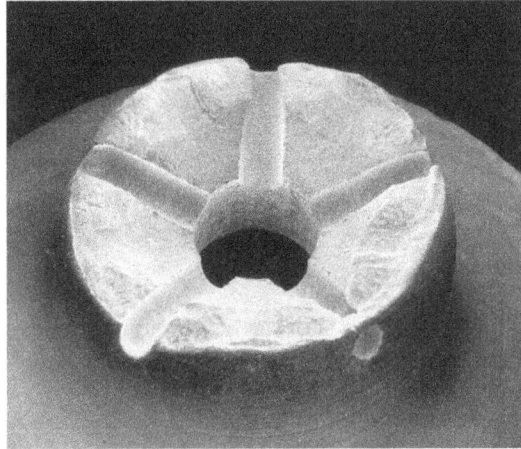

Fig. 6.464 Fatigue fracture and residual overloading, resulting in blown-off nozzle tip.

Example: After about 8500 operating hours, examination of nozzle holder supply lines discovered severe cavitation damage, extending over a length of 120 to 160 mm and 0.3 to 0.65 mm deep (Figs. 6.469, 6.470, and 6.471).

6.5.4 Glow plugs

Cold engines, whether diesel or Otto cycle, are difficult to start; cold lubricating oil is viscous, starter battery capacity falls at low temperatures, piston clearances are larger, and as a result the temperature of compressed charge air is low. In diesel engines, greater ignition delay under cold-start conditions is an added factor. Indirect-injection diesel engines (i.e., those with divided chambers; prechamber

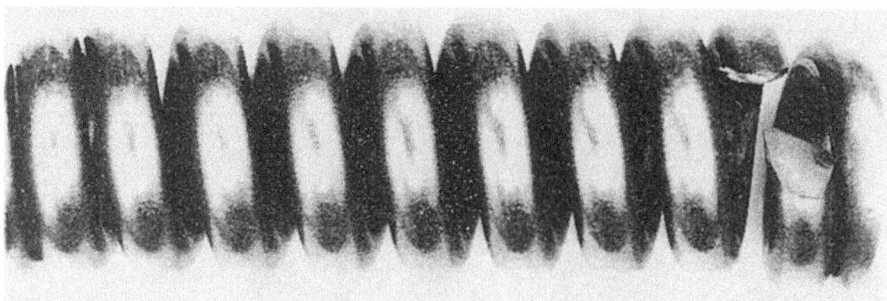

Fig. 6.465 Compression spring with fracture in coil-to-coil contact area of flat-ground spring end, on adjusting shim end.

Fig. 6.466 Fatigue fracture surface with fracture origin in area of coil-to-coil contact.

or swirl chamber) are especially susceptible to starting difficulties. Aside from the difficulty in even getting the engine to start, the starting process was once associated with clouds of soot as well as partially burned and unburned fuel. Old operating manuals hint at the measures once employed to start reluctant engines: "... pour warm water into the radiator, twice if possible; to achieve rapid ignition, lower-boiling fuels such as gasoline or [diethyl] ether ..." (from a Büssing manual). Another example of unusual starting aids was the once familiar method of pre-warming Lanz Bulldog tractors using a blowtorch.

Glow-starting systems have greatly improved this situation. Glow plugs consist of a resistive heating element that heats air in a prechamber or the main cylinder, thereby improving mixture formation of the injected fuel.

As diesel passenger cars became more popular, there was increasingly pressing demand for simpler, faster starting. In the case of indirect-injection (prechamber) engines, glow starting is unavoidable; but even direct-injection engines benefit from preheating, to speed the starting process, and post heating ("post glow") to reduce HC emissions.

Rapid-heating pencil-type glow plug

Pencil-type glow plugs consist of a glow plug body, a glow tube, and a connector thread (Fig. 6.472). The glow tube contains a heating coil and a regulating coil. The heating coil is welded to the tip of the heating tube, while the regulating coil is welded to the threaded connector. Heating and regulating coils are welded together, thereby electrically connecting them in series. When current is applied, the heating coil heats up, and heats the glow tube to 850°C (1560°F). The regulating coil is made of a material with a positive temperature coefficient of resistance, so that its resistance increases with rising temperature. In this way, electrical current is reduced as a function of temperature, and heating is limited. In newer designs, the regulating coil is made of a material whose coefficient rises more steeply above 400°C (750°F), enabling faster heating. To enable a post-start heating capability, glow plugs must be able to withstand higher voltage after starting (about 15.5 instead of 12 V), as well as the added heat of the combustion process. This is achieved by means of a specially designed regulating coil.

Glow plug damage

Depending on operating conditions, pencil-type glow plugs exhibit appearance characteristic of a heating element, which permits corresponding conclusions to be drawn: burnt glow plugs, damaged glow tube tips, or broken resistance coils. Such damage is often the result of injector nozzle problems, overvoltage as a result of jump starting, etc.

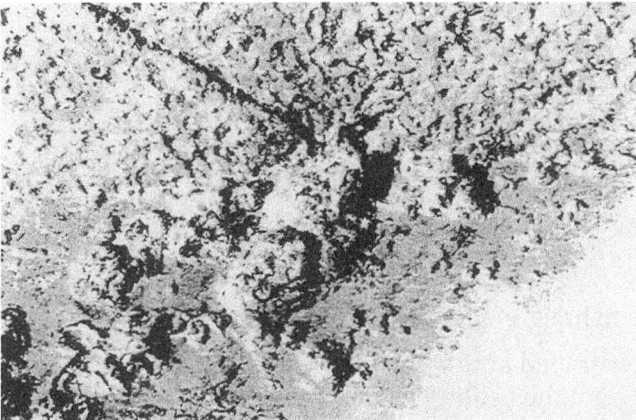

Fig. 6.467 Corrosive attack in crack origination area, and subsequent fracture paths.

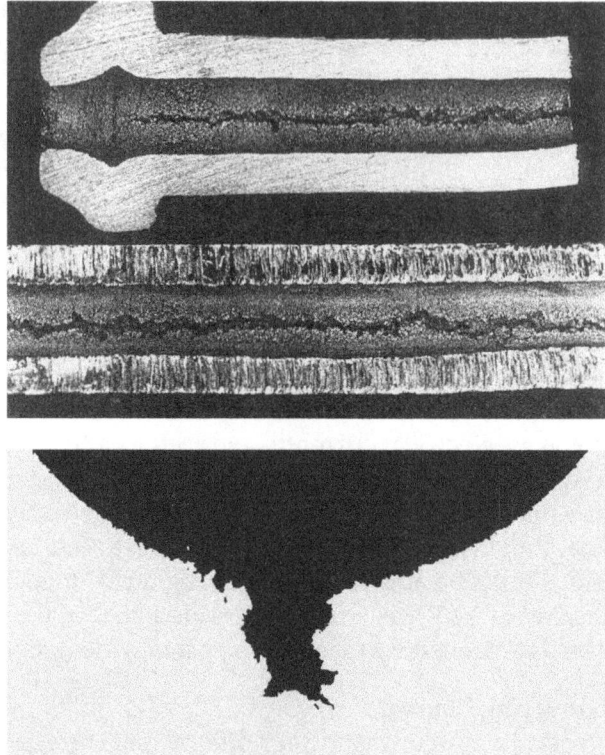

Fig. 6.468 Typical cavitation channel as a result of pressure wave cavitation. Bottom: section through area attacked by cavitation.

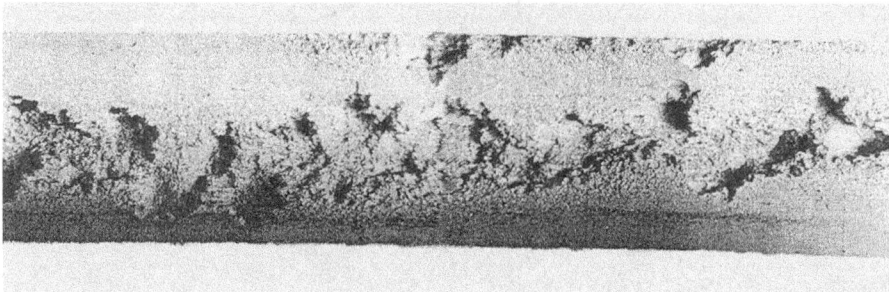

Fig. 6.469 Cavitation attack in fuel supply line.

Damaged glow tube tip

Glow tube overheating leads to damaged glow tube tips (Fig. 6.473). Overheating may have several causes:

- Combustion abnormalities, e.g., excessive injection advance
- In the event of a constricted annular gap between the body and glow tube, heat transfer from the heating element to the body is increased, causing excessive cooling of the regulating coil, which therefore

fails to cut off current to the heating coil (Fig. 6.474). If the annular gap between the body and heating element is closed by excessive tightening torque, too much heat flows from the heating element, causing the regulating element to stay cool, and allowing too much electrical current to the heating coil, which then overheats.

Glow tube shows signs of melting or has melted off, glow tube broken off

Glow tubes (Fig. 6.475) are thermally overloaded by

- Excessive injection advance (too early begin of injection)
- Carbon deposits on nozzles, or nozzles that do not close completely
- Engine faults such as seized pistons, stuck piston rings, or stuck valves

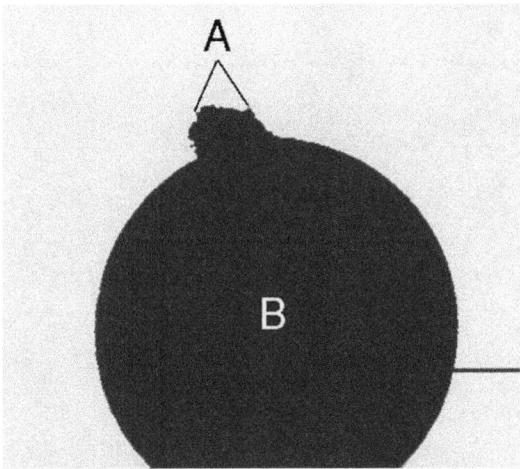

Fig. 6.470 Transverse section through fuel supply line damaged by cavitation (A); B = fuel line bore.

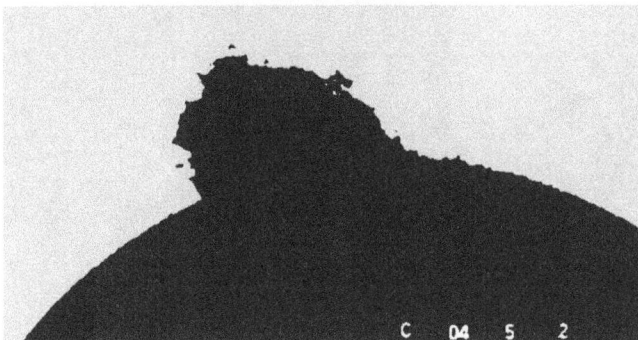

Fig. 6.471 Enlargement of part of previous image; cavitation damage is approx. 0.65 mm deep.

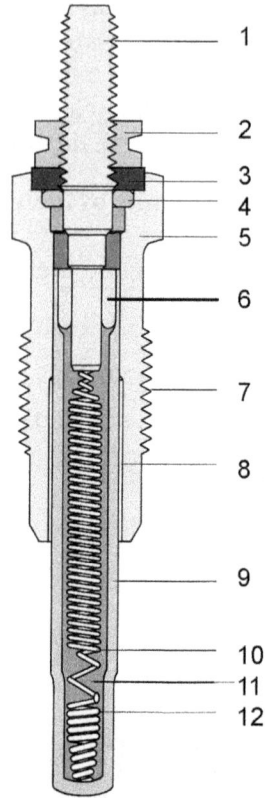

Fig. 6.472 Schematic of pencil-type glow plug: 1, connector; 2, round nut; 3, insulating disc; 4, O-ring; 5, glow plug body; 6, seal; 7, mounting threads; 8, annular gap; 9, glow tube; 10, regulating coil; 11, insulating filling; 12, heating coil. Source: BERU

Fig. 6.473 Damaged glow tube tips. Source: BERU

Fig. 6.474 Glow plug bodies with normal and constricted annular gap. Source: BERU

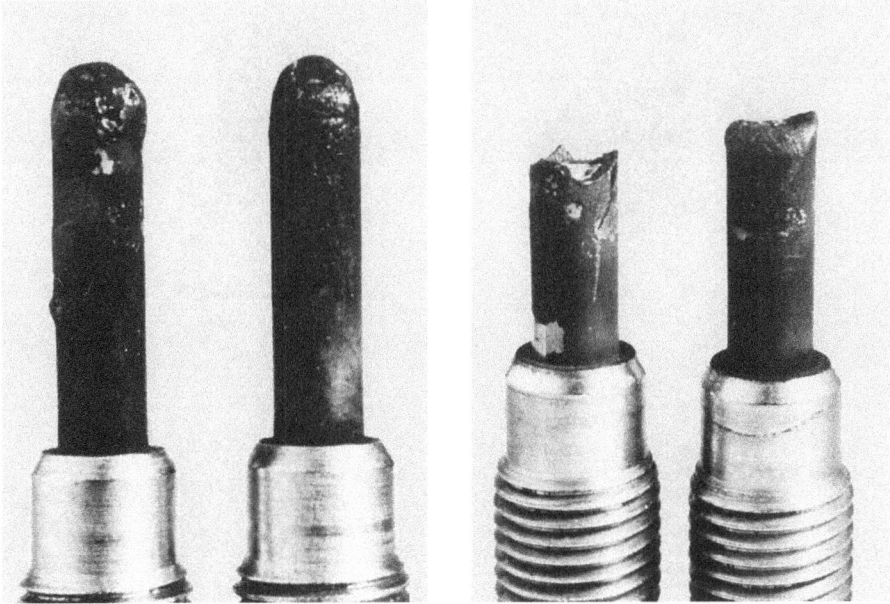

Fig. 6.475 Melted, or melted off, glow tubes; broken glow tubes. Source: BERU

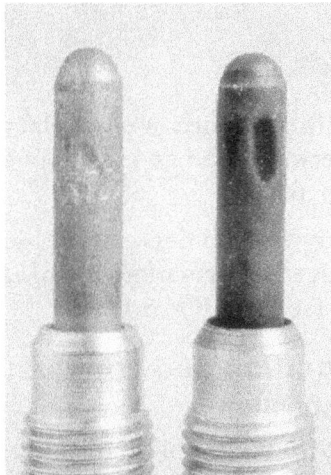

Fig. 6.476 Glow tube creases and dents caused by broken coils. Source: BERU

Creases and dented glow tubes

Creases and dents (Fig. 6.476) are the result of

- Thermal overloading as a result of overvoltage operation, e.g., from external starting assistance (jump starting);

- Excessively long operating duration caused by stuck relays, improper post-glow operation while engine is running, or

- Use of a non-post glow capable glow plug.

Fig. 6.477 Mechanically damaged glow plug (connector broken off). Source: BERU

Fig. 6.478 Mechanically damaged glow plug (damaged wrench flats). Source: BERU

Mechanical damage: connector thread/body wrench flats

- Glow plug with broken connector thread as a result of excessive tightening torque (see also Fig. 6.477).

- Glow plug with damaged body wrench flats caused by use of inappropriate tools. As a result of deformation, the plug has a short between body and round nut (see also Fig. 6.478).

6.5.5 Otto cycle engine ignition and combustion

6.5.5.1 Otto-cycle engine combustion

The idealized thermodynamic cycle representing the Otto-cycle engine is the so-called constant volume process, in which heat is added at constant volume, i.e., theoretically in an infinitely short space of time, in reality at least in a very short time near piston top dead center. In order for combustion to proceed quickly, a homogeneous mixture must be present. Initiated by the spark plug, a flame front spreads outward from the spark plug to encompass the entire combustion chamber volume.

In contrast to diesel engines, the compression temperature within the cylinders of an Otto-cycle engine must remain below the autoignition temperature of the mixture. Mixture in an Otto-cycle engine is ignitable only within a narrow range of relative air/fuel ratios (excess air factor)

λ; $0.8 < \lambda < 1.3$. Ignitability limits are also dependent on the applied spark energy and local mixture composition. Because of this limited ignitability, power output of Otto-cycle engines is regulated by the total amount of mixture, rather than the mixture composition; in other words, power control by means of throttled operation. The resulting throttling losses have a serious negative impact on part-load behavior of Otto-cycle engines.

Even before the onset of combustion, certain fuel components react with oxygen ("cold combustion"). From a chemical perspective, the mixture composition changes prior to regular ignition. The spark plug discharge initiates combustion; a plasma tube is established between the spark plug electrodes, with temperatures up to 6000°C (10,800°F). After a brief ignition delay of 0.5 to 1.0 ms, this initiates combustion of the fuel-air mixture.

The increase in pressure and temperature of the burning mixture, and the resulting displacement action, causes the still-unburned mixture to be compressed and heated even more. This displacement effect is superimposed on gas flow caused by piston motion. The flame core expands outward from the point of ignition and forms a stable flame front, where most of the available energy is converted. Combustion quality determines power, fuel economy, and pollutant formation; it is strongly influenced by mixture composition and ignition timing. Because combustion is intended to take place as quickly as possible near top dead center, and flame speed is limited to about 30 m/s, and the absolute time available to the engine for combustion decreases with increasing speed, ignition timing must be advanced accordingly.

Along with this (highly idealized) normal combustion process, there are also abnormal processes, the most important of which are preignition and detonation.

Preignition

In preignition, combustion is not initiated by the spark plug, but rather by hot engine components such as exhaust valves, spark plug electrodes, or glowing-hot carbon deposits on combustion chamber surfaces. Preignition can only occur if component or deposit temperatures exceed the mixture ignition temperature, and when there is sufficient time available for this type of ignition to take place. This is only possible at low engine speeds, which explains why preignition may occur after the engine has been shut down (a phenomenon known as "dieseling" of an Otto-cycle engine). Carbon deposits with a tendency to glow are primarily formed when too much oil finds its way into the combustion chamber, either through ineffective piston rings or excessive valve guide clearance. Also, if oil sludge from the crankcase reaches the combustion chamber, it first forms a so-called "soft coke," which over time hardens into carbon deposits.

Detonation

Detonation, also known as autoignition, combustion knock, knock, or ping, is an Otto-engine combustion abnormality that has posed a significant obstacle to engine development, and continues to pose a problem today (Fig. 6.479). Detonation is self-ignition of an unburned portion of the fuel-air mixture. After initiation of combustion by the spark plug, the flame front propagates through the combustion chamber at a speed of 10 to 30 m/s. This raises pressure and temperature in the as yet unburned mixture. If critical reaction conditions are satisfied, spontaneous self-ignition takes place in the remaining unburned mixture.

The flame front propagating from this self-ignition travels at several times the speed of the normal flame front: about 300 m/s. This leads to high pressure gradients (pressure increase per unit time) and higher peak pressures than those encountered in normal combustion. High-frequency pressure oscillations are characteristic of the pressure history observed with detonation; the amplitude of these pressure oscillations may exceed the peak pressure of normal combustion by as much as 25% (Fig. 6.480).

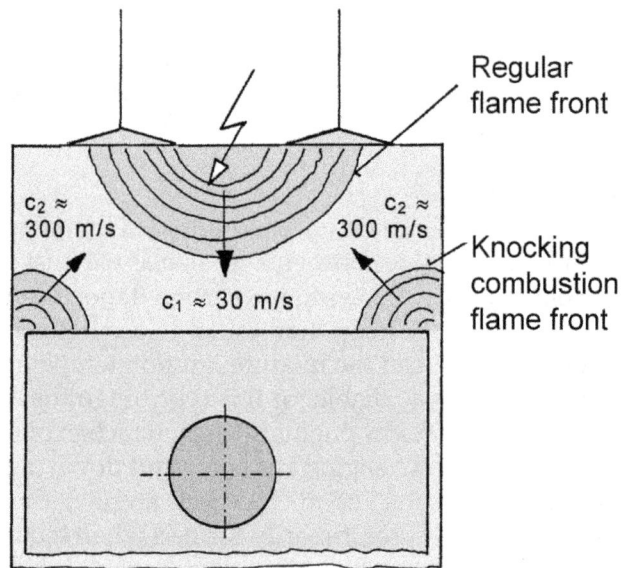

Regular flame front

$c_2 \approx$ 300 m/s

$c_1 \approx 30$ m/s

$c_2 \approx$ 300 m/s

Knocking combustion flame front

Fig. 6.479 Combustion knock (schematic).

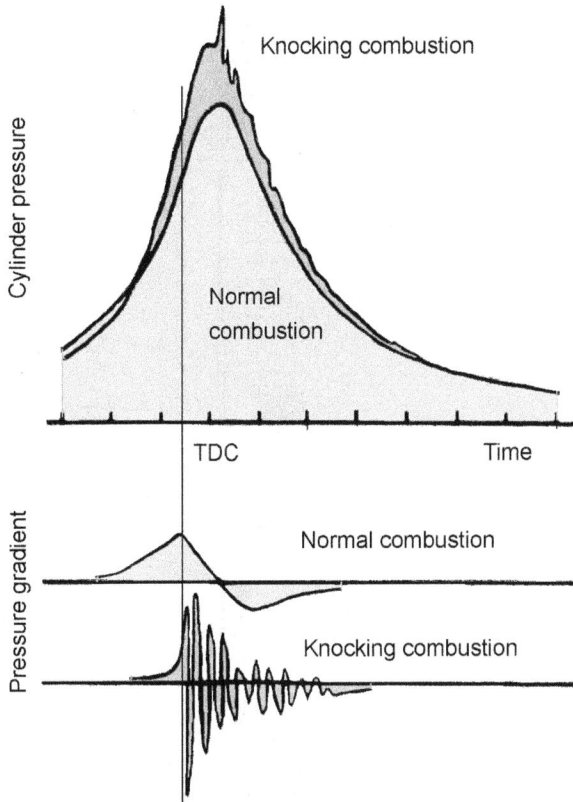

Fig. 6.480 Pressure and pressure rise during combustion knock (schematic).

Upon impact with the combustion chamber surfaces, these pressure waves induce vibration, causing "knock" to be acoustically perceptible as a ringing, knocking noise. The highly turbulent gas motion caused by detonation intensifies heat transfer, causing components bordering the combustion chamber, in particular the piston, to be thermally overloaded. High gas pressures and, worse, the rapid pressure rise ($dp/d\varphi$ or dp/dt) impose mechanical and tribological loads on crank train parts. Due to greater heat losses, detonation also leads to higher fuel consumption and causes an appreciable loss of power. For these reasons, engines must not be allowed to operate in the detonation regime.

Distinction is made between acceleration knock, which occurs while accelerating at full throttle from low engine speeds, and high-speed knock, in the higher engine speed and load ranges. Knock is promoted by a number of factors: fuel type and quality, combustion chamber shape, compression ratio, location of spark plug(s), cooling intensity of components subjected to high thermal loads, and many others. Knock-

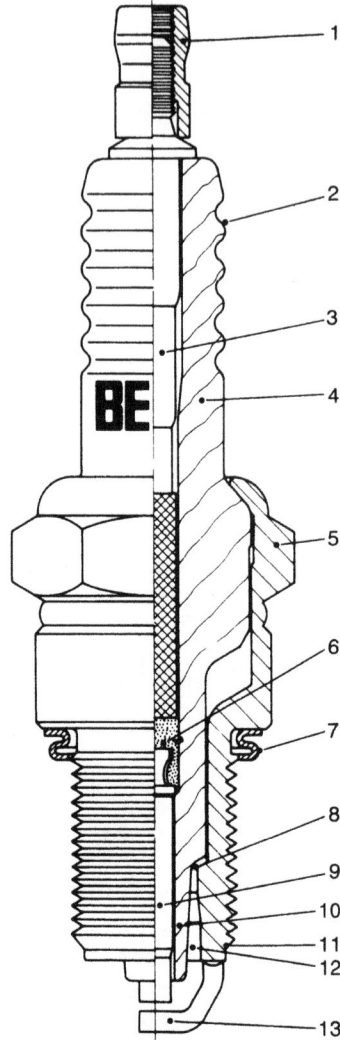

Fig. 6.481 Spark plug construction and terminology: 1, SAE terminal nut; 2, five-ridge profile, current leakage barrier; 3, terminal stud; 4, aluminum oxide insulator; 5, nickel-plated spark plug shell; 6, conductive melted glass seal; 7, captive gasket; 8, internal seal; 9, center electrode; 10, insulator tip; 11, beveled thread end; 12, scavenging volume; 13, ground electrode. Source: BERU

imposed limitations on compression ratio (representing the current state of the art at the time in question) have hampered Otto-cycle engine development to the present day. Fuel octane number serves as the yardstick for knock sensitivity; this number compares the behavior of a given fuel with a reference fuel blend composed of isooctane (knock resistant) and *n*-heptane (knock prone). The octane number represents the volume percent of isooctane in the reference blend. Susceptibility to knock may be countered by reducing compression ratio, but at the price of reduced power output and higher fuel consumption.

6.5.5.2 Ignition and spark plugs

Otto-cycle engines operate with timed external ignition; combustion is initiated by an electrical spark. This spark is created by the spark plug (Fig. 6.481); the spark exists given a sufficiently high voltage provided by an ignition coil or ignition transformer. An arc from the spark plug center electrode to the ground electrode ionizes the intervening gap, and initiates a spark discharge. The resulting energy release ignites the fuel-air mixture. For example, a six-cylinder engine operating at 5000 rpm requires 15,000 spark discharges per minute. Each spark plug fires 2500 times per minute, or more than 40 times per second.

The function and properties of individual spark plug components are described in Table 6.11.

Electrode gap
The space between center and ground electrode is termed the electrode gap (Fig. 6.482). Optimum gap is typically specified by the engine

Table 6.11
Individual Components of a Spark Plug: Properties and Function (see also Fig. 6.481)

1	Solid post (1/4 in) or threaded post (4 mm) terminal enables connection of spark plug to spark plug cable by means of corresponding connector in spark plug boot; conducts ignition voltage to the center electrode
2	Prevents surface leakage by lengthening the leakage path
3	Steel pin, embedded in conductive glass melt to form a gas-tight seal. Establishes electrically conductive path to center electrode
4	Insulates the center electrode from ground, at voltages to 40,000 V. Consists of a special ceramic material with good thermal conductivity, resistance to thermal cycling, as well as high mechanical and electrical durability. The neck of the insulator is glazed to prevent dirt adhesion and surface leakage
5	Is bonded to the insulator by means of swaging and a gas-tight thermal shrink fit; serves to attach the spark plug in the cylinder head; a nickel layer prevents corrosion and seizing into the cylinder head
6	Conductive seal between terminal stud and center electrode
7	Seals against gas leaks and conducts heat away from spark plug
8	Seals insulator against spark plug shell and transfers heat
9	Promotes heat transfer; different electrode materials affect spark plug heat range; enables arcing to ground electrode
10	Portion of insulator extending into combustion chamber; its specific design features significantly determine spark plug heat range
11	Promotes threading spark plug into cylinder head
12	Its shape affects self-cleaning behavior and heat range tolerance
13	Addition of chrome to electrode material improves corrosion and erosion resistance

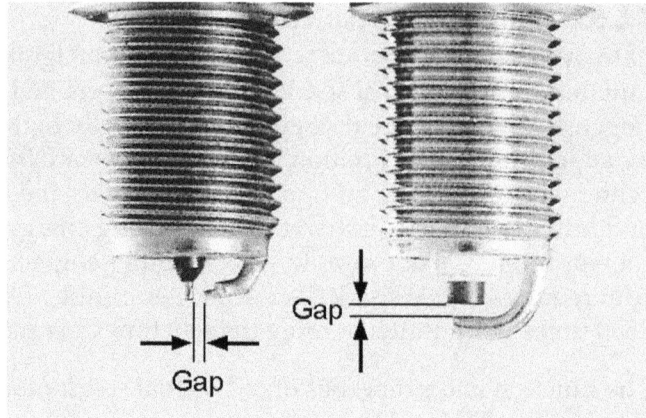

Fig. 6.482 Spark plug electrode gap. Source: BERU

or vehicle manufacturer. An insufficiently large gap may result in misfiring, uneven idling, and poor exhaust gas quality. An excessively large gap will increase ignition voltage demand, resulting in high-speed misfire.

Electrode shapes
Electrode geometry influences wear, heat transfer, ignition voltage demand, and accessibility of the fuel-air mixture (Fig. 6.483).

Electrode materials
Spark plug manufacturers offer various electrode materials (Fig. 6.484). Chrome-nickel alloys with copper cores exhibit good heat transfer and high resistance to corrosion. Depending on design particulars, service life is in the range of 15,000 to 30,000 km (9000 to 18,000 miles). Silver as an electrode material has exemplary heat transfer properties; for this reason, the Beru "Silverstone" spark plug, for example, with its pure silver center electrode, transfers heat more effectively than conventional spark plugs. Moreover, it enjoys a wider heat range, and its high resistance to erosion increases service life to 50,000 km (30,000 miles).

Spark gap, spark orientation
The spark path between the center electrode and ground electrode(s), as well as spark development of a three-pole spark plug, are shown in Fig. 6.485. The location of the spark in the combustion chamber is determined by the spark gap location; two variants are shown in Figs. 6.486 and 6.487.

Top electrode

Top electrode
extended insulator tip

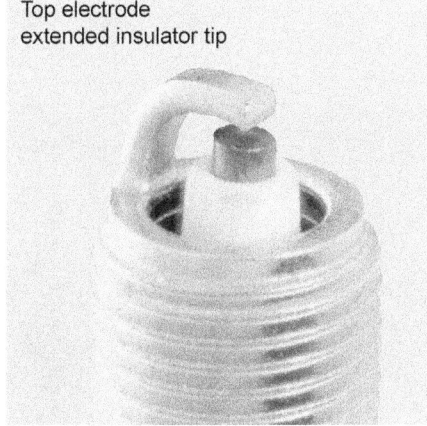

Side electrode

Multi-pole
side electrode

Triangular ground
electrode

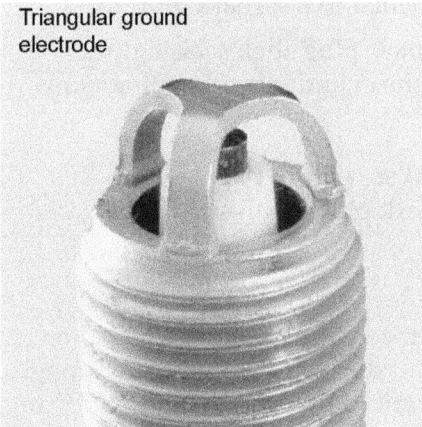

Fig. 6.483 Electrode
configurations.
Source: BERU

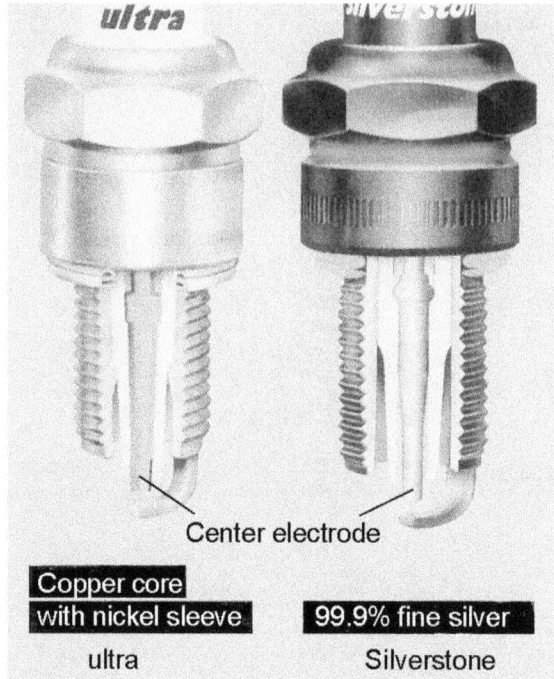

Fig. 6.484 Electrode materials. Source: BERU

Center electrode

Copper core with nickel sleeve
ultra

99.9% fine silver
Silverstone

Spark plug seats and gaskets

Depending on engine configuration, distinction is made between two different configurations for the seal between cylinder head and spark plug (Fig. 6.488):

- Flat seat: a captive gasket ring is attached to the spark plug shell

- Conical seat: a tapered seat on the spark plug shell seals tightly against a mating tapered surface in the cylinder head, without any gasket

Confined installations often employ flat seat spark plugs with smaller wrench flats, or conical seat spark plugs with smaller external dimensions.

Heat ranges

The energy released by combustion in the cylinder imposes a load on the spark plug; an appreciable portion of the heat is conducted through the spark plug to the cylinder head. Spark plug heat range provides an indication of the thermal capacity of individual spark plugs. Plugs with too low a heat range cannot conduct heat to the head quickly enough. At temperatures above about 1000°C (1800°F), this results in preignition, i.e., the fuel-air mixture is not ignited by a spark but rather well ahead of the intended ignition timing by the glowing, thermally overloaded spark plug. This causes a rapid rise in pressure and temperature, which may lead to piston damage. On the other hand, if spark plugs with too high a heat range are installed, the temperature necessary for a self-cleaning effect, about 400°C (750°F), is not reached at low engine speeds, causing soot deposits on the plugs, with resulting misfiring, increased fuel consumption, and elevated exhaust emissions.

Mutual interactions between engine and spark plugs
Abnormal Otto-cycle combustion processes also affect the spark plugs, which gives rise to the impression that the spark plugs are the cause of the resulting engine damage. The causes of damage to spark plugs and/or engine may include spark erosion, corrosion, preignition, combustion knock, and fuel flooding.

Spark erosion and corrosion
Excessive wear of the spark plug electrode material is caused by

* Thermal overloading

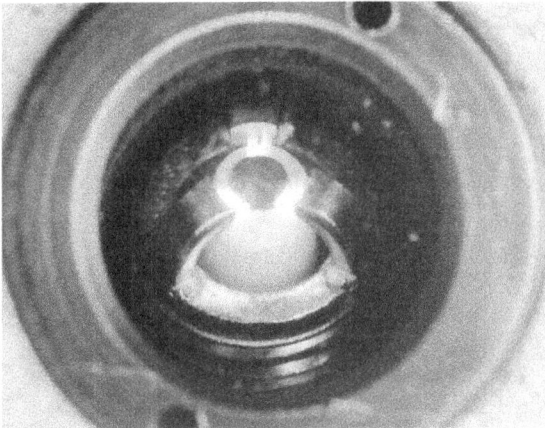

Fig. 6.485 Spark gap. Source: BERU

Fig. 6.486 Normal spark gap location; types A and C, primarily used in older engine designs. Source: BERU

Fig. 6.487 Extended-tip spark gap location; types B and D, modern engines. Source: BERU

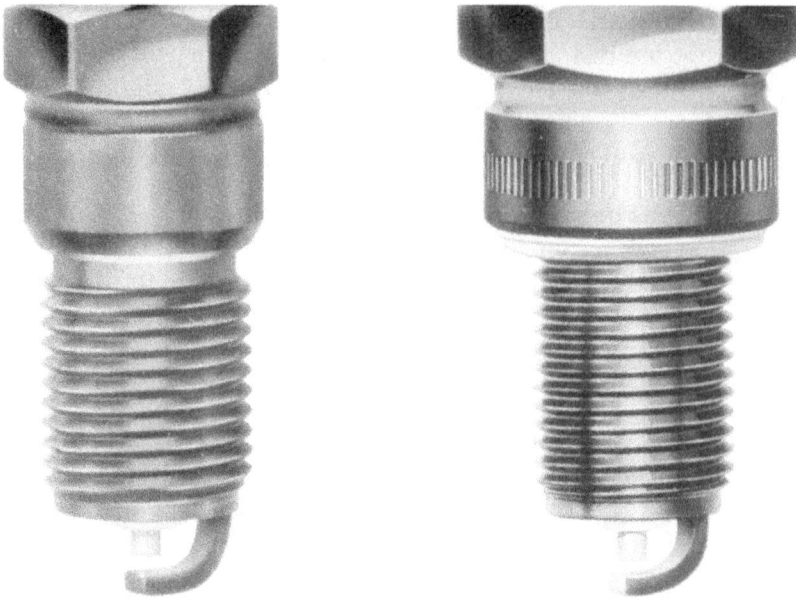

Fig. 6.488 Spark plug seats and gaskets. Left: conical ("tapered") seat plug, no gasket, 16 mm wrench flats. Right: flat seat with gasket ring, 21 mm wrench flats. Source: BERU

- Chemical effects caused by aggressive combustion deposits in the combustion chamber, due to unsuitable fuel, and

- Use of spark plugs with the wrong heat range

The consequences are melted electrodes, preignition, increased electrode gap, and increased ignition voltage demand.

Preignition
If preignition is initiated by deposits on piston crown, cylinder head, valves, or spark plugs, the spark plugs and pistons may be thermally overloaded to the point where holes are burned through the piston heads, usually in the direction of the spark plug axis. Preignition is promoted by spark plugs with the wrong heat range.

Detonation
Detonation causes unacceptable heating of the spark plug, which may in turn cause preignition.

6.5.5.3 Spark plug failures
Figure 6.489 shows an ideal spark plug appearance. The insulator tip is grayish-white or grayish-yellow to a medium reddish brown. Electrode erosion is minimal. The heat range of this spark plug was chosen correctly. The spark plug has not been thermally overloaded, mixture and ignition adjustments are faultless, and the engine is in

perfect running order. The following illustrations show more or less serious deviations from this ideal condition. The examples shown apply only to four-stroke engines. Before evaluating spark plug appearance, the vehicle should be driven for several kilometers at medium load, to prevent soot deposits from distorting the picture. Longer idling periods before spark plug removal, especially with a cold engine, also lead to sooting. These illustrations are taken from literature supplied by spark plug manufacturer Beru.

Sooty spark plug

Appearance
Insulator tip, electrodes, and spark plug shell covered by velvety soot deposits (Fig. 6.490).

Possible causes
- Incorrect mixture adjustment (carburetor or fuel injection system) with excessively rich mixture
- Clogged air filter
- Defective cold-start system (choke)
- Primarily driven on short trips, or
- Spark plug heat range too high

Consequences may include misfiring due to current leakage; poor cold-starting performance.

Oiled spark plug

Appearance
Insulator tip, electrodes, and spark plug shell covered by black or dark oil film (Fig. 6.491).

Fig. 6.489 Normal spark plug appearance. Source: BERU

Fig. 6.490 Sooty spark plug. Source: BERU

Fig. 6.491 Oiled spark plug. Source: BERU

Fig. 6.492 Glazed insulator tip. Source: BERU

Possible causes
- Too much oil in the combustion chamber, engine oil level too high
- Severely worn piston rings, cylinders, and valve guides

Consequences
Misfiring or even spark plug shorting, poor starting performance

Glazing/lead fouling
Appearance (Fig. 6.492)
The insulator tip shows a yellowish-brown glaze in spots; the color may tend toward green.

Possible causes
Ash-like deposits from fuel and engine oil additives

Consequences
At higher loads, these coatings become electrically conductive and cause misfiring.

Deposits
Appearance
- Heavy deposits on insulator tip and ground electrode from oil and fuel additives (Fig. 6.493)
- Slag-like deposits (carbon)

Possible causes
Additives, especially those in oil, may form combustion chamber and spark plug deposits.

Consequences
Preignition with loss of power, even engine damage

Melted center electrode

Appearance
The center electrode has melted (Fig. 6.494); the insulator tip has soft, spongy blisters.

Possible causes
- Thermal overloading as a result of preignition or excessive spark advance
- Combustion chamber deposits, defective valves, defective ignition distributor
- Inadequate fuel quality, possibly spark plug heat range too low

Consequences
Misfiring, loss of power

Melted ground electrode

Appearance
Electrodes have cauliflower-like appearance, possibly deposits of non-spark plug material (Fig. 6.495).

Possible causes
- Thermal overloading as a result of preignition, e.g., as a result of excessive spark advance
- Combustion chamber deposits, defective valves, defective ignition distributor
- Inadequate fuel quality

Fig. 6.493 Deposit-covered insulator tip and electrode. Source: BERU

Fig. 6.494 Melted center electrode. Source: BERU

Consequences

Initially, loss of engine power; ultimately, engine ceases to operate

Severe electrode wear

Possible causes (Fig. 6.496)

- Aggressive fuel and oil additives, unfavorable combustion chamber flow, possibly caused by deposits

- Engine knock

Consequences

Misfiring, especially when accelerating (ignition voltage no longer sufficient for large electrode gap), poor starting behavior

Broken insulator tip

Possible causes (Fig. 6.497)

- Mechanical damage as a result of improper handling; in early stages, often only visible as a hairline crack

- In borderline cases, deposits between center electrode and insulator tip, especially after overextended operating time, may burst the insulator.

Consequences

Misfiring; the spark occurs at locations that are not assured access to fuel/air mixture

Fig. 6.495 Melted electrodes. Source: BERU

Fig. 6.496 Severely worn electrodes. Source: BERU

Fig. 6.497 Spark plug with broken insulator tip. Source: BERU

6.6 Filters

6.6.1 Fundamentals of filtration

When subjected to forces, parts in relative motion to one another undergo wear in the form of progressive loss of surface material. The degree of such wear is a function of

- Condition of materials and surfaces of moving parts
- Force and velocity applied to mutually moving parts
- Type, quantity, and size of particulates in lubricant

To protect the engine, dust and other particulates must be removed from all operating substances (air, fuel, and lubricating oil). In the future, soot particles will also need to be removed from diesel engine exhaust.

Particulates may be removed from a fluid or gas by means of:

- Centrifugal separation (air: cyclone separator); fuel (heavy fuel oil: separators)
- Ejector effect (suction generated by exhaust gas stream used to reject dust particles)
- Binding particulates to fluids (water or oil bath filters)
- Electrical separation: charged particles in the space between two electrodes are accelerated by the electrical field, toward the trapping electrode
- Filtration: separating solid particles from a fluid or gas by means of permeable materials

According to DIN 71450 ("Filters for motor vehicles and internal combustion engines; concepts for filters and components,") filters serve to "contain a suspension or a gas flow containing solid particles, in which only the filter medium, or the filter medium and the filter cake settling upon it, serve to block the flow of these solid particles . . .". Filters are apparatus that remove impurities from liquid or gaseous media by means of a porous component.

The task of filtration is to remove, to as high a degree as possible, wear-causing particles from gases or fluids. Primarily, these particles consist of

- Dust
- Salt
- Soot and combustion products, and
- Metal wear particles

Engine media that are filtered include

- Combustion air
- Fuel
- Lubricant, and to some extent also
- Engine coolant and—to be expected in the future,
- Diesel engine exhaust

Filtration of engine operating materials is based on various effects (Fig. 6.498), primarily on the

- Strainer effect:
  Particles are retained by virtue of the size difference between particles and filter pores.

- Inertia or impact effect:
  When flow is diverted around filter fibers, particles are no longer able to follow, due to inertia; they strike and adhere to these fibers; the effect is strongly dependent on velocity.

- Blocking effect:
  Particles moving along a filter grating at a distance smaller than its (idealized) radius are attracted to the filter; and

- Diffusion separation:
  Very small particles (< 0.5 μm) undergo random motion. If such particles approach within a certain distance of a strainer mesh, the blocking effect comes into play.

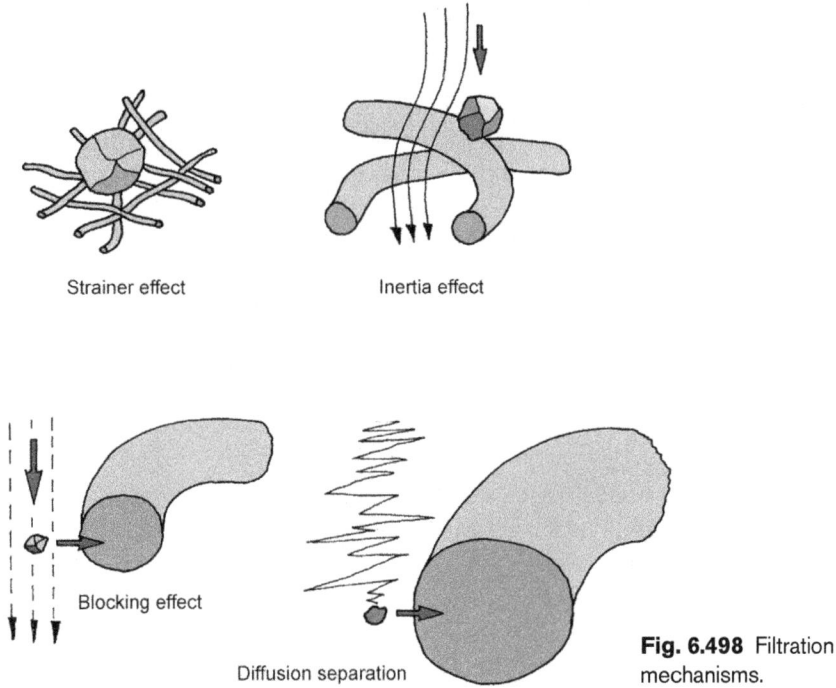

Strainer effect   Inertia effect

Blocking effect

Diffusion separation

Fig. 6.498 Filtration mechanisms.

These filtration mechanisms are superimposed. Operation of individual mechanisms depends on flow velocity, particle size, and location within the filter (at surface or depth). Filtration (probability of a particle touching the filter mesh) depends on:

- *Filter*
  Filter pore size
  Pore count (filter density)
  Filter mesh structure within individual pores

- *Particles to be removed by filtration*
  Particle size
  Particle shape
  Particle density

- *Media to be filtered*
  Viscosity
  Density
  Flow velocity

It is apparent that filtration is determined by the filter itself, the properties of the medium to be filtered (oil, fuel, air), as well as the nature of the contamination.

In principle, filters represent flow restrictions; the more effective they are, and the more separated particles they carry, the greater the restriction. The effectiveness of filters is described by their filter efficiency, which indicates the percentage of particles in the filtered medium removed by filtration. Because of the high degree of filter efficiency (about 99%), a more useful number is their transmittance. This indicates how many particles are allowed through a filter; with a filtration efficiency of 99%, transmittance is 1%.

A key parameter for determining filter capacity is volume flow rate divided by filter area. Dimensionally, this is expressed as a velocity. If one plots filter efficiency and flow resistance as functions of filter capacity, it is apparent that these functions are mutually opposed (Fig. 6.499).

Like any other component, filters can be only loaded to a certain point; their load capacity may be explained by various criteria. To the medium, the filter represents flow resistance, with the result that the fluid experiences a pressure drop in passing through a filter; the magnitude of this pressure drop is a function of the filter itself, the flow velocity (volume flow rate divided by filter area), and the viscosity and density of the fluid. Filter life is determined by the allowable flow resistance. Operational experience as well as laboratory tests have shown that the flow resistance—the pressure differential across the filter—increases hardly at all over a certain time period, but thereafter increases progressively (Fig. 6.500). This point in time is determined

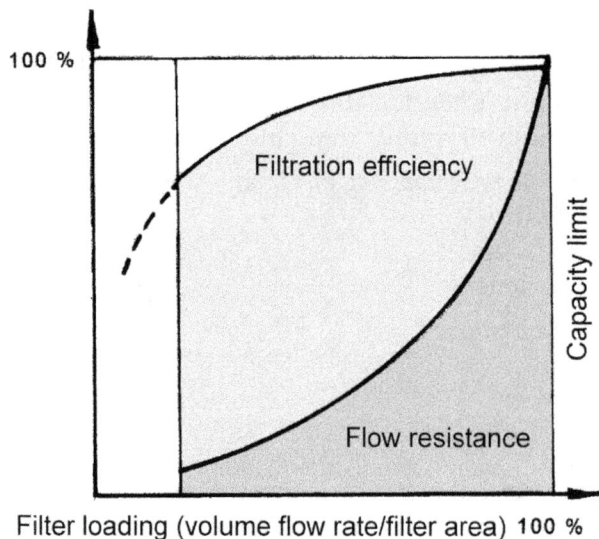

Fig. 6.499 Operational behavior of filters; filtration efficiency and pressure drop (schematic).

Filter loading (volume flow rate/filter area) 100 %

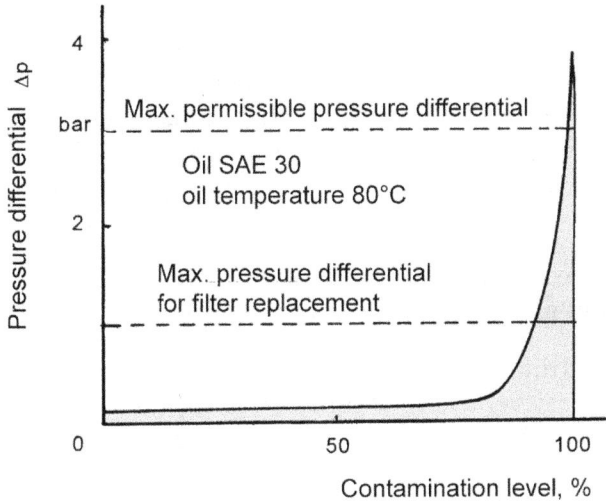

Fig. 6.500 Pressure rise of a pleated-paper filter as a function of contamination. "100%" = 100 g dirt/m² filter area.

by the degree of contamination in the filtered fluid or gas. Filter replacement or cleaning must therefore be carried out prior to this time.

Filter service life is determined by the allowable pressure drop. Filter requirements include:

- High filtration efficiency
- High capacity for separated particles (dirt capacity)
- Low flow resistance
- Long service life
- Mechanical durability
- Fool-proof servicing or replacement
- Environmentally responsible disposal

Demands on engine filters have increased, because air volume flow rates have increased as a result of supercharging; relative to a given engine displacement, more air needs to be filtered. On the other hand, engine oil fill quantities (oilpan capacities) have been reduced, the turnover rate of oil in an operating engine has increased, as have oil temperatures; soot contamination has increased, and filters must be able to withstand aggressive synthetic motor oils. In keeping with extended oil change intervals, filter replacement intervals have also been increased. Overall, installation conditions for automotive filters have suffered.

6.6.2 Air filters

Combustion air contains a multitude of various particles, summarily designated as "dust." This is understood to mean finely divided, airborne solid particles of various shapes, structures, and densities. The size of dust particles, by definition, falls between 0.001 and 0.5 mm (Table 6.12). In addition to dust, other airborne substances pose a threat to combustion engines.

Dust concentration is expressed as (milligrams dust)/m^3 of inducted air. This is strongly dependent on local conditions (Table 6.13).

Off-road vehicles—agricultural equipment, construction equipment, wheeled and tracked military vehicles—are subjected to greater airborne dust concentration, thicker layers of dust on the ground, and the resulting dust clouds churned up by their own wheels or tracks ("self-induced dust," Fig. 6.501).

Table 6.12
Sizes of Airborne Particulates

Type of contamination	Particle size range in μm
Dust	0.1 – 500
Exhaust soot	0.2 – 70
Smog	0.01 – 2
Clouds, haze	2 – 80
Drizzle, fog	80 – 650
Rain	650 – 4000

Fig. 6.501 Agricultural vehicle self-induced dust.

To emphasize the amount of dust that would be ingested by an engine (if it were not protected by a filter), consider the following numbers. A bus, powered by a 180 kW (240 hp) engine, at a speed of 80 mph, draws in about 500 m³/h of air. With a dust content of 0.005 g/m³, this represents 5 g of dust per 100 km (62 miles). After 10,000 km (6200 miles), this engine would have drawn in 500 g (more than a pound) of dust. The consequences are not difficult to imagine. Dust causes engine wear, and not only those components in direct contact with induction air (compressor/turbocharger, cylinders, pistons, and piston rings); dust would find its way into the lubricating oil and so reach every corner of the engine. Wear caused by dust depends on dust concentration in air, the type of dust, the size of dust particles, and the duration of dust exposure; in other words, on the mass of inducted dust.

Air filter flow resistance causes a pressure drop, which determines the filter service life. Individually, pressure drop is a function of:

- Filter load
- Dust concentration
- Duration of dust exposure
- Fineness of the filter
- Location of the air inlet
- Length of the induction tract
- Filter condition

Table 6.13
Dust Exposure in Practice (Source: Mann + Hummel)

Vehicle type & operating conditions	Dust concentration, mg/m³
Operation on normal European roads	0.6
Operation on non-European roads	3
Commercial trucks operating off-road (construction sites)	8
Buses with rear air induction operating on normal European roads	5
Buses with rear air induction operating on non-European roads	30
Earthmoving equipment (wheel loaders, tracked vehicles)	35
Sweepers	8
Agricultural tractors in European service (without significant drought periods)	5
Agricultural tractors (overseas)	15
Combines in individual service	15
Combines operating in teams	30
Military tracked vehicles	100

The consequences of dirty filters are loss of power and increased fuel consumption (Fig. 6.502).

This loss of power is often readily apparent in overloaded vehicles, especially in service outside Europe. It is not unusual for the driver to "remedy" the situation by poking holes through the paper filter element with a screwdriver, to allow more air into the engine.

Dust consists of particles of various sizes ("fractions"); in engine operation, these may cover a hundred-fold range of sizes (largest particles compared to smallest). Because filters, by their operational principle and configuration, are designed for specific grain sizes, various filter designs are employed: coarse filters for preliminary separation, and fine filters for the actual fine particle filtration. Coarse and fine filters may be incorporated in combination filters. The most important engine air filter configurations are:

6.6.2.1 Dry air filters

Cyclone filters (Fig. 6.503)
Inducted air is caused to rotate by guide vanes; dust particles are removed by centrifugal force, and directed to a collection container. Cyclone filters offer up to 85% filtration efficiency for ISO test dust (only larger dust fractions are removed), service life is good, and the cyclone output is cleanable. Cyclone filters are usually employed as prefilters.

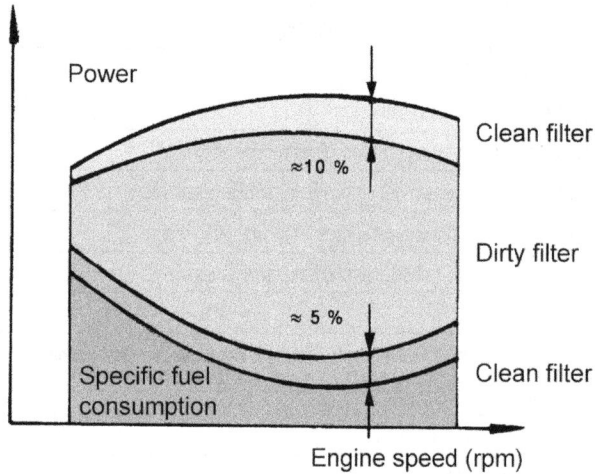

Fig. 6.502 Effect of filter contamination on engine behavior.

Fig. 6.503 Cyclone air filter. Source: MANN+HUMMEL

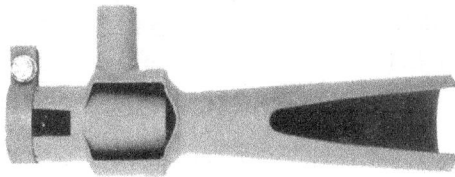

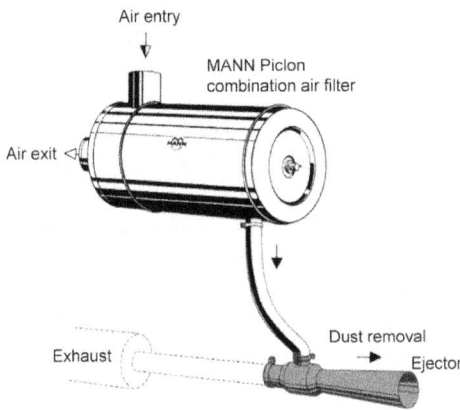

Fig. 6.504 Dust removal by means of ejector. Source: MANN+HUMMEL

Ejectors

In severe dust situations, e.g., tracked military vehicles or harvester applications, pre-filtered dust must be continuously removed. This is achieved by ejectors, which take advantage of the exhaust gas to achieve an ejector effect (Fig. 6.504).

Paper filters

The filter consists of a folded (pleated), often ring-shaped paper element. The large filter surface area (as a rule of thumb, a 400 kW commercial vehicle engine needs an area of about 14 m^2) is achieved by the fold geometry and embossing. Paper filters have very good filtration efficiencies, in excess of 0.995. Their service life, however, is limited, as is their cleanability (Fig. 6.505).

Combination air filters

Combination air filters are used for commercial vehicle engines. These consist of star-pleated paper elements in combination with guide vanes integrated in the filter housing. The guide vanes (Fig. 6.506) induce rotating air flow around the filter element. Centrifugal force causes particle separation.

6.6.2.2 Single-stage air filters

Single-stage air filters usually contain round star-pleated filter elements (Fig. 6.507). In exceptional cases, square filters may be used. Filtration efficiency is up to 0.999. These filters are used because of longer service intervals for truck engines, which may operate for up to 240,000 km (150,000 miles) per year. Fail-safe filter elements are often integrated within the air filters. These prevent contamination of the clean air side while the main filter is being changed. Fail-safe elements usually consist of cylindrical nonwoven elements, and, depending on dust concentration, are only replaced with every third main filter change. In the case of commercial vehicles, the filter element should be replaced when the pressure drop has increased by 40 mbar.

Filters, as such, operate reliably and without problems. Difficulties in air filtration are primarily related to filter peripherals and in filter cleaning or replacement. In many engine applications, whether in motor vehicles, construction equipment, or to power other machinery, the elementary requirement to avoid unnecessary filter loading is disregarded. Vehicles inherently have a zone of self-contamination; every effort should be made, in the design process, to avoid drawing engine air from this zone (Fig. 6.508).

Fig. 6.505 Star-pleated paper filter element (rear) in combination with a fail-safe element (front) to avoid particle contamination while demounting the main element. Under operational condition the fail-safe element is placed inside the main element. Source: MANN+HUMMEL

Fig. 6.506 Combination air filter. Source: MANN+HUMMEL

Fig. 6.507 Single-stage air filter. Source: MANN+HUMMEL

Measurements have shown that airborne dust concentration falls as the height above ground of the induction point is raised. If possible, air should be inducted

- as high, and
- as far forward, as possible.

Furthermore, the induction point must be protected against rain, fog, water splash, and exhaust soot.

The second weak point in engine air supply is the air induction tracts. These tracts have slight vacuum; in the event of any leaks, which occur preferentially at any connection points, dirt and dust will be drawn in. Induction tracts should therefore

- be kept as short as possible, and
- have as few connections as possible.

Because filters are often rigidly mounted in the vehicle or equipment, filter and engine are often connected by flexible tracts such as bellows, which may be excited to vibration by the engine, or by pulsation in the induction air [6-38] (Fig. 6.509).

In particular, if such ducts are installed with bends and curves, the bellows may touch and wear through (Fig. 6.510). The engine then draws in unfiltered air.

Tractor

Zone of ordinary
dust concentration

Zone of five-fold
dust concentration

Zone of ten-fold
dust concentration

Commercial vehicle

B

A

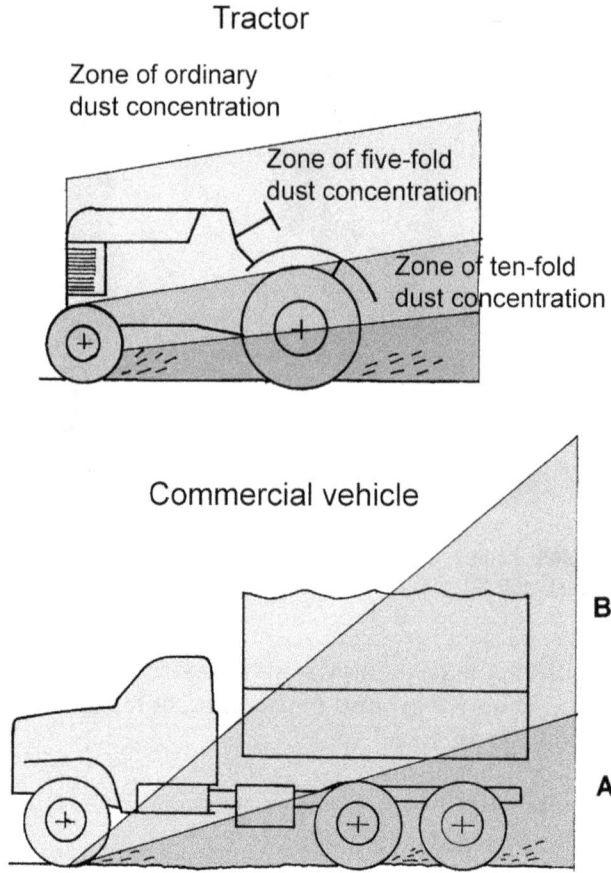

Fig. 6.508 Dust
generated by off-road
vehicles depends in
part on vehicle type and
speed: Commercial
vehicle (truck): A, zone
of self-contamination at
low speeds; B, at higher
speeds.

Filters are only effective if they are mounted upstream of the functional
groups and components they are intended to protect. Some of the
unforeseen events that may occur were described many years ago by
Ernst Mahle in one particular case [6-39]:

"If dust is even capable of penetrating wristwatch cases, and the most
finely divided dust imaginable is floating around in air, it will find its
way into an engine, even by highly unusual means. For example, the
cylinder head was removed from the engine of a car operating in desert
conditions. A small pile of sand was seen atop one piston crown. The
other five pistons were clean. How did sand manage to find its way
to this one piston? Well, on this cylinder, the exhaust valve had hung
open, and sand had blown in—by way of the tailpipe, muffler, exhaust
pipe, and exhaust manifold, following all the twists, turns, and rises of
the exhaust path."

Dull, scored piston skirt areas are clear indicators of dust contamination. Abnormal dust content in combustion air manifests itself as cylinder walls exhibiting partially erased honing marks, a perceptible cylinder ridge at the top ring reversal point, and barrel-shaped wear in the area swept by the rings. The greatest wear may be seen in the upper area of the cylinder wall. Even different wear rates for individual cylinders indicate dust in combustion air, because, depending on cylinder location and intake manifold details, individual cylinders may be more susceptible to dust contamination than their neighbors, as larger dust particles, due to their inertia, are not able to "make the turn" as air is diverted to the upstream cylinders.

Another weak point in filter systems is filter maintenance or replacement. In the case of filter replacement, there is always a danger that dirt will fall into the "clean" side as the filter element or cartridge is removed. From a design standpoint, this situation may be avoided

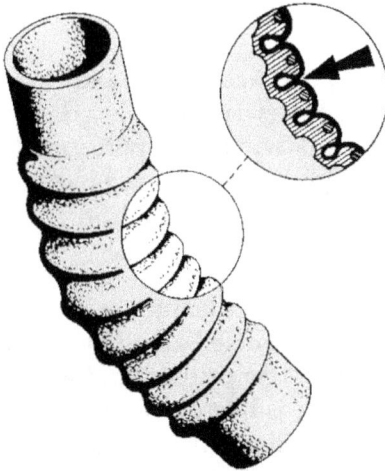

Fig. 6.509 Improper routing of bellows duct. Source: DEUTZ

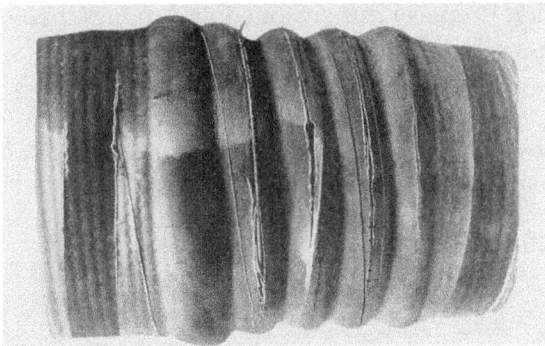

Fig. 6.510 Worn bellows duct. Source: DEUTZ

Fig. 6.511 Location of air filters on an American-made commercial vehicle, operating in Australia.

by placing a protective filter, such as a felt cartridge, downstream of the actual filter element.

Filters must be easily accessible, easily checkable, and readily cleanable (or replaceable) at or near the engine; otherwise, there is a danger that scheduled checks and maintenance work will not be performed as required. Exemplary design in this regard may be found in American commercial vehicles, with air filter(s) mounted on the front fender(s) (Fig. 6.511).

Overloading of oil bath air filters, or filter overheating, will cause the oil viscosity to decrease. The oil may then be carried along by the air flow. This oil then collects on the turbocharger compressor, and within the intake tract. One particular disadvantage is that dirt entrained in this oil (after all, the oil bath is supposed to collect dirt) will cause increased wear. This will primarily affect the cylinder immediately downstream of the air filter, because the dirt-carrying oil stream will take the first available exit to a cylinder. Properly dimensioned and maintained oil bath air filters do not pass oil to combustion air.

6.6.3 Oil filters

From a physical point of view, contaminated fluid (lubricating oil, fuel) may be regarded as a suspension. Engine oil may be contaminated by

- Dirt from the combustion chamber
- Soot
- Oil cracking products

- Mechanical wear particles, and
- Dust as a result of inadequate air filtration

Contamination as a result of so-called initial contamination, i.e., dirt resulting from engine manufacturing (foundry sand, metal chips, fibers from cleaning rags, etc.) are now only rarely encountered in factory-new engines, but are more common after repairs, in particular if these had to be carried out under extreme conditions. Deciding factors are the size of dirt particles in relation to the minimum lubricating film thickness in bearings, and the concentration of dirt particles. For these reasons, engine manufacturers specify limits for oil contamination. Oil is tested and evaluated for

- Solid particle content
- Soot content (diesel engines)
- Fuel content
- Water content

Oil contamination is reported as concentration of fuel and benzene-insoluble components, according to DIN 51588. Contamination is affected by

- Engine configuration: working process, combustion principle, and design features.
- Type of service: power output, loading, operating conditions, driving profile, etc.
- Environmental conditions: ambient air, filtration quality, oil quality.
- Oil consumption and oil change intervals: for high oil consumption rates, dirt content is stabilized by oil replenishment. Oil change then represents a partial cleaning of the oil filter.

The typical ranges of dirt contamination in motor vehicle engine oil are as follows:

- Otto-cycle engines 0.07–0.4 mg/km kW
- Diesel engines 0.25–0.8 mg/km kW

Oil filtration is made more difficult by power increases and the general trend of engine development, because

- Small lubricating film thicknesses demand finer filtration;
- HD oils are more difficult to filter, as their additives bind dirt;

- The volume flows of oil requiring filtration rise as engine performance is increased (e.g., as a result of piston oil cooling), resulting in increased oil velocity relative to filter loading.

The effectiveness of filters is evaluated using the following criteria:

- Contamination based on appearance (visual inspection)
- Appearance of wear components
- Measured wear

The primary function of oil filters is to protect engine sliding components against foreign bodies entrained in lubricating oil. To accomplish this, the filter must be installed immediately upstream of the "consumer" or upstream of a branch to the consumer. A full-flow oil filter is one in which the entire oil flow passes through the filter (Fig. 6.512).

Full-flow filtration is limited with respect to its degree of filtration (filter fineness); filtering smaller particles would impose excessive filter loading, with a risk of damaging the filter medium. Full-flow filters are only intended to retain particles larger than 30 μm, and, with regard to wear protection, at least reduce the number of particles in the 3–30 μm range.

The second function of filters in the lubricating oil circuit is cleaning, i.e., separation of foreign bodies. For this, a portion, perhaps 5 to 10%,

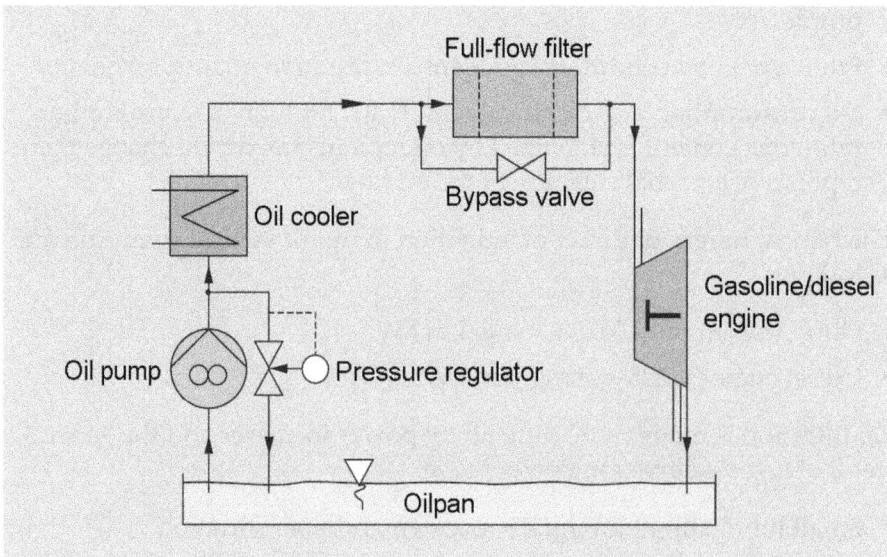

Fig. 6.512 Location of filter in lubricating oil circuit: full-flow circuit. Source: MANN+HUMMEL

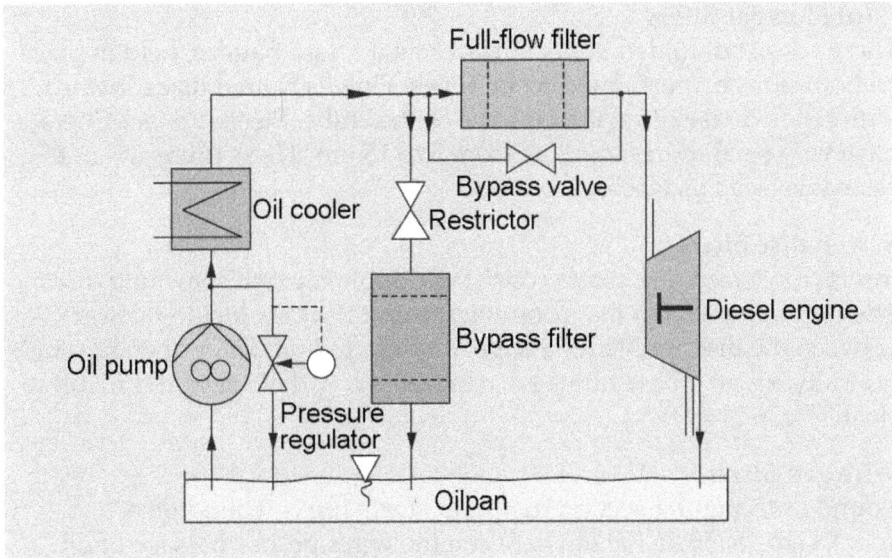

Fig. 6.513 Location of filter in full-flow/bypass flow circuit. Source: MANN+HUMMEL

of the oil flow is diverted and forced through a separate, special filter, and then allowed to return, unpressurized, to the oil pan. Such bypass or partial-flow filters (Fig. 6.513) consist of extremely fine depth filters or free-jet centrifuges (centrifugal oil filters).

The significance of bypass filtration is often overestimated, as the position is advanced that bypass filtration reduces the need to change oil. However, comparative tests have shown that oil change intervals with bypass filtration can only be extended at the cost of significant oil deterioration. Even a bypass filter cannot

- Replace depleted additives, or
- Filter fuel condensate, water, and acids from lubricating oil

Nevertheless, bypass filtration is gaining in importance, because mixture formation and combustion are increasingly being influenced by stricter emissions standards. This leads to added soot contamination of the lubricating oil. To maintain oil change intervals (and, indeed, to extend these), tiny soot particles (size less than 1 μm) must be removed from the oil. This is achievable with compact centrifugal ("free-jet centrifuge") bypass filtration.

There are various oil filter types intended for specific engine sizes and types.

Pleated paper filters

These consist of folded paper surfaces, in a "star" pattern, held in place on both sides by perforated metal tubes. Flow is from outside inward, with cleaned oil exiting through the central tube. Pleated paper filters have very small pores, ranging from 5 to 15 μm. These filters are not cleanable (see Fig. 6.505).

Screen disc filters

Inserts for screen disc filters consist of double-tapered screening discs. They act in parallel to the incoming oil, and flow is outside-in. Seals between the discs prevent passage of unfiltered oil. Filter openings are typically 30 μm. These filters are comparatively delicate and difficult to clean (Fig. 6.514).

Wire gap filters

Round or triangular section rustproof steel wire is wound on a spool. Gaps of 50 or 100 μm between the wires permit passage of oil and perform the filtration function. Oil flows outside-in. By means of a ratchet-actuated scraper, the filter may be cleaned, even while the engine is operating. Wire gap filters are simple and robust, but

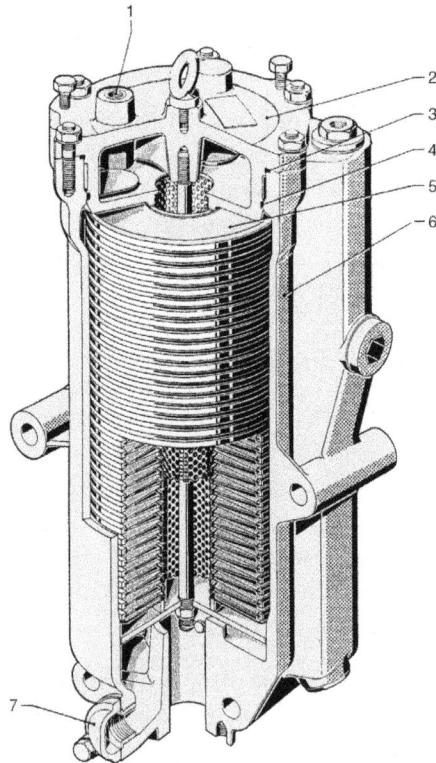

Fig. 6.514 Screen disc filter: 1, Connection for differential pressure shutoff; 2, Filter housing cover; 3,4, Sealing rings; 5, Screen disc cartridge; 6, Filter housing; 7, Sludge drain port. Source: MTU

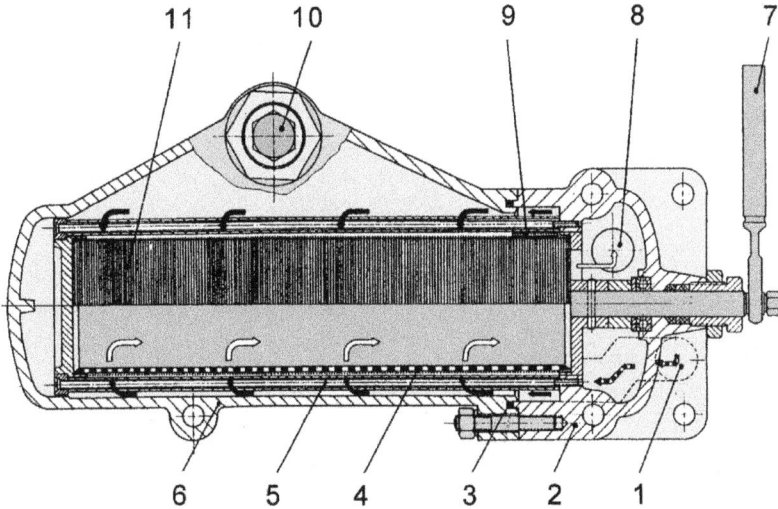

Fig. 6.515 Wire gap filter: 1, Oil inlet; 2, Filter head; 3, Sealing ring; 4, Tube for filter spool; 5, Locking screw; 6, Filter housing; 7, Ratchet handle; 8, Oil outlet; 9, Scraper; 10, Drain plug; 11, Gap filter spool. Source: MTU

relatively coarse (Fig. 6.515). Today, they serve mostly as protective filters and are installed upstream or downstream of the actual filter.

Oil centrifuges

Oil centrifuges (free-jet centrifuges, Fig. 6.516) employ centrifugal force to sling dirt particles contained in oil against the inner wall of the rotor, where they remain. Oil centrifuges are installed as bypass filters. Oil flows through a hollow central shaft into the rotor, and through screens in the upper part of the filter to the drive nozzles. Oil exits the nozzles tangentially; the reaction spins the rotor. After leaving the nozzles, oil flows back to the oilpan unpressurized. Mechanical drive from the engine is not required. A cutoff valve may be used to close the oil supply to the centrifuge, allowing cleaning while the engine is in operation.

Combined full flow and bypass filters

Because automotive filters should be as compact as possible, or even integrated within the engine itself, with a minimum number of interfaces and short flow paths, pleated paper element full-flow filter and centrifugal bypass filter have been combined into a single unit (Fig. 6.517).

The effectiveness of centrifugal bypass filtration may be illustrated by the following example. Old oil from a diesel engine that had been filtered through a conventional full-flow filter was then run through a centrifugal filter on a filter manufacturer's test bench. After 550 hours, this centrifugally removed 500 g (more than 1 lb) of contaminants [6-40] (Fig. 6.518).

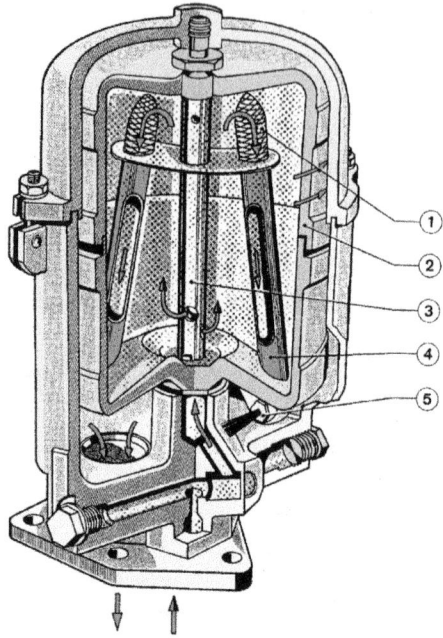

Fig. 6.516 Oil centrifuge (free-jet centrifuge). These are mounted in a bypass branch of the lubricating oil circuit. Oil flows into the centrifuge nozzles at system pressure; as it exits, it imparts rotation to the centrifuge. Centrifugal force slings dirt particles against the inner wall of the rotor, where they adhere: 1, Strainer, 2; Rotor, 3; Hollow shaft, 4; Standpipe; 5; Propulsion nozzle. Source: MTU

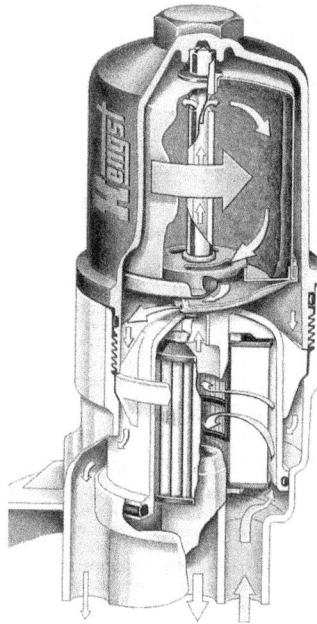

Fig. 6.517 Full-flow oil filter with centrifuge in bypass circuit. Source: Hengst

At low oil temperatures, flow resistance of the filter rises to a level where the filter material might be damaged. To prevent this, bypass valves open at an appropriate pressure differential. Although this lets unfiltered oil into the engine during its warm-up phase, the filter remains undamaged and functional once oil temperature rises to normal levels (Fig. 6.519).

Automotive oil filters have anti-drainback valves that prevent filter draining once the engine has been shut down. These ensure rapid pressure buildup when the engine is restarted (Fig. 6.520).

To absolutely prevent operation with unfiltered oil, medium and large engines add a protective filter downstream of the main filter, for example a wire gap filter. Such filters are installed on general-purpose engines, first because such engines represent a larger investment than road vehicle engines, and second because longer service life is expected

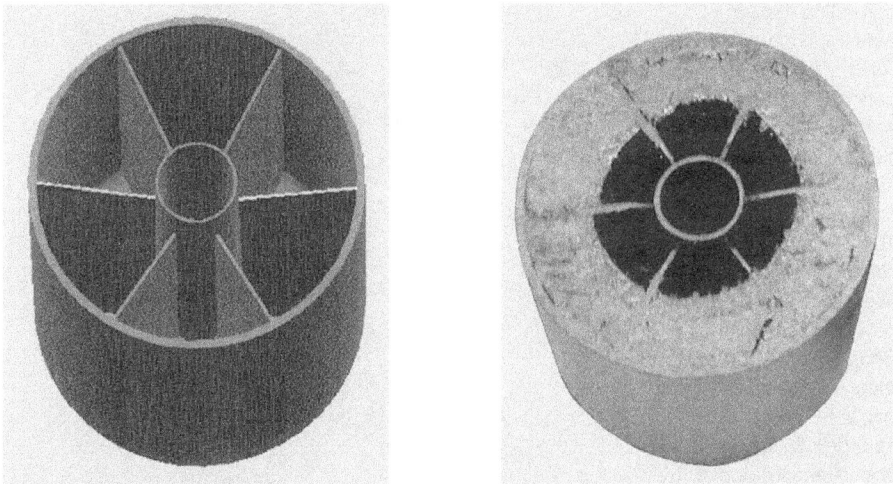

Fig. 6.518 The effectiveness of bypass filtration by means of centrifugal filters is clearly shown by this centrifugally removed contaminant mass (about 500 g). Source: MANN+HUMMEL

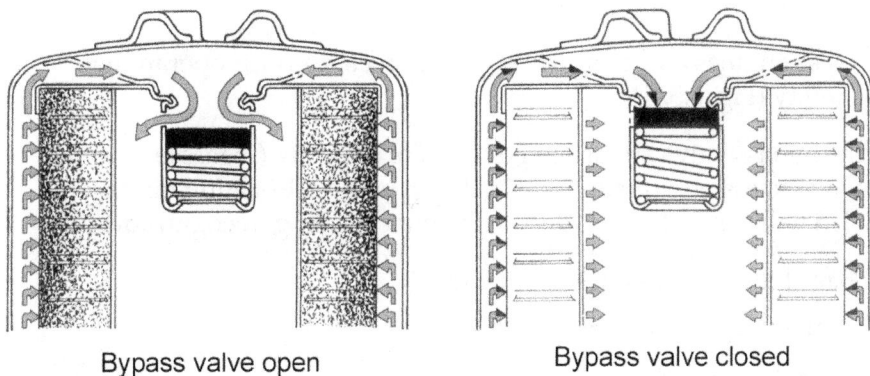

Bypass valve open   Bypass valve closed

Fig. 6.519 Oil filter bypass valve operation. Source: MANN+HUMMEL

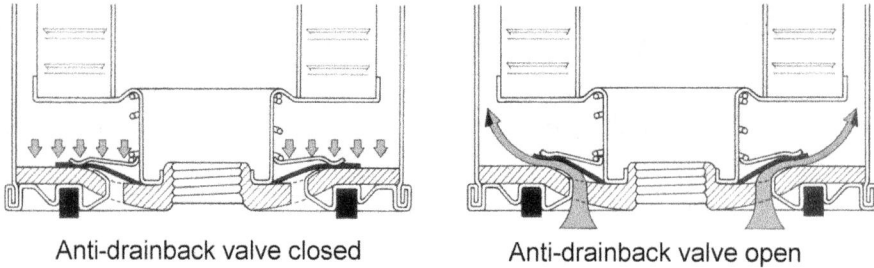

Anti-drainback valve closed   Anti-drainback valve open

Fig. 6.520 Oil filter anti-drainback valve operation. Source: MANN+HUMMEL

Fig. 6.521
Combination oil filter consisting of pleated paper and gap filters: 1, Ratchet handle; 2, Filter cartridge; 3, Carrier; 4, Separating tube; 5, Scraper; 6, Spring; 7, Flange; 8, Round sealing ring; 9, Clearance pin; 10, Filter spool; 11, Oil filter housing; 12, Gap filter cartridge; 13, Filter cover; 14, Disc; 15, Snap ring; 16, Sealing plug; 17, Rotation pin; a, Oil inlet to filter cartridge; b, Oil outlet from filter cartridge; c, Oil passage through carrier; d, Oil inlet to gap filter; e, cleaned oil outlet. Source: MTU

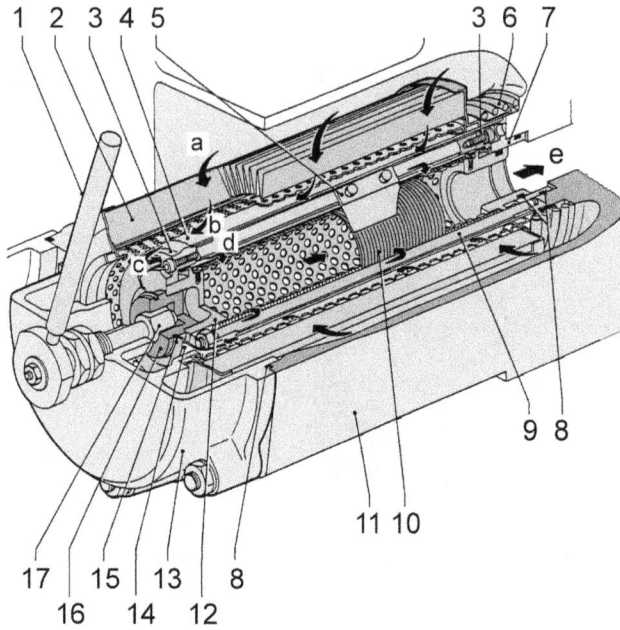

of these engines and there are higher expectations for operational reliability (Fig. 6.521).

If the bypass valve should fail to operate correctly (regardless of the cause), excessive pressure will build up in the lubricating oil system, which will cause deformation of the filter housing, recognizable by:

- Barrel deformation of the housing
- Bursting of the crimped seam
- Gasket extrusion

The two last mentioned items may result in oil loss, usually in association with costly engine damage. Overall, however, lubricating oil filters seldom exhibit any failures; the weakest links are primarily in monitoring and maintenance. In the words of one insurance adjuster's report,

". . . the symptoms indicate that the oil change was performed improperly, due to the cocked installation of the oil filter housing. The vehicle-mounted power unit [author's note: the engine] suffered major damage. This manifested itself as a damaged crankshaft, connecting rods for cylinders 1, 2, and 3, the camshaft, pistons, oil pump, and cylinder walls. Based on the observed damage, repair of the damaged engine is not justifiable on economic grounds."

Severe oil contaminant loading may be the result of negligent maintenance, e.g., pushing dirt into the engine with the dipstick, dirt entry while adding oil, or lack of clean operating practices during engine repair operations. These are made worse if the filter is not easily accessible. Here, again, American practice is exemplary, with generously dimensioned, easily accessible oil filters on commercial vehicle engines (Fig. 6.522).

Manufacturing defects are exceedingly rare, but they do happen.

Example: The gap filter ratchet handle on an emergency power generator could not be turned. Upon removal of the filter, a cracked filter wire was discovered at one of the two wire attachment points to the filter (Fig. 6.523). Furthermore, the wire was wound at an angle over the thread

Fig. 6.522 Readily accessible engine oil filter of a Kenworth tractor.

grooves, and incorrectly wound filter wire showed signs of contact with the wire areas raised by the winding error (Fig. 6.524).

Fractographic analysis of the wire fracture surface indicated a fatigue failure on the tension side of the filter wire wound over the thread groove (Fig. 6.525). The cause of this filter failure was defective winding of the filter wire.

Abnormal amounts of dirt entering the engine oil lead to worn bearing inserts, crankpins, wrist pins, valves, valve guides, etc., in other words, in those areas coated with oil. Cylinder wear occurs primarily in the vicinity of the oil control ring's reversal point; in extreme cases, it may result in the narrow lands of the oil control rings wearing away.

Gas blowby leaking into the crankcase may represent a considerable flow; in automobile engines, this might be 10 to 30 liters/min per cylinder [6-41]. These crankcase ventilation gases contain oil aerosols (0.1 to 2 µm), volatile oil components, and splash oil from the crank train, in other words, droplets of varying sizes. This cannot simply be allowed to escape to the environment; rather, it is directed back into the engine, where it may pool in the induction tract and on the turbocharger compressor. Oil passing through the compressor is caught in the intercooler, turns to varnish, and restricts the intercooler air passages. To prevent all of these undesirable effects, special oil separators are installed, either as fiber-wound or electric separators.

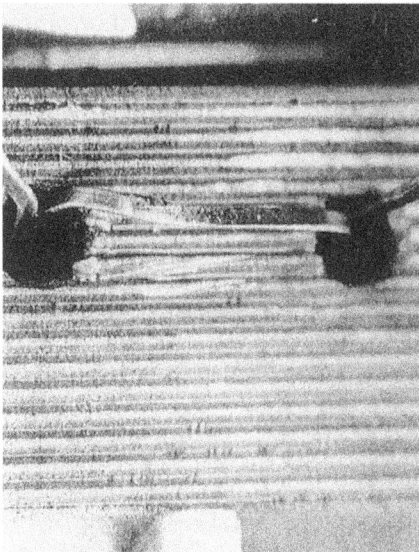

Fig. 6.523 Broken filter wire in the filter body attachment area.

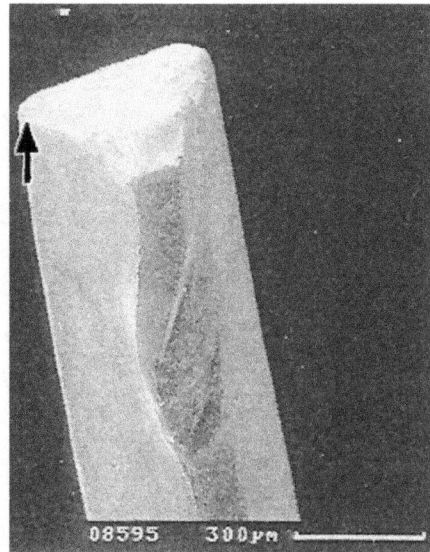

Fig. 6.524 Filter wire severely nicked as a result of crooked winding.

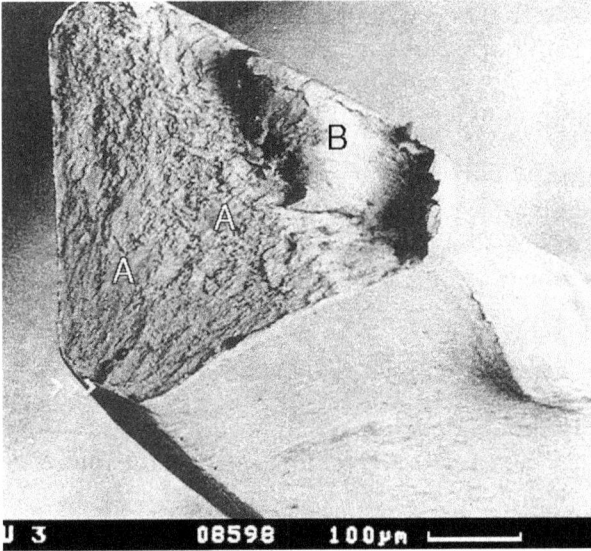

Fig. 6.525 Filter wire with fatigue fracture surface (A) and residual overloading fracture (B).

6.6.4 Fuel filters

Fuel filtration differs from oil filtration insofar as fuel must undergo especially fine filtration, because

- Fuel injection components operate with extremely tight clearances and are exceedingly sensitive to foreign particles. This demands fine filtration.

- Viscosity of Otto-cycle and diesel engine fuel (for commercial vehicles) is lower than that of motor oils, so that lower pressures are encountered in fuel systems. (We will not examine heavy oil operation.)

- Because of the sensitivity of injection system components, fuel filters do not include any bypass valves.

- Fuel filters must include some means of removing water, for one thing because it must be assumed that water will collect in the fuel tank through condensation, for another because fuel injection system elements are sensitive to water in the fuel.

Due to varying fuel storage conditions (at the service station, in the fuel tank, day tank, or vehicle tank), fuels may contain all manner of impurities. One may find particles of rust or paint, sand, and dust—and of course water as a result of contamination or condensation.

In the past, carburetors of Otto-cycle engines were regarded as relatively insensitive to fuel filtration; filter pores on the order of

0.100 mm sufficed. Thanks to fuel injection, this is no longer the case. Screen filters, edge (gap) filters, and, above all, paper filters are employed. Automotive engines generally use especially fine pleated paper filters with 0.001 mm pores (Fig. 6.526).

Diesel fuel injection systems, with component clearances on the order of 0.001 mm, have always demanded extremely fine filtration. In the past, this was accomplished with felt filter elements. Today, paper elements and a meltblown layer for fine filtration and water separation are used.

Modern diesel engine fuel filters consist of a (cleanable) prefilter with a woven strainer inlay, and a subsequent paper element fine filter (Fig. 6.527).

The fine filter automatically drains during maintenance operations. A collector container for coarse dirt and water is integrated in the filter

Fig. 6.526 Fuel filters for passenger car engines. Top: gasoline filter elements; bottom: diesel filter elements. Source: MANN+HUMMEL

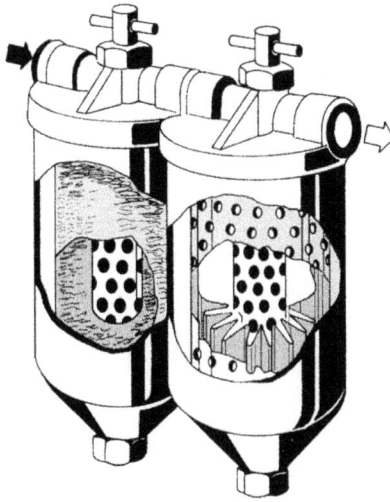

Fig. 6.527 Fuel prefilter and fine filter (MANN+HUMMEL configuration). Source: MANN+HUMMEL

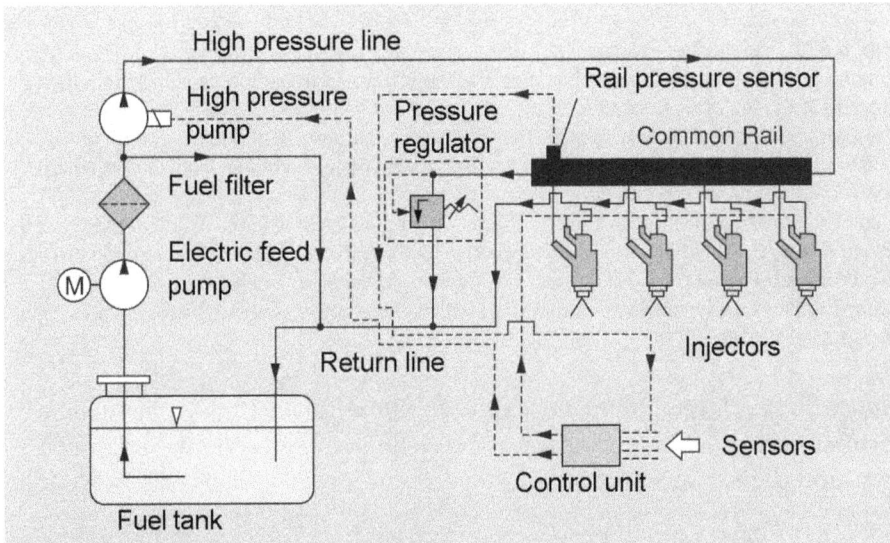

Fig. 6.528 Fuel filter system for engines with timed fuel injection systems. Source: MANN+HUMMEL

housing; the water separator is equipped with a drain. Because there is a slight negative pressure in the line between the fuel tank and fuel supply pump, the prefilter is allowed to create only minimal pressure drop, to avoid cavitation in the supply pump. The fuel prefilter is mounted between the fuel tank (vehicle tank, day tank) and the supply pump; the fine filter between the supply pump and the injection pump (Fig. 6.528).

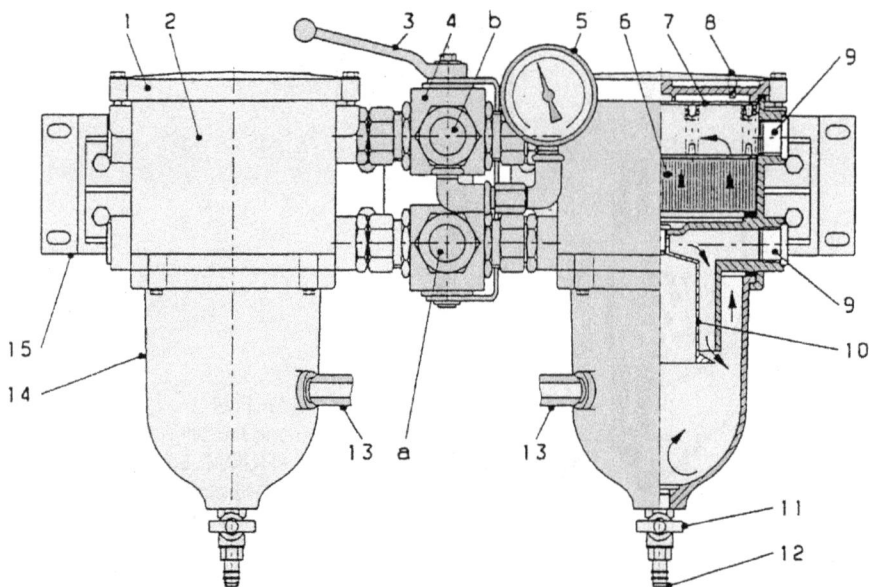

Fig. 6.529 Fuel prefilter (double filter). The fuel prefilter consists of two individual filters (the second filter element serves as a backup). Fuel flows through a centrifuge, which imparts rotation to the fuel. This separates a large portion of the water and dirt , which collects in the filter bowl. Reversing flow through the filter bowl separates an additional portion of the entrained dirt and water. Fuel next flows through a paper filter cartridge, from bottom to top, to filter out any remaining fine dirt particles and water: 1, Filter cover; 2, Filter housing; 3, Selector lever; 4, Ball valve; 5, Indicator gauge; 6, Paper filter cartridge; 7, Spring cassette; 8, Bleed screw; 9, Threaded plug; 10, Centrifuge; 11, Drain/fill valve; 12, Water and dirt drain; 13, Water level sensor; 14, Filter bowl; 15, Mounting bracket. a, Connector for fuel inlet; b, Connector for fuel outlet. Selector handle to left, left filter active; selector handle to right, right filter active. Source: MTU

Medium and large engines use double filters, as prefilters as well as primary filters, so that filters may be cleaned even while the engine is running (Figs. 6.529 and 6.530).

Clogged filters make their presence known by a loss of engine power, as the engine no longer gets enough fuel. The consequences to the injection system of inadequate filtration are shown in Fig. 6.531.

6.7 Heat exchangers and heat transfer devices

Because not all of the heat generated within an engine can be converted into mechanical energy, a portion of that heat, indeed the larger portion, must be removed from the engine. This is accomplished by

• Gas exchange, through exhaust of hot combustion gases

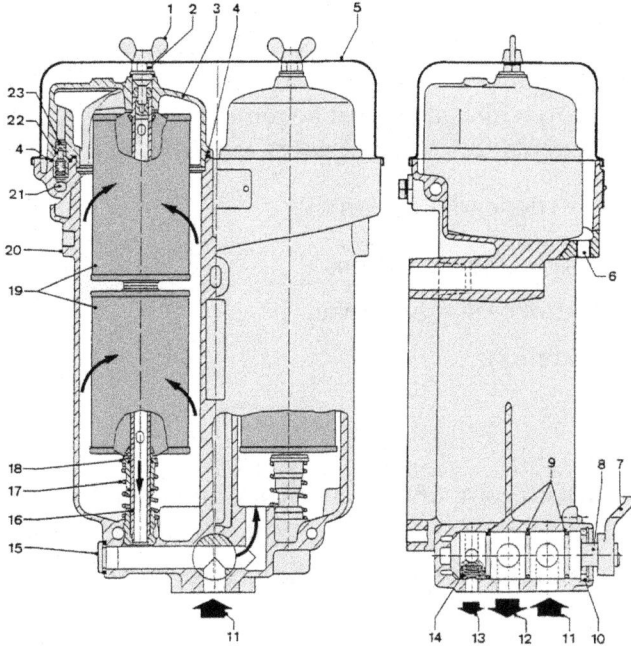

Fig. 6.530 Primary fuel filter (double filter): 1, Wing nut; 2, Central bolt; 3, Cover; 4, Round seal ring; 5, Hood; 6, Connector for drain line; 7, Valve handle; 8, Three-way cock; 9, Seal rings; 10, Disc; 11, Fuel inlet; 12, Fuel outlet; 13, Filter drain; 14, Formed seal; 15, Threaded plug; 16, Drain tube; 17, Compression spring; 18, Spring retainer; 19, Filter cartridge; 20, Housing; 21, Bleed passage; 22, Bleed valve; 23, Restrictor. Source: MTU

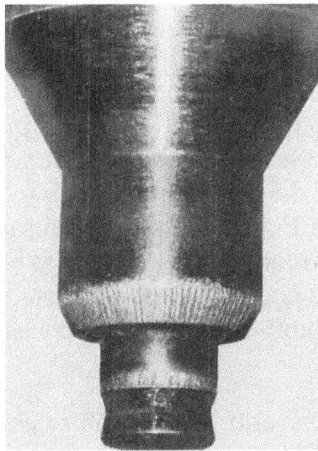

Fig. 6.531 Wear marks on a fuel injector needle as a result of inadequate fuel filtration.

- Heat transfer from combustion gases, through the combustion chamber walls, to the coolant(s); and
- Heat transfer from lubricant to engine parts and coolant

The heat that needs to be removed originates in the nature of engine combustion, as well as heat generated by friction. Furthermore, precompression of air in a compressor converts a portion of the compression work into heat, which also must be removed. In the process of exhaust gas recirculation, exhaust needs to be cooled.

Individually, heat flows are dependent upon the

- Thermodynamic process and its conditions
- Working process (two-stroke vs. four-stroke)
- Engine size and configuration
- Power density
- Design characteristics
- Temperature level at which components and working fluids should be maintained
- Individual ancillary conditions imposed on the engine, its operational behavior, and for safety criteria

The entire engine heat flow that must be removed by coolant amounts to 40 to 100% of rated power (see Table 6.14), in other words 20 to 40% of the heat provided by fuel, depending on engine type and size. For the cooling system as a whole, this means that heat transfer devices (radiators, heat exchangers) and the cooling system are by no means mere "engine accessories."

For any given engine size, the trend toward increased power concentration, above all by means of supercharging, results in ever-increasing heat flow that needs to be removed, while improved engine efficiency reduces relative heat losses. High-pressure supercharging and intercooling, but also further development of engine concepts, have altered engine heat budgets. Even as less heat is removed through engine coolant (cylinder and cylinder head cooling), the heat flow from supercharged combustion air and engine oil has increased significantly. Individual engine types often exhibit dramatic differences in their rejected heat flows.

In engine-transmission installations and, often, motor vehicle engines, waste heat from transmissions and brakes must also be removed, and the cooling system designed accordingly. Components whose mutual arrangement serves to remove heat from a machinery installation are collectively termed the cooling system. Cooling system layout, in principle and in detail, is largely determined by the type, size, application, and service conditions of individual engines. Complex

Table 6.14
Heat Rejection Requirement for Diesel Engines, in Percent of Rated Power
(Source: [6-42])

Engine speed range	Low-speed engines	Medium-speed engines	High-speed engines		
			High-performance engines	Commercial vehicle & industrial engines	
rpm	60 – 200	400 – 1000	1000 – 2000	1800 – 3000	
Engine coolant[1]	14 – 20	12 – 25	30 – 50	30 – 50	50 – 70
Engine oil[2] Engine oil[3]	6 –15 3	10 – 15 –	5 – 15 –	– –	– –
Piston coolant	5 – 6	N.A.	N.A.	N.A.	N.A.
Charge air	20 – 35	20 – 40	10 – 20	15 – 30	N.A.
Coolant heat	40 – 70	40 – 80	45 – 80	45 – 80	50 – 70

[1] Cylinder, cylinder head, and turbocharger cooling
[2] Lubricant and piston cooling oil cooling
[3] Lubricant cooling (water-cooled pistons)

machinery installations, high absolute and specific engine output, and demanding requirements for engine function, reliability, and behavior under varying operating conditions call for targeted, differentiated, and adaptable cooling. The necessary routing and control of coolant may in part only be achievable through considerable design effort. For functional reasons, it is not possible nor desirable to cool certain components directly by means of engine coolant, so that additional heat exchangers must be incorporated in the cooling circuit [6-42]. Components requiring cooling (Fig. 6.532) include the following:

Engine components adjacent to the combustion chamber and which are directly exposed to combustion gases, including

- Cylinder head: valves, pre- or swirl chamber, injector nozzles
- Cylinders
- Pistons
- Exhaust turbocharger (bearings, and possibly turbine housing)
- Exhaust ports to protect personnel in confined spaces (engine rooms of marine vessels, etc.)

Cooled
Turbocharger

Injector nozzle

Valve seat

Exhaust pipe

Cooled
nozzle

Cylinder head

Piston

Cylinder liner

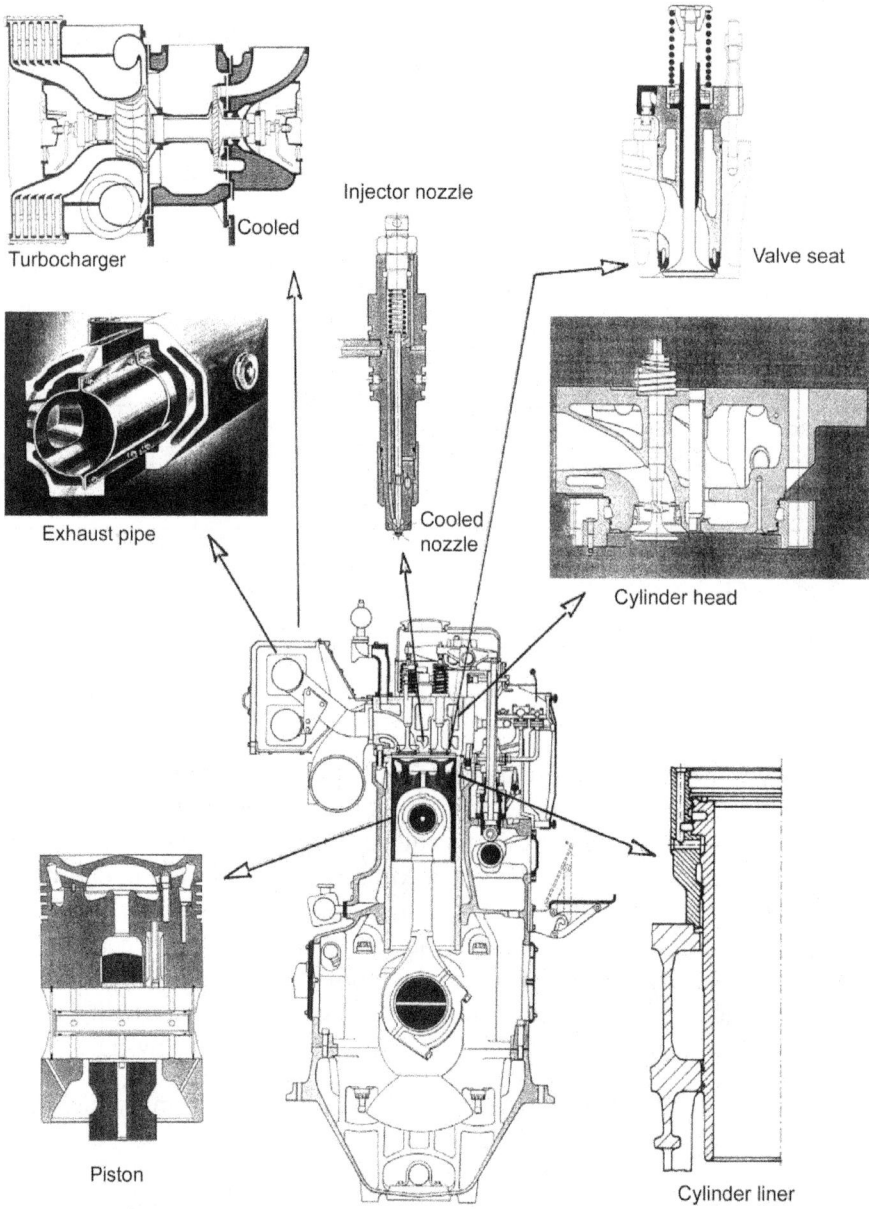

Fig. 6.532 Combustion engine cooling.

Operating fluids:

- Engine coolant
- Charge air/scavenge air
- Lubricating oil
- Fuel
- Exhaust gas, to reduce exhaust gas temperature for exhaust gas recirculation to the cylinders

The type of heat exchanger (colloquially, radiator, cooler, intercooler, heat exchanger) is determined by the

- Physical properties of the cooling media
- Volume flow rates
- Temperature differences, and
- Application

The size of a heat exchanger is a function of

- Heat transfer rate
- Temperature difference between the medium to be cooled and the medium that takes up the rejected heat, and
- Heat exchanger configuration

Engine cooling may employ liquid to liquid, liquid to air, or air to air heat exchangers in the configurations described below.

6.7.1 Shell and tube heat exchangers

Seamless tubing is swaged or brazed to fixed or movable tube sheets. Baffles support the tubes and direct coolant flow. Coolant flows over the outside of the tubes and is deflected by baffles to flow transversely (crossflow heat exchanger) or against the flow (counterflow heat exchanger) within the tubes. Heat transfer is improved by turbulence inducers within the tubes. Configurations include straight tube, U-tube, and Y-tube heat exchangers. Constructively complex, difficult to manufacture concentric-tube heat exchangers have fallen out of favor. Shell and tube heat exchangers are primarily used on medium and large engines (Fig. 6.533).

6.7.1.1 Flat tube or plate heat exchangers

Flat tube or plate heat exchangers operate in parallel on the cooled medium side; the cooling medium flows over the bundle in a crossflow, counterflow, or parallel flow arrangement. Flat tubes have a large

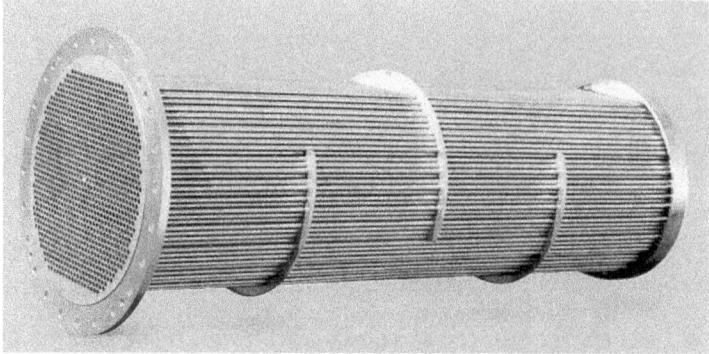

Fig. 6.533 Tube bundle of a shell and tube heat exchanger. Source: Behr

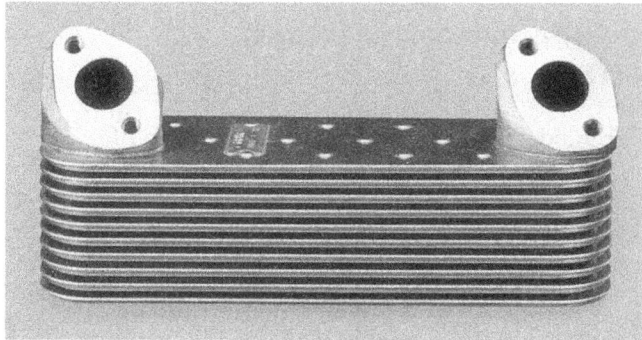

Fig. 6.534 Flat tube or plate heat exchanger. Source: Behr

surface area (relative to their flow cross section) and make possible high-performance coolers occupying small volumes (Fig. 6.534).

6.7.1.2 Stacked plate heat exchangers

Deep-drawn sheet metal plates are stacked together. Their spacing is determined by the interleaved turbulence inducers ("turbulators"). The complete stack is brazed together. Use of standardized components enables cost-effective custom configuration of these heat exchangers for individual applications. Identical turbulence inducers on both sides (medium to be cooled, and coolant), deletion of the coolant housing, and the ability to modify cooling performance by changing the number of plates have led to the rapid rise in popularity of stacked plate heat exchangers (Figs. 6.535 and 6.536).

Air to liquid heat exchangers are used in all types of motor vehicle engines (on-road, off-road, and rail vehicles) to cool engine coolant, to some extent also to cool engine and transmission oil, and as charge air coolers (intercoolers). In principle, they consist of two header tanks on either side of the actual cooling system—in general, a finned tube heat exchanger. Air to air heat exchangers are used as intercoolers in vehicle engines.

Tube and fin systems come in different configurations and with different tube shapes (round, oval, or flat tubes, Fig. 6.537) and fin density (number of fins per unit length) as well as fin form (flat or corrugated fins). High-performance, compact configurations include plate fin separator heat exchangers (Fig. 6.538).

From the laws of heat transfer, it is apparent that cooling performance of a heat transfer device can only be improved by measures applied to the side with the least effective heat transfer, e.g., in an air to water heat exchanger, on the air side, or in a water to oil heat exchanger, on the oil side. This may be accomplished by

- Turbulence inducing inserts in round or flat tubes, or packs, and
- Fins in tube and fin heat exchangers

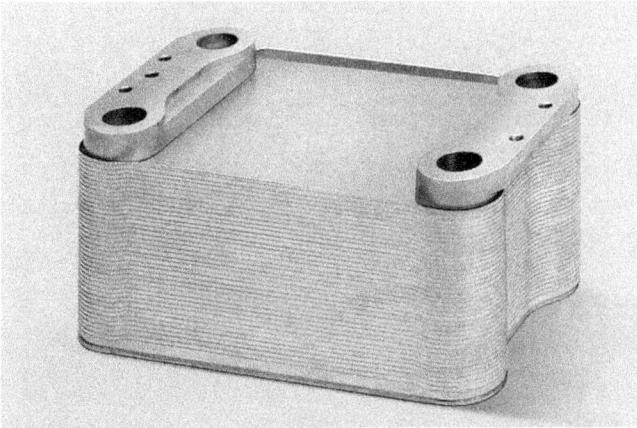

Fig. 6.535 Stacked plate heat exchanger. Source: Behr

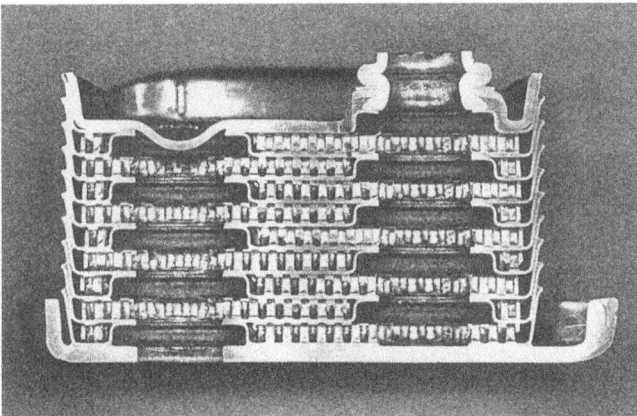

Fig. 6.536 Section through a stacked plate heat exchanger. Source: Behr

The operational behavior of heat exchangers is determined by mass flow rates, and differences in inlet temperature between the medium to be cooled (warm flow) and the cooling medium (cold flow). The heat exchanger effectiveness, ε, is the ratio of actual temperature drop to the theoretically possible temperature drop. The cooling power (heat transfer) of a heat exchanger may be represented by the mass flow rate and temperature difference. If cooling performance is expressed relative to the inlet temperatures, the result is specific cooling power.

Flow velocities largely determine heat transfer, and therefore the heat transfer performance (heat flow, cooling power); however, high flow velocities also result in higher pressure losses, and therefore higher pumping losses. On the other hand, in view of dirt or deposit formation, certain minimum flow velocities must be maintained, their magnitude in keeping with the relevant fluid properties; in an engine,

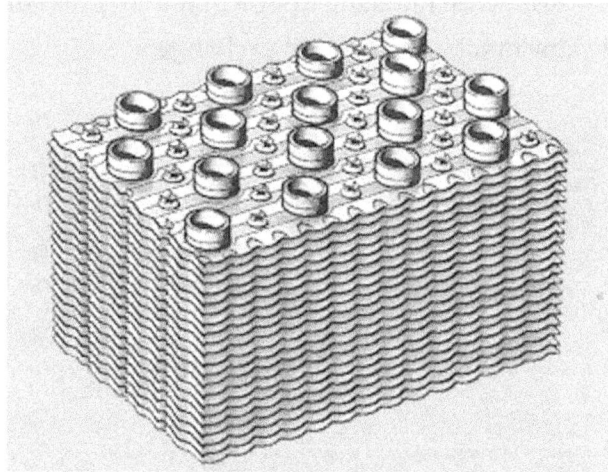

Fig. 6.537 Round tube and fin heat exchanger.

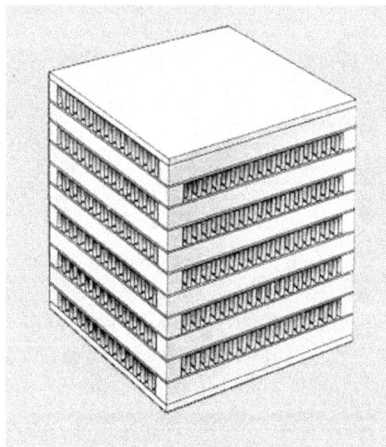

Fig. 6.538 Plate fin separator heat exchanger.

these fluids are water, oil, and air. On the water side, flow velocity should be in the range from 1.5 to 5 m/s; air speeds may go as high as 20 m/s.

Engine coolant is an operating fluid whose condition determines the behavior, functional ability, and service life of components exposed to coolant flow, and therefore also of the engine itself, to a far greater extent than is generally appreciated. For this reason, engine manufacturers issue precise specifications for the water used in engine coolant, and for the necessary additives. Basically, coolant should employ clear, clean water with a pH between 6.5 and 8.0 (at 20°C), and specified maximum content of chloride ions, chlorides, and sulfates. Coolant additives may only be used as approved by the engine manufacturer, and mixtures of various coolant additives are not permissible.

Coolant must satisfy engine manufacturer specifications for

- Overall hardness
- pH
- Chloride content
- Anti-corrosion oil concentration
- Antifreeze concentration
- Concentration of water-soluble corrosion protection

Continuous coolant monitoring is necessary for problem-free engine operation, because anticorrosion oils age and gradually lose their properties over time.

6.7.2 Heat exchanger damage
6.7.2.1 Fouling
Naturally found water (fresh water, salt water, or, worst of all, brackish water) contains dissolved or suspended materials that lead to deposits in water passages (tubes, plates); micro- and macroorganisms may intensify these processes. Thermal conductivity of these deposits is poor, resulting in greater resistance to heat transfer. This deposition process is known as *fouling* (Fig. 6.539).

Fouling consists of various mechanisms:

- Crystallization fouling: precipitation from supersaturated solutions
- Sedimentation: settling of fine particulates
- Reaction fouling: reaction within the fluid, with associated deposition

Fig. 6.539 Inner section of reversal header tank of a charge air intercooler exposed to seawater after short service in tropical waters.

- Corrosion fouling: formation of an oxide layer on passage walls
- Biological fouling (does not apply to gas flows): slime layer and microscopic growth

Because the walls of heat exchangers also experience material loss in addition to deposit formation, fouling is the result of these two opposing processes. Over time, fouling may increase continuously (linearly), at a decreasing rate, asymptotically, or follow a sawtooth curve.

6.7.2.2 Cavitation

Local pressure variations as a result of engine component vibrations are superimposed on coolant system pressure. If local pressure drops below vapor pressure, cavitation will result, often accompanied by significant problems.

6.7.2.3 Damage as a result of design and installation factors

- In designing a heat exchanger, in addition to specific installation and operating conditions, a certain loss of performance due to dirt (about 2 to 5%) must be taken into account.

- Even though modern heat exchangers have largely replaced copper alloys (brass, bronze) with aluminum, cooling systems, with their piping, valves, pumps, and heat exchangers made of different materials, may still be subjected to electrical potential differences, which lead to material loss on less-noble components in the galvanic series. In routing cooling system piping, metallic materials cannot be combined at random.

- Depending on installation conditions, heat exchangers, which often represent considerable mass, are excited to vibration by the engine. This may result in significant vibration amplitudes. These may be transmitted through the heat exchanger attachment points to the exchanger core, with resulting cracking of the outer ends of the heat exchanger tubes. A remedy may be found in modifying the installation conditions (if possible), and/or stiffening the heat exchanger core.

- Corrosion damage may result from inadequate preservation between manufacturing and installation, in particular for replacement parts which are warehoused for long periods. The result is corrosion from condensation. This can be remedied by application of appropriate long-term preservatives.

6.7.2.4 Damage caused by materials and manufacturing

- In the past, heat exchangers were soldered in a salt bath. If the flux was not completely removed, corrosion would result. Today, heat exchangers are vacuum soldered, and this form of corrosion no longer occurs.

- In the event of defective soldering, the exchanger tubes are inadequately supported, so that internal pressure pulsations cause the tubing material to fatigue, in particular in long tubes, leading to fatigue fractures. This may be remedied by redesign, with shorter unsupported tube lengths.

Example a: Leaks were detected in an oil-water heat exchanger. Broken-out material was observed on several tubes, which, however, did not always result in tube penetration. Metallographic examination showed numerous inclusions around the damaged areas; energy-dispersive X-ray analysis identified these as copper oxide. In service, these oxide inclusions were dissolved by coolant, resulting in leaks (Figs. 6.540 and 6.541).

Fig. 6.540 Tube wall penetration as a result of oxide inclusion washout.

Example b: Leaks were detected in flat-tube heat exchanger. After removal, it was observed that several flat tubes were cracked just below the soldered crimp joint (Fig. 6.542). The cause was identified as defective tube support; as a result, the "breathing" effect of varying pressure was amplified, leading to bending fatigue failure.

6.7.2.5 Failures as a result of unfavorable operating conditions and faulty maintenance

Dirt can block coolant and air passages, and create flow patterns that cause corrosion. Inadequate inhibitors may favor corrosion. Air flowing through the heat exchangers of rail vehicles carries with it iron wear particles from brakes, and, in the case of catenary-powered rail vehicles, copper wear particles from overhead lines; these also promote corrosion.

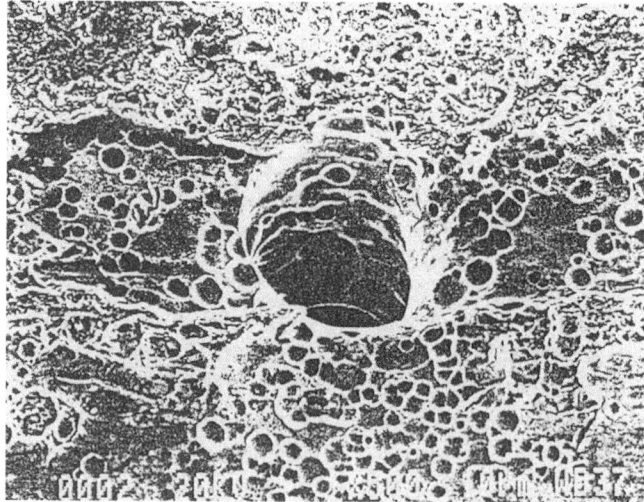

Fig. 6.541 Reflection electron micrograph of the oil-side exit hole of the leak in Fig. 6.540.

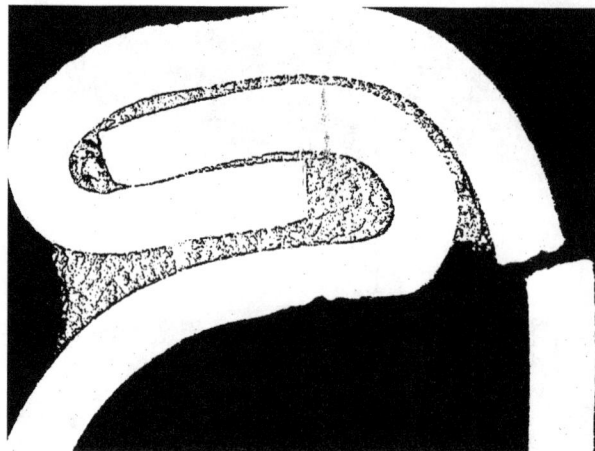

Fig. 6.542 Crack propagation in flat tube below crimped and soldered seal.

Blocked passages restrict coolant flow and air flow, and compromise heat transfer. Cooling system leaks cause component temperatures to rise.

Time and again, marine engines have had their seawater prefilters removed, merely for convenience, to avoid the need for frequent cleaning. This causes cooling system clogging. But even in normal operation, heat exchangers trap dirt, and clog.

Example a: The charge air intercoolers of a marine engine began to leak after about 8000 operating hours. After removing the covers and side panels, both intercoolers exhibited the same appearance: the areas in contact with seawater, especially the water inlet cover, were clogged by various species of marine life, including seashells, seaweed, and trash. The size of individual pieces ranged up to 30 mm. Water flow through the coolers was severely restricted. Within the clogged tubes, stagnant seawater was heated, which increased the rate of corrosion. The overall higher temperature levels in the intercoolers favored corrosion in the as-yet-unclogged tubes. The results were pitting corrosion and tube penetration (Figs. 6.543, 6.544, 6.545, and 6.546).

Example b: After about 7500 operating hours, the charge air intercooler of a marine engine began to leak seawater. Inspection showed deposits of vegetation, foreign objects, and pieces of plastic film in the water cover, several tubes penetrated by pitting corrosion, and fins on the air inlet side with baked-on deposits that restricted air flow. Chemical analysis showed these deposits to be the residue of vaporized seawater. These were formed after extended operation of the intercooler with tubes that leaked seawater. The engine had been drawing its combustion air out of the engine room (Figs. 6.547, 6.548, and 6.549).

Fig. 6.543 Water inlet cover of a charge air intercooler with deposits of marine life (molluscs, vegetation).

Fig. 6.544 Soiled water inlet cover (enlargement of previous image).

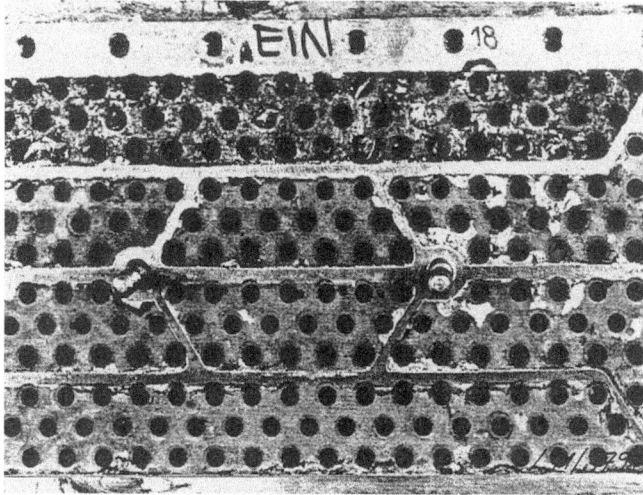

Fig. 6.545 Tube sheet of a soiled charge air intercooler, in which about half of the tubes were blocked by deposits.

- Errors in engine operation, for example neglecting to disengage or shut off certain equipment or functional groups when a vehicle is under tow, may cause cooling system pumps to overspeed and build up extreme pressures, which the heat exchangers are not capable of withstanding (Figs. 6.550, 6.551, 6.552, and 6.553).

6.7.2.6 Damage as a result of changed operating conditions

In all branches of engine technology and engine applications, one may observe, time and again, that engines, functional groups, or components that had never given problems "suddenly" begin to exhibit faults, damage, or failures. The search for a cause then reveals that as a result of altered operating conditions, the nature and magnitude of loads have been changed to the extent that they cause the observed problems.

- Powerful pressure on the development process to reduce fuel
  consumption not only led to long-stroke, slow-turning engines,
  but also, in the case of commercial vehicles, to an engine map
  showing maximum torque at low engine speeds. Because drivers are
  encouraged to apply an energy-efficient driving style [6-43], these
  engines operate primarily in the lower speed range. This, however,
  implies lower oil and water pump speeds, and therefore reduced
  cooling performance.

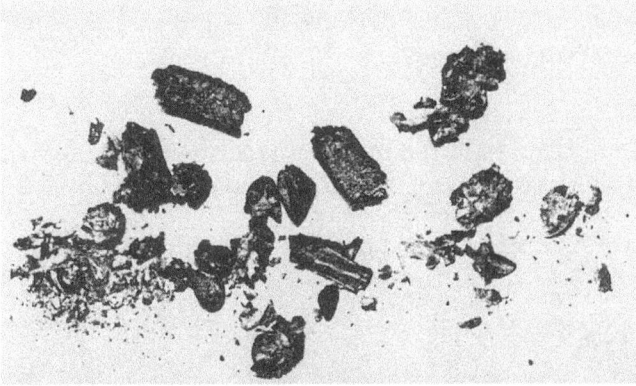

Fig. 6.546 Residue from one tube of the soiled intercooler of Fig. 6.545.

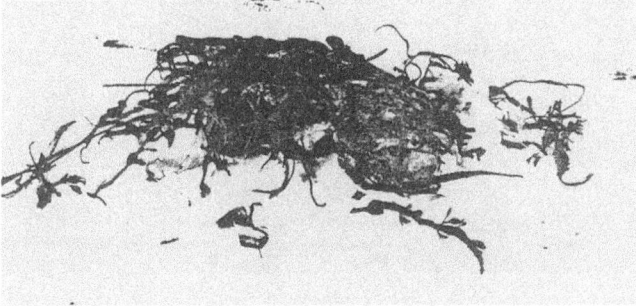

Fig. 6.547 Foreign objects from inlet cover of a charge air intercooler.

Fig. 6.548 Charge air intercooler: inlet face with baked-on deposits.

- Engines in marine ferry applications have always been subject to hard service, because they operate at continuous high loads and, with electric power transmission, spend their entire operating time at full rated engine speed. These conditions have become even more intense in high-speed ferries, which are often catamaran hulls with water jet propulsion. In the tourist season, these engines essentially operate at full power for 18 to 20 hours per day. As a result of such virtually uninterrupted operation, heat exchangers exposed to seawater, such as charge air intercoolers, normally considered seawater resistant, show previously unknown corrosive wear. Due to continuous operation and lack of rest periods, the cupro-nickel alloy has "no time" to form passivating layers.

6.7.2.7 Corrosion

Corrosion may have many causes. What makes this form of damage so problematic is the fact that manufacturers of components and engines

Fig. 6.549 Baked-on deposits in air entry of a charge air intercooler.

Fig. 6.550 Oil heat exchanger cooling system destroyed by excessive internal pressure.

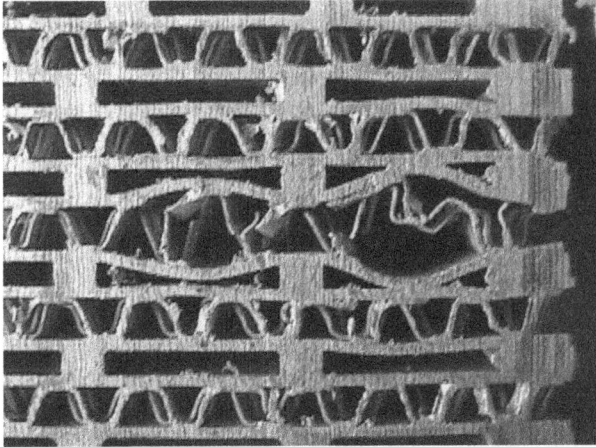

Fig. 6.551 Destroyed oil heat exchanger (detail).

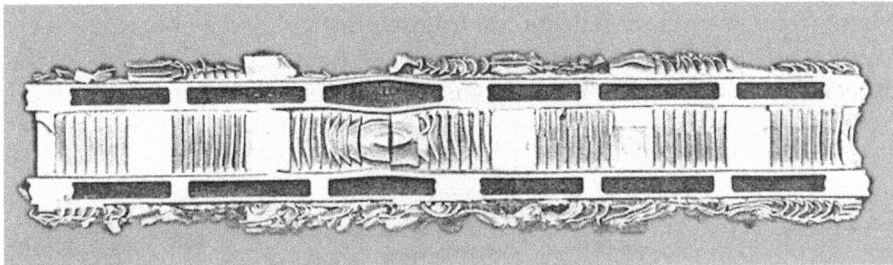

Fig. 6.552 Oil heat exchanger cooling system destroyed by excessive internal pressure.

are essentially unable to influence their products' operating conditions, much less neutralize or eliminate the actual causes of damage.

Example a: A (water) radiator exhibited a 280-mm-long longitudinal crack in its tube sheet bead. Otherwise, the bead area was largely undamaged; the lower tube sheet was nearly free of incipient cracks. More pronounced damage of this type was found in a constructively similar radiator. About 160 mm of the front left, and 130 mm of the front right of the upper tube sheet had broken off at the level of the corrugated slot crimp. Furthermore, after removal of the header, multiple cracks were discovered in the tube sheet material. Reflection electron microscopy of the crack surfaces showed that in all cases, these were intercrystalline cracks. The cracks showed multiple branching and, in some cases, also followed the twin boundaries within crystals.

In this particular material, damage with this manifestation is characteristic of stress crack corrosion. The causative medium was determined to be ammonia (NH_3) from animal excrement, whose dried residue, along with dirt and soil from vehicle operation, was found on the air side of the header tank/tube sheet joints (Figs. 6.554 and 6.555).

Example b: After 140,000 km of service, an aluminum air to water heat exchanger (radiator) showed severe corrosion of its radiator fins. The air inlet side showed considerable soiling, which also affected the radiator face that was masked by the charge air intercooler (an indication of severe turbulence in the air flow). The lower half of the radiator showed pronounced corrosion damage. There, radiator fins crumbled away when merely touched. Damage to the corrugated fins extended through the entire thickness of the radiator. Metallographic examination revealed material disintegration as a result of corrosion. The damage began in the form of intercrystalline corrosive attack, which expanded as pitting. As the damage progressed, the radiator fins loosened as a result of material breakouts and consumption through corrosion. Microprobe examination of the fin fragments provided a clue to the failure-causing substances: road salt may be regarded as the cause of the detected high chlorine content. A sulfur-containing medium also played

Fig. 6.553 Oil heat exchanger expanded by excessive internal pressure.

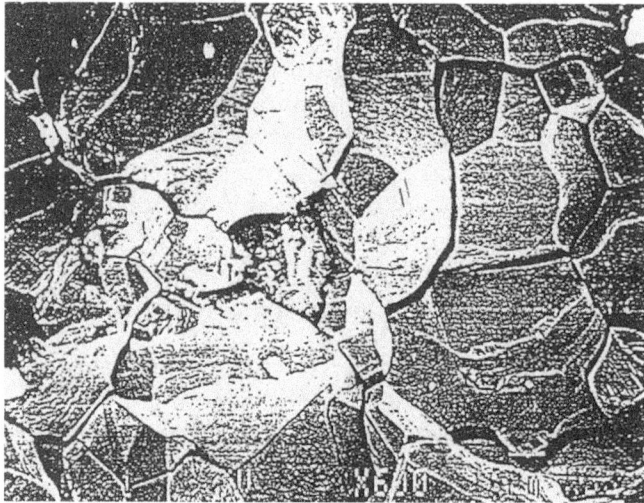

Fig. 6.554 Intercrystalline fracture propagation in tube sheet (600X magnification).

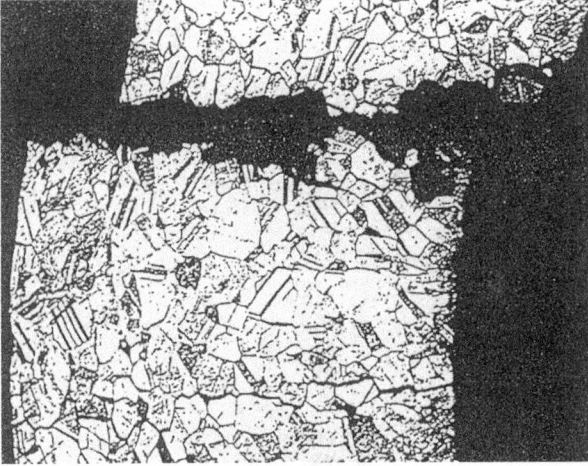

Fig. 6.555 Stress crack corrosion in tube sheet material, propagating from inside of tube sheet bead.

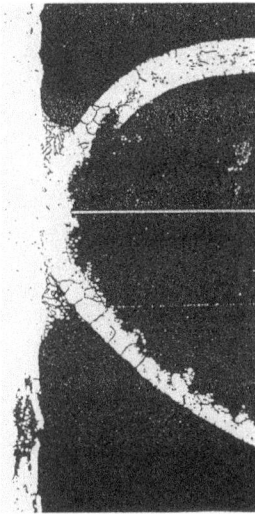

Fig. 6.556 Intercrystalline corrosion at fin and tube material, as well as selective corrosion of the corrugated fin solder joint (100X magnification).

a role, such as the action of a gaseous substance like sulfur dioxide. This corrosion damage could clearly be attributed to outside influences resulting from vehicle operation (Fig. 6.556).

Example c: In an air to water heat exchanger (radiator) of a rail vehicle, formation of oxygenating elements associated with a sulfur-containing medium, presumably sulfur dioxide from combustion products that entered the coolant through a leaking cylinder head gasket, resulted in pitting as well as intercrystalline corrosion, with penetration of the flat tube walls (Fig. 6.557).

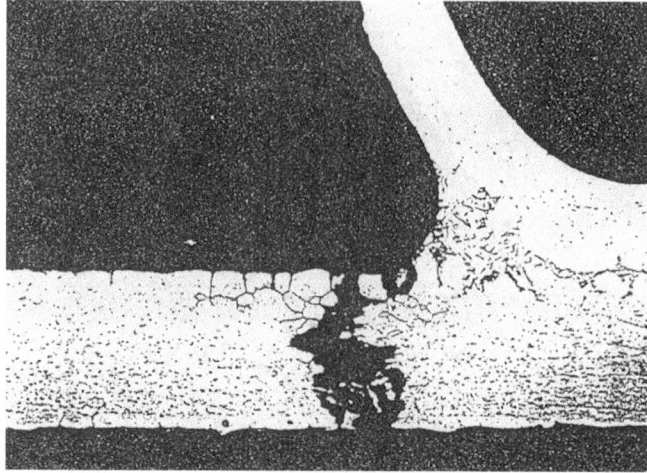

Fig. 6.557 Fractured corrosion area in a flat plate wall of an engine radiator.

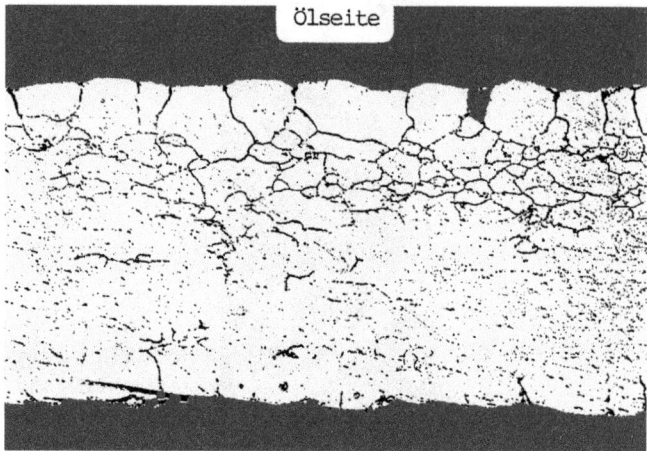

Ölseite

Fig. 6.558 Leakage as a result of intercrystalline corrosion originating from the oil side of a stacked plate oil cooler.

Example d: An oil cooler showed leaks as a result of intercrystalline corrosion originating from the oil side (Fig. 6.558). The plate material itself was damaged as a result. The fact that corrosion damage originated from the oil side, which is normally not subject to corrosion, suggests that corrosion was initiated by moisture inside the flat tubes even before the oil cooler was placed in service. The cooler had been stored for several weeks without preservation, and was only installed in a vehicle more than a year after manufacture. In keeping with customer request, long-term preservation was not applied.

Example e: The oil cooler of a marine engine began to leak after 2500 hours of operation. The plate stack was to some extent severely soiled (with, among other things, seaweed). The oil cooler leak was caused by intercrystalline corrosion of a single plate of the plate stack. Corrosive attack caused a single grain to loosen, leading to the leak (Figs. 6.559, 6.560, and 6.561).

The corrosive attack was caused by locally limited contamination around the grain boundary. Improved prefiltering was installed to eliminate such soiling. A further step was a switch to internal cooling (a central heat exchanger was cooled by seawater; all other heat exchangers received cooled fresh water supplied by the central unit).

6.7.2.8 Mechanical damage

Heat exchangers are assemblies consisting of many individual components. From a manufacturing standpoint as well as from conditions imposed by engine operation (pressure pulsations, flow processes, vibration, dissimilar thermal expansion, etc.) it is understandable that soldered connections and assemblies may loosen, and fretting corrosion and fatigue fractures may occur.

Fig. 6.559 Heavily soiled oil cooler plate stack.

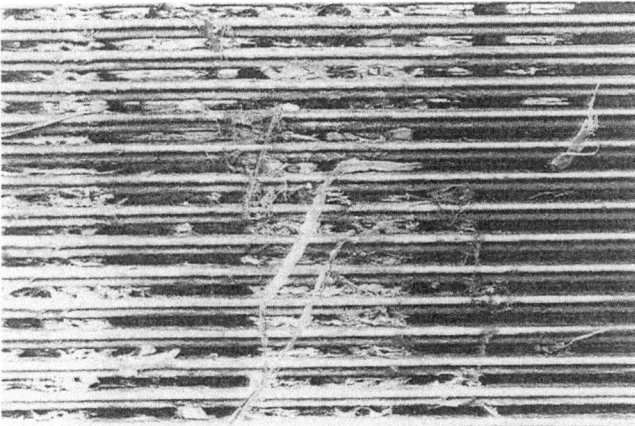

Fig. 6.560 Soiled plate stack (enlarged view).

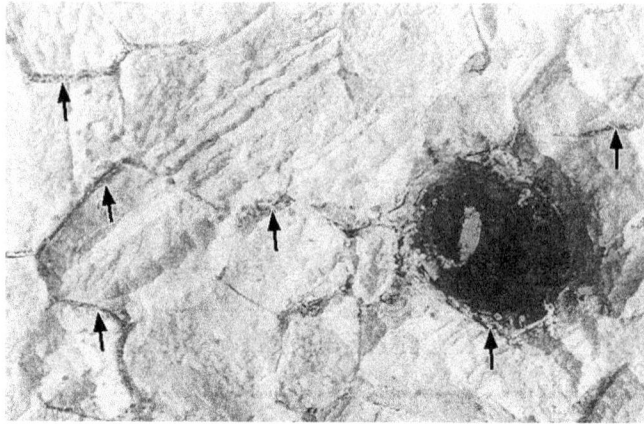

Fig. 6.561 Reflection electron micrograph of the hole causing the leak, due to a single washed-out grain. Surrounding grain boundaries show signs of contamination (arrows).

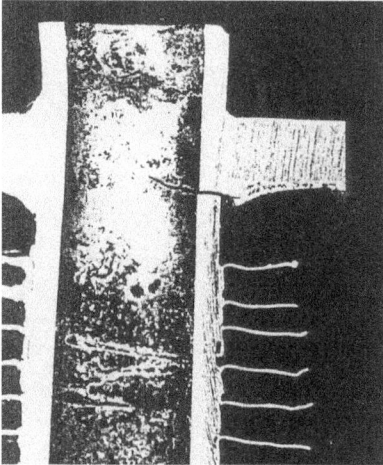

Fig. 6.562 Cracked corner tube of a charge air intercooler; transverse crack below the tube to tube sheet solder joint.

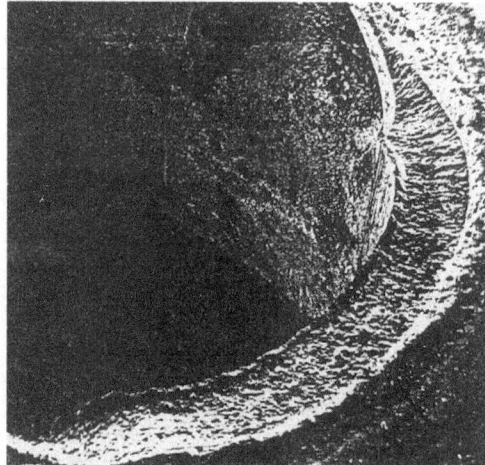

Fig. 6.563 Exposed crescent-shaped crack in a charge air intercooler corner tube.

Example a: After cleaning a severely salt encrusted, leaking corner of a tube and fin charge air intercooler, cracks were discovered on three tubes, from the outside in. The cracks were immediately below the tube/rib solder joint. A fracture had begun on the outside diameter and, consistent with loading, propagated inward in a crescent-shaped pattern. No material or manufacturing defects were found. The cause of these cracks was stress due to thermal cycling, and the resulting differential thermal expansion between tube material and side walls. Design changes were applied to remedy the problem (Figs. 6.562 and 6.563).

Example b: After a relatively short operating time of 500 hours, a tube and fin charge air intercooler showed fin breakouts on the air inlet side. These were fatigue fractures. Induced to vibration by inflowing air, the fins mainly broke on the intercooler face side, near the rigidly soldered joints with the water tubes. The appearance of sometimes staggered fractures between the tube pass-throughs indicate several active fracture fronts, which may have formed after the initial bilateral front-side fin cracks acted as new turbulence generators (Figs. 6.564, 6.565, and 6.566).

6.8 Turbochargers

Combustion engines liberate as useful work only a portion of the energy available in fuel, because the working gas is not allowed to expand completely, i.e., return to ambient temperature and pressure. To even approach this, piston stroke would need to be increased to many times that of current practice. The engine would then become too large and too heavy; furthermore, there would be a considerable increase in friction losses. Instead, exhaust valves are opened while gas pressure inside the cylinder is still well above atmospheric (i.e., before bottom dead center) to facilitate gas transfer. The loss in low-pressure work is accepted as one of the costs.

Alfred Büchi, a Swiss engineer, considered ways to utilize a portion of the "lost" low-pressure work by means of a gas turbine installed in the engine exhaust stream, harnessed to drive a turbine compressor to pre-compress the engine air charge (Fig. 6.567).

Because piston engines work with high pressures and small mass flow rates, while flow machinery works at relatively low pressures and (in comparison to their size) large mass flow rates, *exhaust turbocharging*

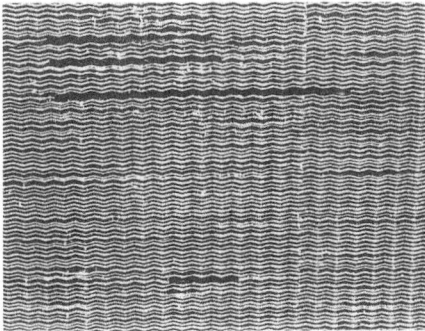

Fig. 6.564 Broken fins on the air entry face of a tube and fin charge air intercooler.

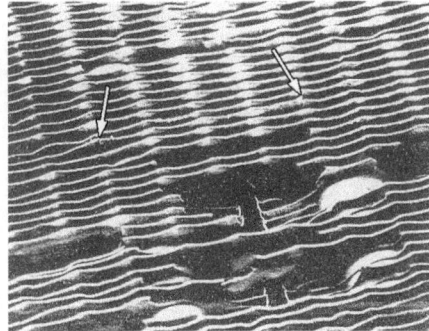

Fig. 6.565 Broken fins in the air inlet duct area showing fin crack propagation (arrows indicate crack initiation sites of a single fin).

combines piston and flow machines to allow each of these very different machines to operate in the regime to which they are best suited (Fig. 6.568). In this way, a major portion of otherwise lost low-pressure work can be recovered.

More air in the cylinder permits more fuel to be burned; this generates more power. Increased power through exhaust turbocharging goes hand in hand with improved thermal efficiency, because the engine's low-pressure work is now used to pre-compress the air charge.

Turbochargers (aka exhaust-driven turbochargers, turbosuperchargers) consist of a rotating section and housing(s). The (radial) compressor impeller is made of an aluminum alloy (depending on configuration, with or without an aluminum bronze alloy first-stage compressor impeller) and is shrink fitted to the turbocharger shaft. The turbine rotors of smaller turbochargers (impeller diameter < 160 mm) are configured as radial turbines (centripetal turbines; Fig. 6.569), while larger diameters (> 300 mm) employ axial turbines (Fig. 6.570). Intermediate sizes may employ either configuration. With few exceptions, "internal" rotating section bearings predominate; smaller turbochargers use journal bearings, larger turbochargers use

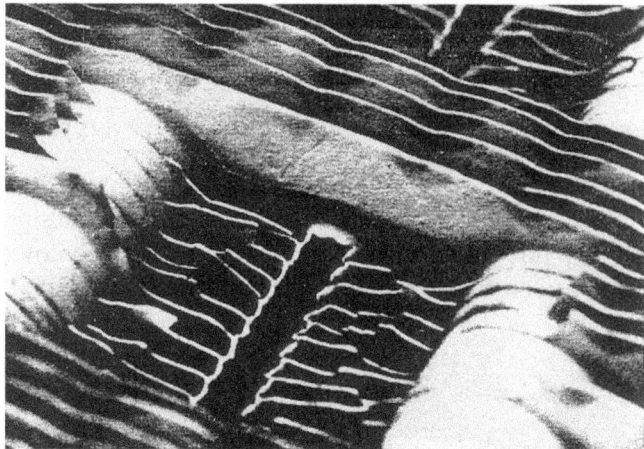

Fig. 6.566 Enlarged view of fin crack propagation.

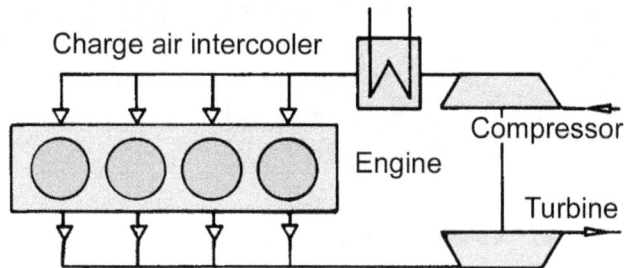

Fig. 6.567 Exhaust turbocharging (schematic).

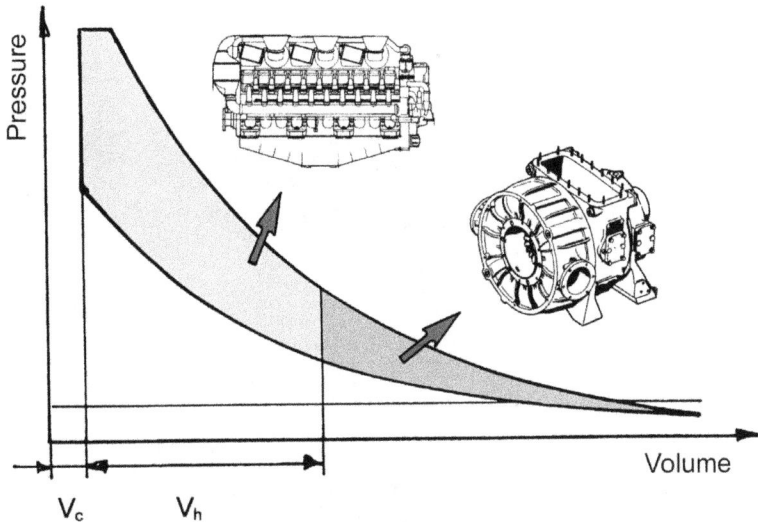

Fig. 6.568 Distribution of high- and low-pressure work between engine and turbocharger.

roller bearings. The bandwidth of turbocharging and their operating conditions matches that of combustion engines themselves; passenger car engines have just about "one handful" of turbocharger, spinning at speeds in excess of 100,000 rpm with glowing red-hot turbine housings, while turbochargers employed on large two-stroke marine engines are bigger than a man, turn at about 10,000 rpm, and are subject to the abuses of heavy-oil operation.

In compressing air, part of the compression work is converted into heat, which raises charge air temperature, with the result that air density is increased less than the theoretically possible amount. By employing an intercooler (aka charge air intercooler, charge air cooler) to cool the charge air after it leaves the compressor and before it enters the engine, one achieves higher air density at the same boost pressure, with lower thermal loads and improved exhaust emissions (NO_x).

One advantage is that engine and turbine are connected "elastically" by means of the gas flow, providing a certain degree of self-regulation. As the engine is loaded, increased energy content in the exhaust stream provides more energy to the turbine, which in turn allows the engine, now receiving more boost pressure, to develop more power. This is, however, combined with increased mechanical, thermal, and tribological stresses. As a very rough first approximation, piston engines exhibit speed-independent behavior; i.e., torque is constant over speed. Flow machines, on the other hand, have a pronounced speed-dependent characteristic; their torque increases with the square

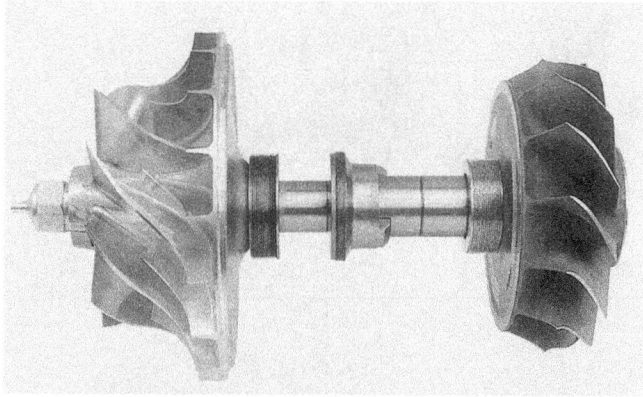

Fig. 6.569
Turbocharger rotating
section, with radial-flow
turbine. Source: MAN

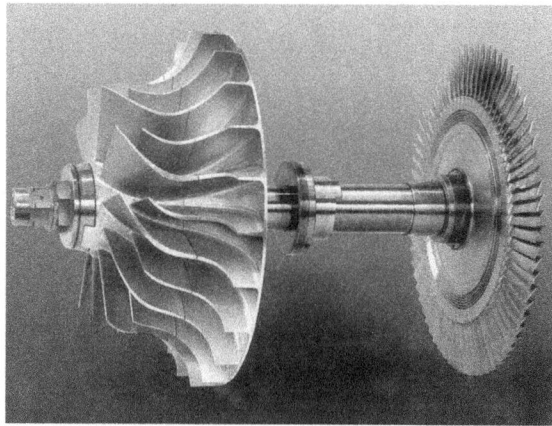

**Fig.
6.570** Turbocharger
rotating section, with
axial-flow turbine.
Source: MAN

of rotational speed, their power as the cube of speed. If one combines a
piston engine with a turbocharger, engine torque is not raised uniformly
across the speed range, but rather only in the higher engine speed
range. The usable portion of the turbocharger map becomes so narrow
that engine and turbocharger maps are no longer matched, especially
in the case of operational modes that deviate from the propeller curve.
The greater the increase in specific work (mean effective pressure) by
means of turbocharging, the greater the tendency of the turbocharger
to impose flow machine characteristics on the engine. This is readily
apparent if one compares the engine maps of engines with different
levels of turbocharging; the engine map begins to resemble that of a
propeller curve. Engine and turbocharger must therefore be carefully
matched to permit them to work together effectively. This means
that the characteristic curves of the "consumer," the engine, must fall
within the compressor map: pressure ratio $p_2/p_1 = f$(volume flow rate
$\dot{V}$). The compressor map (Fig. 6.571) is limited on one side by the *surge
limit* (aka stall line, pumping limit, stability limit) and on the other by

the *choke limit* (choke line). The choke limit establishes the maximum volume flow rate that the compressor can handle; if operated beyond the surge limit, the turbocharger becomes unstable, and begins to "surge." This is understood to mean oscillating flow, back and forth, of the volume flow from the compressor, readily recognizable by its distinctive sound and pressure impulses.

The interaction of two very different machines, such as piston engine and turbocharger, with their vastly different operational behavior, working together, poses difficulties, because the available energy at low engine power and low engine speeds is insufficient to achieve adequate turbine speeds; at the engine's upper range, on the other hand, the available energy is too great. Passenger car turbochargers are therefore configured for operation in the lower engine speed ranges. At higher engine speeds, a pressure diaphragm actuated by turbocharger boost pressure controls a wastegate (aka bypass) that allows a portion of the exhaust stream to circumvent the turbine, thereby limiting turbine speed to the desired level. Recently, variable geometry turbochargers have been introduced (Fig. 6.572). This is understood to mean turbochargers that employ a means of adjusting the turbine-side guide vanes. By enlarging or restricting the exhaust gas flow cross sections to the turbine, these provide optimum exhaust gas speed entering the turbine, largely independent of exhaust gas flow rate.

Turbochargers have reached such a state of development that not all of the available turbine energy is needed for engine supercharging.

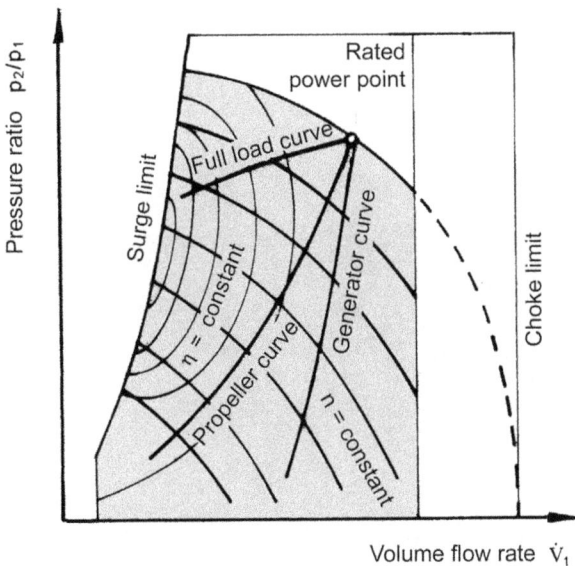

Fig. 6.571 Compressor map with choke limit lines (schematic).

A technique known as *turbocompounding* may be employed to take advantage of this otherwise wasted energy; a working turbine is installed downstream of the exhaust turbocharger, and delivers mechanical energy to the crankshaft by means of a clutch and power transmission (Fig. 6.573).

High output engines, so called because of their high specific output (power/engine displacement or power/engine physical volume), achieve their power by high-pressure supercharging (pressure ratios of 4 or 5:1). Such engines employ *sequential supercharging:* engine combustion air is no longer supplied by one or two large turbochargers, but rather by four or five small chargers that may be added or taken

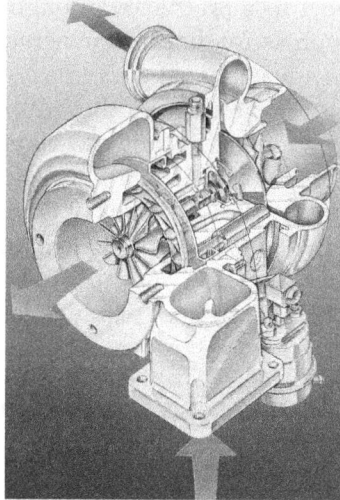

Fig. 6.572 Variable turbine geometry turbocharger.

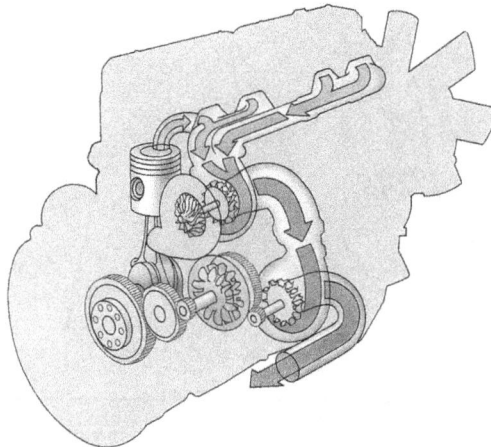

Fig. 6.573 Turbocompounding. Source: Scania

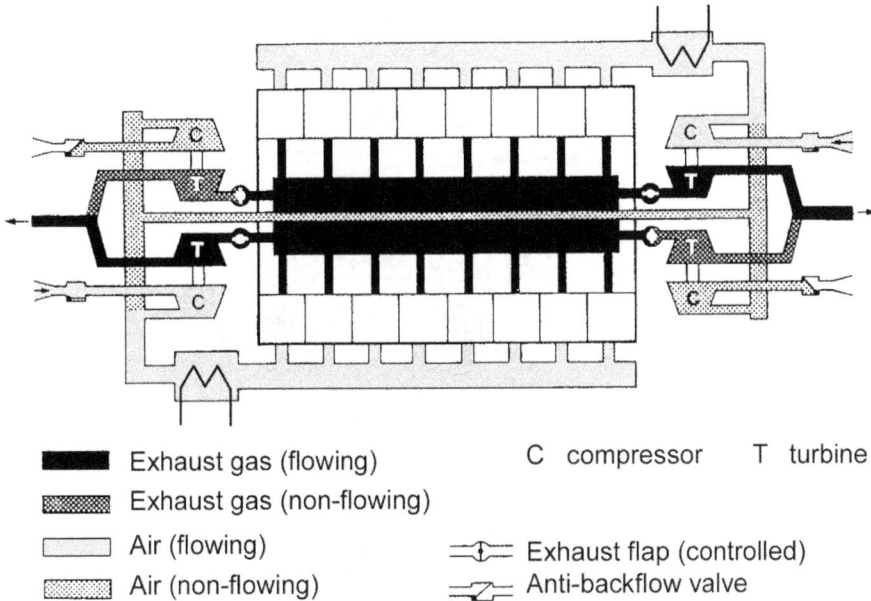

■■■ Exhaust gas (flowing)	C compressor	T turbine
▓▓▓ Exhaust gas (non-flowing)		
☐ Air (flowing)	═╤═ Exhaust flap (controlled)	
░░░ Air (non-flowing)	═╤═ Anti-backflow valve	

Fig. 6.574 Sequential turbocharging (schematic). Source: MTU

offline as needed (Fig. 6.574). In this way, these engines no longer employ large turbochargers operating at low part-load efficiency, but rather an appropriate number of smaller turbochargers operating at higher efficiency. With sequential turbocharging, the compressors of all turbochargers are connected to common, shared air supply ducts, the turbines to the shared exhaust pipe(s). Switching individual chargers on or off is accomplished by automatic opening or closing of flaps or butterflies mounted ahead of the turbines.

Further increases in boost pressure may be achieved with two-stage supercharging, with two turbines and compressors each operating in series, with the high-pressure and low-pressure stages coupled to each other. Intercooling permits an appreciable improvement in compressor efficiency (Fig. 6.575).

The combination of two-stage supercharging and sequential supercharging represents an additional improvement, albeit requiring complex design solutions. This is used on large, fast ferry vessels. In such schemes, exhaust turbocharging carries an ever larger share of the total developed power. This is readily apparent in the external appearance of the engine installation: the supercharging package, with exhaust turbocharger, intercooler, air induction system, exhaust, and

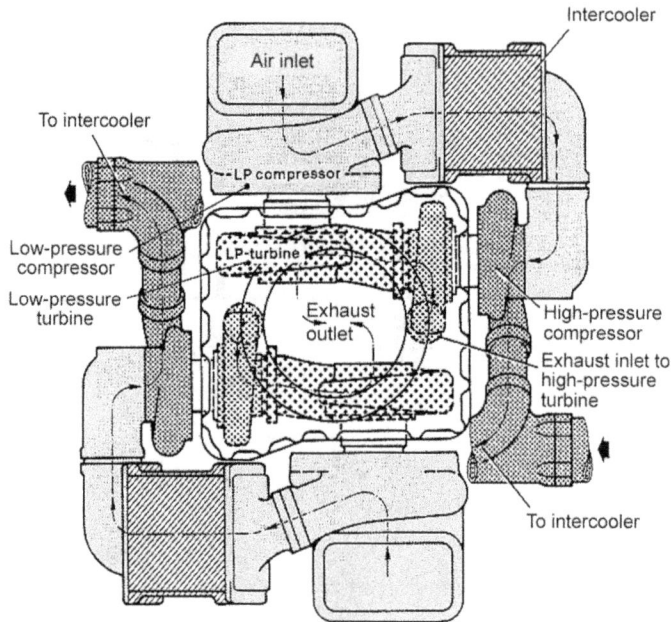

Fig. 6.575 Two-
stage turbocharging
(schematic).
Source: MTU

various ductwork and plumbing carrying air and coolant, takes up
nearly as much space as the actual engine itself.

6.8.1 Turbocharger damage

Because of high tip speeds (rotational speed × diameter), turbochargers
are sensitive to external influences, to operational irregularities of both
the turbocharger itself and the engine.

- Increased flow restriction due to dirty charge air intercoolers or
  exhaust mufflers, or severe dirt deposits on compressor impeller or
  turbine rotor, may result in *compressor surge.*

- Fuel injection problems, dirty charge air intercoolers, or inadequate
  cooling cause exhaust temperatures to rise, imposing added thermal
  loads on the turbine.

- Foreign bodies can damage compressor impellers or turbine rotors.

6.8.1.1 Material defects

The fundamental problem of completely inspecting three-dimensional
components with regard to their material condition (microstructure,
inclusions, contamination, etc.) also arises in the case of turbochargers.

Example: On the compressor impellers of several turbochargers, a
segment of the outer diameter in the vicinity of a single blade was
broken away (Figs. 6.576 and 6.577). The fracture surface showed that

this was a fatigue fracture, originating from an oxide surface inclusion nearly as thick as the vane itself. Modification of the casting process and stricter quality control measures provided a remedy. In addition, the impeller disk was reinforced to positively eliminate the negative effects of qualitative material differences.

6.8.1.2 The effects of foreign bodies

Compressor side

- Compressor impeller damage caused by foreign objects (nuts, bolts, cleaning rags, etc.) is a common occurrence that may be attributed to operating conditions, and therefore may not be curable by the turbocharger development process. Depending on how and where foreign objects strike the impeller vanes, they may break off smaller or larger fragments, which in turn amplify the damage. Usually, this damage proceeds "benignly," even as the damage continues to progress, without any great drama. This may be seen from the fact that in an advanced stage of damage, the vanes have been ground away, down to the impeller hub (Figs. 6.578, 6.579, and 6.580).

Fig. 6.576 Broken compressor vanes.

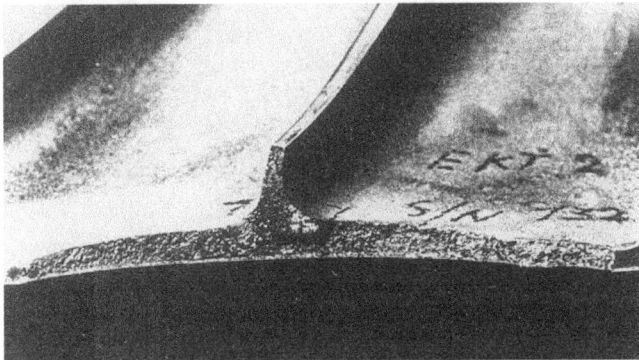

Fig. 6.577 Vane fracture surface from Fig. 6.576.

- If an engine is operated with a damaged air filter, or indeed without any filter at all, dirt particles will enter the compressor housing, where they lodge throughout the entire unit (Figs. 6.581 and 6.582). Furthermore, at points of high flow velocity, they have an erosive effect.

- Crankcase ventilation to the intake tract can result in oil drenching within the charge air duct and carbon deposits on the compressor blades.

Turbine side

The turbine rotor (Figs. 6.583 and 6.584) may be damaged by

- Fragments from the combustion chamber (originating from valves, piston rings, or damaged pistons)

- Carbon particles

- Casting residue (foundry sand) in the exhaust manifold or turbine housing

- Improper turbocharger installation, e.g., if the shaft nut is not tightened according to specifications and therefore comes loose, or if the compressor housing is removed incorrectly during maintenance, damaging the impeller

6.8.1.3 Imbalance

High-speed flow machinery such as exhaust turbochargers must meet high standards for rotational balance, i.e., exhibit very small residual imbalance. Accordingly, turbochargers are balanced by their manufacturers. As a result of

- Damage to rotating components or the turbine-side damping wire (to protect against vibration, vanes of axial compressors are joined by a damping wire)

Fig. 6.578 Damaged vane leading edges. Source: BorgWarner

Fig. 6.579 Ground-down vanes. Source: BorgWarner

Fig. 6.580 Compressor impeller: deformed vane leading edges caused by foreign object in turbocharger.

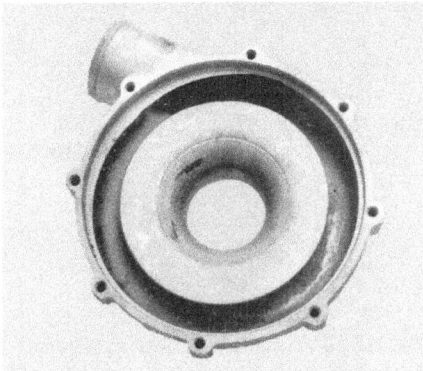

Fig. 6.581 Dirt deposits in compressor housing snail. Source: BorgWarner

Fig. 6.582 Dirt deposits in compressor housing. Source: BorgWarner

- Deposits on the compressor impeller and/or turbine rotor
- Defective bearings, or
- Asymmetric wear

The center of mass of the rotating section may shift, resulting in imbalance. The measure of imbalance is the vibration speed (the product of center of mass displacement from the ideal shaft axis, and mass of the rotating section). Imbalances manifest themselves as additional rotating bearing loads, compromise the kinematics of the rotating section, and allow the compressor impeller or turbine rotor to scrape against the housing. An out-of-balance rotating section may be recognized by:

- Vibration during operation

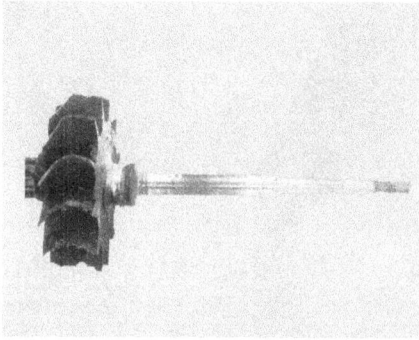

Fig. 6.583 Shaft section with damaged turbine wheel. Source: BorgWarner

Fig. 6.584 Damaged turbine leading edges. Source: BorgWarner

Fig. 6.585 Asymmetrically worn radial bearing. Source: BorgWarner

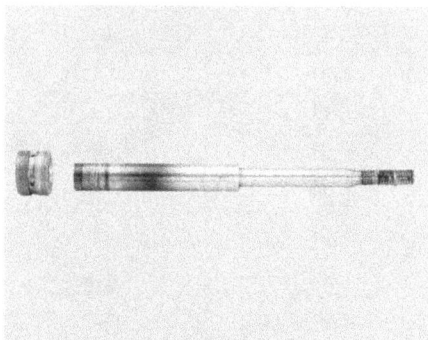

Fig. 6.586 Asymmetrical bearing material deposit on turbocharger shaft. Source: BorgWarner

- Worn radial bearings and bearing areas in the bearing housing (Fig. 6.585), or

- Asymmetrical bearing metal deposits on the shaft surface (Fig. 6.586)

Example: On a diesel engine (cylinder bore ca. 170 mm), turbocharger damaged was discovered after about 1000 hours of operation. Upon disassembly, it was discovered that the tips of the turbine blades had worn down as a result of scraping against the turbine housing. Furthermore, the back of the rotor and the turbine blades showed asymmetrical carbon deposits, forming a layer about 0.7 to 1.5 mm thick. This carbon layer contained oil-specific elements S, Ca, P, and Zn. The turbine rotor shaft was broken, the result of a one-sided bending fatigue failure with a residual fracture of about 40% at the transition to the turbine-side shaft step. The cause of this failure was imbalance due to heavy carbon deposits, but a manufacturing-related imbalance cannot be ruled out (Figs. 6.587 and 6.588). The appreciable loss of material strength in the turbine-side radial bearing seat was presumably caused by wear-related heating.

6.8.1.4 Erosion

Turbine blades are eroded by particles in the exhaust stream.

6.8.1.5 Fractures

Fatigue fractures result in fragments being broken out of compressor impellers and turbine rotors, in worst cases to burst compressor impellers (Figs. 6.589 and 6.590). The actual cause of this failure is the combination of static preload and high dynamic stresses (fracture in the creep regime). This type of failure occurs primarily during full-load engine operation. Such fatigue failures of compressor impellers and turbine rotors are initiated by:

- Miniscule pre-existing damage by
  - Microstructural inhomogeneities in the compressor and turbine materials (e.g., oxide inclusions) or

Fig. 6.587 Heavy carbon deposits on vanes and back side of turbine rotor.

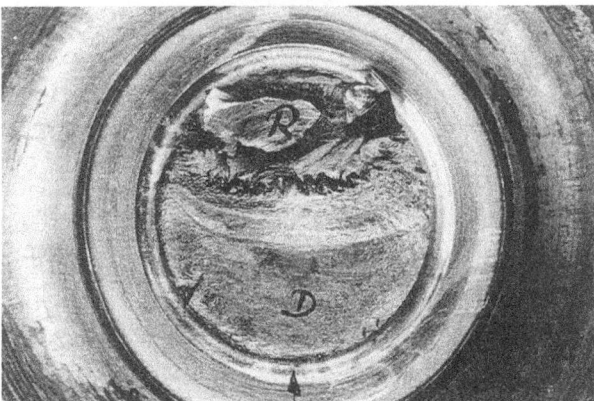

Fig. 6.588 One-sided bending fatigue fracture of a turbine shaft: D, fatigue fracture; R, residual fracture.

- Pre-existing damage of the compressor impeller by improper assembly after service/maintenance operations (compressor housing not pulled off straight)

- Corrosion scars on the compressor impeller caused by seawater
- Effects of foreign bodies
- Imbalance
- Vibration: high cycle failures (HCF) caused by self-resonance

Occasionally, one might see a broken turbocharger shaft,

- Primarily as a result of imbalance
- Secondarily by scoring of the bearing oil seal areas, and the turbine rotor contacting the housing, as well as

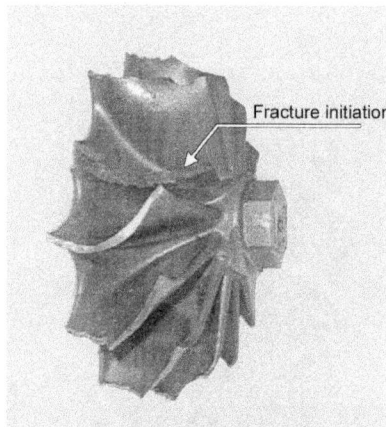

Fig. 6.589 Fracture of a compressor vane. The remaining vanes are damaged or bent.

Fig. 6.590 Compressor vane fracture (HCF failure): vane tip deformation and breakouts on vane leading edges.

- A loose shaft assembly. A rotating bending fatigue fracture may be recognized by severe wear marks on the shaft and the mating surfaces of components stacked on the shaft.

6.8.2 Lubrication inadequacies

6.8.2.1 Oil contamination

Dirt in lubricating oil causes

- Scoring over the entire inner and outer circumference of radial bearings, the shaft surface in the area of the radial bearings, and also causes washouts (erosion) of axial and radial bearing oil supply passages (Fig. 6.591)
- Clogged axial bearing oil supply passages; and
- Evenly distributed scoring over the circumference of radial bearing areas in the bearing housing (Fig. 6.592)

The causes of such oil contamination might include:

- Exceeded oil change intervals, or
- An oil filter that does not meet the engine manufacturer's specifications.

6.8.2.2 Oil starvation

In turbochargers, as anywhere else, oil starvation causes the usual tribological damage:

- Wear to the axial (thrust) bearing tapered land surfaces. In advanced stages, the land surfaces are no longer discernible (Figs. 6.593 and 6.594).

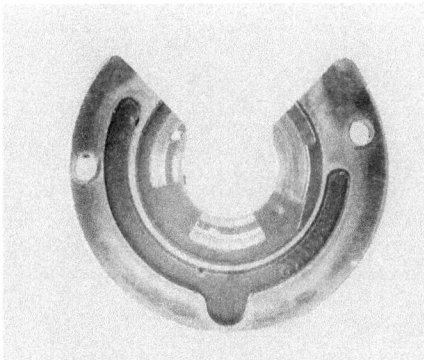

Fig. 6.591 Scoring of axial (thrust) bearing tapered land surfaces. Source: BorgWarner

Fig. 6.592 Scoring on outside circumference of radial bearing. Source: BorgWarner

- Wear to the radial load-carrying surfaces of the bearing body (to inside and outside diameters); oil grooves and oil supply passages smeared shut.

- Wear to the radial bearing areas in the bearing housing. Evenly distributed bearing metal deposits on the shaft circumference in the radial bearing area (Figs. 6.595 and 6.596). In an advanced stage, the radial bearings may seize firmly on the shaft or in the bearing housing.

Causes include:

- Oil supply failure (e.g., defective oil pump, broken oil line, pinched oil supply line, clogged supply lines)

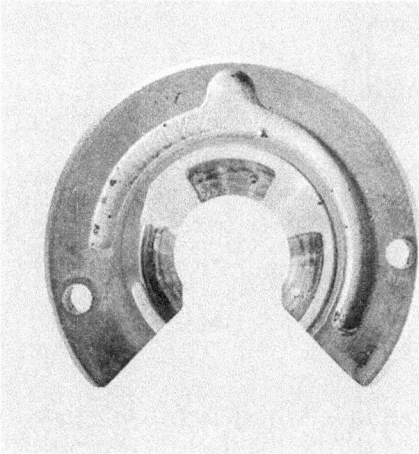

Fig. 6.593 Worn axial (thrust) bearing tapered land surfaces. Source: BorgWarner

Fig. 6.594 Axial bearing tapered land surfaces and turbine-side bearing boss. Source: BorgWarner

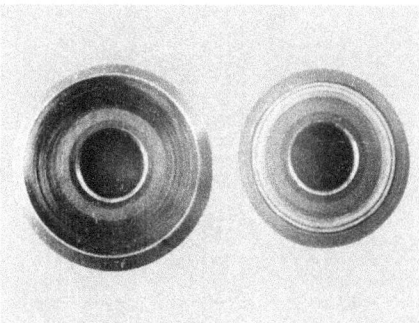

Fig. 6.595 Oil deflector ring (left) and bearing boss. Source: BorgWarner

Fig. 6.596 Evenly distributed bearing metal deposits on turbocharger shaft. Source: BorgWarner

- Insufficient oil pressure upstream of the turbocharger
- Sudden revving of the turbocharger before adequate oil supply has been established

Example: The turbocharger of a diesel engine exhibited axial bearing damage that was traced to inadequate oil supply during the engine startup phase. In some situations, turbine run-up was accelerated by blown-in air during engine start, while the oil pressure required for adequate oil supply to the bearings only built up slowly. This led to mixed lubrication (and resulting bearing damage). In this case, the severely worn-out axial bearing led to axial contact of the rotating section against the axial thrust washer/compressor housing and turbine rotor boss/bearing sleeve. The bearing sleeve and oil seal area of the shaft were damaged (Fig. 6.597). The increased axial clearance as a result of advanced wear resulted in contact of the turbine rotor against the bearing housing boss. Subsequently, the bearing cracked in the oil passage area, and caused the shaft to fracture at the inertia (friction) weld. Axial bearing damage resulted during run-in and acceptance tests on the dynamometer, during which the engine had not yet been equipped to match production engines. In its production configuration, actively controlled check valves were provided, to ensure adequate oil supply even during engine start.

6.8.2.3 Carbon deposit formation

Carbon deposits on hot components harden over time, forming so-called varnish (Fig. 6.598 and 6.599).

- Oil varnish formation in the axial bearing oil pocket.
- Oil varnish formation in the oil groove/oil supply area of a turbine-side axial bearing. In an advanced state, heavy carbon deposits

Fig. 6.597 Turbine rotor: fractured shaft in inertia-welded area. The turbine rotor scoring resulted from contact with the bearing housing boss.

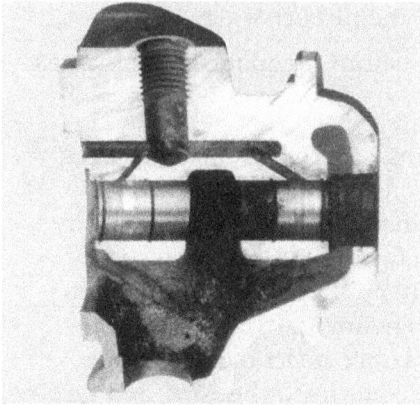

Fig. 6.598 Carbon deposits in bearing housing. Source: BorgWarner

Fig. 6.599 Oil varnish in axial (thrust) bearing.

would be visible on the back of the turbine rotor and turbine-side oil seal rings.

Possible causes:

- Unsuitable oil
- Oil sludge or over-aged oil

6.8.2.4 Oil leaks

Oil leaks may occur at various locations, and for various reasons (Fig. 6.600):

- Oil leaks on the compressor or turbine side due to
    - Incorrect oil seal ring installation
    - Turbocharger operation in zero boost pressure regime
    - Excessive crankcase pressure
    - Oil drain clogged or impaired
    - Excessive oil level in oilpan, or
    - Seal bushings damaged or worn out
- Oil leaks between compressor housing backplate and bearing housing due to
    - Loose compressor housing backplate, or
    - Defective O-ring seal between backplate and bearing housing
    - Oil leaks at oil supply flange or drain flange due to damaged gaskets

6.8.3 Turbocharger housing leaks

Cracks due to thermal stresses, as a result of

- Material defects

- Inadequate coolant supply, or
- Contaminated cooling system (in the case of water-cooled housings)

6.8.4 Turbocharger operation in zero pressure regime

Under certain engine operating conditions, it is possible for oil vapor to enter the compressor housing by way of the compressor-side bearing seal rings. This may occur if the operating point in the compressor map lies below the zero-pressure line; i.e., the engine is operated for an extended period in the lower part of the breakaway curve (Fig. 6.601).

Fig. 6.600 Heat shield with carbon deposits.

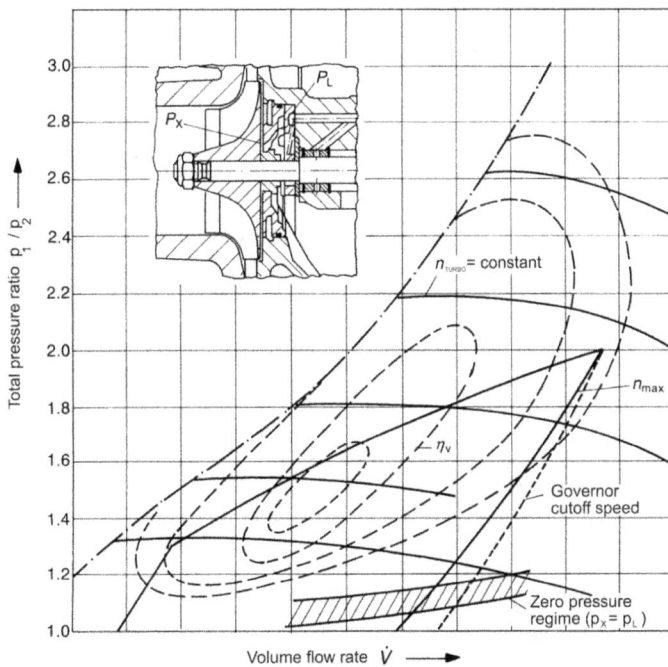

Fig. 6.601 Compressor map showing turbocharger zero-pressure regime. Source: BorgWarner

Below this zero-pressure line, oil pressure on the compressor side bearing oil seals is higher than the pressure behind the compressor impeller. This pressure difference may cause oil vapor from the bearing housing to be forced past the oil seals and into the compressor, forming a film in the compressor housing and in the air ducts. Experience has shown that compared to overall oil consumption, the added oil consumption at this operating point is negligible.

6.8.5 Noise complaints

The typical sound of a turbocharged engine is caused by the turbocharger's rotational frequency. These flow noises are more or less audible, depending on noise damping and duct insulation. These sounds are more pronounced during gear shifts (load changes). If complaints are lodged against a particular turbocharger but no faults or defects are detected upon examination (dynamic balance, compressor and turbine clearances, axial and radial clearances, rotating section out-of-roundness), the cause is often a defective seal or gasket in an air duct or exhaust pipe.

Particular attention should be given to connections between

- Compressor outlet and air plenum/intercooler
- Exhaust manifold and engine
- Exhaust manifold and turbine housing inlet flange, as well as
- Turbine housing, outlet flange, and exhaust pipe

Chapter 7
Preventing Combustion Engine Damage

7.1 Preliminary remarks

If one attempts to obtain more precise statistics regarding the types and causes of engine damage in *road vehicles* (motorcycles, passenger cars, trucks), one will generally be disappointed. The evaluations compiled by automobile clubs and their roadside assistance services are, by their very nature, non-specific. The conclusions derived within this framwork have already been addressed in Section 6.1. By contrast, the most interesting, detailed compilations and overviews are, if anything, only maintained internally by manufacturers and are generally not available to the public. However, in addition to updating operating handbooks and repair manuals, such statistics mainly serve the continuing engine development process, in order to make these engines as fault and damage resistant as possible. Added to this, the fiscal extent of road vehicle engine damage or failure is, by industrial standards, relatively minor. In a worst case, if repair is technically impossible or uneconomical, a replacement engine will be required. Accordingly, covering road vehicle engine damage is not a topic of interest for insurers. However, the situation is quite different for large vehicles, heavy equipment, and transport (e.g., in open-pit mining service), for agricultural and construction equipment, large engines in rail or marine service, stationary installations, etc. In addition to very high repair or replacement costs, engine failures in these fields can incur immense costs. These costs arise from mere damage to total loss of the vehicle (e.g., ship sinking), loss of life, personal injury, loss or spoilage of cargo, loss of contracts, and many others. Extent and causes of engine failures in these fields are therefore primarily collected and evaluated by insurance firms.

If one seeks recommendations and guidelines for preventing engine damage, these will most likely be available from property insurance firms. To minimize their risks, these have a vital interest in preventing engine damage from happening in the first place.

The following statistics are primarily based on the work of Kalina and Muffert[1], on behalf of one of Europe's largest insurance firms, with worldwide operations. They have evaluated damage reports generated over an extended timeframe [7-1]. Their emphasis is on diesel engines, including industrial applications. Nonetheless, insights and conclusions gained from these failure statistics may, to a large degree, be extended to Otto-cycle and diesel engines for road vehicles. This applies at least insofar as these do not share the service, layout, or design-specific peculiarities of the diesel engine, in particular stationary applications. If one keeps this limitation in mind, the material offered is of undiminished importance and value.

7.2 Introduction

Diesel engines used as prime movers for a great variety of purposes are to be found all over the world. This results mainly from their reliability, robustness, and favorable thermodynamic efficiency.

Diesel engines operate in two-stroke or four-stroke mode and can, in the case of both modes, be of the trunk piston type or the crosshead type. In general, however, four-stroke engines are of the trunk piston type and large two-stroke engines of the crosshead type. Diesel engines are designed preferably with vertical cylinders. Because of better economy of space, the horizontal type is used in road-bound or track-bound vehicles as a so-called underfloor engine. However, the disadvantage of this type is that it requires a larger surface area and develops increased one-sided wear. The following advice and examples of instances of damage are based on the knowledge and experience the insurer gained through numerous cases of damage. This knowledge is passed on, above all, with the intention of preventing or correcting those faults and defects that recur frequently and lead, time and again, to damage and breakdowns.

---

[1] *Klaus Kalina*, Dipl.-Ing. [M.Eng.], formerly Senior Engineer, Allianz Insurance Co., Engineering Insurance, Claims Dept., Berlin. *Karl-Heinz Muffert*, Dipl.-Ing. [M. Eng.], formerly Senior Engineer, Allianz Insurance Co., Head Office, Engineering Insurance, Claims Dept., Munich.

Practical experience shows that a high percentage of instances of damage is attributable to factors which, without any doubt, can be controlled. Technical shortcomings of the design, defects of fabrication and incorrect assembly, unsuitable materials, as well as operating errors or insufficient maintenance are often the cause of extensive and costly damage.

7.3 Loss statistics

The statistical results and brief explanations convey a general idea of the frequency and typical areas where damage occurs. Examples of instances of damage and information on prevention of damage will be dealt with subsequently. The evaluations were based on 676 instances of damage, involving diesel and gas engines, and which occurred in the course of 7 years. Only such cases were taken into account where both the cause and the course of events had clearly been established, that is to say damaging events whose cause was "unknown" were not taken into account.

Table 7.1 shows a breakdown of the investigated causes of damage, according to the area/manner of application of the internal combustion engines. When considering the 351 cases where diesel engines used with earth-moving machinery were damaged, it must be taken into account that the engines of such equipment are much more likely to be insured than, for example, those of water-craft or track vehicles and that, therefore, the frequency of damaging events in the different areas of applications cannot be compared directly.

Insured construction site vehicles are mainly dump trucks, truck mixers, and mobile concrete pumps [7-2].

Table 7.1
Breakdown of Damage to Internal Combustion Engines
According to the Area of Application

Area of application	Number	%
Earth-moving machinery	351	51.9
Current generation	95	14.0
Water-craft (ship propulsion)	67	9.9
Construction site vehicles	54	8.0
Fork lift trucks	45	6.6
Track vehicles	32	4.8
Compressors, pumps	18	2.7
Other applications	14	2.1
	676	100.0

Table 7.2 shows a compilation of cases of damage to internal combustion engines according to the components mainly involved. It must be noted that in almost all cases covered in this analysis, several components were involved, and that these components are not only those where the initial damage occurred. Table 7.2 shows that in 50% of all cases of damage the driving parts (pistons, connecting rods, crankshaft) and the cylinder liners were damaged.

Table 7.2
Main Areas Where Damage Occurred

Location of damage	Number	%
Piston and connecting rod	316	24.1
Crankshaft	217	16.5
Cylinder liners	211	16.1
Bearings	195	14.9
Housing	126	9.6
Cylinder heads	100	7.6
Timing gears	76	5.8
Other locations	71	5.4
	1312	100.0

Crankshaft housings are often damaged as a result of the seizing of pistons, for example when a broken connecting rod or other knocked-off parts penetrate the wall of the housing. Table 7.3 outlines the primary causes of damage. The table shows which primary causes are classified under product faults, operational faults, and external influences. Errors of design constitute the largest proportion of product faults, and are followed by those of incorrect assembly and defective materials. Although the extremely large proportion of incorrect maintenance and operating errors, classified under the general term of operational faults, has slightly decreased over the last few years [7-2], its value of almost 66% is still very high. From the insurer's point of view, these facts permit the conclusion that the level of knowledge or attention of the operating and the maintenance personnel to this could still be improved. The attendant circumstances that, in the end, caused the damage are listed in Table 7.4.

In nearly one-third of all cases of damage, lubricant shortage was the secondary cause of damage, i.e., the consequence of a primary cause of damage. Taken together, lubricant shortage and cooling water shortage represent a proportion of causes of damage exceeding 50%.

Table 7.3
Primary Causes of Damage

Primary causes of damage	Number	%
Product faults:		
Design errors	39	5.8
Incorrect assembly	37	5.6
Faulty material	28	4.1
Incorrect repairing/overhauling	22	3.2
Defect of fabrication	20	2.9
	146	21,6
Operational faults:		
Incorrect maintenance	270	39.9
Operating error	175	25.9
	445	65.8
External influences:		
Foreign bodies	44	6.5
Sabotage or other	41	6.1
influences	85	12.6

Table 7.4
Attendant Circumstances During the Occurrence of the Damaging Event

Secondary causes of damage	Frequency [%]
Lubricant shortage ('starvation')	31.6
Cooling water shortage	26.1
Ingress of water	13.2
Loosening	6.3
Oil drain facility damaged	5.6
Damage to combustion air inlet	4.4
Frost	3.5
Other causes	9.3

Incorrect or improper maintenance is shown in Table 7.3 as representing a proportion of almost 40% of the causes of damage that occurred through mainly irregular checking of oil or cooling water levels, and shows at the same time that the causes of excessive oil consumption or loss of cooling water are not always investigated.

7.4 Advice for the prevention of damage by product faults

7.4.1 Planning and design

At the planning stage, basic faults and damage arising out of them can be prevented by ensuring that the conception meets the requirements for future use [7-3], [7-4]. Depending on the area of application

- Use in road vehicle/water-craft or rail traction
- Use in stationary drive units (generator, pumps, etc.)
- Use in intermittent drive units (construction machinery, drilling rigs, etc.)

Existing main and secondary conditions must be taken into account, and an optimization study for the engine design must be carried out.

Determining conditions, with information on any problems and influences, are listed below:

- Space requirement (overall length and height, accessibility)
- Mass (sturdiness)
- Type (trunk piston type or crosshead type; the use of heavy fuel oil on trunk piston engines is to be taken into account when choosing a lubricating oil and considering lubricating oil maintenance care)
- Kind of duty (short-time duty or continuous duty, peak load)
- Use (availability, readiness, overload capacity)
- Operating mode (two-stroke mode or four-stroke mode with and without supercharging; the two-stroke engine is subjected to greater thermal stress and, therefore, requires a more expensive cooling system)
- Governor type and operation (different controlled quantities, e.g., marine engines—rotational speed; construction machinery—torque; generators—constant rotational speed, degree of irregularity)
- Cooling mode (water cooling or air cooling)
- Environmental conditions (climate: condensation water, danger of frost; altitude: barometric pressure, supercharging; location: dust filter, air washing units, protection against corrosion)
- Maintenance (personnel, automation, redundancy, repairability, service)
- Environmental protection (noise prevention and control of exhaust emission)

Experience gained from numerous cases of damage shows that certain weak points in the construction lead time and again to destruction. Cylinder heads and pistons are machine parts that are subjected to great dynamic and thermal stress. That is why the designer must ensure that local material accumulation is kept to a minimum, for otherwise the consequences of insufficient cooling and uncontrollable material expansion would be detrimental [7-5]. Particular care must be taken at the design stage to achieve favorable flow conditions in the cooling chambers where shortcomings may result in cavitation and damage to cooled machine parts.

Time and again, abrupt changes of sectional area of both fixed and moving parts cause damage. Rounding radii are to be in accordance with DIN 250 [7-6]. A further preventive measure is relieving, which is also standardized, of the risk areas or the use of tapered transition pieces. Another important point for the designer to take into account is accessibility to all areas requiring maintenance. If there is no access, certain maintenance tasks are likely to be omitted.

Examples of instances of damage
Diesel engine used for current generation
A diesel engine (370 kW at 1500 rpm) was installed at the top station of a funicular at an altitude of over 2500 m. The engine was meant to supply electric power to the unfinished station, including a mountain hotel, during the construction period and subsequently, following the connection to the public mains supply, be used as an emergency power-generation set. After no more than 570 hours of operation, the engine was severely damaged in winter by the seizing of a piston, causing extensive consequential damage. Four pistons as well as the associated connecting rods were completely destroyed.

The engine housing was ruptured in several places and needed replacing while the crankshaft was re-used after eliminating the scores by grinding and, as a precaution, checking for cracks. The investigation into the cause of the damage revealed the following facts: In view of the very low ambient temperatures frequently prevailing at the top station, the manufacturer had equipped the engine with electric preheaters for the cooling water and the lubricating oil. However, because during the construction period, the engine was the only unit available for the generation of current, and no other source of electric supply was yet available for the preheaters, there was no preheating when the engine was out of operation. On starting the engine after a long shut-down period, both the cooling water and the lubricating oil had cooled down almost to outdoor temperatures. According to the user, the engine was started with the (charge-air) compressor intercooler frozen up. Owing to a grave error at the planning stage, the need for engine preheating by means of another source of energy, e.g., an electric battery or an oil heater, was overlooked; 24,000 Euros were required to repair the damage.

Diesel Engine Used for the Propulsion of a Ship
On a 250-kW diesel engine used for the propulsion of a ship on inland waterways, the hexagon head cap screws securing the camshaft became loose. The camshaft stopped abruptly, leaving some valves in the open position so that all pistons were damaged. As a result, the cylinder liners as well as the main bearings and connecting rod bearings were damaged. The investigation into the damage revealed that the design did not include screw locking devices; the repair cost amounted to approximately 13,000 Euros.

Aircraft Piston Engine
After approximately 400 hours of operation, a connecting rod of an aircraft piston engine broke, causing extensive damage to the piston, cylinder liner, and housing.

The examination of the connecting rod in the Allianz Centre for Technology revealed that the connecting rod had broken, because of dynamic stress, where the sectional area of the material was smallest, with the fracture starting from chromium deposits in the connecting rod bore. The fracture through vibration started from cracks in the built-up chromium deposit. The chromium deposits (cold weldings) were presumably caused by relative movement between the bearing shell and the connecting rod, resulting from insufficient rigidity.

7.4.2 Fabrication and assembly

One must, on no account, go below the radii of curvature specified in the drawings where there is a change in cross-section. Particular care is required regarding surface finish where there is a change in cross-section area, as any tool marks, scratches, grooves, or indentations cause local stress peaks leading to incipient cracks and fatigue fractures. Incorrect heat treatment, including hardening, also causes stress in the workpiece and may result in cracks or fractures. This fault is mainly experienced on crankshafts.

As for cast parts, it is essential that molding sand remainders are removed. This applies particularly to parts containing hollow spaces, such as cylinder heads. These remainders may cause damage by blocking pipes or impeding heat transmission. Furthermore, in such a case, increased wear and thus reduced service life is to be expected. Before restarting engines damaged in this way, cooling water circuits or oil lines, after draining, are to be thoroughly cleaned. The engine is to be refilled with the required amount of new or cleaned fluids.

Engines vibrate during operation, which may cause certain parts to break. This is the case, for example, with tubes and linkages that are secured to the housing. Therefore, tube mountings must be designed with great care and checked regularly.

Faults in the material, such as inclusions, impurities, and forging faults, can be detected by non-destructive test methods, i.e., ultrasonic, radiation, dye penetrant, or magnetic particle testing, before the parts are further processed or at least before they are fitted. Such faults are mainly found in parts of the engine casing, crankshafts, connecting rods, rockers, and valves.

During assembly, often individual components are omitted or not fitted in the correct position. Intensive training of the mechanics as well as detailed assembly instructions, including check lists, help to prevent such errors.

Necked-down bolts (expansion bolts with reduced shanks according to DIN 2510 [7-7]) as well as other essential screws must be tightened by means of a correctly set torque wrench. The required torque must be shown clearly in the manufacturer's assembly instructions.

A further cause of damage is that securing devices for screws, nuts, etc. are used several times. In particular, split pins, tab or tongued washers, or similar components must only be used once. After dismantling, these parts must always be replaced by new ones.

Time and again, foreign bodies in the engine (welding beads [globules], cleaning wool remainders, cloths, nuts, and the like) cause damage. Before shutting off hollow spaces (e.g., crankcase, cylinder heads, pipes) it is absolutely essential to carry out a visual check for foreign bodies. In inaccessible areas, an endoscope (borescope) may be used. Furthermore, flushing oil has been used successfully to clean interiors of engines and remove such foreign bodies.

Example of a damage incident: On a 200-kW diesel engine used for driving an earth-moving machine, a loud noise suddenly developed after approximately 5000 hours of operation. During the following investigation, a crack was found in one cylinder liner in the area of the bottom sealing ring groove. This described an arc of about 270° and then, on one side, sloped down to the rim. Four clearly distinguishable traces of rubbing were found on the liner, each offset by 90°, though meeting in the area of the crack. After breaking open the crack, it was established that it was longer on the inside of the liner. This means that the crack, caused by vibration, traveled from inside to outside. Diametrically opposite the area of the start of the crack, was an incipient crack in the groove of the collar of the cylinder liner. The edges of the circular crack in the liner caused damage to the piston and detrimental touching. From the findings, it can be concluded that the cylinder liner was distorted by improper fitting. The increase in stress at the bearing points exceeded the fatigue strength of the material and caused the crack.

7.5 Advice for loss prevention by operational faults

To ensure trouble-free operation and prevent damage, adequate training of the operating personnel is imperative. In this connection, unusual operating conditions should be simulated. The operating instructions should be handed to the operating personnel.

To avoid fluid-hammering, diesel engines must be cranked with the indicator cocks open, especially after long shutdown periods. In the case of smaller engines, the decompression cocks are to be opened for this purpose or, if there are no decompression cocks, the injection valves must be removed. After the start, the load of the engine should be increased gradually, if possible, to achieve a uniform temperature rise in the driving parts and the engine.

During operation, pressures and temperatures must be checked regularly. Hot engines should not be started by using compressed air because of the risk of cracks forming as a result of thermal shock.

The power output, the combustion, and thus the operating condition can be monitored [7-8] by means of compression and firing pressures, inlet gas temperatures, performance curves, and exhaust pressure and gas analyses. Comparative data facilitates the detection of imminent defects or damage.

However, even without extensive technical resources, trained personnel can recognize defects at an early stage:

- Knocking noise is caused by advanced or delayed ignition, excessive bearing clearance, dry running, and seizing pistons. In the case of the single-acting two-stroke engine, excessive bearing clearance does not cause knocking as there is no change in pressure in the bearing shell.

- Exhausts emitting dense smoke indicate incomplete combustion (CO formation among other things) and may cause fire in the exhaust gas duct.

- In a supercharged plant, an exhaust gas temperature increase is attributable to dirt accumulation on the supercharger intercooler. Apart from that, an increase in exhaust gas temperature indicates that the injection parts, exhaust valves, and control systems are defective.

Example of a damage incident: After no more than 200 hours of operation, the crankshaft of a diesel engine of a ship used on inland waterways broke aft of the second main bearing. Furthermore, several crankshaft bearing seals had either superficial cracks or were broken. An examination of the crankshaft revealed that bending stress had caused the fracture of the crankshaft through vibration. As there was no indication of any manufacturing defects, it must be assumed that external forces, acting on the crankcase that was bolted to the ship's bottom, were the cause of the crack.

Particular features of the operation of emergency power-generating sets

There are a few points in which the mode of operation of emergency sets is basically different from that of other engines:

- Long shut-down periods between test runs
- Test run-up with immediate power output
- Short running periods
- No cooling water preheater on older sets

It is, therefore, absolutely necessary to ensure that the design of new emergency sets includes both cooling water and lubricating oil preheaters. This applies in particular to open-air sets or sets installed in unheated areas. If possible, already existing sets should be retrofitted with preheaters. Emergency sets in particular fail time and again, as the fast run-up with immediate power output leads to a closure of clearance between piston and cylinder liner and consequently to the pistons seizing. Experience has shown in many cases that running the emergency set slowly to working temperature at the beginning of the weekly test runs and then stopping the engine before starting it for the actual test run itself, is worthwhile.

7.6 Engine cooling

The cooling system covers the pistons, cylinder liners, cylinder head cover, exhaust valves, circulating oil, intake air, and supercharging equipment. On larger engines, the exhaust stack and exhaust pipes as well as silencers are also cooled.

For liquid cooling, fresh water or, less frequently, sea water is used. Fixed engine installations on dry land are mainly cooled with fresh water. If suitable cooling water is not available in sufficient quantities, a water re-circulation system with cooling facilities is used. This system is efficient, as the only water loss to be made up is that caused by leakage and evaporation. If sea water is used as coolant, there is a risk of salts and minerals being deposited, so cooling water temperatures must be limited to approximately 45°C. If fresh water is used, the risk is much smaller so that operation at higher temperatures is possible.

The operation at higher cooling water temperatures is advantageous from the point of view of economy and loss prevention:

- Better economy because of lower fuel consumption.
- Prevention of low-temperature corrosion as the temperatures are above the dew-point.
- Injected fuel does not condense on the walls of the combustion chamber. By maintaining the lubricating film on cylinder liners and piston rings, wear is reduced.
- Warmer engines start more easily—their operation is therefore more reliable.

The influence cooling has on an engine's availability is often underestimated. Trouble in the cooling system can cause extensive damage through overheating. Liquid-cooled engines require particular care regarding the use of properly treated cooling water.

7.6.1 Information on cooling water treatment

If insufficiently treated cooling water is used in an engine, during use, and depending on the water hardness, furring will occur, which impairs the transmission of heat. The resulting overheating often causes damage to cylinder heads (cracked webs), cylinder liners, and pistons. Furthermore, if the cooling water is too hard (above 10° dH ≡ 8 grain

CaCO$_3$/gal)[2], the slushing oil may separate, become ineffective, and form a coating with heat insulating properties on the cooling surfaces, resulting in a further reduction of heat transmission. Fresh water should meet the following specification:

Hardness: 4 to 10° dH
pH-value: 7 to 9
Chloride content: ≤ 150 mg/L

Several simple devices for measuring these values are on the market (tablets for measuring hardness, and indicator paper for measuring the pH-value).

The above-mentioned values are average values. As specifications from various machine manufacturers differ, it is their specifications that are applicable. A water analysis can be obtained from the appropriate water authority. Water which is not hard enough, such as rain water or condensate, is unsuitable as it favors corrosion.

Water that is too hard must be treated before use. Calcined trisodium phosphate is a well-tried agent for this treatment, whereby 4 g added to 100 L of cooling water leads to a decrease of 10 dH, and the addition of 40 to 50 g to 100 L of cooling water/slushing oil emulsion results in an increase of 1 pH unit.

The slushing oil content during the initial filling, as well as after cleaning the cooling oil system, should be approximately 1.5% by volume (i.e., 15 cm^3/L). It will drop to approximately 0.5 to 0.6% by volume after wetting the surface of the walls. The oil content can be checked any time by means of an emulsion tester obtainable from special shops.

Slushing oil is added to the cooling water before it is introduced into the engine. An adequately clean container should be used with the oil added slowly to the water, the latter being stirred continuously. Not vice versa. The water temperature should be between 20°C and 70°C.

If antifreeze is used, the manufacturer's specifications must be strictly adhered to. In temperate zones, as a rule, frost must be expected from the beginning of October until the end of April. In order to clean dirty

---

[2] 1° dH (German degree of water hardness) is equivalent to 0.8 English degrees; 1 degree English is equivalent to 1 part CaCO$_3$ in 70,000 parts H$_2$O or 1 grain CaCO$_3$ (≡ 0.0648 g) CaCO, in 1 English gallon (= 4.543 L); for American degree of water hardness the American gallon (≡ 3.785 L) has to be used.

cooling chambers, the engine must be run until the cooling water has reached its working temperature and all loose scale particles have been stirred up. After stopping the engine, the cooling water must be drained off quickly. The cooling system is then filled with a mixture of 10% naphtha (kerosene) and 90% hot soda solution (60 g of soda in 1 liter of water). After running the engine for about 30 min, the liquid must be drained off. The cooling system should be flushed several times with clean water. After filling with cooling water as specified, the engine will be operative again. If proprietary solvents are to be used for cleaning the cooling system, the directions for use must be followed to the last detail. If the outer surface of the honeycombs and tubes of surface radiators are dirty, heat transmission is impaired. Depending on the degree of dirt accumulation, radiators are to be cleaned with a brush and water and then blown through with compressed air. The radiators cleaned in this way must be submitted to a leak test. The predominant causes of damage to air-cooled diesel engines are

- Dirt accumulation on the outside cooling surfaces
- Lack of cooling air

7.6.2 Cooling water shortage

Experience shows that the causes of damage through cooling water shortage are largely attributable to the carelessness of operating personnel in not checking the cooling water level or the condition of the cooling system. For example, damage to engines through insufficient cooling can be caused if the operating personnel does not ensure, before starting and at regular intervals during operation, that the cooling water level is sufficient, and the general condition of the cooling system is satisfactory. Unfortunately, there are still many engines in operation that are not equipped with cooling water-level monitoring facilities.

7.6.3 Examples of damage incidents

Diesel Engine Emergency Power-Generating Set

After a diesel engine emergency set had been in operation for several years, a cooling water pipe connected to the underside of the engine broke. The drainage of the cooling water went unnoticed. The cooling water temperature monitor was located in a connection piece that had become empty of water, and no cooling water level monitor was fitted. The ensuing overheating of the engine caused seizing of a piston, which extensively damaged the engine. An examination of one of the cylinder heads, which were also damaged, revealed a crack between the valve ports. The cost of repairing the engine was EUR 65,000.

Diesel engine damaged by corrosion

On inspecting a 410-kW diesel engine, the cooling-water side of the cylinder liners was found to have coatings of approximately 2 mm thickness. An analysis of these coatings revealed that they consisted mainly of corrosion products. Cooling water residues were also found albeit in small proportions. An inquiry revealed that the engine had been run with untreated tap water.

7.7 **Engine lubrication** [7-9], [7-10]

For reliable operation, only stable oils whose firing (burning) point (ignition temperature) varies little with the operating conditions, should be used. They offer the following advantages over instable oils:

- They impede crank case explosions in trunk piston engines in case of a blow-by of gas.
- They do not participate in the combustion as easily, preventing dry running of the cylinders and thus breakdowns.

As lubricating oils have higher coke (carbon residue) and ash contents than good fuels, the amount of oil that is burnt should be as small as possible. This means that you should not lubricate more than absolutely necessary! Contamination of the oil by abraded metal particles (fine debris), oil coke, ash, rust, water, and air leads to premature aging of the oil, i.e., the deterioration of its lubricating properties, and consequently to damage:

- Metals in combination with atmospheric oxygen make the oil acid, cause corrosion, and favor the formation of lacquer-like deposits. Copper has a particularly strong effect and should not be used anywhere in the oil circuit.
- The combination of carbon black, lubricating oil resins, and oil coke leads to the formation of lacquer-like deposits at higher temperatures. As a result, piston rings stick, and the higher surface pressure produced causes increased wear. If this leads to the piston rings breaking, the gastight seal could be lost with the risk of crankcase explosions.
- Contamination through abraded metal particles, oil coke, water, etc. leads to the formation of oil sludge and may cause constrictions or obstructions of oil pipes.

To avoid such damage, the use of heavy-duty (HD) circulating oils and "single phase" cylinder oils as well as the continuous mechanical purification by means of strainers, filters, separators, etc., is recommended.

These measures are facilitated by sufficiently large oil systems with a low number of oil circulation ratios. Clever design can increase the dwell time period of the oil in the tank of the circulation system.

Usual circulating oil quantities in L/kWh:

	with	without
		cooling of pistons	
High-speed engines	35 to 40	30 to 35
Medium-speed engines	30 to 35	15 to 20
Large engines		10 to 15
Number of circulation cycles per hour ("circulation ratio"):		
	gearbox medium sized diesel engines large diesel engines high-speed engines	1.5 4 to 6 8 to 10 ~ 25

These are reference values. However, they should not be exceeded because for larger circulation ratios the renewal of the oil is required at shorter intervals.

To obtain an indication of the effectiveness of the lubrication and prevent damage, the pressure, temperature, oil level, and the degree of contamination are to be monitored continuously.

- Decreasing oil pressure, with the pump output remaining constant, is an indication of oil dilution or excessive bearing clearance.

- Increasing pressure before and decreasing pressure after the filter, or increasing differential pressure, are indications of filter choking.

- In order to prevent cooling water from entering the oil circuit, the oil pressure should be higher than the water pressure.

- At regular intervals, the filters should be cleaned or exchanged, and oil samples taken and analyzed.

7.8 Engine fuel

To ensure economical and trouble-free operation, only fuels containing the smallest possible amount of undesirable admixtures such as water, ash, sulfur, and coke should be used. In particular, ash and coke cause increased wear and reduce the service life of the pistons, piston rings, cylinder liners, and valves. While the standard qualities of diesel oil available remain largely the same, there are in general no precise quality specifications for heavy fuel oils. The composition and thus properties of heavy oils depend on their particular treatment and may, therefore, be quite different. While the sulfur content of diesel oil may be a maximum of 1%, the amount of sulfur contained in heavy fuel oils may be as much as 5%. Experience shows that the wear of cylinder liners and cylinders naturally depends on the ash content, purification by means of filters, settling tanks, separators, etc. The rate of piston and cylinder liner wear, when heavy oil is used, is four to five times higher than with diesel operation.

High sulfur contents have a corrosive effect because of the formation of sulfuric acid (H_2SO_4). Even more unpleasant is the fact that the presence of sulfur causes the precipitation of asphaltenes (I.P. spirit insolubles), and the formation of coke (carbonaceous deposits) and ash. The amounts of these undesirable substances increase with the sulfur content. A high asphalt content in heavy oil causes a slow, incomplete combustion and favors the carbonization (graphite formation) of the nozzles, piston rings, exhaust valves, and ports. The precipitation of asphalt is increased by low engine temperatures.

The unfavorable properties of heavy oils can be counteracted by chemical additives or the use of intermediates. Information on this is available from the oil industry.

To summarize, use of heavy oils, in comparison with diesel oils, requires more extensive supply and treatment equipment and makes greater demands on the personnel.

7.9 Combustion air

The reliability and service life of an engine depend to a large extent on the cleanliness of the combustion air supplied to the engine. That is why the air supplied must be as free of dust as possible. The whole combustion system, including associated pipes, flexible tubes, and expansion bellows, must therefore be checked carefully at regular intervals. The maintenance instructions from the engine manufacturers contain information on this. However, it must be taken into account that, depending on the conditions of operation, the maintenance intervals may have to be shortened considerably. When checking the air system, all screwed connections as well as pipe clamps are to be checked for tightness. Damaged parts must be exchanged. The combustion air is cleaned by means of dry-type filters or oil bath air cleaners. Both filter systems can also be combined and, if required, preceded by a preliminary filter for crude dirt extraction.

When dry-type filters are used, the filter cartridge must be removed during maintenance and carefully cleaned by tapping. At the same time, the cartridges are to be checked for tears. Furthermore, all parts of the filter housing are to be cleaned externally and internally.

When oil bath air cleaners are used, the oil must be changed at or before the end of the specified period. Only fresh motor oil must be used for filling purposes, never used oil.

Example of a damage incident: During a collision, the driver of a wheel loader had the oil bath air cleaner damaged so that unfiltered air was supplied to the engine. After a short period of further operation, the pistons were damaged. The repair cost amounted to approximately EUR 7000.

7.10 Maintenance and inspection

7.10.1 Maintenance

The maintenance instructions from the manufacturers should be followed to ensure trouble-free operation. These checking and overhauling tasks are carried out by rotation, depending on the hours of running of the engine, thus ensuring reliability of operation, increasing the useful life, and finally, preventing damage. To keep to the maintenance schedule, engines should be provided with service-hour indicators, if possible. A log of maintenance and repair tasks carried out should be kept, showing the history of the engine [7-9]. Unscheduled maintenance tasks must be carried out when the diesel engines have been in heavy-duty operation for a longer period (e.g., in dusty surroundings, at very high or very low temperatures).

The Allianz Information Sheet No. 7 [7-9] contains information on trouble-free operation of diesel engines.

7.10.2 Inspection

In general, inspections are only carried out on larger engines, and involve the dismantling of the engine to such an extent that all components are accessible for close examination. Before dismantling, however, it is advisable to check the connecting rod bearing and main bearing clearance, and to check the crankcase bottom for foreign bodies.

Crankshaft: Crankweb breathing is to be measured in forward and reverse directions. In the case of marine engines, it must be ensured that the ship is without cargo and its keel on even level. Any remaining loads and weights must be taken into account. Crank pins and journals must be checked for run-out and cracks.

Checking cracks is only possible by means of a dye penetrant test. If cracks are detected, their depth must be measured or established by grinding out. Depending on the strength of the remaining shaft diameter, the shaft is to be reused, repaired (e.g., sprayed), or replaced.

Main and connecting rod bearings: In the case of plain bearings, their bearing surface must be checked for abnormal appearance and seizing marks. If the bearing clearance is excessive and the surface appearance unsatisfactory, the bearings should be replaced.

Roller bearings should be replaced, as a precaution, without any further examination. If any bearings have been replaced, an oil pressure test should be carried out after the inspection, with the engine hatch doors open. This test will show whether or not the lubricating oil supply to the bearings and pistons is satisfactory.

Connecting rod and piston rod: In areas of small cross-sections and abrupt changes of cross-section, a check for cracks should be carried out.

Piston: The piston diameter should be measured at three pre-determined points in the transverse and the longitudinal direction, and compared with specified values. Slight seizing marks on the piston skirt can be removed by polishing them out. As a precaution, the piston rings are to be exchanged without further examination. In principal, the reworking of worn-out piston ring grooves is possible. However, the expenditure involved must be considered. Highly stressed parts, such as piston crowns and piston pin bosses, should be checked carefully for cracks.

Cylinder liner: The liner diameter should be measured at three predetermined points in the transverse and the longitudinal direction and compared with specified values. A check for cracks by means of a dye penetrant test is advisable. The water side of the liner must be checked for deposits, corrosion, and cavitation. If pistons or cylinder liners are replaced, the cylinders of the engine must be checked for constancy of temperature on the resumption of operation.

Cylinder cover: The underside of the cylinder cover and the webs must be checked for cracks. Any scale deposits can be removed chemically. The valve seats in the cylinder head are to be reworked, and the valves, depending on the degree of wear, refaced or replaced.

Fuel pump and injection valves: Without any further examination, these parts should be reconditioned by the manufacturer.

Frame: Cooling water chambers are to be checked for deposits, corrosion, and cracks, the latter particularly at the liner guide and at sealing points. Check the tension of the tension rod. If the tension has altered, the crankweb clearance must be rechecked.

Exhaust gas supercharger: Check the housing for cracks and the rotor for deformation. As a preventive measure, the bearings should be replaced. Deposits in the cooling water chambers are to be removed chemically.

Control system: Check gear and chain wheels for incipient cracks and pressure marks. Camshafts and cams as well as tappet rollers, sliding tappets, and tappet push rods must be checked for wear.

Rocker arms should be checked for deformation caused by bending and valve springs should be checked for fatigue. As a precaution, needle bearings should be replaced. Parts of the governor are to be checked for wear, deformation by bending, and breaks.

The play in the control devices is to be adjusted according to the manufacturer's instructions.

Cooling water and lubricating oil pumps: The rotor and housing must be checked for breakouts (chipping, flaking, etc.) and cavitation. Bearings, scaling devices, and other parts subject to wear are to be replaced as a preventive measure.

Oil cooler, fresh water cooler, and supercharger intercooler: Tubular coolers or plate coolers are to be checked for tightness.

Shut-off devices: Valves and slide valves must be dismantled, cleaned, and examined. Rework valve seats and repack glands. Existing safety devices on discharging and draining devices must be inspected.

Instruments: Safety devices, thermometers, manometers, etc. must be submitted to an operational test and checked for accuracy of indication and, if necessary, adjusted or replaced.

Pipes: Pipes and filters must be checked for erosion, corrosion, and dirt accumulation. The scavenging air duct and the exhaust gas duct are to be checked for dirt accumulation.

Example of a damage incident: The engine of a sea-going vessel broke down after a longer period of sub-load operation. The investigation revealed several piston seizures and damage to two cylinder heads. The engine was completely dismantled in the manufacturing plant, where further damage to other parts was found. Both supercharger intercoolers had about 75% of the cooling surface on their water side covered with foreign bodies such as plastic parts, wooden pieces, metallic pieces, stones, etc. The decrease of the cooling effect had caused the above mentioned extensive damage. The seawater filter, located before the supercharger intercooler, must have been either completely destroyed or missing.

Appendix

Engine output

The output of any engine is obtained from its dimensions (bore, stroke, number of cylinders), engine speed, and number of strokes per working cycle, as well as a measure which indicates its level of technical development: specific work. On the other hand, engine output, in general, is the product of torque and engine speed. This allows us to express a relationship between these general and engine-specific dimensions.

Power (P) is (generally) work (W) over time (t):

$$P = \frac{W}{t}$$

or

$$P = M \cdot \omega$$

(where $\omega = 2 \pi n$, and M is mass)

Accordingly, power output of an individual engine cylinder is calculated as

$$P_{cyl} = \frac{W_e}{t}$$

Engine-related concepts of work and time are applied:

effective (actual) work of a cylinder, W_e

cycle time (duration of cycle), T

Because cycle time is the inverse of engine speed,

$1/T$ = frequency = engine speed n

Output of an individual engine cylinder is therefore

$$P_{Cyl} = W_e \cdot n$$

Now one must consider that depending on the working principle, work is produced with every crankshaft rotation (two-stroke cycle) or with every-other rotation (four-stroke cycle). Instead of engine speed n, we therefore insert the working stroke frequency n_a in the power equation. This working stroke frequency is obtained from the engine speed n and stroke number a:

$n_a = n/a$

$a = 1$ for two-stroke engines

$a = 2$ for four-stroke engines

$$P_{cyl} = W_e \cdot n_a$$

The work available from the cylinder over one working stroke depends on the mass of fuel which can be burned. Fuel mass, in turn, requires a certain mass of air for combustion, which in turn represents a certain piston displacement volume for a given air density. Therefore one obtains the following functional chain, Fig. A 1:

In equation form, the work of a single cylinder, W_e, may be expressed as

$$W_e = \eta_{ideal} \cdot \eta_{eng} \cdot \eta_{mech} \cdot \frac{H_L \cdot \rho_O}{\lambda \cdot L_{sto}} \cdot V_D$$

where

W_e is the effective work of a single cylinder

η_{ideal} is the efficiency of an ideal engine cycle

η_{eng} is the actual engine efficiency

η_{mech} is the engine mechanical efficiency

H_L is the lower heat value of the fuel

ρ_O is the air density upstream of the cylinder

λ is the relative air/fuel ratio (relative to stoichiometric = 1.0)

L_{sto} is the stoichiometric air/fuel ratio

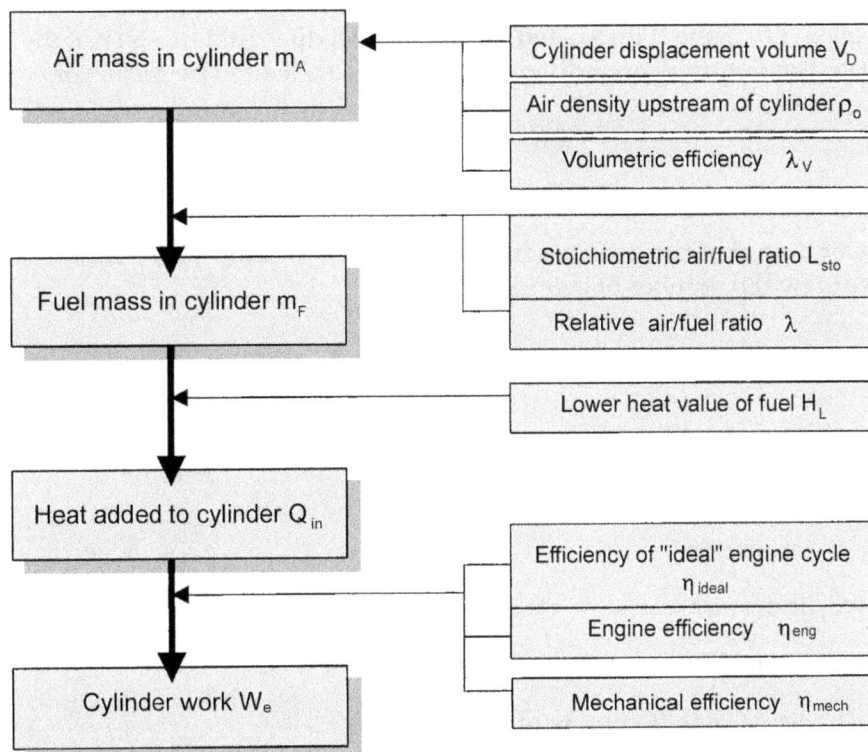

Fig. A 1 Factors affecting power output of a single engine cylinder.

V_D is the piston displacement volume

If this mechanical work is expressed relative to the piston displacement volume V_D, one obtains the specific work w_e.

$$w_e = \frac{W_e}{V_D}$$

Dimensions:

W_e N·m, J, or kJ

w_e kJ/dm³ or J/dm³

Because the expression for specific work contains the various efficiencies which engine designers and test engineers are able to influence by means of the engine conception and design process, modification of combustion and heat transfer processes, reduction of thermodynamic, mechanical, or hydraulic losses, this quantity is well suited for evaluation of an engine's development level. It is also a measure of engine thermal and mechanical loading, and mastery of the same. The output of an individual cylinder is measured by means of specific work:

$$P_{cyl} = w_e \cdot V_D \cdot n_a$$

Power output of the entire engine is obtained by multiplying cylinder output by the number of cylinders:

$$P = z \cdot w_e \cdot V_{cyl} \cdot n_a$$

P engine output

z number of cylinders

d cylinder bore

s cylinder stroke

A_p piston area

V_{cyl} displacement volume of single cylinder

V_{eng} displacement volume of entire engine

w_e specific work

n_a frequency of working strokes

n engine speed

a revolutions per cycle

$a = 1$, two-stroke engine

$a = 2$, four-stroke engine

$$A_p = \frac{\pi}{4} \cdot d^2$$

$$V_{cyl} = \frac{\pi}{4} \cdot d^2 \cdot s$$

$$V_{eng} = z \cdot V_{cyl}$$

Dimensional analysis:

$$P = z \cdot w_e \cdot V_{cyl} \cdot n_a$$

1 kJ/s = 1 kW

1 kJ/dm³ = 1 kWs/dm³

1 min = 60 s

P [kW] = z[-] · V_{cyl} [dm³] · w_e [kJ/dm³] [kW/kJ/s] · n [min⁻¹]· (1/60) · [min/s] · (1/a)

By substitution into the power equation, one obtains

$$P = z \cdot \frac{\pi}{4} \cdot d^2 \cdot s \cdot w_e \cdot \frac{n}{a}$$

Because the piston of a reciprocating engine executes nonuniform motion over the course of a single crankshaft rotation, in which it twice accelerates from rest to maximum speed and then decelerates to a stop, one may use the mean piston speed, which is the product of total piston travel for a single crankshaft revolution (in other words, twice the piston stroke) and engine speed:

$$c_m = 2 \cdot s \cdot n$$

Mean piston speed is just as powerful an indicator as specific work. It is a yardstick for loads on an engine and its components, imposed by inertia forces, mechanical wear, and thermal loading; it is also an index for mastery of gas exchange, and of friction and lubrication processes. Using c_m, the power equation may be expressed as

$$P = \frac{\pi}{8 \cdot a} \cdot z \cdot d^2 \cdot w_e \cdot c_m$$

Therefore engine output is the product of:

- an engine constant, C = [$\pi/(8a)$] · z · d²
- and two variable quantities,
    - specific work w_e and
    - average piston speed c_m

If one equates engine output as the product of engine speed and torque, it becomes apparent how much of the engine's output is due to torque, and how much to speed.

$$P = V_{eng} \cdot w_e \cdot \frac{n}{a} = M \cdot \omega = M \cdot 2 \cdot \pi \cdot n$$

$$w_e = \frac{2 \cdot \pi \cdot a}{V_{eng}} \cdot M$$

Specific work w_e is proportional to torque (divided by engine displacement), and may be regarded as an "engine size independent" expression for torque.

Crank angle—a measure of time and distance

The crank train, more precisely the piston motion, controls the engine's thermodynamic processes. The instantaneous position of the piston is defined by the crank angle—the angle between the crank throw and the cylinder axis (Fig. A 2). During a single crankshaft revolution, the piston travels through its stroke twice. At the extreme upper limit of the crank train travel (top dead center, or TDC), the crank angle is defined as 0°;

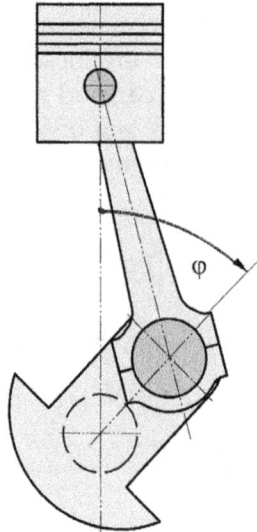

Fig. A 2 Crank angle and piston stroke.

at its lower limit (bottom dead center, BDC), 180°. The distance covered by the piston in the course of a single crankshaft rotation is described by the piston travel equation. However, the relationship between crank angle and piston travel is nonlinear; at a crank angle of 90°, the piston has already covered slightly more than half of its total stroke (Fig. A 3a).

At the moment the piston reverses its motion at top or bottom dead center, piston speed is zero. Then the piston accelerates to its maximum speed, and decelerates back to zero at the opposite dead center point. A mean piston speed is obtained if total piston travel over a single crankshaft rotation is divided by the time required for the piston to execute this motion.

The rotation period is T, its inverse is $1/T$ = frequency = engine speed n. Dividing by rotation period is synonymous with multiplying by speed.

$$c_m = 2 \cdot s \cdot n$$

If one substitutes stroke, in meters, for s and speed in min⁻¹ (rpm), one obtains the following numerical value equation:

$$c_m[m/s] = \frac{s[m] \cdot n[\min^{-1}]}{30}$$

(Numerical value equations indicate the relationships between numerical values expressed in dissimilar units; therefore, the applicable units [dimensions] must be specified.) The time required for the piston to travel a certain distance or crank angle is a function of engine speed (Fig. A 3b). The following relationship holds for engine speed, time, and crank angle:

In one minute, the crank train makes n revolutions; one revolution represents 360 crankshaft degrees, therefore it rotates through

$$\varphi = n \cdot 360°$$

crankshaft degrees. In one second, it covers 1/60 of this, or

$$\varphi = n \cdot 360/60 = n \cdot 6$$

In t seconds, this is

$$\varphi \text{ [crankshaft degrees]} = t[s] \cdot n[\min^{-1}] \cdot 6$$

The time required to cover φ crankshaft degrees is

$$t[s] = \frac{\varphi[\deg\,crankangle]}{n[\min^{-1}] \cdot 6}$$

Time specifications in crankshaft degrees are relative; these indicate how much time is required for a process, such as fuel injection or mixture formation, relative to the working stroke, but they say nothing about the absolute duration of a process. Absolute duration is obtained in conjunction with engine speed.

Example: If combustion takes place over 60° of crankshaft rotation, then at an engine speed of 2000 rpm, this occurs in the space of 0.005 s (5 ms); at 6000 rpm, however, this takes only 0.00167 s (1.67 ms).

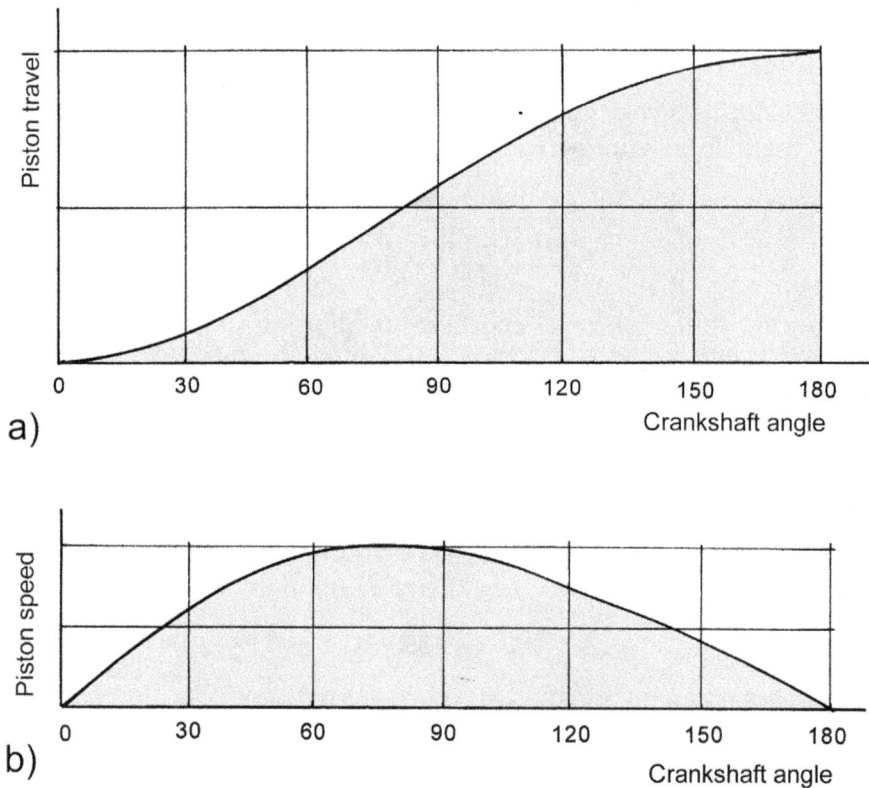

a)

b)

Fig. A 3 (a) Piston travel as a function of crankshaft angle. (b) Piston speed as a function of crankshaft angle.

List of Acronyms

ADAC   *Allgemeiner Deutscher Automobil-Club* [German auto club]

ATZ   *Automobiltechnische Zeitschrift* [German Journal for Automotive Engineering]

ATDC   After top dead center

BTDC   Before top dead center

DIN   *Deutsches Institut für Normung/Deutsche Industrie-Norm* [German Standards Association/German Industrial Standard]

ECU   Electronic control unit

HCF   High cycle fatigue

IVK   *Institut für Verbrennungskraftmaschinen und Kraftfahrzeugbau* [Institute for Internal Combustion Engines and Automotive Vehicle Construction, Vienna Technical University]

ISO   International Standards Organization

KS   *Kolbenschmidt Pierburg AG* [Piston manufacturing company, Neckarsulm, Germany]

LCF   Low cycle fatigue

LDO   Long-distance oil

MAN   *Maschinenfabrik Augsburg-Nürnberg AG* [Augsburg-Nuremberg Machine Manufacturing Company, Munich, Germany]

MTU   *Motoren- und Turbinen-Union Friedrichshafen* [Motors and Turbines Union, Friedrichshafen/Lake Constance, Germany]

MTZ   *Motortechnische Zeitschrift* [Journal for Engine Engineering]

OT   *Oberer Totpunkt* [top dead center] (in German combustion engine diagrams)

PDE   *Pumpe/Düse-Einheit* [pump/nozzle unit]

SHPD   Super high performance diesel (oil)

UT   *Unterer Totpunkt* [bottom dead center] (in German combustion engine diagrams)

VDI   *Verein Deutscher Ingenieure* [Association of German Engineers]

W   *Wartungsstufe* [Maintenance echelon]

ZF   *Zahnradfabrik Friedrichshafen AG* [Friedrichshafen Gear Manufacturing Co., Friedrichshafen/Lake Constance, Germany]

References

[2-1]   MTU Friedrichshafen, Die Motoren der Baureihe 396 – Betriebserbebnisse in Schienenfahrzeugen, company publication, Signet [registration mark] VTS/VKB 09045 ZPW 10/90 A.

[2-2]   "Der Druckwellenlader Comprex – eine ganz einfache Maschine für einen hochkomplexen Strömungsmaschinenprozeß," BBC-Technik Vol.74, 1987.

[2-3]   "Neue Diagnose- und Wartungssysteme: Garanten für einen kostenoptimierten Einsatz von Motoren," MTU Report No. 1, 1995.

[2-4]   Kuhn, C., Wartungskonzepte für die Bahnanwendung," MTU Report No.1, 1996.

[2-5]   MTU Friedrichshafen, Die Motoren der Baureihe 396 – Betriebsergebnisse in Schienenfahrzeugen, MTU company publication, Signet [registration mark] VTSNKB 09045 (52 2E) 1/90.

[2-6]   Haußmann, G., "Entwicklung eines neuen Wartungskonzeptes anhand der Betriebserfahrungen mit den Motoren der Baureihe 396," MTZ Motortechnische Zeitschrift, special issue "Die neuen Motorbaureihen 2000 und 4000 von MTU und DDC," 1997.

[2-7]   Bockelmann, W., Gerve, A., Kehrwald, B., and Willenbockel, O., "Optimierung des Verschleißverhaltens am Ventiltrieb des Opel 3,0/24V-Motors," MTZ Vol. 52, No. 2, 1991.

[2-8]   Reinhardt, G. P., "Anforderungen an Schmierstoffe durch den modernen Fuhrpark," in: Reinhardt, G. P. et al., *Schmierung von Verbrennungskraftmaschinen*, Expert-Verlag, Ehningen, 1992.

[3-1]   Eichler, C., *Instandhaltungstechnik*, 5th ed., Verlag Technik, Berlin, p. 27.

[3-2]   Schnittier, M., Heinrich, R., and Treutlein, W., "Mercedes-Benz Baureihe 500 – Lebensdauerabsicherung und Wartungskonzept der neuen V-Motoren für schwere Nutzfahrzeuge," MTZ Motortechnische Zeitschrift Vol. 57, No. 10, 1996.

[4-1]   DIN 31051 Instandhaltung – Begriffe und Maßnahmen, Deutsches Institut für Normung, Berlin, January 1985.

[4-2]   Vögtle, G., "Anforderungen heutiger und zukünftiger Großmotoren an den Schmierstoff," in: Reinhardt, G. P. et al., *Schmierung von Verbrennungskraftmaschinen*, Expert-Verlag, Ehningen, 1992.

[4-3]   Vliet, R. van, "Erfahrungen mit MaK M 32-Motoren während der ersten 15 000 Betriebsstunden," MaK-Toplaterne No. 73, October 1998.

[4-4]   Duden, *Deutsches Universalwörterbuch*, 2nd edition, Bibliographisches Institut, Mannheim.

[4-5]   Grupp, H., "Analyse von Schmierölen aus stationären Großmotoren und deren Bewertung zur Schadenverhütung," Der Maschinenschaden Vol. 64, No. 3, 1991.

[4-6]   Dhillon, B. S., *Zuverlässigkeitstechnik*, VHC Verlagsgesellschaft, Weinheim, 1988; p. 48.

[5-1]   Czichos, H. (Ed.), *Hütte – Die Grundlagen der Ingenieurwissenschaften*, 30th ed., Section D 65, Springer-Verlag, Berlin, 1996.

[5-2]   Engel, L. and Klingele, "Rasterelektronenmikroskopische Untersuchung von Bruchflächen und Oberflächenschäden" in: VDI-Berichte 243, Methodik der Schadensuntersuchung, VDI Verlag, Düsseldorf, 1975.

[5-3]   Roßmann, A., "Untersuchung von Schäden als Folge thermischer Überbeanspruchung," in: Grosch, J. et al., *Schadenkunde im Maschinenbau*, 2nd edition, Expert-Verlag, Renningen, 1995.

[5-4]   Eichler, C., *Instandhaltungstechnik*, 5th ed., Verlag Technik, Berlin, 1990; p. 43.

[5-5]   MTU (ed.), *Taschenbuch der Luftfahrtantriebe*, 4th ed., MTU, Munich, 1981.

[5-6]   Czichos, H. and Habig, K.-H., *Tribologie-Handbuch*, Vieweg-Verlag, Wiesbaden, 1992; p. 109.

[5-7] Rieger, H., *Kavitation und Tropfenschlag*, Werkstofftechnische Verlagsgesellschaft, Karlsruhe, 1977.

[5-8] Grein, H., Kavitation – eine Übersicht, Sulzer-Forschungsheft, 1974.

[5-9] Schegk, C.-D. and Gerdes, G., "Heavy Oil Fueling and Solid Particle Erosion in Turbocharger Turbines," I. Mech. Eng., 1994, C 484/028.

[6-1] Kalina, K. and Muffert, K.-H., "Verbrennungsmotoren," in: Allianz (Ed.), *Allianz-Handbuch der Schadenverhütung*, 3rd ed., VDI-Verlag, Düsseldorf, 1984; pp. 469-470.

[6-2] ADAC Pannenstatistik: Bis zu 10 Jahre alte Fahrzeuge, ADAC, Munich, 1997/2007.

[6-3] Auto – Motor – Sport, 1999, No. 8, p. 189.

[6-4] ADAC-Motorwelt, 1999, No. 5, p. 14, and 2007, No. 5, p. 12.

[6-5] Kolben-Seeger (Ed.), KS Technisches Handbuch, Chapter 1, Kolben-Seeger, Steinbach, p. 13.

[6-6] Mahle, E., MAHLE-Kolbenkunde, Heft 2, Funktion des Kolbens, 1984 issue, MAHLE GmbH, Stuttgart, 1984, pp. 28-33.

[6-7] Mahle, E., Kolbenschäden – Ursache und Abhilfe, Signet [registration mark] Mahle 7142/1 Dr. 1. 97.

[6-8] Bäumler, H. and Wiemann, L., Brandspurversuche in einem Fahrzeug-Dieselmotor, (company publication), Goetze-Werke, Burscheid, Druckschrift K1, Signet [registration mark] 786/7.67.

[6-9] Bäumler, H., Messungen und Einflussgrößen zum Ölverbrauch in Verbrennungsmotoren, 2nd ed., (company publication), Goetze-Werke, Burscheid, Druckschrift K2, Signet [registration mark] 809/8.75/2.

[6-10] OLG Hamm, Urteil vom 31.5.1989 (20 U 328/88), [Provincial Supreme Court Hamm/Westphalia, decision of 05/31/1989 (file reference no. 20 U 328/88)].

[6-11] Glyco (Ed.), Betriebsspuren in Gleitlagerschalen von Verbrennungsmotoren und deren Bedeutung, 1979 issue, (company publication), Glyco-Metallwerke, Wiesbaden, 1979.

[6-12] Miba (ed.), *Miba-Gleitlager-Handbuch*, Miba AG, Laakirchen, Austria, 1985.

[6-13] BHW-Beurteilungskriterien für Haupt- und Pleuellager in mittelschnellaufenden Verbrennungsmotoren, Zollern BHW Gleitlager GmbH, Braunschweig, Signet [registration mark] TKB 03.91.2 D-2.

[6-14] Cummins Engine Company, Inc. (Ed.), Analysis and Prevention of Bearing Failures, Bulletin No. 3810387-00 8/88.

[6-15] Cummins Engine Company, Inc. (Ed.), Failure Analysis – Bearings, Bulletin No. 985573 5-69.

[6-16] Caterpillar, Inc. (ed.), Engine Bearings – Applied Failure Analysis Reference Book, Signet [registration mark] SEB V0544.

[6-17] Gläser, H., *Schäden an Gleit- und Wälzlagerungen*, Verlag Technik, Berlin, 1990.

[6-18] Bartz, W. J. et al., *Gleitlager als moderne Maschinenelemente*, Expert-Verlag, Ehningen, 1993.

[6-19] Affenzeller, J. and Gläser, H., *Lagerung und Schmierung von Verbrennungsmotoren*, series *Die Verbrennungskraftmaschine – Neue Folge*, Vol. 8, Springer-Verlag, Vienna, 1996.

[6-20] Cummins Engine Company, Inc. (Ed.), Analysis and Prevention of Bearing Failures, Bulletin No. 3810387-00 8/88.

[6-21] Nemec, K. J., "Erkenntnisse zur Problematik der Zverlässigkeit- und Lebendauererhöhung von Dieselmotoren-Gleitlagern," Maschinenbautechnik Vol. 24, 1975, No. 3.

[6-22] Kreutzer, R., Schäden an Gleitlagern, ed. by Allianz-Zentrum für Technik, Institut Industrielle Technik, Signet [registration mark] TI-DE-28.

[6-23] Aral (Ed.): Ölfibel, No. 1-4; 10th ed., Aral Aktiengesellschaft, 1989-1991.

[6-24] Wedepohl, E. and Hildebrandt, U., "Schlammbildung in Dieselmotoren," Erdöl & Kohle, Erdgas, Petrochemie Vol. 37, No. 6, 1984.

[6-25] Dahm, W. and Daniel, K., "Entwicklung der Ölwechselintervalle und deren Beeinflußbarkeit durch Nebenstrom-Feinölfilterung," MTZ Vol. 57, No. 6, 1996.

[6-26] deutz antriebe, No. 41, 1999.

[6-27] Aral (Ed.), Ölfibel, No. 2, 10th ed., Aral Aktiengesellschaft ,1989.

[6-28] Horn, P. and Leidenroth, V., *Qualität von Schraubenfedern*, Dr. Riederer Verlag, Stuttgart, 1987.

[6-29] TRW (Ed.), *TRW-Thompson-Handbuch*, 6th ed., TRW Automotive, Barsinghausen, 1983.

[6-30] Umland, E. and Ritzkopf, M., "Ventilkorrosion in Dieselmotoren," MTZ Vol. 36, No. 7/8, 1975.

[6-31] mtu heute, No. 2, 1999, p. 27.

[6-32] Wächter, K. (ed.), *Konstruktionslehre für Maschineningenieure*, VEB Verlag Technik, Berlin, 1987.

[6-33] ContiTech-Praxistips: Keilrippenriemen und Keilriemen (ContiTech company publication).

[6-34] Konzern-Norm ZFN 201 – Zahnradschäden: Begriffsbestimmung, ZF Friedrichshafen.

[6-35] Stumpp, G., "Einspritzsysteme für moderne Pkw-Motoren," in: Essers, U. (Ed.): *Dieselmotorentechnik 98*, Expert-Verlag, Renningen, 1998.

[6-36] Ickiewitz, J. and Jeszke, T., "Einfluß des Verschleißes der Einspritzpumpe auf die Charakteristik des Dieselmotors," MTZ Vol. 40, No. 4, 1979.

[6-37] Bosch-Druckschrift VDT AKP 6/1: Düsen und Düsenhalter (Bosch company publication).

[6-38] Staubverschleiß (company publication), Deutz AG, Cologne, Signet [registration mark] W 0999-102 50-4-74 pa.

[6-39] Mahle, E.: "Fortschritte im Kolbenbau," ATZ Vol. 45, No. 21, 1942.

[6-40] MANN+HUMMEL technical papers.

[6-41] Batram, B., Brunsmann, L., and Knickmann, K.-U., "Ölabscheider – the next generation!", MTZ offprint, 1997; p. 12.

[6-42] Mollenhauer, K. (Ed.) *Handbuch Dieselmotoren*, Section 11, Kühlsysteme und Kühlmittel bei Flüssigkeitskühlung, Springer-Verlag, Berlin, 1997.

[6-43] Fahr + Spar-Training, ed. by MAN Nutzfahrzeuge
        Aktiengesellschaft, Vertrieb, Marketing-Training, March issue,
        1995.

[7-1]   *Allianz Handbook of Loss Prevention*, ed. by Allianz AG, Section 11,
        Internal Combustion Engines, VDI-Verlag, Düsseldorf, 1997, pp.
        495-508; by courtesy of Allianz AG, Munich.

[7-2]   Muffert, K.-H., "Ursachen und Beispiele von Schäden an
        Verbrennungsmotoren" [Causes and examples of damage to
        combustion engines; in German with English summary], *Der
        Maschinenschaden*, Vol. 53, No. 2, 1980, pp. 95-100.

[7-3]   Kraemer, O. and Jungbluth, G., *Bau und Berechnung von
        Verbrennungsmotoren – Hubkolben- und Rotationskolbenmaschinen*
        [Design, Calculation and Construction of Internal Combustion
        Engines – Reciprocating and Rotary Engines; in German], 5th ed.,
        Springer Publ. Berlin, Heidelberg, New York, Tokyo, 1983.

[7-4]   Alcraft, D. A., "Ensuring the reliability of diesel engine
        components," in: *Design and Application in Diesel Engineering*
        (ed. by S. D. Haddad and N. Watson), Ellis Horwood Publ. and
        Halsted/Wiley, Chichester, 1984, pp. 317-336 [Chapter 10].

[7-5]   Pavlin, M. and Pirš, J., "Die Untersuchung von Stegrissen in
        Zylinderköpfen von Dieselmotoren" [The Investigation of Bridge
        Cracks in the Cylinder Heads of Diesel Engines; in German with
        English summary], *Der Maschinenschaden*, Vol. 48, No. 5, 1975, pp.
        149-154.

[7-6]   DIN 250, Rounding radii, Deutsches Institut für Normung (DIN),
        Berlin, July 1972.

[7-7]   DIN 2510, Part 1, Bolted connections with reduced shank; survey;
        range of application and examples of installation, Deutsches
        Institut für Normung (DIN), Berlin, September 1974.
        DIN 2510 Part 2, Bolted connections with reduced shank;
        metric thread with large clearance; nominal values and limits,
        Deutsches Institut für Normung (DIN), Berlin, August 1971.
        DIN 2510, Part 3, Bolted connections with reduced shank; stud
        bolts, Deutsches Institut für Normung (DIN), Berlin, August
        1971.
        DIN 2510, Part 4, Bolted connections with reduced shank; studs,
        Deutsches Institut für Normung (DIN), Berlin, August 1971.
        DIN 2510, Part 5, Bolted connections with reduced shank;
        hexagon nuts, Deutsches Institut für Normung (DIN), Berlin,

August 1971.

DIN 2510, Part 6, Bolted connections with reduced shanks; cap nuts, Deutsches Institut für Normung (DIN), Berlin, September 1971.

DIN 2510, Part 7, Bolted connections with reduced shank; extension sleeves, Deutsches Institut für Normung (DIN), Berlin, August 1971.

DIN 2510, Part 8, Bolted connections with reduced shank; threaded holes for studs, Deutsches Institut für Normung (DIN), Berlin, August 1971.

[7-8]   Haddad, S. D., "Condition Monitoring and Fault diagnosis in Diesel Engines," in: *Principles and Performances in Diesel Engineering* (ed. by S. D. Haddad and N. Watson), Ellis Horwood Publ. and Halsted Press/Wiley, Chichester, 1984, pp. 246-277 [Chapter 8].

[7-9]   Allianz Information Sheet No. 7 Guidance for satisfactory operation of Diesel engines, 12 pages, Allianz, Munich, Berlin, 1983, Order No. TI-E4-7.

[7-10] Everett, P. N., "Diesel Engine Maintenance," in: *Diesel Engine Principles and Practice*, (ed. by C. C. Pounder), G. Newnes Publ., London, 1962, pp. 26-1 to 26-28 [= Section 26]. (See also below: *Diesel Engine Reference Book*).

Bibliography

Machine elements (general)

DUBBEL, *Taschenbuch für den Maschinenbau*, 19th ed., Springer-Verlag, Berlin, 1997.

Meißner, M. and Wanke, K., *Handbuch Federn*, VEB Verlag Technik, Berlin, 1988.

Wächter, K., *Konstruktionslehre für Maschineningenieure*, VEB Verlag Technik, Berlin, 1987.

Engine technology (general)

Beier, R., et al., *Verdrängermaschinen – Hubkolbenmotoren*, Vol. 2/11, Energie (ed. by Bohn, T.), Techn. Verlag Resch, Munich, 1983.

Bosch (Ed.), *Dieselmotor-Management*, 2nd ed., Vieweg-Verlag, Wiesbaden, 1998.

Bosch (Ed.), *Kraftfahrtechnisches Taschenbuch*, 23rd ed., Vieweg Verlag, Wiesbaden, 1999.

Bosch (Ed.), *Ottomotor-Management*, 1st ed., Vieweg-Verlag, Wiesbaden, 1998.

BUSSIEN: *Automobiltechnisches Handbuch*, 18th ed., Verlag Walter de Gruyter, Berlin, 1965.

BUSSIEN: *Automobiltechnisches Handbuch. Ergänzungsband zur 18. Aufl*, Berlin: Verlag Walter de Gruyter, 1979.

Essers, U. (Ed.), *Dieselmotorentechnik 98*, Expert-Verlag, Renningen, 1998.

Gerigk, Bruhn, Endruschat, Göbert, Gross, Komoll, *Kraftfahrzeugtechnik*, 2nd ed., Westermann-Verlag, Braunschweig, 1995.

Köhler, E., *Verbrennungsmotoren*, Vieweg-Verlag, Wiesbaden, 1998.

Küntscher, V. (Ed.), *Kraftfahrzeugmotoren*, 3rd ed., Verlag Technik, Berlin, 1995.

Lilly, L.R.C. (Ed.), *Diesel Engine Reference Book*, Butterworth, London, 1984.

Mau, G., *Handbuch Dieselmotoren im Kraftwerks- und Schiffsbetrieb*, Vieweg-Verlag, Wiesbaden, 1984.

Mettig, H., *Die Konstruktion schnellaufender Verbrennungsmotoren*, Verlag Walter de Gruyter, Berlin, 1973.

Mollenhauer, K. (Ed.), *Handbuch Dieselmotoren*, Springer-Verlag, Berlin, 1997.

Neumeister, O. and Erbling, H.-G., *Betrieb von Schiffsmotorenanlagen*, VEB Verlag Technik, Berlin, 1989.

Pucher, H. (Ed.), *Gasmotorentechnik*, Expert-Verlag, Sindelfingen, 1986.

Shell-Lexikon Verbrennungsmotor, ATZ/MTZ supplement, ed. by. van Basshuysen, R. and Schäfer, F.

Sperber, R., *Technisches Handbuch Dieselmotoren*, 4th ed., VEB Verlag Technik, Berlin.

Urlaub, A., *Verbrennungsmotoren*, 2nd ed., Springer-Verlag, Berlin, 1994.

Zima, S., *Kurbeltriebe*, Vieweg-Verlag, Wiesbaden, 1998.

Engine installations

Englhard, O., *Dieselmotorenanlagen*, Vogel-Buchverlag, Würzburg, 1999.

MAN Nutzfahrzeuge AG (Ed.), Technische Information für den Einbau von MAN-Dieselmotoren in Fahrzeuge und selbstfahrende Arbeitsmaschinen, company publication, 1990.

MAN Nutzfahrzeuge AG (Ed.), Planung von stationären Dieselmotorenanlagen, Signet [registration mark] 50.99497-8074.

MAN, Schiffsdieselmotoren-Einbauanleitung, Signet [registration mark] 50.99497-8065.

Experimental methods and measurement technology

Bohl, W. and Mathieu, W., *Laborversuche an Kraft- und Arbeitsmaschinen*, Carl Hanser Verlag, Munich, 1975.

Grohe, H., *Messen an Verbrennungsmotoren*, Vogel-Buchverlag, Würzburg, 1977.

Kuratle, R., Motorenmeßtechnik, Vogel-Buchverlag, Würzburg, 1995.

Machine failure, general (fundamentals)

Broichhausen, J., *Schadenskunde*, Hanser Verlag, Munich, 1995.

Cummins Engine Co, Inc. (Ed.): Failure Analysis – Steel Parts Fractures, Bulletin No. 3387210-R.

Czichos, H. (Ed.), *HÜTTE – Die Grundlagen der Ingenieurwissenschaften*, 30th ed., Section D 65, Springer-Verlag, Berlin, 1996.

Czichos, H. and Habig, K.-H., *Tribologie Handbuch*, Vieweg-Verlag, Wiesbaden, 1992, p. 109.

DIN 31051, "Instandhaltung – Begriffe und Maßnahmen," Deutsches Institut für Normung (DIN), Berlin, January 1985.

DIN 31661, "Gleitlager – Begriffe, Merkmale und Ursachen von Veränderungen und Schäden," Deutsches Institut für Normung (DIN), Berlin.

DIN 50320, "Verschleiß – Begriffe, Systemanalyse von Verschleißvorgängen, Gliederung des Verschleißgebietes", Deutsches Institut für Normung (DIN), Berlin.

DIN 50321, "Verschleiß-Meßgrößen," Deutsches Institut für Normung (DIN), Berlin.

DIN 50323-2, "Tribologie," Deutsches Institut für Normung (DIN), Berlin.

Eichler, C., *Instandhaltungstechnik*, 5th ed., Verlag Technik, Berlin, 1990.

Grein, H., Kavitation – eine Übersicht, Sulzer-Forschungsheft, 1974.

Grosch, J. et al., *Schadenskunde im Maschinenbau*, 2nd ed., Expert-Verlag, Renningen, 1995.

Heckel, K., *Einführung in die technische Anwendung der Bruchmechanik*, 3rd ed., Hanser Verlag, Munich, 1991.

Held, G., "Betriebsbewährung von Schiffsantriebsanlagen," *Der Maschinenschaden*, Vol. 53, No. 1, 1980.

Kalina, K. and Muffert, K.-H., "Verbrennungsmotoren," in: Allianz (Ed.), *Allianz-Handbuch der Schadenverhütung*, 3rd ed., VDI Verlag, Düsseldorf, 1984, pp. 469-470.

Muffert, K.-H., Ursachen und Beispiel von Schäden an Verbrennungsmotoren, Technische Information, ed. by Allianz Versicherungs-AG, Order No. WBAlTI-DE6-161.

Naumann, F. K., *Das Buch der Schadensfälle*, Dr. Riederer-Verlag, Stuttgart, 1976.

Nolting, E., Erfahrungen beim Betrieb von Dieselaggregaten und Blockheizkraftwerken, Technische Information, ed. by. Allianz Versicherungs-AG, Order No. WBA/TI-DE6-161.

Pfender, M., "Werkstoffbeurteilung bei mechanischer Beanspruchung," *Konstruktion*, Vol. 5, No. 12, 1953.

Pohl, E. J. and Bark, R., *Wege zur Schadensverhütung im Maschinenbetrieb*, ed. by Allianz Versicherungs-AG, 1964.

Pohl, E. J., *Das Gesicht des Bruches metallischer Werkstoffe*, Vol. I/II, ed. by Allianz Versicherungs-AG, 1956.

Pohl, E. J., *Das Gesicht des Bruches metallischer Werkstoffe*, Vol. III, ed. by Allianz Versicherungs-AG, 1956.

Rieger, H., *Kavitation und Tropfenschlag*, Werkstofftechnische Verlagsgesellschaft, Karlsruhe, 1977.

Schittler, M., Heinrich, R., and Treutlein, W., "Mercedes-Benz Baureihe 500 – Lebensdauerabsicherung der neuen V-Motoren für schwere Nutzfahrzeuge," *MTZ Motortechnische Zeitschrift*, Vol. 57, No. 10, 1996.

Schmitt-Thomas, K. G., *Integrierte Schadenanalyse*, Springer-Verlag, Berlin, 1999.

VDI 3822, Blatt 1, "Schadenanalyse – Grundlagen, Begriffe und Definitionen, Ablauf einer Schadenanalyse," Verein Deutscher Ingenieure (VDI), Düsseldorf.

VDI 3822, Blatt 2, "Schadenanalyse – Schäden durch mechanische Beanspruchungen," Verein Deutscher Ingenieure (VDI), Düsseldorf.

VDI 3822, Blatt 3, "Schadenanalyse – Schäden durch Korrosion in wäßrigen Medien," Verein Deutscher Ingenieure (VDI), Düsseldorf.

VDI 3822, Blatt 4, "Schadenanalyse – Schäden durch thermische Beanspruchungen," Verein Deutscher Ingenieure (VDI), Düsseldorf.

VDI 3822, Blatt 5.1, "Schadenanalyse – Schäden durch tribologische Beanspruchungen," Verein Deutscher Ingenieure (VDI), Düsseldorf.

VDI 3822, Blatt 5.2, "Schadenanalyse – Schäden durch tribologische Beanspruchungen," Verein Deutscher Ingenieure (VDI), Düsseldorf.

VDI 3822, Blatt 6, "Schadenanalyse – Erfassung und Auswertung von Schadenanalysen," Verein Deutscher Ingenieure (VDI), Düsseldorf.

VDI-Berichte 243, Methodik der Schadensuntersuchung, VDI-Verlag, Düsseldorf, 1975.

Wendler-Kalsch and Gräfen, H., *Korrosionsschadenkunde*, Springer-Verlag, Berlin, 1999.

Engines, engine components, and engine component failures

Engines (general)

Englisch, C., *Verschleiß, Betriebszahlen und Wirtschaftlichkeit von Verbrennungskraftmaschinen.* (Series: List, *Die Verbrennungskraftmaschine*, Vol. 14), Springer-Verlag, Vienna, 1952.

Haussmann, G., "Entwicklung eines neuen Wartungskonzeptes anhand der Betriebserfahrungen mit den Motoren der Baureihe 396," in: Die neuen Motorbaureihen 2000 und 4000 von MTU und DDC, *MTZ Motortechnische Zeitschrift*, special issue, 1997.

Klaver, P., *Motorschades*, Technische Leergang, ed. by Nederlandse Ingenieursvereniging, Delta Press BV, Amerongen-Overberg, The Netherlands, 1999.

Kuhn, C., "Wartungskonzepte für die Bahnanwendung," *MTU Report* 1/96.

Muffert, K. H., "Ursachen und Beispiele von Schäden an Verbrennungsmotoren," *Der Maschinenschaden*, Vol. 53, No. 3, 1980.

Belt drives

Keilrippenriemen und Keilriemen, ContiTech-Praxistips, Continental, Hannover.

Zahnriemen und Zahnriemenkits, ContiTech-Praxistips, Continental, Hannover.

Connecting rods

Caterpillar Reference Book: Connecting Rods – Applied Failure Analysis, Signet [registration mark] SEBV0546.

Coolant

Wissenswertes über Kühlmittel, published by MTU-Friedrichshafen Service, Signet [registration mark] A060630/OOD.

Crankcases

Pflug, E. and Piltz, H.-H., "Korrosion und Kavitation in den Kühlwasserräumen von Hochleistungs-Dieselmotoren im Eisenbahnbetrieb – Über Ursachen und Schutzmaßnahmen," *MTZ Motortechnische Zeitschrift*, Vol. 26, No. 3, 1965.

Crankshafts

Caterpillar Reference Book, Crankshafts – Applied Failure Analysis, Signet [registration mark] SEBV0548.

Gassner, E. and Schütz, W.,"Zur Dauerfestigkeit von Fahrzeug-Kurbelwellen," *MTZ Motortechnische Zeitschrift*, Vol. 22, No. 8, 1961.

Oppitz, A., "Kurbelwellen: Brüche, Ursachen und Folgerungen," *MTZ Motortechnische Zeitschrift*, Vol. 17, No. 5, 1956.

Cylinders

Collins, H. H., "Pitting of Diesel Cylinder Liners I/II," *The Oil Engine*, Feb./March 1960.

Dück, G., "Zur Laufflächengestaltung von Zylindern und Zylinderbuchsen," *MTZ Motortechnische Zeitschrift*, Vol. 28, No. 3, 1967.

Dück, G., Zur Laufflächengestaltung von Zylindern und Zylinderlaufbuchsen, Druckschrift K 9, Goetze-Werke, Burscheid, Signet [registration mark] 944/9.73.

Guertler, R. W., "Excessive Cylinder Wear and Bore Polishing in Heavy-Duty Diesel Engines: Causes and Proposed Remedies," SAE Paper No. 860165, Society of Automotive Engineers, Warrendale, PA, USA, 1986.

Handbuch Kolbenringe und Zylinderbuchsen, ed. by Sealed Powers Technologies Europe, Barsinghausen, 1993.

Häusler, F., Flüssigkeits- und luftgekühlte Zylinder in Verbrennungsmotoren, Nüral Technischer Hinweis, Aluminiumwerke Nürnberg, Signet [registration mark] E 2755/26.

Immisch, H., Erfahrungen mit Nüral-Perimatic-Regelkolben in Dieselmotoren zur Verhütung von Kavitationsschäden an Zylinderlaufbuchsen, Technische Mitteilung K 46, Aluminiumwerke Nürnberg.

Oetz, H., "Einfluß der Honbearbeitung von Zylinderlaufbuchsen auf die innere Grenzschicht und den Einlaufverschleiß," *MTZ Motortechnische Zeitschrift*, Vol. 30, No. 12, 1969.

Pflaum, W., "Zur Kavitation an Zylinderbuchsen von Dieselmotoren," *MTZ Motortechnische Zeitschrift*, Vol. 30, No. 2, 1969.

Pirs, J., "Zylinderbuchsen eines Dieselmotors durch ungeeignetes Kühlwasser korrodiert," *Allianz Report*, Vol. 66, No. 3, 1993.

Teetz, C., "Einfluß des Brennbeginns auf den Zwickelverschleiß im Dieselmotor," *MTZ Motortechnische Zeitschrift*, Vol. 44, No. 12, 1983.

Wiemann, L., "Die Bildung von Brandspuren auf den Laufflächen der Paarung Kolbenring – Zylinder in Verbrennungsmotoren," *MTZ Motortechnische Zeitschrift* , Vol. 32, No. 2, 1971.

Zürner, H.-J., Schibalsky, W., and MÜLLER, H., "Kavitation und Korrosion an Zylindern von Dieselmotoren," *MTZ Motortechnische Zeitschrift*, Vol. 49, No. 9, 1988.

Exhaust gas turbochargers

ABB Turbolader Betriebshandbuch, VTR 564-32.

Caterpillar Reference Book: Turbochargers – Applied Failure Analysis, Signet [registration mark] SEBVO550.

Cummins Engine Co, Inc. (Ed.), Turbocharger Failure Analysis, Bulletin No. 3387125-R.

KKK-Turbolader, Katalog zur Schadensbeurteilung, ed. by Aktiengesellschaft Kühnle, Kopp & Kausch.

Mayer, M., *Abgasturbolader*, 3rd ed., Verlag Moderne Industrie, Landsberg/Lech, 1997.

Zinner, K., *Aufladung von Verbrennungsmotoren*, 2nd ed., Springer-Verlag, Berlin, 1980.

Filters

Baumann, D. and Brunsmann, L., "Ölfiltertechnik – Impulsgeber der Systemintegration," *MTZ Motortechnische Zeitschrift*, Vol. 59, No. 9, 1998.

Blumenstock, K.-U., *Motorenfilter*, Verlag Moderne Industrie, Landsberg/Lech, 1994.

Dahm, W. and Daniel, K., "Entwicklung der Ölwechselintervalle und deren Beeinflußbarkeit durch Nebenstrom-Feinölfilterung," *MTZ Motortechnische Zeitschrift*, Vol. 57, No. 6, 1996.

Erdmannsdörfer, H., "Leistungsmöglichkeiten von Papierfiltern zur Reinigung der Ansaugluft von Dieselmotoren," *MTZ Motortechnische Zeitschrift*, Vol. 32, No. 4, 1971.

Erdmannsdörfer, H., "Trockenluftfilter für Fahrzeugmotoren – Auslegungs- und Leistungsdaten," *MTZ Motortechnische Zeitschrift*, Vol. 43, No. 7/8, 1982.

Rose, K. and Kemmer, U., "Vorgänge bei der Kraftstoffiltrierung," *Bosch Techn. Berichte*, Vol. 4, No. 7, 1974.

Taufkirch, G., "Papierluftfilter in der Einsatzpraxis von Nutzfahrzeugen," *MTZ Motortechnische Zeitschrift* , Vol. 58, No. 4, 1997.

Fuel injection technology

Bosch (Ed.), Technische Unterrichtung: Diesel-Einspritzausrüstung 1; Diesel-Einspritzausrüstung 2; Diesel-Einspritztechnik im Überblick, Signet [registration mark] 1987722038; Diesel-Radialkolben-Verteilereinspritzpumpen; Diesel-Speichereinspritzsystem Common Rail.

Brucker, E., "Die Entwicklung des Common-Rail-Einspritzsystems für die Baureihe 4000," *MTZ Motortechnische Zeitschrift*, special issue, 1997.

Cummins Deutschland, Das Wichtigste über Select, Bulletin No. 3666967-00.

Huber, E. W. and Schaffitz, W., "Kavitationsverschleiß in Kraftstoff-Einspritzanlagen," *MTZ Motortechnische Zeitschrift*, Vol. 32, No. 10, 1971.

Ickiewicz, J. and Jeszke, T., "Einfluß des Verschleißes der Einspritzpumpen auf die Charakteristik des Dieselmotors," *MTZ Motortechnische Zeitschrift* , Vol. 40, No. 4, 1979.

Ickiewicz, J., Jeszke, T., and Puczynski, W., "Die Abhängigkeit der Kraftstoffverluste vom Abnutzungsgrad bei Einspritzsystemen von Dieselmotoren," *MTZ Motortechnische Zeitschrift*, Vol. 40, No. 4, 1979.

Ivosevic, S., "Untersuchungen der Kavitationserosion an Einspritzpumpen-Steuerventilen von langsamlaufenden Dieselmotoren," *MTZ Motortechnische Zeitschrift*, Vol. 34, No. 12, 1973.

Kasedorf, J., *Benzineinspritzung und Katalysatortechnik*, Vogel-Buchverlag, Würzburg, 1995.

Krieger, K., "Diesel-Einspritztechnik für Pkw-Motoren – Überblick über Verfahren und Ergebnisse," *MTZ Motortechnische Zeitschrift*, Vol. 60, No. 5, 1999.

Olszewski, J. D., "Die Verhütung der Kavitationserosion in Einspritzleitungen durch eine neuartige Gleichdruckentlastung," *MTZ Motortechnische Zeitschrift* , Vol. 29, No. 5, 1968.

Prescher, K., and Schaffitz, W., "Verschleiß von Kraftstoff-Einspritzdüsen für Dieselmotoren infolge Kraftstoffkavitation," *MTZ Motortechnische Zeitschrift*, Vol. 40, No. 4, 1979.

Stan, C., *Direkteinspritzsysteme für Otto- und Dieselmotoren*, Springer-Verlag, Berlin, 1999.

Gears

Caterpillar Reference Book: Gears – Applied Failure Analysis, Signet [registration mark] SEBV0561 (1990).

Niemann, G., and Bötsch, H., "Neuere Versuchsergebnisse zur Zahnflanken-Tragfähigkeit von Stirnrädern aus Vergütungsstahl," *Konstruktion*, Vol. 18, No. 12, 1966.

ZFN 201, "Zahnradschäden: Begriffsbestimmung, Bezeichnungen und Ursachen," Konzernnorm [concern standard], ZF Friedrichshafen AG.

Heat exchangers

Herr, A., "Einfluß der Kühlanlage auf Motorschäden bei flüssigkeitsgekühlten Panzermotoren," *Soldat und Technik*, No. 3, 1983.

Jenz, S., "Entwicklungen in der Nutzfahrzeug-Motorkühlung," *MTZ Motortechnische Zeitschrift*, Vol. 54, No. 10, 1993.

Wilken, H., "Derzeitiger Stand der Motorkühlung," *ATZ Automobiltechnische Zeitschrift*, Supplement in No. 12, 1980.

Journal bearings

Achilles, M., "Gleitlagerschäden an Verbrennungsmotoren – Ursachen und Erscheinungsformen," *Schmierungstechnik*, Vol. 9, No. 2, 1978.

Affenzeller, J. and Gläser, H., *Lagerung und Schmierung von Verbrennungsmotoren*, Series *Die Verbrennungskraftmaschine – Neue Folge*, Vol. 8, Springer-Verlag, Vienna, 1996.

Austauschkriterien für Stahl-Leichtmetall-Dreistofflager, ed. by Miba AG, Laakirchen, Austria, Signet [registration mark] AK 0586.

Austauschkriterien für Stahlleichtmetall-Rillenlager, ed. by Miba AG, Laakirchen, Austria, without Signet [registration mark].

Bartz, W. J., et al., *Gleitlager als moderne Maschinenelemente*, Expert-Verlag, Grafenau, 1993.

Betriebsspuren in Gleitlagerschalen von Verbrennungsmotoren und deren Bedeutung, company publication, Glyco-Metallwerke Daelen & Hofmann KG, Wiesbaden, 1979.

Beurteilungskriterien für Haupt- und Pleuellager in mittelschnellaufenden Verbrennungsmotoren, ed. by BHW Braunschweiger Hüttenwerk, Signet [registration mark] TKB 03.91.2 D-4.

Beurteilungskriterien für Haupt- und Pleuellager in mittelschnellaufenden Verbrennungsmotoren, ed. by BHW Braunschweiger Hüttenwerk, Signet [registration mark] TKB 03.91.2 D-2.

Cummins Diesel Engine Company, Inc., Failure Analysis - Bearings Bulletin 9855735-69.

Cummins Engine Company, Inc., Analysis and Prevention of Bearing Failures, Bulletin No. 3810387-00 8/88.

DIN 31661, "Gleitlager – Begriffe, Merkmale und Ursachen von Veränderungen und Schäden," Deutsches Institut für Normung (DIN), Berlin.

Eckardt, C., "Einfluß von Fremdkörpern im Schmiermittel auf das Betriebsverhalten von Motorengleitlagern," *MTZ Motortechnische Zeitschrift*, Vol. 44, No. 10, 1983.

Ederer, U. G., "Lager für hohe Belastungen in Zweitakt- und Viertakt-Dieselmotoren," *MTZ Motortechnische Zeitschrift*, Vol. 44, No. 11, 1983.

Ederer, U. G. and Kirsch, H., "Neue Gleitlagertechniken für höhere Lebensdauer," *MTZ Motortechnische Zeitschrift*, special issue, 1997.

Ederer, U., "Neue Gleitlagerbauarten für gesteigerte Anforderungen," Miba-Symposium 1990.

Engel, U., "Schäden an Gleitlagern in Kolbenmaschinen," Ingenieurbericht No. 8/87, Glyco-Metallwerke, Wiesbaden.

"Erkenntnisse zur Problematik der Zuverlässigkeits- und Lebensdauererhöhung von Dieselmotoren-Gleitlagern," *Maschinentechnik*, Vol. 24, No. 3, 1975.

Gläser, H., *Schäden an Gleit- und Wälzlagerungen*, Verlag Technik, Berlin, 1990.

"Gleitlagerschäden und ihre Beurteilung auf dem Gebiet der Kolbenmaschinen," seminar report, BHW Braunschweiger Hüttenwerk, Signet [registration mark] 5204951 DE-2.

Glyco-Metallwerke (Ed.), Betriebsspuren in Gleitlagerschalen von Verbrennungsmotoren und deren Bedeutung, issue 1979.

Glyco-Metallwerke (Ed.), Technischer Lehrgang Gleitlager für Verbrennungsmotoren, Vieweg-Verlag, Wiesbaden, 1992.

Glyco-Metallwerke, Ingenieur-Bericht No. 1/91.

Glyco-Metallwerke, Sputterlager, Signet [registration mark] 5/1997.

Grobuschek, F., "Konsequenzen für die Gleitlagerungen aus den zu erwartenden Anforderungen," Miba-Symposium, 1990.

Grobuschek, F., "Lagerung im Dieselmotor," Kolloquium Kolbenmaschinen, MTU Friedrichshafen, lecture No. 15, October 27, 1987.

Grobuschek, F., "Tieftemperaturverhalten von Dieselmotoren-Lagerungen," *MTZ Motortechnische Zeitschrift*, Vol. 47, No. 11, 1986.

Grobuschek, F., Neue Gleitlagerbauarten, Miba company publication no. 845091.

Grünthaler, K.-H., Lucchetti, W., and Schopf, E., "Gleitlager für höchste Beanspruchungen in Verbrennungsmotoren," *MTZ Motortechnische Zeitschrift*, Vol. 59, No. 4, 1998.

Haller, R., Wollfarth, M., and Heumer, H., "Die experimentelle Simulation von Kavitationsschäden im Gleitlager," *MTZ Motortechnische Zeitschrift*, Vol. 56, No. 1, 1995.

Harbordt, J., "Spannungen und Materialermüdung in mehrschichtigen Schalen von Gleitlagern," *Z-VDI*, Vol. 118, No. 2, 1976.

Klumpp, G., "Gleitlagerschäden," *Krafthand*, No. 5/6, 1976.

Koroschetz, E., "Diffusionsvorgänge in galvanisch abgeschiedenen Pb-Sn-Cu-Gleitlagerlaufschichten," *Mikrochimica Acta*, Vienna, Suppl. 9, 1981, pp. 139-152.

Koroschetz, E., "Gleitlagerwerkstoffe in Entwicklung, Produktion und Einsatz," Miba-Symposium 1990.

Koroschetz, E. and Gärtner, W., Neue Werkstoffe und Verfahren zur Herstellung von Gleitlagern, Miba company publication.

Kreutzer, R., Schäden an Gleitlagern, ed. by Allianz-Zentrum für Technik GmbH, Institut Industrielle Technik, Signet [registration mark] TI-DE-28.

Kreutzer, R., Schäden an Gleitlagern, ed. by Allianz-Zentrum für Technik, Signet [registration mark] TI-DE-28.

Lang, O. R. and Steinhilper, W., *Gleitlager*, Springer-Verlag, Berlin, 1978.

Maass, H., "Modellbetrachtungen zur Gleitlagerkavitation," *Technica*, No. 3 and No. 7, 1978.

Miba (Ed.), *Gleitlager-Handbuch*, 2nd ed., Miba AG, Laakirchen, Austria, 1986.

Nemec, K. J., "Erkenntnisse zur Problematik der Zuverlässigkeit- und Lebensdauererhöhung von Dieselmotoren-Gleitlagern," *Maschinenbautechnik*, Vol. 24, No. 3, 1975.

Roemer, E., "Gleitlagerschäden und ihre Verhütung," offprint from: Technische Akademie Wuppertal, Berichte 9, Vulkan-Verlag, Essen, pp. 21-52.

Schopf, E., Hochbelastbare Gleitlager in Verbrennungsmotoren, company publication, Glyco-Metallwerke Daelen & Hofmann KG, Wiesbaden.

Schopf, E., "Notlaufeigenschaften von Gleitlagern aus metallischen Werkstoffen," offprint from: *Antriebstechnik*, No. 5, 1983.

Steeg, M., Engel, U., and Roemer, E., "Hochleistungsfähige metallische Mehrschicht-Verbundwerkstoffe für Gleitlager," *Glyco Ingenieur-Berichte*, No. 1, 1991.

Wlodraski, J. K., "Reibkorrosion in Kurbelwellenlagern mittelschnellaufender Schiffsmotoren," *MTZ Motortechnische Zeitschrift* , Vol. 50, No. 5, 1989.

Lubrication and lubricants

Aral, Ölfibel, Heft 1-4, 10th ed., Aral AG, 1989-1991.

Bartz, W. J., *Handbuch der Betriebsstoffe für Kraftfahrzeuge, Teil 2, Schmierstoffe*, Expert-Verlag, Grafenau, 1983.

Castrol (Ed.), *Technischer Lehrgang Schmierstoffe und Motoren*, Vieweg-Verlag, Wiesbaden, 1992.

Cummins Engine Company, Inc., Oil Consumption Analysis, Bulletin No. 3387148-00 7-80.

Cummins Engine Company, Inc., Technical Overview of Oil Consumption, Bulletin No. 3379214-00 8-80.

Gupp, H., "Analyse von Schmierölen aus stationären Großmotoren und deren Bewertung zur Schadenverhütung," *Der Maschinenschaden*, Vol. 64, No. 3, 1991.

Häusler, F., Erhöhter Ölverbrauch bei Verbrennungsmotoren, Alcan Aluminiumwerk Nürnberg GmbH, Signet [registration mark] E 27511/26.

Paehr, G., "Zustand und Veränderung der Schmierstoffe im Betrieb und deren Auswirkungen auf den Verbrennungsmotor," *MTZ Motortechnische Zeitschrift*, Vol. 53, No. 7/8, 1992.

Reinhardt, G. P. et al., *Schmierung von Verbrennungskraftmaschinen*, Expert-Verlag, Ehningen, 1992.

Wissenswertes über Motorenöle, published by MTU-Friedrichshafen GmbH Service, Signet [registration mark] A060481/10D.

Pistons

Aeberli, K. and Lustgarten, G.-A., "Verbessertes Kolbenlaufverhalten bei langsamlaufenden Sulzer-Dieselmotoren," *MTZ Motortechnische Zeitschrift*, Vol. 50, No. 5/12, 1989.

Caterpillar Reference Book: Pistons, Rings & Liners – Applied Failure Analysis, Signet [registration mark] SEBV0553.

Cummins Diesel Failure Analysis Piston and Liners Bulletin No. 985579.

Kolbenschmidt AG (Ed.), Aus der Praxis – Für die Praxis, Signet [registration mark] D 1667 1, 5 5/08 C.

Kolbenschmidt AG (Ed.): Kolbenschäden, in: KS-Kolbenhandbuch, Heft 14.

MAHLE, Kolbenschäden – Ursache und Abhilfe, Signet [registration mark] Mahle 6836 a IX.81.

MAHLE, Kolbenschäden – Ursache und Abhilfe, Signet [registration mark] Mahle 7142/1 Dr.L97.

Moebus, H., "Vermeidung von Kolbenschäden bei mittelschnellaufenden Dieselmotoren," *MTZ Motortechnische Zeitschrift*, Vol. 39, No. 5, 1978).

Nüral (Ed.), *Nüral-Kolbenhandbuch*, ed. by Aluminiumwerk Nürnberg GmbH, Nürnberg 1983.

Röhrle, M., "Ermittlung von Spannungen und Deformationen am Kolben." *MTZ Motortechnische Zeitschrift*, Vol. 31, 1970.

Röhrle, M., *Kolben für Verbrennungsmotoren*, Verlag Moderne Industrie, Landsberg/Lech, 1994.

Piston rings

Bäumler, H. and Wiemann, L., Brandspurversuche in einem Fahrzeug-Dieselmotor, Druckschrift K 1, Goetze-Werke, Burscheid, Signet [registration mark] 786/7.67.

Bäumler, H., Messungen und Einflußgrößen zum Ölverbrauch in Verbrennungsmotoren, 2nd ed., Druckschrift K 2, Goetze-Werke, Burscheid, Signet [registration mark] 809/8.75/2.

Englisch, C., *Kolbenringe*, Vols. 1 and 2, Springer-Verlag, Vienna, 1958.

Flattern und Brechen von Kolbenringen, discussion paper dated November 16/17, 1967, Goetze-Werke, Burscheid, without registration mark.

Goetze, A. E., *Kolbenring-Handbuch*, A. E. Goetze GmbH, Burscheid, 1995.

Groth, K. and Beherns, R., "Korrosionsverschleiß an Kolbenringen von Dieselmotoren," *MTZ Motortechnische Zeitschrift*, Vol. 51, No. 11, 1990.

Handbuch Kolbenringe und Zylinderbuchsen, ed. by Sealed Powers Technologies Europe Barsinghausen, 1st ed., 1993.

Jöhren, P., Ölabstreifringe: Ein Vergleich bewährter und neuer Ringkonstruktionen, Druckschrift K 23, Goetze-Werke, Burscheid, Signet [registration mark] 154/5.78.

Truhan, J. J. and Covington, C. B., "Der Einfluß von Filtration auf den Verschleiß von Kolbenringen," *MTZ Motortechnische Zeitschrift*, Vol. 54, No. 7/8, 1993.

Wiemann, L., "Die Bildung von Brandspuren auf den Laufflächen der Paarung Kolbenring – Zylinder in Verbrennungsmotoren," *MTZ Motortechnische Zeitschrift*, Vol. 32, No. 2, 1971.

Spark plugs, glow plugs, and flame start systems

Beru (Ed.), Technische Information Nr. 01, Flammstartanlagen, Signet [registration mark] 5100006006.

Beru (Ed.), Technische Information Nr. 02, Zündkerzen, Signet [registration mark] 5001006005.

Beru (Ed.), Technische Information Nr. 0, Alles über Stabglühkerzen, Signet [registration mark] 5100008001.

Bosch (Ed.), *Autoelektrik – Autoelektronik*, 3rd ed., Vieweg-Verlag, Wiesbaden, 1998.

Bosch (Ed.), Technische Unterrichtung, Zündkerzen.

Endler, M., Schlanke Glühkerzen für Dieselmotoren mit Direkteinspritzung, offprint from *MTZ Motortechnische Zeitschrift*, Vol. 59, No. 2, 1998.

Springs

Barthold, G., "Spring Failures and Their Causes," *Springs – The Magazine of Spring Technology*, Vol. 37, No. 3, 1998.

Hempel, H., "Untersuchungen an Eindraht-Schraubenfedern," *Konstruktion*, Vol. 5, No. 12, 1953.

Hora, P., "Schraubenfederbrüche beurteilen," *Materialprüfung*, Vol. 39, No. 10, 1997.

Hora, P. and Leidenroth, V., *Qualität von Schraubenfedern*, Dr. Riederer-Verlag, Stuttgart, 1987.

Krickau, G. and Huhnen, H., "Federbrüche und ihre Beurteilung," *Draht-Fachzeitschrift*, No. 10, 1972.

Meissner, M. and Wanke, K., *Handbuch Federn*, VEB Verlag Technik, Berlin, 1988.

Valves

Caterpillar Reference Book: Engine Valves – Applied Failure Analysis, Signet [registration mark] SEBV0551.

Hesse, A., "Verhinderung der Hochtemperaturkorrosion an Auslaßventilen von Dieselmotoren bei der Verwendung von Mischkraftstoffen," *MTZ Motortechnische Zeitschrift*, Vol. 43, No. 3, 1982.

Marx, W. and Müller, R., "Ein Beitrag zum Einlaßventilsitzverschleiß in aufgeladenen Viertakt-Dieselmotoren," *MTZ Motortechnische Zeitschrift*, Vol. 29, No. 6, 1968.

Milbach, R., Ventilschäden und ihre Ursachen, ed. by TRW Thompson GmbH.

Schönau, H., "Probleme und Lösungswege an Aus- und Einlaßventilen hochbelasteter Verbrennungsmotoren," *MTZ Motortechnische Zeitschrift*, Vol. 42, No. 9/10, 1981.

Siebert, W., "Auslaßventile für zukünftige Schweröle," *MaK Toplaterne*.

TRW Thompson (Ed.), *Technischer Lehrgang Ventile: Schäden und ihre Ursache*, Vieweg-Verlag, Wiesbaden, 1992.

Umland, E. and Ritzkopf, M., "Ventilkorrosion in Dieselmotoren," part 2, *MTZ Motortechnische Zeitschrift*, Vol. 43, No. 9, 1982.

Umland, F. and Ritzkopf, M., "Ventilkorrosion in Dieselmotoren," *MTZ Motortechnische Zeitschrift*, Vol. 36, No. 7/8, 1975.

Valve train

Bockelmann, W., Gerve, A., Kehrwald, B., and Willenbockel, O., "Optimierung des Verschleißverhaltens am Ventiltrieb des Opel 3,0 1/24-V-Motors," *MTZ Motortechnische Zeitschrift*, Vol. 52, No. 2, 1991.

Cummins Diesel, Technical Overview of Camshaft Durability, Bulletin No. 3379031-OIR.

Cummins Diesel: Failure Analysis – Cylinderhead and Valve Train, Bulletin No. 985586.

Fuhrmann, W., "Die Einlauf-Oberfläche. Untersuchungen an Nocken und Stößeln," *MTZ Motortechnische Zeitschrift*, Vol. 41, No. 6, 1980.

Körner, W. D., "Untersuchungen über die Elastizität der Ventilsteuerung bei untenliegender Nockenwelle," *MTZ Motortechnische Zeitschrift*, Vol. 23, No. 3, 1962.

Further readings to Chapter 7

Dorinson, A. and Ludemar, K. C., *Mechanics and Chemistry in Lubrication* (Tribology Series, 9), Elsevier Publ. Co., Amsterdam, Oxford, New York, Tokyo, 1984.

Jones, M. H. and Scott, D., *Industrial Tribology – The Practical Aspects of Friction, Lubrication and Wear*, (Tribology Series, 8), Elsevier Publ. Co., Amsterdam, Oxford, New York, 1983.

"Performance Testing of Lubricants for Automotive Engines and Transmissions," ed. C. F. McCue; J. C. G. Cree; R. Tourret [Proceedings of a Symposium organised by the Institute of Petroleum and held at Montreux, Switzerland, 2nd to 6th April, 1973], Applied Science Publisher Ltd., Barking, and The Institute of Petroleum, London, 1974.

Kates, E. J., *Diesel and High-compression Gas Engines – Fundamentals*, 2nd edition, Chapter 20, Operation and Maintenance, American Technical Society, Chicago, 1965, pp. 417-440.

Oil Engine Manual, ed. by R. C. Southerton, Chapter 8, Maintenance, Temple Books, London, 1964, pp. 192-223.

Landsdown, A. R., *Lubrication – A Practical Guide to Lubricant Selection*, Pergamon Press, Oxford, New York, Toronto, 1982.

Diesel Engine Reference Book, ed. by L.C. R. Lilly, Butterworths Publ. Co., London, Boston, Durban, 1984. (Mainly successor to C. C. Pounder's Diesel Engine Principles and Practice = Ref. [7-10].)

ISO 4113-1978, Road vehicles, Calibration fluid for diesel injection equipment, International Organization for Standardization (ISO).

ISO/TR 4011-1976 Road vehicles, Apparatus for measurement of opacity of exhaust gas from diesel engines, International Organization for Standardization (ISO).

ISO 7342-1982 Road vehicles, Diagnostic systems, Equipment for ignition system testing, International Organization for Standardization (ISO).

BS 6380, Guide to low temperature properties and cold weather use of diesel fuels and gas oils (classes A1, A2 and D of BS 2869), 1983.

Illustration Credits

Allianz Global Corporate & Specialty AG, Munich, Germany: Figure 5.2; Figure 5.3; Figure 5.7; Figure 5.8; Figure 6.1; Figure 6.2; Figure 6.100; Figure 6.102; Figure 6.235; Figure 6.236; Figure 6.237; Figure 6.334; Figure 6.335; Figure 6.336; Figure 6.350; Figure 6.351; Figure 6.352; Figure 6.359; Figure 6.381; Figure 6.382; Figure 6.383.

Behr GmbH & Co. KG, Stuttgart, Germany: Figure 6.533; Figure 6.534; Figure 6.535; Figure 6.536.

BERU Aktiengesellschaft, Ludwigsburg, Germany: Figure 6.472; Figure 6.473; Figure 6.474; Figure 6.475; Figure 6.476; Figure 6.477; Figure 6.478; Figure 6.481; Figure 6.482; Figure 6.483; Figure 6.484; Figure 6.485; Figure 6.486; Figure 6.487; Figure 6.488; Figure 6.489; Figure 6.490; Figure 6.491; Figure 6.492; Figure 6.493; Figure 6.494; Figure 6.495; Figure 6.496; Figure 6.497.

BorgWarner Turbo Systems, Kirchheimbolanden, Germany: Figure 6.578; Figure 6.579; Figure 6.581; Figure 6.582; Figure 6.583; Figure 6.584; Figure 6.585; Figure 6.586; Figure 6.591; Figure 6.592; Figure 6.593; Figure 6.594; Figure 6.595; Figure 6.596; Figure 6.598; Figure 6.601.

BOSCH: Robert Bosch GmbH, Automotive Aftermarket, Plochingen, Germany: Figure 6.433.

Caterpillar Marine Power Systems, Hamburg, Germany: Figure 6.247; Figure 6.420.

ContiTech AG, Hannover, Germany: Figure 6.385; Figure 6.386; Figure 6.388; Figure 6.389; Figure 6.390; Figure 6.391; Figure 6.392; Figure 6.393; Figure 6.394; Figure 6.395; Figure 6.396; Figure 6.397; Figure 6.398; Figure 6.399; Figure 6.400.

Cummins Deutschland GmbH & Co. KG, Gross-Gerau, Germany: Figure 6.434.

Daimler AG, Archive & Sammlung, Stuttgart, Germany: Figure 2.21; Figure 2.22; Figure 6.220; Figure 6.221; Figure 6.222; Figure 6.357; Figure 6.419.

DEUTZ AG, Cologne, Germany: Figure 6.308; Figure 6.509; Figure 6.510.

Federal-Mogul Burscheid GmbH (ex Goetze-Kolbenringe), Burscheid, Germany: Figure 6.115; Figure 6.116; Figure 6.122; Figure 6.123; Figure 6.301.

Federal-Mogul Wiesbaden GmbH (ex Daelen & Loos, Glyco-Metallwerke), Wiesbaden, Germany: Figure 6.232.

Hengst GmbH & Co. KG, Münster/Westf., Germany: Figure 6.517.

IVK, TU Wien: Vienna Technical University, Institute for Internal Combustion Engines and Automotive Vehicle Construction (Prof. Dr. techn. B. Geringer), Vienna, Austria: Figure 2.17.

iwis motorsysteme GmbH & Co. KG, Munich, Germany: Figure 6.401; Figure 6.402; Figure 6.404a-e; Figure 6.405a-c; Figure 6.406.

Klaver Engines and Engineering, Rolde, The Netherlands: Figure 6.286; Figure 6.287.

KS: Kolbenschmidt Pierburg AG, Neckarsulm, Germany: Figure 6.18; Figure 6.19; Figure 6.20; Figure 6.21; Figure 6.29; Figure 6.30; Figure 6.31; Figure 6.32; Figure 6.34; Figure 6.38; Figure 6.39; Figure 6.40; Figure 6.43; Figure 6.50; Figure 6.51; Figure 6.61; Figure 6.62; Figure 6.70; Figure 6.71; Figure 6.91; Figure 6.104; Figure 6.105; Figure 6.107; Figure 6.110; Figure 6.111; Figure 6.298; Figure 6.330.

L'Orange GmbH, Stuttgart, Germany: Figure 6.435; Figure 6.438.

MAHLE Engine Systems UK Ltd., Rugby, United Kingdom: Figure 6.285.

MAHLE International GmbH, Stuttgart, Germany: Figure 6.4; Figure 6.5a-f; Figure 6.10; Figure 6.13; Figure 6.22; Figures 6.23–6.27; Figure 6.33; Figures 6.35–6.37; Figure 6.41; Figure 6.42; Figures 6.44–6.49; Figures 6.52–6.55; Figure 6.59; Figure 6.60; Figures 6.67–6.69; Figures 6.74–6.89; Figures 6.92–6.99; Figure 6.103; Figure 6.106; Figure 6.108; Figure 6.109; Figure 6.121; Figure 6.124; Figures 6.314–6.316.

MAN Nutzfahrzeuge AG, Munich, Germany: Figure 4.2; Figure 4.3; Figure 5.16; Figure 6.332; 6.424; Figure 6.427; Figure 6.430; Figure 6.569; Figure 6.570.

MANN+HUMMEL GMBH, Ludwigsburg, Germany: Figure 6.503; Figure 6.504; Figure 6.505; Figure 6.506; Figure 6.507; Figure 6.512; Figure 6.513; Figure 6.518; Figure 6.519; Figure 6.520; Figure 6.526; Figure 6.527; Figure 6.528.

Miba Gleitlager GmbH, Laakirchen, Austria: Figure 6.180; Figure 6.181; Figure 6.196; Figure 6.197; Figure 6.203; Figure 6.218; Figure 6.219; Figure 6.233; Figure 6.252; Figure 6.257; Figure 6.259; Figure 6.261; Figure 6.263; Figure 6.268; Figure 6.271; Figure 6.274; Figure 6.276; Figure 6.277; Figure 6.281; Figure 6.282; Figure 6.283; Figure 6.284.

MTU: From: Zima, S., *Entwicklung schnellaufender Hochleistungsmotoren in Friedrichshafen*, [Diss.], Series: Technikgeschichte in Einzeldarstellungen, Vol. 44, VDI-Verlag, Düsseldorf, Germany, 1987, pp. 652-806, by courtesy of MTU Motoren- und Turbinen-Union Friedrichshafen GmbH, Friedrichshafen, Germany. Treue, W., and Zima, S., *Hochleistungsmotoren, Karl Maybach und sein Werk*, VDI-Verlag, Düsseldorf, Germany, 1992, pp. 358-383, by courtesy of MTU Motoren- und Turbinen-Union Friedrichshafen GmbH, Friedrichshafen, Germany. Figure 2.7; Figure 2.23; Figure 2.28; Figure 2.29; Figure 4.1; Figure 6.289; Figure 6.306; Figure 6.329; Figure 6.331; Figures 6.343–345; Figure 6.358; Figure 6.387; Figure 6.407; Figure 6.431; Figures 6.514–6.516; Figure 6.521; Figure 6.529; Figure 6.530; Figure 6.574; Figure 6.575.

Scania Engines AB, Södertälje, Sweden: Figure 6.73; Figure 6.319.

SCHERDEL GmbH, Marktredwitz, Germany: Figure 6.346; Figure 6.444; Figure 6.445.

TRW Automotive GmbH, Barsighausen, Germany: Figure 6.361; Figure 6.374; Figure 6.378.

VDI-Verlag GmbH, Düsseldorf, Germany, from: Arbeitsmappe Kfz-Technik, ed. by Robert Bosch GmbH, Vol. 1, VDI-Verlag, Düsseldorf, Germany, 1991, Section 6, pp. 16-21, by courtesy of Robert Bosch GmbH, Stuttgart, Germany: Figure 6.422; Figure 6.425; Figure 6.428; Figure 6.429; Figure 6.432; Figure 6.436; Figure 6.437.

Woidich: Ingenieurbüro Woidich u. Kollegen, Mainz-Kastel, Germany: Figure 4.4.

ZF Friedrichshafen AG, Friedrichshafen, Germany: Figure 6.408; Figure 6.409; Figure 6.410; Figure 6.411; Figure 6.412; Figure 6.413; Figure 6.414; Figure 6.415; Figure 6.416; Figure 6.417.

Zollern BHW Gleitlager GmbH, Braunschweig, Germany: Figure 6.266.

Index

Abrasion, 77
Abrasive wear of cylinders, 301–303
Acronyms, 525–526
Adhesion, 78
Adhesive wear of cylinders, 300
Air filters
    combination, 436
    cyclone, 434
    dry, 434–436
    ejectors, 435
    overview of, 432–434
    paper, 435–436
    single-stage, 436
Airborne particulates, sizes of, 432*f*
Appendix, 517–522

Bathtub curve, 46*f*
Bearing force, 226–227
Boundary lubrication, 80–81
Break-in, 55–56
Brittle coating method, 69, 71
Brittle overload failures, 64–65

Camshaft and cam followers, 342–347
    camshaft wear, 345
    gray staining (flecking,
        micropitting), 346–347
Cavitation, 305–309
    crack, 307
    cracks and fractures, 307–309
    in crank train bearings, 243–248
    exit, 246–247
    flow, 247
    in heat exchangers, 464
    in injection pumps, 394–395

    in injector nozzles, 400
    suction, 245–246
    throttling, 247–248
Chain drives
    broken bushings, 364–365
    broken rails, 364
    chain breakage, 365
    chain lengthening, 363
    damage, 360–363
    guide and tension rail failure,
        364
    overview of, 357–360
    wear and surface damage,
        363–364
Cold start, 57
Combustion, 20
Connecting rods
    angled journal bearing cap, 191*f*
    assembly defects, 200, 202
    beam force, 194*f*
    deformation, 194*f*
    effect of surface finish on load
        limits, 193*f*
    failures caused by engine
        operation, 202–203
    fracture zones, 196*f*
    fretting corrosion, 198–200
    journal bore out-of-roundness
        as a function of bolt tension,
        195*f*
    manufacturing defects, 197
    overview of, 190–196
    positions during crankshaft
        revolution, 191*f*
    terminology for, 190*f*
Corrosion, 42–43
    of crank train bearings, 250–251

Corrosion (*continued*)
    of cylinder heads, 315–316
    engine failure through, 42–43, 73–76
    erosive, 76
    failure from, 73–76
    fatigue crack, 75
    fretting, 75, 78, 80–81
        of connecting rods, 198–200
        of crank train bearing back, 269
        of valves, 337
    hot, 340–342
    of heat exchangers, 470–475
    of injection pumps, 395
    of injector nozzles, 400–401
    in piston skirt area, 130
    selective, 75
    spark plug, 421–422
    stress crack, 75
    surface, 74
    of valves, 339–349
Cracks, 61–62
Crank angle, as measure of time and distance, 522–524
Crank train, 20
    forces, 22*f*
    failures
        connecting rods, 190–203
        crank train bearings, 215–275
        crankshafts, 203–215
        engine oil, 275–284
        piston rings, 178–190
        pistons, 99–178
Crank train bearings
    anomalies in geometry, 257–259
    bearing and bearing housing press fit, 221*f*
    bearing complex, 216*f*
    bushings, 218*f*
    cavitation, 243–248
    clearance terminology, 218*f*
    clearance tolerances, 219*f*
    comb wear, 264
    contamination, 237–240
    corrosion, 250–251
    crankpin bearing force, 225*f*

damage, 231–233
deformation of connecting rod bore from bearing installation, 221*f*
edge wear, 259–264
    alternating wear, 260–261
    contact marks in center of bearing, 263
    narrow wear-free zones at edges, 263
    single-ended, 259–260
    two-ended, 263
effect of engine operating mode on minimum lubricating film thickness, 227*f*
electrical arcing, 251–253
erosion, 249
fatigue, 240–243
friction coefficient as function of sliding speed, 224*f*
insufficient bearing shell interference, 269–275
    bearing shell fatigue fracture, 270–275
    fretting corrosion of bearing back, 269
load-carrying mechanism in, 223*f*
manufacturing defects, 254–256
materials in, 228*t*
measurement of shell, 222*f*
overview of, 215–231
radial clearances, 220*f*
shaft displacement path, 226*f*
sleeve bearing terminology, 217*f*
surface wear, 264–268
    bearing cap offset, 264–266
    out-of-round housings, 266–268
trimetal, 229*f*
wear, 233–237
Crankcase
    cracks and fatigue fractures, 287–289
    damage and failure, 287–291
    fracture from excessive force, 289
    overview of, 284–287

relief valves, 38
wear, 289–291
Crankshaft
bearing failures, 210–211, 213
bending fatigue failures, 209
causes of failure, 208
crank throw deformation, 207*f*, 208*f*
crankpins, 206*f*
failures, 209–215
nomenclature for, 204*f*
overview of, 203–208
tangential and radial forces, 205*f*
torsion fatigue failures, 209–210
Cylinder head, 309–312
blow-through at threads or injector mount, 318
cracks and fractures, 313–315
damage, 312–318
erosion and corrosion, 315–316
face distortion, 316
gasket leaks, 317–318
temperatures, 28*f*, 311*t*
valve guide wear, 316–317
valve seat wear, 312
Cylinder liners, 291–296
Cylinders, 299–305
abrasive wear, 301–303
adhesive wear, 300
honing of, 297–299
overview of, 291–299
ring reversal wear, 300
scoring, 303
seizing, 303–305
wear, 299–300
see also Cylinder head

Damage
describing, 90
determining relevant data, 87
evaluating, 86–87
Declared power, 17
Decompression valves, 38
Design flaws, 49–52
Diesel engines, 92, 498
afterburning, 371–376

begin of delivery and begin of injection, 376–377
controlled combustion, 370
cracked piston heads, 161
cylinder head temperatures, 311*t*
engine cooling damage examples, 510–511
erosion damage to piston head and fire land, 157
example design failures, 503–504
exhaust valve and valve seat temperatures, 329*t*
fuel matching, 380
full load, 381
heat rejection requirement for, 457*t*
idling and part load operation, 380
ignition delay, 369
injection duration and injection rate, 377
injection pressure, 377–379
injection system matching, 380
mixture formation and combustion, 369–381
piston burn through, 146
piston head and fire land melting, 150–151
ring land failure from combustion problems, 141
seized fire lands on pistons, 133–134
starting, 379
starting assistance, 380
uncontrolled combustion with steep pressure rise, 369–370
Droplet impact wear, 82
Ductile fractures, 62–63
Duration of developed power, 17

Emergency air shutoff valves, 38
Engine coolant, 463
Engine failures
by application, 95*f*
breakdown statistics, 95–99

Engine failures (*continued*)
    causes of, 47–60
    corrosion in aqueous media, 73–76
    course of events, 88–89
    definitions and concepts, 41–46
    from design (planning) flaws, 49–52
    evaluating, 44–46
    fatigue fractures, 65–71
    history of, 91–92
    human errors, 59–60
    introduction to, 1–7
    by location, 99*f*
    from manufacturing defects, 53
    from materials defects, 52–53
    from mechanical loading, 61–62
    overload failure, 62–65
    overview of, 91–99
    from technical (product) defects, 49–53
    thermal damage, 71–73
    through corrosion in aqueous media, 42–43
    through mechanical loads, 41–42
    through thermal loading, 43
    by tribological loading, 44, 76–85
    types of, 61–85
    from wear and tear, 47–49
Engine map, 21–25
Engine oil
    acidification, 280–281
    loading from unfavorable operating conditions, 36*f*
    oil consumption, 282
    oil loss, 281
    overview of, 275–284
    properties of, 277*t*
    sludge formation, 279–280
    vaporization losses, 280
    viscosity increase, 280
    viscosity reduction, 280
Engines
    basic concepts of, 11–14
    bearing temperature map, 26*f*
    characteristics as functions of speed and torque, 23*f*
    combustion air, 513–514
    cooling, 508–511
        cooling water shortage, 510
        cooling water treatment, 508–510
        damage examples, 510–511
    damage by application area, 499*t*
    damage by component area, 500*t*
    damage prevention, 497–516
        combustion air, 513–514
        for emergency power-generating sets, 507
        engine cooling, 508–511
        engine fuel, 512–513
        engine lubrication, 511–512
        examples of, 503–504
        from fabrication and assembly, 504–506
        introduction to, 498–499
        loss statistics, 499–502
        by maintenance and inspection, 514–516
        from operational faults, 506–507
        from planning and design, 502–504
        from product faults, 502–506
    fuel for, 512–513
    gas and inertia forces, 21*f*
    heat budget of, 16*f*
    heat lost to coolant, 28*f*
    historical overview, 91–92
    ignition pressure map, 26*f*
    inspection of, 514–516
    introduction to, 1–7
    life expectancy as function of load profile and power output, 36*f*
    loading of, 31*f*, 32*f*, 33*f*
    lubrication of, 511–512
    maintenance of, 514
    oil loading from unfavorable operating conditions, 36*f*

oil pressures, 232*f*
operating conditions of, 11–20
operating errors, 53–59
operational behavior of, 20–40
output, 517–522
power generators, 13, 93
power output and power
    reduction, 14–20
primary causes of damage,
    501*t*
properties and peculiarities of,
    9–11
protective measures to prevent
    damage, 37–39
secondary causes of damage,
    501*t*
specific fuel consumption, 24*f*
speed range of, 12*t*
tractive effort vs. road speed,
    15*f*
see also Engine failures
Erosion, 81–82
    of crank train bearings, 249
    of cylinder heads, 315–316
    of fire land, 143
    of piston head and fire land,
        157–158
    of spark plugs, 421–422
    of turbochargers, 489
    of valves, 342
Erosive corrosion, 76
Exit cavitation, 246–247

Failure analysis, 86–90
    course of events, 88–89
    describing damage, 90
    determining damage-relevant
        data, 87
    on-site inspection, 86
    securing damaged parts, 86–87
Fatigue crack corrosion, 75
Fatigue fractures, 65–71
Filters, 427–454
    air, 432–440
    airborne particulates, sizes of,
        432*f*
    dust exposure, 433*t*, 437–439

filtration efficiency, 430*f*
filtration mechanisms, 429*f*
fuel, 451–454
fundamentals of filtration,
    427–431
oil, 440–450
pressure rise as function of
    contamination, 431*f*
Fire land damage, 130–144, 149, 149–
    151, 157–160
Flow cavitation, 247
Fracture mechanisms, 61, 63*f*
Fretting corrosion, 75, 78, 80–81
    of connecting rods, 198–200
    of crank train bearing back,
        269
    of valves, 337
Fuel filters, 451–454
Fuel injection and ignition systems,
    369–426
    diesel engine mixture formation
        and combustion, 369–381
    fuel injection systems, 381–392
        damage, 392–405
    glow plugs, 405–412
    Otto-cycle engine ignition and
        combustion, 412–426
    common rail injection systems,
        390
    damage, 392–405
    fuel injection line damage,
        404–405
    injection pump damage, 394–
        396
        cavitation, 394–395
        corrosion, 395
        injection pump cams, 396
        injector springs, 395–396
        plunger pumps, 394–395
        seizing, 395
    injector nozzle damage, 398–404
        cavitation, 400
        compression springs,
            403–404
        corrosion, 400–401
        from dirt, 398
        from excessive operating
            temperatures, 401

Fuel injection and ignition systems
(*continued*)
        uncapping, 402–403
        wear, 398–400
    distributor pump, 386–389
    individual-plunger pump,
        383–384
    nozzles and injectors, 391–392
    overview of, 381–383
    pressure generation in nozzle,
        389
    unit injectors, 385–386
Full-load curve, 13

Gears, 365–369
    damage, 366–369
    galling, 367
    micropitting (flecking, gray
        staining, microspalling),
        368–369
    pitting, 366
    tooth breakage, 366–367
    wear, 367
Generator curve, 13
Glow plugs, 405–412
    creased and dented tubes, 411
    mechanical damage, 412
    tube melting or tube broken off,
        409
    tube tip damage, 408–409
    rapid-heating pencil-type, 407
Governors, 39

Heat exchangers
    cavitation, 464
    combustion engine cooling, 458*f*
    corrosion, 470–475
    damage, 463–477
        from changed operating
            conditions, 468–470
        from design and installation
            factors, 464–465
        from unfavorable operating
            conditions, 466–468
    faulty maintenance, 466–468

    flat tube or plate, 459–460
    fouling, 463–464
    heat rejection requirement for
        diesel engines, 457*t*
    materials and manufacturing
        damage, 465–466
    mechanical, 475–477
    overview of, 454–459
    shell and tube, 459–463
    stacked plate, 460–463
Heat lost in charge air intercooler, 28*f*
Hydraulic locking, 55

Idling, 58

Knock, 91, 414–416, 506

Loading , 25–27, 31
    brittle overload, 64–65
    effect of connecting rods surface
        finish on load limits , 193*f*
    fracture mechanisms for
        various external loads, 63*f*
    full-load curve, 13
    load-carrying mechanism in
        crank train bearings, 223*f*
    mechanical, 41–42, 61–62,
        138–140
    of engines, 31*f*, 32*f*, 33*f*, 36*f*
    oil loading from unfavorable
        operating conditions, 36*f*
    overloading, 53–54, 62–65
    piston overloading, 112, 138–
        140
    thermal, 43
    tribological, 44
    valve overloading, 332–334
Lubrication, boundary, 80–81

Manufacturing defects, 53
Materials defects, 52–53
Mechanical loads, 41–42
    failures from, 61–62

Oil filters
    centrifuges, 445
    combined full flow and bypass,
        445–450
    overview of, 440–443
    pleated paper, 444
    screen disc, 444
    wire gap, 444–445
Operating errors, 53–59
    break-in, 55–56
    cold start, 57
    engineering progress, 58–59
    external forces, 58
    long idling periods, 58
    and operating conditions, 54–55
    overloading, 53–54
    unsuitable working materials,
        57–58
Otto-cycle engines
    combustion, 412–416
    cracked piston heads, 161
    detonation, 414–416, 422
    erosion damage to piston head
        and fire land, 157
    ignition and combustion,
        412–426
    ignition and spark plugs,
        417–422
    piston burn through, 145–146
    piston head and fire land
        melting, 150
    preignition, 413–414, 422
    ring land failure from
        combustion problems,
        140–141
Overload failure, 62–65

Piston rings
    burned, 186–187
    compression, 179f
    contact pressure distribution,
        182f
    faulty assembly, 184–185
    high oil consumption, 189–190
    oil control rings, 180f, 181f
    overview of, 178–184

    pressure relief, 183f
    ring and fire land damage,
        130–144
    ring breakage, 188
    ring flutter, 187–188
    stuck rings, 188
    terminology for, 178f
Pistons
    basic construction of, 101f
    clearance, 110f
    configurations of, 103f
    contact patterns, 114f
    damage symptoms, 113t
    failure areas, 116f
    failures in power transmission
        area and wrist pin bearings,
        164–178
      broken or loose wrist pin
        retainers, 172–173
      broken wrist pin boss, 171–
        172
      broken wrist pin, 175–176
      broken-out wrist pin boss,
        174–175
      cracks in wrist pin bores,
        164–166
      damaged wrist pin
        retainers, 173–174
      piston noise (running
        noise), 177–178
      split-piston and wrist pin
        boss fractures, 164–166
      transverse fracture through
        lower part of built-up
        pistons, 167
      wrist pin bore seizing,
        168–171
      wrist pin boss cracking
        inside wrist pin bores,
        166
    failures in skirt area, 115–130
      crooked piston–
        asymmetrical contact
        pattern, 123
      diagonal seizing marks on
        skirt near wrist pin boss,
        121–122

Failure in skirt area (*continued*)
    long-term dry seizing,
      126–128
    one-sided seizure marks
      from oil starvation,
      118–120
    scuffing as a result of fuel
      flooding, 128–129
    seizing from distorted
      (warped) cylinders,
      123–124
    seizing from overheating,
      117–118
    seizure of major and minor
      thrust sides, 120–121
    skirt corrosion, 130
    skirt seizing, 115–118
    wear caused by dirt, 125–
      126
fatigue strength diagram, 109*f*
faulty kinematic relationships,
    111–112
force transfer in, 106*f*
functional zones of, 100*f*
gas temperature history, 107*f*
head failures, 144–163
    ablated edge of head and
      damaged fire land,
      158–160
    bowl cracking, 163
    burn through, 144–148
    burned head and fire land,
      149
    cracking, 161–163
    damage from valve
      interference, 153–154
    deformation (valve or
      cylinder interference),
      151–153
    erosion damage to head
      and fire land, 157–158
    foreign body on piston
      head, 155–156
    head and fire land melting,
      149–151
    recession, 154–155
heat flow, 107*t*
and normal force history, 104*f*

offset, 105*f*
overview of, 99–112
    failures, 111–112
quench height, 112*f*
ring and fire land damage,
    130–144
    broken fire land (two-stroke
      Otto cycle), 144
    burned piston rings, 131–
      132
    fire land erosion damage,
      143
    from skewed piston due
      to jammed piston rings,
      130–131
    piston burning (Otto-cycle
      engines), 132–133
    ring and ring groove wear,
      136–137
    ring land failure from
      combustion problems,
      140–142
    ring land failure from
      mechanical overloading,
      138–140
    ring land failure in Otto-
      cycle engines, 137–138
    ring zone damaged by
      broken piston rings,
      134–136
    seized fire lands on diesel
      engine pistons, 133–134
    seized pistons from extreme
      overheating, 142–143
temperatures, 108*f*
thermal, mechanical, and
    tribological overloading, 112
wear measurement of, 100*f*
see also Piston rings
Pit formation, 79*f*
Power, 31–32
    engine life expectancy as
      function of power output, 36*f*
    power output and power
      reduction, 14–20
    reference conditions for, 20*t*
    specification, 17–18
    transmission failures, 164–178

Power generators, 13, 93
Propeller curve, 13

Relative air/fuel ratio, 29*f*
Rillenlager bearings, 230

Sankey diagram, 16*f*
Selective corrosion, 75
Skirt, piston, failures in, 115–130
Smoke number, 29*f*
Spark plugs
    broken insulator tip, 426
    components of, 417*t*
    deposits, 424–425
    electrode gap, 417–418
    electrode materials, 418
    electrode shapes, 418
    erosion and corrosion, 421–422
    failures, 422–426
    glazing/lead fouling, 424
    heat ranges, 420–421
    melted center electrode, 425
    melted ground electrode,
      425–426
    mutual interactions between
      engine and spark plugs, 421
    oiled plug, 423–424
    severe electrode wear, 426
    sooty, 423
    spark gap, spark orientation,
      418
    spark plug seats and gaskets,
      420
Sputtered bearings, 230
Stress crack corrosion, 75
Stribeck curve, 224*f*
Suction cavitation, 245–246
Surface corrosion, 74
Surface disruption, 77

Technical (product) defects, 49–53
Temperature
    bearing temperature map, 26*f*
    of components bordering
      combustion chamber, 72*f*

cylinder head temperatures, 28*f*,
    311*t*
of exhaust valve and valve seat,
    329*t*
gas temperature history, 107*f*
injector nozzle damage from
    excessive, 401
piston temperatures, 108*f*
Thermal damage, 71–73
Thermal loading, 43
Throttling cavitation, 247–248
Timing belts
    backing worn in roots, 354
    cracks in belt backing, 356
    damage, 354–356
    edge wear, 354
    overview of, 351–354
    separated (torn) belts, 356
    teeth and reinforcing material
      separated from backing, 355
    tooth root breakage and shear,
      354
    wear marks on the tooth side,
      356
Torsional fatigue fractures, 68–71
Tribochemical reactions, 78
Tribological loading, 44
    failure from, 76–85
Turbochargers
    carbon deposit formation,
      493–494
    compressor map, 481*f*
    damage, 484–491
      compressor side, 485–486
      turbine side, 486
    distribution of work between
      engine and turbocharger,
      479*f*
    effects of foreign bodies, 485–
      586
    erosion, 489
    exhaust turbocharging, 478*f*
    fractures, 489–491
    housing leaks, 494–495
    imbalance, 486–489
    lubrication inadequacies,
      491–494
    material defects, 484–485

Turbochargers (*continued*)
  noise complaints, 496
  oil contamination, 491
  oil leaks, 494
  oil starvation, 491–493
  operation in zero pressure
    regime, 495–496
  overview of, 477–484
  sequential turbocharging, 483*f*

Valve springs, 322–325
Valve train
  camshaft and cam followers,
    342–347
  chains, 347–348, 357–365
  gears, 347–348, 365–369
  overview of, 318–322
  terminology, 320*f*
  timing belts, 347–356
  valve springs, 322–325
    failures, 323–325
  valves, 325–342
Valves, 325–342
  clearance adjustment, 329*f*
  corrosion, 339–349
  crankcase relief, 38
  decompression, 38
  deposits, 338–339
  emergency air shutoff, 38
  erosion, 342
  exhaust valve and valve seat
    temperatures, 329*t*
  failures, 332–342
  fretting corrosion, 337
  hot corrosion, 340–342
  introduction, 325–332
  mechanical overloading, 333–
    334

reciprocating motion of, 321*f*
rotator details, 330*f*
rotator faults, 342
standard valve materials, 331*f*
stuck, 335–336
terminology, 327*f*
thermal overloading, 332–333
wear, 335
V-belts, 349–351

Wear, 47–49
  basic mechanisms of, 80*f*
  of camshaft, 345
  causes of, 80–82
  of chain drives, 363–364
  of crank train bearings, wear,
    233–237, 259–268, 289–291
  of cylinder heads, 312, 316–317
  of cylinders, 299–300
  of cylinders, 300–303
  droplet impact, 82
  effects of, 83, 85
  of gears, 367
  of injector nozzles, 398–400
  of piston rings and ring
    grooves, 136–137
  in piston skirt area, 125–126
  of pistons, measurement of, 100*f*
  preconditions for, 76–77
  rates of, 80*f*
  of spark plug electrode, 426
  speed of, 78
  of timing belts, 354, 356
  of valves, 335
  zones of, 85*f*
Wrist pin bearings, failures in, 164–
  178

About the Authors

Ernst Greuter (1922-1995). After graduation from high school, Ernst Greuter completed training as a motor vehicle mechanic. After passing his journeyman and master's exams, he attended engineering school. In 1950, he joined his father's business, "Motoren-Greuter," an engine overhaul shop located in Saarbrücken, Germany. In that same year, he set up a small subsidiary in Algiers, Algeria, specializing in cylinder boring and crankshaft grinding, with a small sales office. In 1951 he established additional engine overhaul branches in Idar-Oberstein and Mainz, Germany. In 1952 and 1953, Greuter served as an engineer trainee with several vehicle and engine component manufacturers in Germany and abroad. In 1955, he was given overall technical responsibility for all of the family firm's engine overhaul facilities. Also beginning in 1955, Greuter worked as a combustion engines consultant (gasoline and diesel). He participated in research, development, and material testing activities with various engine component manufacturers in Germany and abroad. Greuter conducted tests of new designs and materials on diesel and gasoline engine test stands. In 1964, he established additional subsidiaries in Central and South America, and began license manufacture of ocean-going diesel engines and manufacture of exchange engines. In addition to his work as technical director and managing director of "Motoren-Greuter" GmbH, beginning in 1975 Greuter operated an engineering and consulting office which specialized in planning and establishing engine repair facilities, exchange engine manufacturing, and production of engine components in Germany and overseas. Additional specialties included personnel training, and combustion engine seminars for the motor vehicle repair trade and for engine specialists. Periodically Greuter served as a visiting instructor at various educational facilities overseas. From 1987 to 1997, he had an independent consulting service, specializing in combustion engines, engine overhaul, and the outfitting of engine and motor vehicle repair facilities.

Stefan Zima (1938-2004). Stefan Zima studied mechanical engineering at the Technical University of Berlin (majoring in engines and machines). He came into contact with high-speed, heavy-duty engines for special applications when, after obtaining his intermediate diploma with Deutsche Bundesbahn [German Federal Railways], he underwent training to become an engine driver on diesel locomotives of the V 60, V 1002, V 160, and V 2001 series. After his examination in humanistics, in 1967 he gained his full diploma with a thesis on experimental investigations into a heat-transfer problem. Subsequently, Zima worked as assistant engineer on merchant-navy ships, where he acquired a thorough knowledge of the operation, maintenance, and repair of marine propulsion systems, including those powered by high-speed engines. From 1968 to 1977, Dipl.-Ing. Stefan Zima worked in the R&D division of Maybach-Mercedes-Benz Motorenbau (now: MTU Friedrichshafen GmbH), and was last employed as deputy head of a design department. In 1977 he joined the Giessen-Friedberg University for Applied Sciences as a lecturer in piston engines and head of the IC laboratory. He was appointed professor in 1979. Also during his service as a university lecturer, Zima maintained contact with the everyday world of engine technology in a variety of ways, e.g., by trips on motor ships or by driver training and testing on tracked vehicles (Leopard 1 A 4 combat tank). In 1985 he was awarded his doctorate in engineering from the Technical University of Berlin with a thesis on *"The Development of High-Speed Heavy-Duty Engines in Friedrichshafen"* (VDI-Verlag, Düsseldorf, 1987). He published a number of books on internal combustion engines, e. g., *Hochleistungsmotoren. Karl Maybach und sein Werk*, [Heavy-Duty Engines. Karl Maybach and His Work], VDI-Verlag, Düsseldorf, 1992 and 1995 (co-author); *Kurbeltriebe*, [Development of the Crank Assembly], Vieweg-Verlag, Wiesbaden, 1999; and *Ungewöhnliche Motorkonstruktionen* [Unusual Engine Designs], Vogel-Verlag, Würzburg, 2003. Dr. Zima unexpectedly passed away in April 2004.

www.ingramcontent.com/pod-product-compliance
Lightning Source LLC
Chambersburg PA
CBHW062010190326
41458CB00009B/3028